Renewable Energy Resources

Renewable Energy Resources

Edited by **George Thomson**

SYRAWOOD
PUBLISHING HOUSE
New York

Published by Syrawood Publishing House,
750 Third Avenue, 9th Floor,
New York, NY 10017, USA
www.syrawoodpublishinghouse.com

Renewable Energy Resources
Edited by George Thomson

International Standard Book Number: 978-1-68286-335-0 (Hardback)

Printed in China.

Contents

Preface IX

Chapter 1 **Advances in surface passivation of c-Si solar cells** 1
Mohammad Ziaur Rahman, Shahidul Islam Khan

Chapter 2 **Metal-organic vapor-phase epitaxial growth of InGaN and InAlN for multi-junction tandem solar cells** 12
A. Yamamoto, K. Sugita, A. G. Bhuiyan, A. Hashimoto, N. Narita

Chapter 3 **Dislocation reduction in heteroepitaxial $In_xGa_{1-x}N$ using step-graded interlayer for future solar cells** 21
Md. Arafat Hossain, Md. Rafiqul Islam, M. K. Hossain, A. Hashimoto, A. Yamamoto

Chapter 4 **Strontium and iron-doped barium cobaltite prepared by solution combustion synthesis: exploring a mixed-fuel approach for tailored intermediate temperature solid oxide fuel cell cathode materials** 35
Francesca Deganello, Leonarda F. Liotta, Giuseppe Marcì Emiliana Fabbri, Enrico Traversa

Chapter 5 **Processing and characterizations of a novel proton-conducting $BaCe_{0.35}Zr_{0.50}Y_{0.15}O_{3-\delta}$lectrolyte and its nickel-based anode composite for anode-supported IT-SOFC** 49
Ashok Kumar Baral, Sungmin Choi, Byung Kook Kim, Jong-Ho Lee

Chapter 6 **Oxidation pathways towards Si amorphous layers or nanocrystalline powders as Li-ion batteries anodes** 58
Kamal Annou, Mathilde Pelosi, Gregory Gershinsky, Frederic Favier, Yvan Cuminal, Monique Tillard, David Zitoun

Chapter 7 **High-temperature thermoelectric properties of $Cu_{1.97}Ag_{0.03}Se_{1+y}$** 67
Tristan W. Day, Kasper A. Borup, Tiansong Zhang, Fivos Drymiotis, David R. Brown, Xun Shi, Lidong Chen, Bo B. Iversen, G. Jeffrey Snyder

Chapter 8 **Modified graphite and graphene electrodes for high-performance lithium ion hybrid capacitors** 74
Ji Hoon Lee, Weon Ho Shin, Soo Yeon Lim, Byung Gon Kim, Jang Wook Choi

Chapter 9 **Processing of pristine carbon nanotube supported platinum as catalyst for PEM fuel cell** 82
Chanchal Gupta, Priyanka H. Maheshwari, S. Sasikala, R. B. Mathur

Chapter 10 **Hydrogen selectivity and carbon behavior during gasoline steam reforming over nano-Al_2O_3 catalysts** 93
Dalia R. Abd El-Hafiz, Mohamed A. Ebiad, Radwa A. El-salamony

Chapter 11 **Novel and efficient texturing approach for large-scale industrial production line of large-area monocrystalline silicon solar cell** 106
U. Gangopadhyay, S. Jana, S. Das, P. Ghosh

Chapter 12 **An experimental investigation of mesoporous MgO as a potential pre-combustion CO_2 sorbent** 115
Sushant Kumar, Surendra K. Saxena, Vadym Drozd, Andriy Durygin

Chapter 13 **Phase change materials (PCMs) integrated into transparent building elements: a review** 123
Paris A. Fokaides, Angeliki Kylili, Soteris A. Kalogirou

Chapter 14 **Electrochemical and in situ ATR-FTIR studies of ethanol electro-oxidation in alkaline medium using PtRh/C electrocatalysts** 136
E. H. Fontes, R. M. Piasentin, J. M. S. Ayoub, J. C. M. da Silva, M. H. M. T. Assumpção, E. V. Spinacé, A. O. Neto, R. F. B. de Souza

Chapter 15 **Effects of acid concentration, temperature, and time on recycling of post-vehicle-application lithium-ion batteries of varying chemistries** 146
Haikun Li, Lindsay M. Corneal, Charles R. Standridge

Chapter 16 **Investigations on thermal storage systems containing micron-sized conducting particles dispersed in a phase change material** 151
S. Jegadheeswaran, S. D. Pohekar, T. Kousksou

Chapter 17 **Effect of Ni concentration on the structural and hydrogen storage characteristics of Zr–Mn based laves phase system** 167
Vinod Kumar, D. Pukazhselvan, A.K. Tyagi, S.K. Singh

Chapter 18 **Possibilities to decrease the absorber thickness reducing optical and recombination losses in CdS/CdTe solar cells** 175
L.A. Kosyachenko

Chapter 19 **Feasibility of an oxygen-getter with nickel electrodes in alkaline electrolysers** 195
Daniel Symes, Connie Taylor-Cox, Leighton Holyfield, Bushra Al-Duri, Aman Dhir

Chapter 20 **Self-standing electrochemical double layer capacitors for operation in severe temperature conditions** 202
C. Lecoeur, B. Daffos, R. Lin, L. Divay, P. Le Barny, M. Pham Thi, P.-L. Taberna, P. Simon

Chapter 21 **Betalain pigments as natural photosensitizers for dye-sensitized solar cells: the effect of dye pH on the photoelectric parameters** 209
Kasim Uthman Isah, Umar Ahmadu, Adamu Idris, Mohammed Isah Kimpa, Uno Essang Uno, Muhammed Muhammed Ndamitso, Noble Alu

Chapter 22 **A review on the application of iron(III) fluorides as positive electrodes for secondary cells** **214**
Donato E. Conte, Nicola Pinna

Chapter 23 **Trials for oxide photo-thermoelectrics** **236**
Ichiro Terasaki, Ryuji Okazaki, Partha Sarathi Mondal, Yu-Chin Hsieh

Chapter 24 **Materials science aspects of zinc–air batteries: a review** **243**
Vincenzo Caramia, Benedetto Bozzini

Chapter 25 **A new method of pretreatment of lithium manganese spinels and high-rate electrochemical performance of $Li[Li_{0.033}Mn_{1.967}]O_4$** **255**
A. V. Potapenko, S. I. Chernukhin, S. A. Kirillov

Permissions

List of Contributors

Preface

This book was inspired by the evolution of our times; to answer the curiosity of inquisitive minds. Many developments have occurred across the globe in the recent past which has transformed the progress in the field.

This book provides a detailed study on renewable energy resources covering the full range of renewable energy technologies and their implementations. Renewable energy mitigates pollution and climate changes. These benefits are analyzed and illustrated with case studies and examples in this text. The book includes reviews of methods and techniques for applications of renewable energy such as electrical power, etc. The chapters included herein have undergone a tremendous reviewing process and aims to serve as valuable reference for graduates, researchers and professionals.

This book was developed from a mere concept to drafts to chapters and finally compiled together as a complete text to benefit the readers across all nations. To ensure the quality of the content we instilled two significant steps in our procedure. The first was to appoint an editorial team that would verify the data and statistics provided in the book and also select the most appropriate and valuable contributions from the plentiful contributions we received from authors worldwide. The next step was to appoint an expert of the topic as the Editor-in-Chief, who would head the project and finally make the necessary amendments and modifications to make the text reader-friendly. I was then commissioned to examine all the material to present the topics in the most comprehensible and productive format.

I would like to take this opportunity to thank all the contributing authors who were supportive enough to contribute their time and knowledge to this project. I also wish to convey my regards to my family who have been extremely supportive during the entire project.

Editor

1

Advances in surface passivation of c-Si solar cells

Mohammad Ziaur Rahman · Shahidul Islam Khan

Abstract In order to avoid an unacceptably large efficiency loss when moving towards thinner silicon materials, the near-term challenge in the c-Si PV industry is to implement an effective passivation method for both cell surfaces. This paper discussed several suitable passivation schemes available. While the efficiency potential of industrially produced thin film poly-Si cells on foreign substrates cannot yet reliably be predicted, it is clear that wafer-based c-Si solar cells will allow to maintain (or even improve) today's efficiency levels while at the same time reducing the consumption of (expensive) crystalline silicon by up to 50 %. Given the trend towards these Si materials, the most promising surface passivation methods are identified to date. The key issues to be considered are cost-effectiveness, added complexity, additional benefits, reliability, and efficiency potential. The efficiency increase for best cells is around 0.5–0.6 %abs and the current efficiency potential already demonstrated for all technologies is around 19.0 %. Average efficiencies in industrial mass production for selected technologies are 18.5–18.6 % for Cz and 17.1 % for mc-Si.

Keywords Solar cell · Passivation · Emitter · Efficiency · c-Si

M. Z. Rahman (✉)
Department of Electrical and Electronic Engineering,
Ahsanullah University of Science and Technology,
141-142 Love Road, Tejgaon I/A,
Dhaka 1208, Bangladesh
e-mail: ziaeee_083@yahoo.com

S. I. Khan
Department of Electrical and Electronic Engineering,
Bangladesh University of Engineering and Technology,
Dhaka 1000, Bangladesh

Introduction

The global photovoltaic (PV) market is booming over the years. Solar cell producers have to concentrate on the cost per watt peak (Wp) of their product, especially as the competition between the manufacturers becomes more and more important. Therefore, most cell manufacturers try to optimize their solar cell processes concerning higher efficiencies while not significantly increasing costs. There are four different ways to improve c-Si solar cell efficiency:

- Increased light trapping effect by improving surface structuring and texturization.
- Redistributing the emitter profile on the front surface.
- Upgrading or changing metallization processes to get thinner contacts with excellent electrical properties.
- Optimizing the passivation layer on both surfaces to reduce the recombination losses.

This study focused on the present status of passivation of p-type c-Si solar cells with a special emphasis on passivation by AL_2O_3 due to its superiority over other passivation methods available.

The surface of c-Si represents the largest possible disturbance of the symmetry of the crystal lattice and hence, due to non-saturated ('dangling') bonds, a large density of defects ('surface states') within the bandgap exists at the surface of the crystal. These dangling bond defects can also be divided into intrinsic and extrinsic defects. There are typically additional processing-related extrinsic surface defects, for instance, due to dislocations or chemical residues and metallic depositions on the surface. These defects are highly prone to different recombination mechanisms, namely surface recombination, auger recombination and Shockley–Read–Hall (SRH) recombination (occurs via defects level within the bandgap) [1, 2]. The recombination processes reduce the minority carrier life time and are responsible for lower conversion efficiency of Si solar cells. One way to keep the recombination loss at tolerable minimum is to passivate the silicon surfaces electronically [3, 4].

High-efficiency silicon solar cells require both surfaces (front and rear) to be well passivated. In this regard, electrical surface passivation of crystalline silicon (c-Si) solar cells has emerged as a key issue involving both conversion efficiency and fabrication cost. A low thermal budget (≤400 °C) passivation technique that remains stable against ultraviolet photons of sunlight for longer period (>20 years) is considered to be the more industrially convenient passivation technique.

In the following paragraphs, we will discuss the advances in surface passivation for high-efficiency c-Si solar cells.

c-Si solar cell surface passivation techniques

Surface passivation has become an essential factor for translating high-efficiency crystalline silicon solar cell concepts into industrial production schemes. In photovoltaics, a widespread method to determine the surface recombination is to measure the effective charge carrier lifetime from the photoconductance of symmetrically passivated silicon wafers in transient or quasi-static mode [5]. An excellent electrical interface quality is essential for many devices relying on the bulk electronic properties of semiconductors [6]. Electrical losses at a semiconductor interface or surface should be minimized in photonic devices based on group III–V or group IV semiconductors when radiative recombination should be the dominant process [7, 8]. Moreover, electronic losses at the crystalline Si (c-Si) surface have become increasingly important in the field of c-Si solar cells due to the trend toward thinner c-Si wafers used as base material. Consequently, the reduction of recombination losses at semiconductor interfaces is a prime concern for numerous semiconductor applications.

Recombination losses at a semiconductor interface or surface can be reduced by two different passivation strategies. As the recombination rate is directly proportional to the interface defect density, the first strategy is based on the reduction in the number of defect states at the interface. The interface defect density can be reduced significantly by the passivation of under coordinated atoms (dangling bonds) by, e.g., atomic H or by a thin dielectric or semiconductor film. This strategy is commonly referred to as chemical passivation. For example, the midgap interface defect density of c-Si can be as low as 1×10^9 eV^{-1} cm^{-2} after the growth of a high-quality thermal SiO_2 film and a subsequent alneal in a H_2 atmosphere, e.g., a FGA [9].

The second strategy to reach surface passivation is based on a significant reduction of the electron or hole concentration at the semiconductor interface by means of a built-in electric field. As recombination processes require both electrons and holes, the highest recombination rate is obtained when the electron and hole concentration at the interface are approximately equal in magnitude (assuming identical capture cross sections for electrons and holes). In other cases the recombination rate scales with the minority carrier concentration at the surface. In the so-called field-effect passivation, the electron or hole concentration at the semiconductor interface is altered by electrostatic shielding of the charge carriers through an internal electric field present at the interface. This internal electric field can either be obtained by a doping profile below the interface or by the presence of fixed electrical charges at the semiconductor interface. Consequently, the application areas of field-effect passivation are limited, but the effect can be employed successfully in devices such as light emitting diodes and solar cells [10].

There are two fundamentally different types of surfaces in a c-Si solar cell: metallized and non-metallized surfaces. Metal silicon interfaces feature very high SRVs and hence need to be carefully designed to avoid excessively large recombination losses. Similarly, to ensure a good blue response of the cell, the illuminated non-metallized surface regions need to be well passivated and not too heavily doped to avoid the formation of a dead layer. In the case of laboratory c-Si cells, the importance of the passivation of both cell surfaces is well recognized.

Fundamental theory of surface recombination

The surfaces or interfaces of a silicon substrate represent a severe discontinuity in its crystalline structure. The large numbers of partially bonded silicon atoms give rise to many dangling bonds, and therefore, a large density of defect levels are found within the bandgap near the semiconductor surface. Even if the silicon surface is not bare, say due to a native oxide, the presence of silicon-oxygen bonds can strain the crystal structure at the surface, which again introduces many defect states.

For a single defect at the surface, the rate of surface recombination, U_s is given by [11]

$$U_s = \frac{n_s p_s - n_i^2}{\frac{n_s + n_1}{S_{p0}} + \frac{p_s + p_1}{S_{n0}}} \tag{1}$$

where n_s and p_s are the concentrations of electrons and holes at the surface, and S_{p0} and S_{n0} are related to the density of surface states per unit area, N_{ts}, the capture cross-sections, σ_n and σ_p, and thermal velocity, v_{th} for the specific defect

$$S_{n0} \equiv \sigma_n v_{th} N_{ts} \quad \text{and} \quad S_{p0} \equiv \sigma_p v_{th} N_{ts} \tag{2}$$

In reality, defect levels are so numerous that they can be considered to be continuously distributed throughout the bandgap, and both their density and capture cross-sections will be dependent on their energy level.

Similar to the definition of the recombination lifetime $\left(\tau(\Delta n) = \frac{\Delta n}{U(\Delta n)}\right)$, the surface recombination velocity, S is defined as

$$U_s \equiv S\Delta n_s \tag{3}$$

where Δn_s is the excess carrier density at the surface.

Hence, the surface recombination velocity can be related to the fundamental properties of the surface defects through (3). For the case of flatband conditions at the surface and negligible carrier trapping, the excess densities of electrons and holes are equal ($\Delta n_s = \Delta p_s$), giving [3]

$$S(\Delta n_s) = \frac{n_0 + p_0 + \Delta n_s}{\frac{n_0 + n_1 + \Delta n_s}{S_{p0}} + \frac{p_0 + p_1 + \Delta n_s}{S_{n0}}} \tag{4}$$

It is the surface recombination velocity, S, that is typically used for quantifying surface recombination processes. In general, the surface passivation is controlled by the reduction of interface defect states (chemical passivation) and by shielding of minority charge carriers from the surface (field-effect passivation). This is illustrated in Eq. 1. The driving force of the surface recombination process is the term $(n_s p_s - n_i^2)$ which describes the deviation of the system from the thermal equilibrium under illumination. The relative importance of either chemical passivation or field-effect passivation depends on the passivation scheme used.

Front surface passivation

Front surface passivation was attained with thermal SiO_2 and TiO_2 fabricated by atmospheric chemical vapor deposition. Thermal growth of silicon oxide is the most effective c-Si surface passivation technique for solar cell, but this technique is not suitable for low-cost industrial processes as it requires very high temperature (above 1,000 °C) treatment. High temperature degrades the bulk life time significantly as well as the stability of passivated surface [12]. Moreover, SiO_2 has very low refractive index which is not suitable for good antireflection performance. On the other hand, TiO_2 does not provide any electronic surface passivation. In the past two decades, various research efforts have been devoted to the development of a more industrially convenient solution, in particular, on passivation layers deposited at low temperature (≤400 °C) onto the c-Si surface. SiN_x front side passivation using plasma-enhanced chemical vapor deposition (PECVD) has been evolved as an alternative to high-temperature oxidation of silicon [13–15]. Record low effective surface recombination velocity (SRV) of 4 cm/s has been obtained on 1 Ω cm p-Si wafers for SiN_x using remote PECVD or high-frequency (13.56 MHz) direct PECVD [14]. SiN_x passivation for front surface of c-Si solar cells is deemed to be superior over other passivation techniques such as SiO_2, TiO_2, etc. due to its (1) field effect passivation provided by positive interface charges, (2) properties of capture cross-sections of dominant defects, (3) adjustable refractive index as antireflection coating (ARC) and (4) hydrogenated passivation of bulk defects [3, 12, 16, 17].

Rear surface passivation

Usage of thinner wafers and an effective reduction of surface recombination losses are increasingly important for low-cost highly efficient silicon solar cells. Hence the recent trend in silicon-wafer-based PV industries is toward thinner wafers (<200 μm). But the use of thinner wafers increases the rear surface recombination [12] and consequently necessitates a well-passivated rear surface to keep the rear surface recombination loss in acceptable minimum. Moreover, the increasing demands for optical quality require a higher internal reflection as well. Therefore, all designs for high-efficiency silicon solar cells today use a dielectric passivation layer on the rear to meet the requirements of passivation quality and optical properties. Keeping in mind the above attributes a good number of rear surface passivation techniques were developed over the years; among them atomic layer deposition (ALD)-assisted Al_2O_3, wet oxidation process using pyrogenic steam, passivation by stack of a-Si:H/SiO_2 or Al_2O_3/SiN_x and passivation by phosphorus-doped a-SiC_xN_y:H(n) alloys are the most notable rear side passivation techniques for crystalline silicon solar cells. Recent research results on these rear surface passivation techniques are aggregated below.

Rear passivation by Al_2O_3

In the case of mc-Si wafers, thermal processes above 900 °C typically lead to a significant degradation of bulk lifetime [18]. Al_2O_3 rear passivation by ALD has been evolved as an alternative of low-temperature surface passivation for high-efficiency silicon solar cells which have comparable properties as that of the alnealed SiO_2 [19–24]. The excellent level of surface passivation by Al_2O_3 films is reflected by SRVs <5 cm/s for low-resistivity p-type c-Si, after alnealing at 400 °C in N_2 [22, 25–27]. A detailed study on ALD deposited Al_2O_3 has been reported by Schmidt et al. [28].

Recently, it was demonstrated that an exceptional high level of c-Si surface passivation can be achieved for SiO_2

synthesized at low temperatures when combined with a very thin Al_2O_3 capping film [29]. The passivation was virtually independent of the SiO_2 thickness and showed an exceptional firing stability. *C–V* measurements revealed that the SiO_2/Al_2O_3 stacks yielded a high level of chemical passivation after alnealing due to a low interface defect density. Preliminary results using an alternative method to synthesize the low-temperature SiO_2 films, and (similar) results for thermally grown SiO_2 [30], have suggested that the effective hydrogenation under influence of the Al_2O_3 capping film may be general. Moreover, preliminary data on the long-term stability showed a low S_{eff} (<4 cm/s) for a stack measured after 250 days. The excellent efficiency values (up to 21.3–21.5 %), the high voltages (680 mV), the excellent long-wavelength quantum efficiency, and the high short-circuit currents of these cells (40 mA/cm^2) are a proof for the low rear-SRV and excellent internal rear-surface reflection [31]. Moreover, Al_2O_3 passivation overcomes the so-called 'parasitic shunting' effects which is very common for SiN_x passivated rear [32]. Low-temperature-synthesized SiO_2/Al_2O_3 passivation schemes can provide a realistic alternative for thermally grown SiO_2 in next-generation solar cells.

The key differentiator of Al_2O_3 in comparison with other passivation schemes is its high fixed *negative* charge density Q_f (10^{12}–10^{13} cm^{-2}) located at the Al_2O_3/Si interface which produces effective field-effect passivation by shielding electrons from the interface [33–35]. The origin of the negative Q_f in Al_2O_3 deposited on c-Si has been attributed to intrinsic and extrinsic defects in Al_2O_3. Matsunaga et al. [36] calculated the energetics of intrinsic vacancies and interstitials in Al_2O_3 from first principles. These calculations showed that each intrinsic point defect is most stable in their fully ionized form. Hence, Al vacancies and O interstitials exhibit a negative charge and Al interstitials and O vacancies exhibit a positive charge in good agreement with the ionic nature of the Al_2O_3 [36]. Extrinsic H has also been proposed as origin for the negative fixed charges in Al_2O_3. Peacock and Robertson [37] calculated that interstitial H acts as a deep trap site for electrons in Al_2O_3. H is, for example, a common constituent in Al_2O_3 synthesized by ALD because H-containing precursors such as $Al(CH_3)_3$ and H_2O are used in the deposition process. Hence, Al vacancies, O interstitials, and interstitial H are proposed as the origin of the negative Q_f in Al_2O_3. Based on the ionic nature of Al_2O_3, Lucovsky [38] postulated that Al_2O_3 consists of tetrahedrally coordinated Al in AlO_{4-} units and octahedrally coordinated Al_{3+} in a ratio of 3:1 to assure charge neutrality. Kimoto et al. [39] demonstrated that both tetrahedrally and octahedrally coordinated Al are present in Al_2O_3 grown by thermal ALD on H-terminated c-Si. However, tetrahedrally coordinated Al was found to be dominant at the interface. This dominance was attributed to the fact that Si in the interfacial SiO_x film also has a tetrahedral coordination [39]. Consequently, the interfacial SiO_x film could fulfill an important role in the origin of the negative Q_f that is found in Al_2O_3 films grown on c-Si by inducing a high-density of negatively charged Al vacancies close to the interface. This hypothesis is in good agreement with the location of the negative Q_f extracted from thickness-dependent *C–V* measurements by various authors [40–42].

The Al_2O_3 passivation mechanism has been elucidated by contactless characterization of c-Si/Al_2O_3 interfaces by optical second-harmonic generation (SHG). SHG has revealed a negative fixed charge density in 10^{11} cm^{-2} that increased to as-deposited Al_2O_3 on the order of 10^{12}–10^{13} cm^{-2} upon alnealing, causing effective field-effect passivation. In addition, multiple photon-induced charge trapping dynamics suggest a reduction in recombination channels after alneal and indicate a c-Si/Al_2O_3 conduction band offset of 2.02 ± 0.04 eV [43]. This makes Al_2O_3 especially suited for the passivation of p-type silicon and p + emitters, while Al_2O_3 is expected to be less suitable for n-type (due to parasitic shunting) and n+ type silicon (yet this still remains to be proven). The thermal stability of Al_2O_3 during firing and the low-temperature deposition of the films are compatible with industrial solar cell processes [44]. Al_2O_3 is transparent (bandgap ~9 eV) and stable under UV illumination [45, 46], but the refractive index of ~1.64 makes it less suitable for single layer ARC on the front side as compared with *a*-SiN_x. As a back reflector, however, Al_2O_3 is superior. Also the fact that very thin Al_2O_3 films (down to 5 nm) can be used is another benefit of Al_2O_3 and enables flexibility in the design of surface passivation schemes and stacks [47]. As a consequence of these (unique) properties, Al_2O_3 is now considered an important candidate for surface passivation in industrial solar cells, either for front and/or rear side passivation. A remaining challenge in this respect is the development of deposition tools for Al_2O_3 films compatible with the requirements, especially in terms of throughput, of the solar cell industry [48, 49].

Wet oxidized rear passivation

A wet oxidation process using pyrogenic steam at 800 °C has been applied on 250 μm thick boron-doped (0.5 Ω cm) high-quality float zone (FZ) silicon to study its suitability for rear side passivation at Fraunhofer ISE [50, 51]. In the long-wavelength region of incident photons, this wet oxidized rear passivation has high internal reflectance which leads to an absolute gain of 6 % in both J_{sc} and V_{oc} compared with Al-back surface field (BSF) cells though there is about 3 % absolute loss in FF due to increase in series resistance for point-contacts [52].

Wet oxides grown from purified steam have been demonstrated to provide a high level of surface passivation that at least is comparable to wet oxides grown by the standard process, where the water vapor is introduced by the pyrolytic generation from high-purity gases (H_2, O_2) [53]. The rear side of laboratory type laser-fired contact (LFC) solar cells has been passivated by a wet oxide grown from purified steam at a temperature of 850 °C. As a reference also a standard wet oxide grown at the same temperature as well as a dry oxide grown at 1,050 °C has been applied for the passivated emitter rear cell (PERC) type solar cells. For all cells, conversion efficiencies well above 20 % could be reached. The average open-circuit voltages for the different rear side passivation layers lie in a very close range of 648.4 mV for the wet oxide grown from purified steam to 650.6 mV for the dry oxidation. The rear SRV that was extracted from the measured quantum efficiency for all oxides is in the range of 80–100 cm/s. Thus, the wet oxides grown from purified steam seem to be well suited for the rear side passivation of PERC-type solar cells. The one-sun parameters of the PERC-type solar cells featuring the rear side passivation by the different dry and wet thermal oxides are summarized in Table 1.

In addition, this wet oxidation process has about one order of magnitude faster growth rate than the traditional high-temperature (1,050 °C) dry oxidation process and can close the gap between laboratory and industrial application by overcoming the longer high thermal budget of dry oxidation [54].

Rear passivation by stack of a-Si:H/SiO_2

A PECVD double layer of amorphous silicon-rich oxynitride and amorphous silicon nitride (SiN_x) can provide a nearly perfect passivation after subsequent rapid thermal process (RTP) and light soaking. The resulting effective minority carriers' lifetime (τ_{eff}) is close to the modeled maximum on p-type as well as on n-type c-Si. Restrictions on the RTP of passivated surfaces, typical of other common passivation schemes (e.g., amorphous Si), are relieved by this double layer. Harsher thermal treatments can be adopted while still obtaining salient passivation. It is shown that the strong quality of surface passivation is ensured by a mechanism that emits electrons from shallow donor states in the passivation layer system and therefore creates a positive field effect [55].

Very low SRVs <6 and <11 cm/s were obtained for SiO_x/*a*-SiN_x:H stacks synthesized by PECVD on low-resistivity n- and p-type c-Si, respectively. The stacks induced a constant effective lifetime under low illumination, comparable to Al_2O_3 on p-type Si. Compared with single layer *a*-SiN_x:H, a lower positive fixed charge density was revealed by SHG measurements, while field-effect passivation was absent for a reference stack comprising thermally grown SiO_2. The results indicate that hydrogenation of interface states played a key role in the passivation and remained effective up to alnealing temperatures >800 °C [56]. The best achieved energy conversion efficiency of 21.7 % for c-Si solar cells (that was confirmed by the Fraunhofer ISE CalLab) was reported in [57, 58].

Passivation by phosphorus-doped a-SiC_xN_y:H(*n*) alloys

Amorphous silicon nitride (a-SiN_x:H) deposited by PECVD was considered as the most feasible alternative to thermally grown silicon oxide on c-Si surface due to its ability to serve as both passivation and anti-reflection coating. Recently, a study conducted on amorphous silicon carbide (a-SiC_x:H) proved that it could be the best alternative to a-SiN_x:H since its optical characteristics can be tuned by the carbon content of the film [14, 59].

Phosphorus-doped amorphous silicon carbide (a-SiC_x:H(*n*)) demonstrated the ability to passivate p-type c-Si substrates and highly doped n-type emitters in solar cells [60–63]. Hydrogenated amorphous silicon carbide (a-SiC:H) provides excellent electronic surface passivation for crystalline silicon solar cells. The hydrogen and carbon content of the passivation layers control the surface passivation depending on hydrogen bonding and alnealing temperature [64, 65]. The carbon content C_c of the amorphous layers varies depending on the methane-to-silane gas flow ratio during deposition. The electronic passivation quality exhibits best thermal stability for an optimum $C_c = 2.3$ at. %. Alnealing this sample under forming gas atmosphere up to TFG = 550 °C enables excellent effective minority carrier lifetimes of 1.2 ms [66]. Hydrogen effusion measurements relate this result to an increase in H-content with rising C_c and to a simultaneous shift of the

Table 1 One-sun parameter (best values) and surface recombination velocity, SRV of LFC solar cells with different rear surface passivations: (1) dry SiO_2, (2) wet SiO_2 (purified steam), (3) wet SiO_2 (pyrolytic generation of water vapor)

	V_{oc} (mV)	J_{sc} (mA/cm^2)	FF (%)	η (%)	SRV (cm/s)
Dry Oxidation (1,050 °C, 97 nm)	652.4	39.9	80.5	21	90
Wet oxidation, purified steam (850 °C, 110 nm)	650.7	40.1	79.1	20.6	80
Wet oxidation, pyrolytic generation (850 °C, 95 nm)	651.7	39.9	80.5	20.9	100

effusion peaks to higher temperatures [65, 67]. Higher carbon content reduces the diffusion of atomic hydrogen out of the amorphous layers. The Si–H bonding configurations in the amorphous layers, analyzed from infrared absorption spectroscopy, reveal that a-SiC:H layers with lower carbon content have a higher density. Increasing C_c induces voids and microvoids in the amorphous structure, favoring the diffusion of molecular hydrogen out of the a-SiC:H layers [68, 69].

Very silicon-rich films yielded effective SRVs at 1 sun-illumination as low as 3 and 2 cm s^{-1} on 1 Ω cm p- and n-type crystalline silicon substrates, respectively [70]. An efficiency of 20.2 % was reported for PERC following the approach in [34, 63]. The reported values had been obtained without forming gas alneal (FGA) and the stability in surface passivation had been verified for a 2-month period. Such stacks have the additional benefit of being more dielectric than silicon-rich films, thus avoiding the degradation of field effect passivation due to shunting effects [32]. This result demonstrates that a-SiC_xN_y:H(n) films can be useful for designing high-efficiency c-Si solar cells.

Passivation by Al_2O_3/SiN_x-layer stack

Despite the merits of Al_2O_3, it was shown that the samples which are passivated by a single layer of Al_2O_3 without the protecting SiN_x layer show a largely destroyed Al_2O_3 layer after metallization etch-back [71]. Investigations on lifetime samples showed a 2.5-fold increase in effective lifetime and an efficiency gain of 0.7 % absolute (max. efficiency 18.6 %) for surfaces passivated by an Al_2O_3/SiN_x stack compared with fully metallized Al-BSF rear sides. An increase in the infrared spectrum of the internal quantum efficiency is determined as the source of this gain. However, this cell exhibited a reduction in FF due to an increase in series resistance locally. For Al_2O_3 and Al_2O_3/a-SiN_x:H stacks on 2 Ω cm n-type c-Si, ultralow SRVs of $S_{eff} < 3$ cm/s were obtained and the passivation proved sufficiently stable ($S_{eff} < 14$ cm/s) against a high-temperature "firing" process (>800 °C) used for screen printed c-Si solar cells [72].

Discussion and outlook

Surface recombination is very effectively suppressed by means of silicon dioxide (SiO_2) grown in a high-temperature (≥900 °C) oxidation process. Very low SRVs are in particular realized at the lightly doped rear surface, where the combination of a thermally grown SiO_2 layer with an evaporated film of Al give—after an additional alnealing treatment at 400 °C (the so-called 'alneal')—SRVs below 20 cm/s on un-metallized low-resistivity (~1 Ω cm) p-type silicon wafers. In addition, the SiO_2/Al stack at the cell rear acts as an excellent reflector for near bandgap photons, significantly improving the light-trapping properties and hence the short-circuit current of the cell.

One of the main reasons why high-temperature oxidation has not been implemented into the majority of industrial cell processes up to now is the high sensitivity of the silicon bulk lifetime to high-temperature processes. In particular in the case of multi-crystalline silicon wafers, thermal processes above 900 °C typically lead to a significant degradation of the bulk lifetime. Hence, low-temperature surface passivation alternatives are required for future industrial high-efficiency silicon solar cells, which should have comparable properties as the alnealed SiO_2.

One intensively investigated low-temperature surface passivation alternative to thermal oxide is silicon nitride (SiN_x) grown by PECVD at ~400 °C, which has proven to give comparably low SRVs as thermal SiO_2 on low-resistivity p-type silicon [14, 73]. However, when applied to the rear of passivated emitter and rear cell (PERC)-type solar cells the short-circuit current density is strongly reduced compared with the SiO_2-passivated cell rear [74]. This effect has been attributed to the large density of fixed positive charges within the SiN_x layer, inducing an inversion layer in the crystalline silicon underneath the SiN_x. The coupling of this inversion layer to the base contact leads to a significant loss in the short-circuit current density. This detrimental effect is known as 'parasitic shunting' [32]. Another alternative low-temperature passivation scheme resulting in comparable SRVs as alnealed SiO_2 is intrinsic hydrogenated amorphous silicon (a-Si) deposited by PECVD in the temperature range between 200 and 250 °C [75]. Despite the fact that no parasitic shunting occurs in the case of a-Si passivated cell rear, new problems arise from the high sensitivity of the a-Si passivation to thermal processes. More recently, it was shown that thin films of aluminum oxide (Al_2O_3) grown by ALD provide an excellent level of surface passivation on lowly doped (~1 Ω cm) p- and n-type silicon wafers [25, 28, 76] as well as on highly boron-doped p+ emitters [75].

Figure 1 shows a comparison of the injection dependent effective SRVs measured on 1–2 Ω cm p-type silicon wafers passivated by (1) SiN_x deposited by remote- PECVD [78], (2) intrinsic a-Si deposited in a parallel-plate PECVD reactor [75], and (3) Al_2O_3 deposited by means of plasma-assisted ALD [79]. The measurements are shown in the injection range between 10^{12} and 10^{15} cm^{-3}, which is the most relevant range for the rear of one-sun silicon solar cells. Within this injection range it becomes obvious from Fig. 1 that the Al_2O_3 passivation outperforms the well-optimized SiN_x as well as the a-Si passivation. Importantly, the Al_2O_3 passivation results in a very weak injection-level

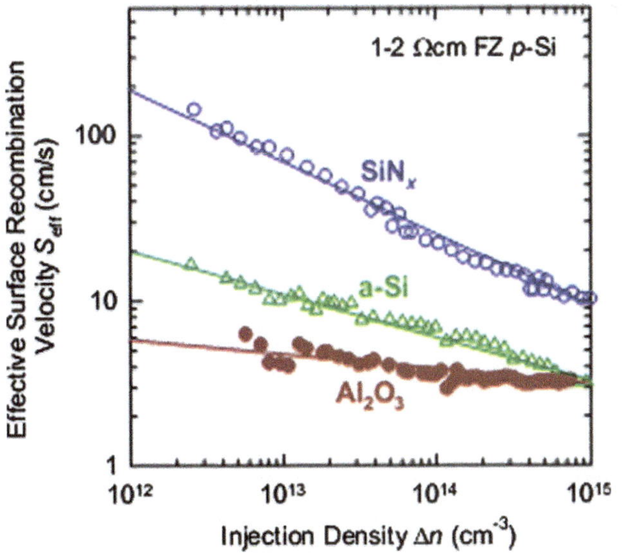

Fig. 1 Comparison of the injection-dependent effective SRVs S_{eff} (Δn) measured on 1–2 Ω cm p-type floatzone silicon wafers passivated by (1) SiN_x deposited by remote-PECVD [78], (2) intrinsic a-Si deposited in a parallel-plate PECVD reactor [75] and (3) Al_2O_3 deposited by means of plasma-assisted ALD [79]. The *lines* are guides to the eye

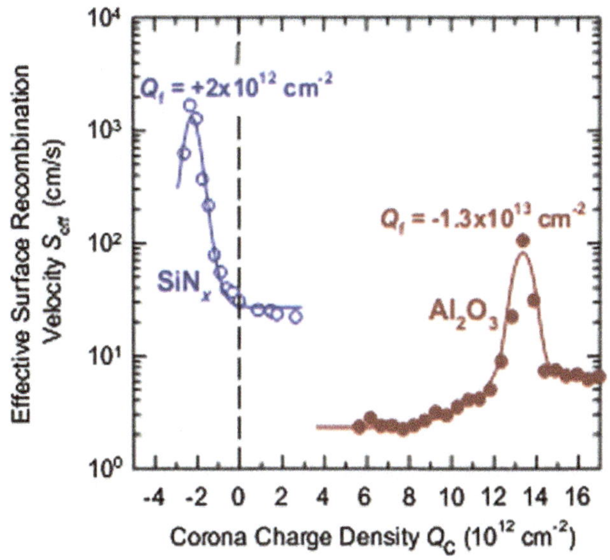

Fig. 2 Measured effective SRV S_{eff} as a function of the corona charge density Q_C deposited onto dielectric passivation layers of SiN_x and Al_2O_3. The *lines* are guides to the eye

dependence in the injection range between 10^{12} and 10^{15} cm^{-3}, and the effective SRV stays well below 10 cm/s even at low injection densities $<10^{13}$ cm^{-3}. The strong injection-level dependence of the effective SRV measured on SiN_x-passivated p-Si surfaces can be attributed to recombination within the space charge region induced by the high fixed positive charge density within the SiN_x layer [73].

Figure 2 shows the measured effective SRV as a function of the corona charge density deposited onto dielectric passivation layers of SiN_x and Al_2O_3. In the flatband case when the deposited Corona charge density equals the fixed charge density within the dielectric layer, the recombination rate shows a maximum. From this maximum the charge density within the layer can be deduced. As can be seen from Fig. 2, the fixed charge density within the SiN_x layer is positive and amounts to $Q_f = +2 \times 10^{12}$ elementary charges/cm^2, whereas the fixed charge density within the Al_2O_3 layer is negative and amounts to $Q_f = -1.3 \times 10^{13}$ cm^{-2}. The fixed negative charge density within the Al_2O_3 layer induces an accumulation layer at the p-type silicon surface in contrast to the SiN_x layer inducing and inversion layer. As a consequence, the Al_2O_3 provides a very effective field-effect passivation without the additional contribution of recombination in the space charge region, which is the dominant recombination channel in the case of the SiN_x passivation of p-type silicon surfaces at low injection densities [73, 77]. In addition, due to the formation of an accumulation layer instead of an inversion layer at the p-type silicon surface, the above-mentioned parasitic shunting effect at the solar cell rear is not expected for an Al_2O_3-rear-passivated cell. In combination with its very high transparency for near-bandgap photons, ALD-deposited Al_2O_3 should hence be an optimal choice for a dielectric layer at the silicon solar cell rear.

In addition to the outstanding surface passivation provided by ALD-deposited Al_2O_3, the deposition process itself is also beneficial from an application point of view. In contrast to the conventionally applied PECVD, ALD consists of two self-limiting half-reactions, which implies several important advantages: (1) ALD gives highly conformal coatings, which allows to deposit and passivate, e.g., deep trenches or even pores in silicon, (2) pin-hole and particle-free deposition is achieved, (3) as ALD is a self-limiting process, uniform films can be deposited over large areas with mono-layer growth control, and (4) very low impurity concentrations of deposited films and hence very high film quality is achieved.

The main disadvantage of ALD for photovoltaic applications is its relatively low deposition rate. However, this disadvantage can be overcome by depositing ultrathin (2–30 nm) ALD-Al_2O_3 films and capping them with a thicker film of, e.g., PECVD-SiO_x, SiN_x or SiC_x [80]. Apart of the advantageous optical properties of these stacks, the passivation quality of such ALD/PECVD stacks can even be superior to that of single layers of Al_2O_3, which could be attributed to the hydrogenation of interface states at the Al_2O_3/Si interface during deposition of the hydrogen-rich PECVD layer [28]. Combination of ALD and PECVD might hence be a key technology for future industrial high-efficiency solar cells.

Table 2 Surface passivation methods of c-Si solar cells that were developed in the past 30 years

Cell concepts/types	Types of passivation used	Remarks	Efficiency (%)
Screen-printed cells	p–n junction passivation	Larger shading loss, poor blue response, modest short-circuits currents	15–16
Buried-contact cells	Thermal oxide passivation	Reduced shading loss by 50 %	17–18
PERL/LBSF cells	Thermal oxide passivation	Lower surface recombination	24.7
Backside-contact concentrator solar cells	Thermal oxide passivation	Eliminate metal shading loss, facilitating the interconnection of individual cells	22.7–26
Bifacial cells	Silicon nitride passivation	Both surfaces of the cell are passivated, minimize reflection losses, etc	18.1–20.1
MIS inversion-layer cells	Silicon nitride passivation	Low temperature treatment, promising candidates for cost-effective generation of PV electricity	18.5
HIT cells	Double-layer stack of a-Si	Low SRV, junction is formed at only 200 °C	20.0
Thin-film polycrystalline silicon cells	PECVD deposited silicon of only 2 μm	Very thin	10.1

Efficiencies shown in the table are the theoretical limit

Aggregated results of efficiencies of different cell types after exploiting different passivation schemes are shown in Table 2.

Conclusions

Solar cells made from crystalline silicon have lower conversion efficiency and hence optimization of each process step is very important. Increasing the efficiency of crystalline silicon solar cells relies on the understanding and optimization of each individual processing step as well as of the interplay between the material properties and the processing conditions. Our focus was to review the advances in existing surface passivation techniques in an industrial process line as well as in the research laboratories over the world. Every c-Si solar cell fabricated to date features one or more of these surface passivation methods. With regard to solar cell applications, it is important to make sure that the surface passivation is long-term stable (>20 years) and stable against the UV photons of sunlight.

In order to increase the efficiency of silicon-wafer-based solar cells in production well above 20 %, it is indispensable to improve the currently applied level of surface passivation at the front as well as at the rear of the cells. In the past, low surface recombination velocities (<20 cm/s) have been achieved on low-resistivity ($\sim 1\ \Omega$ cm) p-type silicon using plasma-enhanced chemical-vapor-deposited (PECVD) silicon nitride (SiN_x) as well as amorphous silicon (a-Si). However, the high density of fixed positive charges within the PECVD-SiN_x layer induces an inversion layer at the rear of p-type Si cells, producing a detrimental parasitic shunting, which reduces the short-circuit current density by up to 3 mA/cm^2. The passivation quality of a-Si, on the other hand, is very temperature sensitive. More recently it has been shown that atomic-layer-deposited (ALD) aluminum oxide (Al_2O_3) provides an outstanding level of surface passivation, which can be attributed to its extremely high *negative* fixed charge density in combination with the very gentle deposition technique ALD, leading to low interface state densities. The application of these ALD-Al_2O_3 layers to the rear of p-type solar cells shows that this new passivation scheme is indeed suitable for high efficiencies and that due to the large negative fixed charge density no parasitic shunting occurs. In a direct comparison with other passivation schemes, it is found that Al_2O_3 even outperforms optimized thermally grown SiO_2 and opens the possibility of achieving very large open-circuit voltages up to $V_{oc} = 740$ mV.

However, whenever a particular process is required to be optimized, the amount of extra steps should be kept to an absolute minimum and ideally the general cell line concept should not be changed drastically to make the approach cost-effective and easy to implement.

References

1. Hall, R.: Electron-hole recombination in germanium. Phys. Rev. **87**, 387 (1952)
2. Shockley, W., Read, W.: Statistics of the recombinations of holes and electrons. Phys. Rev. **87**, 835–842 (1952)

3. Aberle, A.: Crystalline Silicon Solar Cells: Advanced Surface Passivation and Analysis. Centre for Photovoltaic Engineering, University of NSW, Australia (1999)
4. http://www.iea.org
5. Seiffe, J., Hofmann, M., Rentsch, J., Preu, R.: Charge carrier trapping at passivated silicon surfaces. J. Appl. Phys. **109**, 064505 (2011)
6. Röder, T., Eisele, S., Grabitz, P., Wagner, C., Kulushich, G., Köhler, J., Werner, J.: Add-on laser tailored selective emitter solar cells. Prog. Photovolt. Res. Appl. **18**, 505–510 (2010)
7. Queisser, H.J., Haller, E.E.: Defects in semiconductors: some fatal, some vital. Science **281**, 945–950 (1998)
8. Boroditsky, M., Gontijo, I., Jackson, M., Vrijen, R., Yablonovitch, E., Krauss, T., Cheng, C.C., Scherer, A., Bhat, R., Krames, M.: Surface recombination measurements on III–V candidate materials for nanostructure light-emitting diodes. J. Appl. Phys. **87**, 3497 (2000)
9. Jin, H., Weber, K.J., Dang, N.C., Jellett, W.E.: Defect generation at the Si–SiO_2 interface following corona charging. Appl. Phys. Lett. **90**, 262109 (2007)
10. Hoex, B., Gielis, J.J.H., van de Sanden, M.C.M., Kessels, W.M.M.: On the c-Si surface passivation mechanism by the negative-charge-dielectric Al_2O_3. J. Appl. Phys. **104**, 113703 (2008)
11. Rein, S.: Lifetime Spectroscopy: A Method of Defect Characterization in Silicon for Photovoltaic Applications. Springer, Berlin (2004)
12. Aberle, A.: Surface passivation of crystalline silicon solar cells: a review. Prog. Photovolt. Res. Appl. **8**, 473–487 (2000)
13. Hoex, B., Heil, S., Langereis, E., Sanden, M., Kessels, W.: Ultralow surface recombination of c-Si substrate passivated by plasma-assisted atomic layer deposited Al_2O_3. Appl. Phys. Lett. **89**, 1–3 (2006)
14. Lauinger, T., Schmidt, J., Aberle, A.G., Hezel, R.: Record low surface recombination velocities on 1 ohm-cm p-silicon using remote plasma silicon nitride passivation. Appl. Phys. Lett. **68**, 1232–1234 (1996)
15. Leguijt, C., Lölgen, P., Eikelboom, J.A., Weeber, A.W., Schuurmans, F.M., Sinke, W.C., Alkemade, P.F.A., Sarro, P.M., Marée, C.H.M., Verhoef, L.A.: Solar energy materials. Solar Cells **40**, 297–345 (1996)
16. Nagel, H., Aberle, A., Hezel, R.: Optimisedantireflection coatings for planar silicon solar cells using remote PECVD silicon nitride and poroud silicon dioxide. Prog. Photovolt. Res. Appl. **17**, 245–260 (1999)
17. Kimura, K.: Recent developments in polycrystalline silicon solar cells. In: Proceedings of the 1st International Photovoltaic Science and Engineering Conference, Kobe, Japan, pp. 37–42, 13–16 November 1984
18. Stocks, M., Cuevas, A., Blakers, A.: Minority carrier lifetimes of multicrystalline silicon during solar cell processing. In: Proceedings of the 14th European Photovoltaic Solar Energy Conference, Barcelona, Spain, pp. 770–773, 30 June–4 July 1997
19. Agostinelli, G., Delabie, A., Vitanov, P., Alexieva, Z., Dekkers, H.F.W., De Wolf, S., Beaucarne, G.: Very low surface recombination velocities on p-type silicon wafers passivated with a dielectric with fixed negative charge. Sol. Energy Mater. Sol. Cells **90**, 3438 (2006)
20. Hoex, B., Heil, S.B.S., Langereis, E., van de Sanden, M.C.M., Kessels, W.M.M.: Ultralow surface recombination of c-Si substrate passivated by plasma-assisted atomic layer deposited Al_2O_3. Appl. Phys. Lett. **89**, 042112 (2006)
21. Hoex, B., Schmidt, J., Bock, R., Altermatt, P.P., van de Sanden, M.C.M., Kessels, W.M.M.: Excellent passivation of highly doped p-type Si surfaces by the negative-charge-dielectric Al_2O_3. Appl. Phys. Lett. **91**, 112107 (2007)
22. Hoex, B., Schmidt, J., Pohl, P., van de Sanden, M.C.M., Kessels, W.M.M.: Silicon surface passivation by atomic layer deposited Al_2O_3. J. Appl. Phys. **104**, 044903 (2008)
23. Schmidt, J., Merkle, A., Brendel, R., Hoex, B., van de Sanden, M.C.M., Kessels, W.M.M.: Surface passivation of high efficiency solar cells by atomic layer deposited Al_2O_3. Prog. Photovolt. **16**, 461 (2008)
24. Benick, J., Hoex, B., van de Sanden, M.C.M., Kessels, W.M.M., Schultz, O., Glunz, S.: High efficiency n-type Si solar cells on Al_2O_3-passivated boron emitters. Appl. Phys. Lett. **92**, 253504 (2008)
25. Chen, M.J., Shih, Y.T., Wu, M.K., Tsai, F.Y.: Enhancement in the efficiency of light emission from silicon by a thin Al_2O_3 surface-passivating layer grown by atomic layer deposition at low temperature. J. Appl. Phys. **101**, 033130 (2007)
26. Dingemans, G., Kessels, W.M.M.: Recent progress in the development and understanding of silicon surface passivation by aluminum oxide for photovoltaics. In: Proceedings of the 25th European Photovoltaic Solar Energy Conference, Valencia, Spain (2010)
27. Grant, N.E., McIntosh, K.R.: Silicon surface by silicon dioxide grown in nitric acid. IEEE Electron Device Lett. **30**, 922 (2009)
28. Schmidt, J., Merkle, A., Brendel, R., Hoex, B., Sanden, M., Kessels, W.: Surface passivation of high efficiency solar cells by atomic layer deposited Al_2O_3. Prog. Photovolt. Res. Appl. **16**, 461–466 (2008)
29. Dingemans, G., van de Sanden, M.C.M., Kessels, W.M.M.: Excellent Si surface passivation by low temperature SiO_2 using an ultrathin Al_2O_3 capping film. Phys. Status Solidi RRL **5**(1), 22–24 (2011).
30. Saint-Cast, P., Benick, J., Kania, D., Weiss, L., Hofmann, M., Rentsch, J., Preu, R., Glunz, S.W.: High-efficiency c-Si solar cells passivated with ALD and PECVD aluminum oxide. IEEE Elec. Dev. Lett. **31**(7), 695–697 (2010)
31. Dingemans, G., et al.: Hydrogen induced passivation of Si interfaces by Al_2O_3 films and SiO_2/Al_2O_3 stacks. Appl. Phys. Lett. **97**, 152106 (2010)
32. Dauwe, S., Mittelstädt, L., Metz, A., Hezel, R.: Experimental evidence of parasitic shunting in silicon nitride rear surface passivated solar cells. Prog. Photovolt. Res. Appl. **10**, 271–278 (2002)
33. Saint-Cast, P., Kania, D., Hofmann, M., Benick, J., Rentsch, J., Preu, R.: Very low surface recombination velocity on p-type c-Si by high-rate plasma-deposited aluminum oxide. Appl. Phys. Lett. **95**, 151502 (2009)
34. Dingemans, G., van de Sanden, M.C.M., Kessels, W.M.M.: Influence of the deposition temperature on the c-Si surface passivation by Al_2O_3 films synthesized by ALD and PECVD. Electrochem. Solid-State Lett. **13**, H76 (2009)
35. Werner, F., Veith, B., Zielke, D., Kühnemund, L., Tegenkamp, C., Seibt, M., Brendel, R., Schmidt, J.: Electronic and chemical properties of the c-Si/Al_2O_3 interface. J. Appl. Phys. **109**, 113701 (2011)
36. Matsunaga, K., Tanaka, T., Yamamoto, T., Ikuhara, Y.: First-principles calculations of intrinsic defects in Al_2O_3. Phys. Rev. B **68**, 085110 (2003)
37. Peacock, P.W., Robertson, J.: Behavior of hydrogen in high dielectric constant oxide gate insulators. Appl. Phys. Lett. **83**, 2025 (2003)
38. Lucovsky, G.: A chemical bonding model for the native oxides of the III–V compound semiconductors. J. Vac. Sci. Technol. **19**, 456 (1981)
39. Kimoto, K., Matsui, Y., Nabatame, T., Yasuda, T., Mizoguchi, T., Tanaka, I., Toriumi, A.: Coordination and interface analysis of atomic-layer-deposition Al_2O_3 on Si(001) using energy-loss near-edge structures. Appl. Phys. Lett. **83**, 4306 (2003)

40. Aboaf, J.A., Kerr, D.R., Bassous, E.: Charge in SiO_2–Al_2O_3 double layers on silicon. J. Electrochem. Soc. **120**, 1103 (1973)
41. Johnson, R.S., Lucovsky, G., Baumvol, I.: Physical and electrical properties of noncrystalline Al_2O_3 prepared by remote plasma enhanced chemical vapor deposition. J. Vac. Sci. Technol. A **19**, 1353 (2001)
42. No, S.Y., Eom, D., Hwang, C.S., Kim, H.J.: Property changes of aluminum oxide thin films deposited by atomic layer deposition under photon radiation. J. Electrochem. Soc. **153**, F87 (2006)
43. Gielis, J.J.H., Hoex, B., van de Sanden, M.C.M., Kessels, W.M.M.: Negative charge and charging dynamics in Al_2O_3 films on Si characterized by second-harmonic generation. J. Appl. Phys. **104**, 073701 (2008)
44. Dingemans, G., et al.: Firing stability of atomic layer deposited Al_2O_3 for c-Si surface passivation. In: Proceedings of the 34th IEEE Photovoltaics Specialists Conference, Philadelphia, USA (2009)
45. Schmidt, J., et al.: Progress in the surface passivation of silicon solar cells. In: Proceedings of the 23rd European Photovoltaic Solar Energy Conference, Valencia, Spain (2008)
46. Dingemans, G., Seguin, R., Engelhart, P., Einsele, F., Hoex, B., van de Sanden, M.C.M., Kessels, W.M.M.: Stability of Al_2O_3 and Al_2O_3/a-SiNx:H stacks for surface passivation of crystalline silicon. J. Appl. Phys. **106**, 114907 (2009)
47. Dingemans, G., Seguin, R., Engelhart, P., van de Sanden, M.C.M., Kessels, W.M.M.: Silicon surface passivation by ultrathin Al_2O_3 films synthesized by thermal and plasma atomic layer deposition. Phys. Status Solidi RRL **4**, 10 (2010)
48. Terlinden, N.M., et al.: Role of field-effect on c-Si surface passivation by ultrathin (2–20 nm) atomic layer deposited Al_2O_3. Appl. Phys. Lett. **96**, 112101 (2010)
49. Dingemans, G., Engelhart, P., Seguin, R., Mandoc, M.M., van de Sanden, M.C.M., Kessels, W.M.M.: Comparison between Al_2O_3 surface passivation films deposited with thermal ALD, plasma ALD and PECVD. In: 35th IEEE PVSC, Honolulu, Hawaii, 20–25 June 2010
50. Stocks, M., Cuevas, A.: Surface recombination velocity of thermally oxidized multicrystalline silicon. In: Proceedings of the 2nd World Conference on Photovoltaic Energy Conversion, Vienna, Austria, pp. 1623–1626, 6–10 July 1998
51. Schultz, O., Mette, A., Hermle, M., Glunz, S.: Thermal oxidation for crystalline silicon solar cells exceeding 19 % efficiency applying industrially feasible process technology. Prog. Photovolt. Res. Appl. **16**, 317–324 (2008)
52. Catchpole, K., Blakers, A.: Modelling of the PERC structure with strip and dot back contacts. In: Proceedings of the 16th European Photovoltaic Solar Energy Conference, Glasgow, UK, pp. 1719–1722, 1–5 May 2000
53. Benick, J., Zimmermann, K., Spiegelman, J., Hermle, M., Glunz, S.W.: Rear side passivation of PERC-type solar cells by wet oxides grown from purified steam. Prog. Photovolt. Res. Appl. **19**, 361–365 (2011)
54. Schultz, O., Glunz, S., Goldschmidt, J., Lautenschlager, H., Leimenstoll, A., Schneiderlöchner, E., Willeke, G.: Thermal oxidation processes for high efficiency multicrystalline silicon solar cells. In: Proceedings of the 19th European Photovoltaic Solar Energy Conference, Paris, France, pp. 604–607, 7–11 June 2004
55. Seiffe, J, Gautero, L, Hofmann, M, Rentsch, J, Preu, R, Weber, S, Eichel, R.A.: Surface passivation of crystalline silicon by plasma-enhanced chemical vapor deposition double layers of silicon-rich silicon oxynitride and silicon nitride. J. Appl. Phys. **109**, 034105 (2011)
56. Dingemans, G., Mandoc, M.M., Bordihn, S., van de Sanden, M.C.M., Kessels, W.M.M.: Effective passivation of Si surfaces by plasma deposited SiO_x/a-SiN_x:H stacks. Appl. Phys. Lett. **98**, 222102 (2011)
57. Hofmann, M., Schmidt, C., Kohn, N., Rentsch, J., Glunz, S.W., Preu, R.: Stack system of PECVD amorphous silicon and PECVD silicon dioxide for silicon solar cell rear side passivation. Prog. Photovolt. Res. Appl. **16**, 509–518 (2008)
58. Glunz, S., Grohe, A., Hermle, M., Hofmann, M., Janz, S., Roth, T., Schultz, O., Vetter, M., Martin, I., Ferré, R., Bermejo, S., Wolke, W., Warta, W., Preu, R., Willeke, G.: Comparison of different dielectric layers for application in industrially feasible high-efficiency crystalline silicon solar cells. In: Proceedings of the 20th European Photovoltaic Solar Energy Conference, Barcelona, Spain, pp. 572–577, 6–10 June 2005
59. Martín, I., Vetter, M., Orpella, A., Voz, C., Puigdollers, J., Alcubilla, R.: Characterization of p-type c-si surface passivation by n-doped a-SiC_x:H films In: Proceedings of the 17th European Photovoltaic Solar Energy Conference, Munich, Germany, pp. 2954–2957, 22–26 October 2001
60. Ferre, R., Martín, I., Vetter, M., Garín, M., Alcubilla, R.: Effect of amorphous silicon carbide layer thickness on the passivation quality of crystalline silicon surface. Appl. Phys. Lett. **87**, 202109–202111 (2005)
61. Janz, S., Riepe, S., Hofmann, M., Reber, S., Glunz, S.: Phosphorous-doped SiC as an excellent p-type Si surface passivation layer. Appl. Phys. Lett. **88**, 133516–133517 (2006)
62. Martín, I., Vetter, M., Orpella, A., Puigdollers, J., Cuevas, A., Alcubilla, R.: Surface passivation of p-type crystalline Si by plasma enhanced chemical vapor deposited amorphous SiC_x:H films. Appl. Phys. Lett. **79**, 2199–2201 (2001)
63. Ferre R, Martín, I., Ortega, P., Vetter, M., Torres, I., Alcubilla, R.: n-Type emitter surface passivation in c-Si solar cells by means of antireflective amorphous silicon carbide layers. J. Appl. Phys. **100**, 073703–073710 (2006)
64. Folsch, J., Robel, H., Schade, H.: Change in bonding properties of amorphous hydrogenated silicon-carbide layers prepared with different gases as carbon sources. Appl. Phys. Lett. **61**, 3029–3031 (1992)
65. Street, R.A.: Hydrogen chemical potential and structure of a-Si:H. Phys. Rev. B. **43**, 2454–2457 (1991)
66. Ehling, C., Bilger, G., Schubert, M.B.: Thermally stable a-Si1-xCx:H passivation layers. In: Proc. 24th European Photovoltaic Solar Energy Conference, Germany, 2009
67. Langford, A.A., Fleet, M.L., Nelson, B.P., Lanford, W.A., Maley, N.: Infrared absorption strength and hydrogen content of hydrogenated amorphous silicon. Phys. Rev. B. **45**, 13367–13377 (1992)
68. Iqbal, Z., Vepfek, S.: Raman scattering from hydrogenated microcrystalline and amorphous silicon. J. Phys. C Solid State Phys. **15**, 377–392 (1982)
69. Gall, S., Muske, M., Sieber, S., Nast, O., Fuhs, W.: Aluminum-induced crystallization of amorphous silicon. J. Non Cryst. Solids **299**, 741–745 (2002)
70. Ferre, R., Orpella, A., Munoz, D., Martín, I., Recart, F., Voz, C., Puigdollers, J., Cabarrocas, P., Alcubilla, R.: Very low surface recombination velocity of crystalline silicon passivated by phosphorus-doped a-SiC_xN_y:H(n) alloys. Prog. Photovolt. Res. Appl. **16**, 123–127 (2008)
71. Lauermann, T., Lüder, T., Scholz, S., Raabe, B., Hahn, G., Terheiden, B.: Enabling dielectric rear side passivation for industrial mass production by developing lean printing based solar cell processes. In: 35th IEEE PVSC, Honolulu, 20–25 June 2010
72. Dingemans, G., Engelhart, P., Seguin, R., Einsele, F., Hoex, B., van de Sanden, M.C.M., Kessels, W.M.M.: Stability of Al_2O_3 and Al_2O_3/a-SiN_x:H stacks for surface passivation of crystalline silicon. J. Appl. Phys. **106**, 114907 (2009)
73. Schmidt, J., Moschner, J.D., Henze, J., Dauwe, S., Hezel, R.: Proc. 19th EUPVSEC, Paris, France, p. 391 (2004)
74. Dauwe, S., Mittelstädt, L., Metz, A., Schmidt, J., Hezel, R.: Low-temperature rear surface passivation schemes for >20% efficient

silicon solar cells. In: Proceedings of the 3rd WCPEC, Osaka, Japan, p. 1395 (2003)
75. Dauwe, S., Schmidt, J., Hezel, R.: Very low surface recombination velocities on p- and n-type silicon wafers passivated with hydrogenated amorphous silicon films. In: Proceedings of the 29th IEEE PVSEC, New Orleans, USA, pp. 1246–1249 (2002)
76. Agostinelli, G., Delabie, A., Vitanov, P., Alexieva, Z., Dekkers, H.F.W., De Wolf, S., Beaucarne, G.: Very low surface recombination velocities on p-type silicon wafers passivated with a dielectric with fixed negative charge. Sol. En. Mat. Sol. Cells **90**, 3438 (2006)
77. Hoex, B., Schmidt, J., Bock, R., Altermatt, P.P., van de Sanden, M.C.M., Kessels, W.M.M.: Excellent passivation of highly doped p-type Si surfaces by the negative-charge-dielectric Al_2O_3. Appl. Phys. Lett. **91**, 112107 (2007)
78. Dauwe, S.: Low-temperature rear surface passivation of crystalline silicon solar cells.Ph.D. thesis, ISFH, University of Hanover, Germany (2003)
79. Hoex, B., Schmidt, J., Pohl, P., van de Sanden, M.C.M., Kessels, W.M.M.: Silicon surface passivation by atomic layer deposited Al_2O_3. J. Appl. Phys. **104**, 044903 (2008)
80. Schmidt, J., Merkle, A., Hoex, B., van de Sanden, M.C.M., Kessels, W.M.M., Brendel, R.: Atomic-layer-deposited aluminium oxide for the surface passivation of high-efficiency silicon solar cells. In: Proceedings of the 33rd IEEE PVSC, San Diego, USA (2008)

2

Metal-organic vapor-phase epitaxial growth of InGaN and InAlN for multi-junction tandem solar cells

A. Yamamoto · K. Sugita · A. G. Bhuiyan · A. Hashimoto · N. Narita

Abstract Metal-organic vapor-phase epitaxial (MOVPE) growth of InGaN and InAlN has been studied to prepare a wanted band-gap from 0.65 to 2.5 eV for multi-junction tandem solar cells. The main subjects in the growth of InGaN are the suppression of phase separation and metallic In incorporation and the control of composition in grown films. Both phase separation and metallic In segregation can be avoided by choosing the appropriate substrate position on the susceptor. By optimizing growth temperature and TMI/(TMI + TEG) molar ratio, InGaN films with full composition range are successfully grown. The Mg-doping behavior of MOVPE InGaN (In composition 0.1–0.4) is also studied using Cp_2Mg as a Mg source. The growth behavior of InAlN is studied with the dominant parameters such as growth pressure, TMI/(TMI + TMA) molar ratio and substrate position on the susceptor. The major difficulty in the InAlN growth is found to be the adduct formation by the parasitic reaction of TMA and NH_3. By employing the atmospheric-pressure growth, adduct-free InAlN films are grown with a reasonable growth rate ($\sim$1 μm/h). This enables us to grow InAlN films with an In content from 0.3 to 1, corresponding to band-gaps from 3.6 to 0.65 eV. In order to demonstrate an ability to prepare these different alloys sequentially, InAlN/InGaN hetero-structures are prepared and the photo-response is observed for the first time for an n-InAlN/p-InGaN hetero-junction.

Keywords Multi-junction tandem solar cell · InGaN · InAlN · MOVPE · n-InAlN/p-InGaN hetero-junction

A. Yamamoto (✉) · A. Hashimoto
Graduate School of Engineering, University of Fukui, 3-9-1 Bunkyo, Fukui 910-8507, Japan
e-mail: ayamamot@u-fukui.ac.jp

K. Sugita
HISAC, University of Fukui, 3-9-1 Bunkyo, Fukui 910-8507, Japan

A. G. Bhuiyan
Khulna University of Engineering and Technology, Khulna 9203, Bangladesh

N. Narita
Energy Use R&D Centre, The Kansai Electric Power Co., Inc, 3-11-20 Wakoji, Amagasaki, Hyogo 661-0974, Japan

Introduction

Photovoltaic power generation is expected to be one of the key technologies for the realization of a low-carbon society. For this purpose, we need to develop low cost and high conversion efficiency solar cells. The most effective way to realize an high conversion efficiency is to fabricate a multi-junction tandem solar cell, where many sub-cells composed of a different band-gap material are stacked perpendicularly and electrically connected in series. Figure 1 shows the relationships between band-gap energy and lattice parameter for typical semiconductor materials applicable to solar cell. In order to realize a high-efficiency multi-junction cell, semiconductor materials with band-gap energies from 0.6 to 2.5 eV are required [1]. Since the band-gap of InN was found to be about 0.7 eV [2], InN-based nitride semiconductor alloys, such as InGaN and InAlN, have had much attention as materials for multi-junction solar cells [1, 3]. This is mainly because a wide range of band-gaps can be realized by changing only their composition; from 0.65 to 3.4 eV with InGaN and from 0.65 to 6.2 eV with InAlN. Figure 1 shows that there is no such a material except for the InN-based materials. The authors have shown that, using InGaN or InAlN films with a band-gap energy from 0.7 to 2.5 eV, a multi-junction tandem solar cell with a conversion efficiency more than 50 % is expected to be realized [1]. Thus, the InN-based alloys are ones of a few of the promising materials for multi-junction solar cells.

Since 2003, the studies on InN-based solar cells, especially InGaN cells, have been started using Ga-rich materials [1, 3]. Single-junction cells composed of GaN/InGaN heterojunction and InGaN homojunction have been extensively studied. However, their efficiencies and performances are still poor, mainly due to theoretical limits or transparency loss for high bandgap energy InGaN cells as a result of low In composition and poor crystalline quality of the grown InGaN cells. Although various efforts have been carried out toward the goal of high-efficiency solar cells from different viewpoints, however, still there are lots of challenges to achieve a high-efficiency cell.

The most important challenging task for tandem cell fabrication is to grow high-quality InGaN or InAlN materials with wide range of In content. Among the III-nitrides, GaN is the most extensively studied material and comparatively has matured, while the lower band-gap InGaN or InAlN alloys, that are more useful for device application, are still a topic of fundamental research. The difficulties in growing high-quality InGaN and InAlN materials can be attributed to a number of problems: for instance, the large difference in interatomic spacing between InN and GaN or AlN results in a solid phase miscibility gap [4, 5] and the relatively high vapor pressure of InN as compared to the vapor pressure of GaN or AlN leading to low indium incorporation in these alloys [6]. In addition, the difference in formation enthalpies for InN and GaN or AlN causes a strong indium surface segregation on the growth front [7]. Nevertheless, single-crystalline InGaN films with full composition range have been successfully grown by MBE [8]. The growth of high-quality InAlN seems to be more difficult compared with InGaN growth because the optimum growth temperatures for InN and AlN are largely different. The growth of the InAlN film in all composition regions has been realized with MBE [9], while In-rich InAlN with an In content >32 % grown by MOVPE was reported to show phase separation [10]. It is also expected that the MOVPE growth of InAlN is hindered by the parasitic reaction of TMA and NH_3, which was observed in the MOVPE growth of AlN [11].

In this work, the metal-organic vapor-phase epitaxial (MOVPE) growth of InGaN and InAlN is studied to grow high-quality InGaN or InAlN materials with wider range of In content. It is shown that InGaN films with a full range of In compositions can be grown using MOVPE by optimizing growth temperature and TMI/(TMI + TEG) molar ratio. The Mg-doping behavior and p-type conduction of MOVPE InGaN are also studied using Cp_2Mg as a Mg source. As a result of detailed investigation of the growth behavior of MOVPE InAlN including the adduct formation, a single-crystalline InAlN film with an In content of 0.3–1 are successfully grown using the atmospheric-pressure MOVPE. Based on these achievements, an n-InAlN/p-InGaN hetero-structure is successfully prepared and its photo-response is confirmed for the first time.

Experimental procedures

The growth of InGaN and InAlN is performed using a MOVPE system with a horizontal reactor. As sources, trimethyl-indium (TMI), triethyl-gallium (TEG), trimethyl-

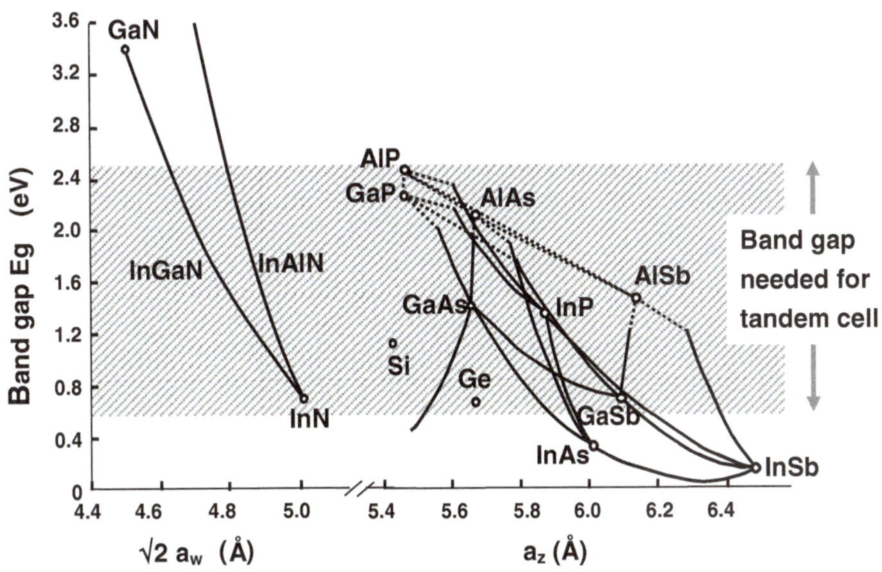

Fig. 1 Relationship between band-gap energy and lattice parameter for typical semiconductor materials. a_z is lattice parameter for diamond-structure or zincblende materials and a_w is a-lattice parameter for wurtzite materials. For nitrides, $\sqrt{2}a_w$ is used for the horizontal axis for the convenience of the combination of nitrides and diamond-structure or zincblende materials

aluminium (TMA) and ammonia (NH_3) are used. As substrates, α-Al_2O_3(0001) and GaN/α-Al_2O_3(0001) templates are used. A 20 nm thick GaN layer grown at 550 °C is used as a buffer for the growth on α-Al_2O_3(0001). In the growth of InGaN, growth pressure is fixed at 150 Torr and growth temperature is varied from 600 to 900 °C. In the growth of InAlN, growth temperature is fixed at 600 °C and growth pressure is varied from 76 to 730 Torr. TMI/(TMI + TEG) molar ratio is varied from 0 to 1 for the InGaN growth, while TMI/(TMI + TMA) molar ratio is varied from 0.3 to 1 for the InAlN growth. The substrates are placed at different positions on the 150-mm long carbon susceptor. The composition of grown films is determined using X-ray diffraction ($2\theta/\omega$) patterns. Full width at half maximum (FWHM) of (0002) X-ray rocking curve, tilt fluctuation, is also measured for grown InGaN and InAlN. The Mg-doping of MOVPE InGaN is performed using bis-cyclopentadienyl magnesium (Cp_2Mg) as a Mg source. Carrier concentration is measured using the Hall effects with the Van der Pauw method. As an ohmic contact to p-InGaN, Ni/Au is used. The Mg concentration in InGaN is measured with the secondary ion-mass spectrometer (SIMS) (outsourced to Toray Research Center, Inc., Japan).

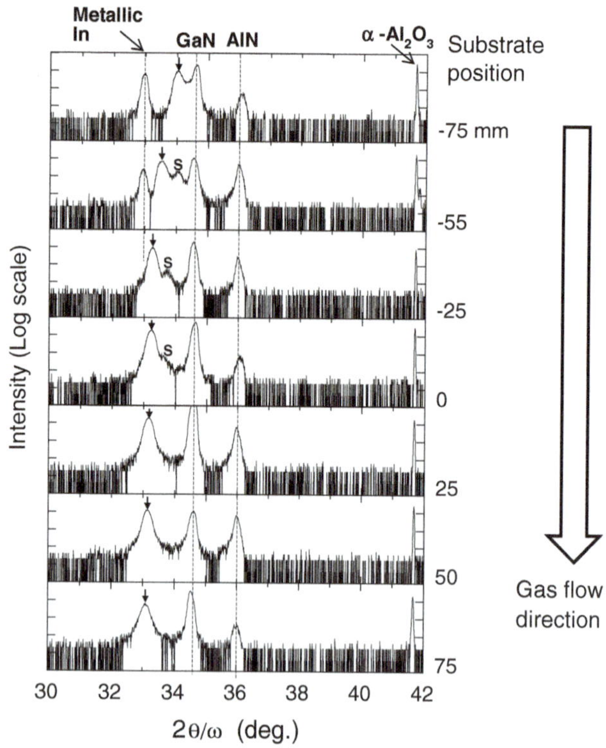

Fig. 2 X-ray diffraction $2\theta/\omega$ profiles for InGaN grown at a different substrate position on the 150-mm long susceptor (along the gas flow direction) in the horizontal reactor

MOVPE growth of InGaN with full composition range

Growth behavior and composition control of InGaN

Non-doped InGaN films are grown at a growth temperature from 600 to 900 °C with a different TMI/(TMI + TEG) molar ratio. Figure 2 shows the X-ray diffraction $2\theta/\omega$ profiles for films grown at 700 °C at a different substrate position on the 150-mm long (along the gas flow direction) susceptor in the horizontal reactor. The substrate used here is a GaN/α-Al_2O_3(0001) template with an AlN buffer. Metallic In segregation is found for the films grown near the upstream end of the susceptor. In addition to the main peak marked by an arrow, a sub-peak denoted by "S" is found for the films grown near the upstream end of the susceptor, indicating phase separation. It is noted that that both phase separation and metallic In segregation can be suppressed by choosing the substrate position near the downstream end of the susceptor, as shown in Fig. 2. As reported in literatures, there are many parameters that affect phase separation in InGaN, such as growth temperature [12], strain in growing films [12, 13], film thickness [14], growth rate [14] or impurity incorporation [15]. We believe that, through such parameters, the gas flow near the upstream side indirectly affects the phase separation. Further investigation will be needed to clarify the thermodynamics for phase separation and metallic In segregation near the upstream side.

The In composition in single-phase InGaN is changed by varying growth temperature and TMI/(TMI + TEG) molar ratio. Figure 3 shows the X-ray diffraction $2\theta/\omega$ profiles for InGaN with a different In composition. The films are grown on α-Al_2O_3(0001) with a GaN buffer. As can be seen in this result, all samples show a distinct single peak of InGaN (0002), indicating that they have no phase separation and no metallic In segregation. Even for the InGaN with In content of 0.7, no obvious phase separation is observed, although its diffraction peak is relatively broad. The films with In contents up to 0.4 were grown by changing growth temperature with a constant TMI/(TMI + TEG) molar ratio 0.45. Those with an In content from 0.4 to 1.0, on the other hand, were grown by changing TMI/(TMI + TEG) molar ratio at the same temperature 600 °C. Figure 4 shows the growth temperature dependence of In composition with a parameter of TMI/(TMI + TEG) molar ratio. At 600 °C, the In composition in solid InGaN is very close to that in the gas phase (TMI/(TMI + TEG) molar ratio), showing that the In composition in solid is controlled by In/(In + *Ga*) ratio in gas phase. At a growth temperature higher than 700 °C, on the other hand, In composition in solid is almost independent on TMI/(TMI + TEG) molar ratio and is gradually decreased with increasing growth temperature. For example, the InGaN grown at 700 °C has In composition of

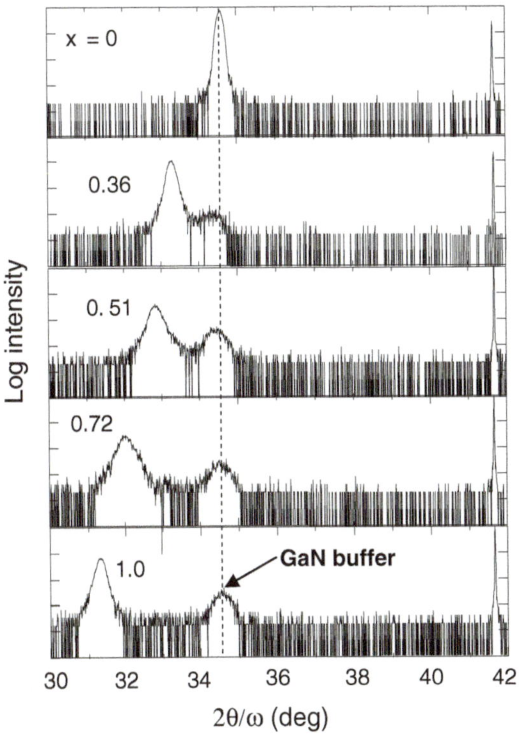

Fig. 3 X-ray diffraction $2\theta/\omega$ profiles for $In_xGa_{1-x}N$ with a different In content x

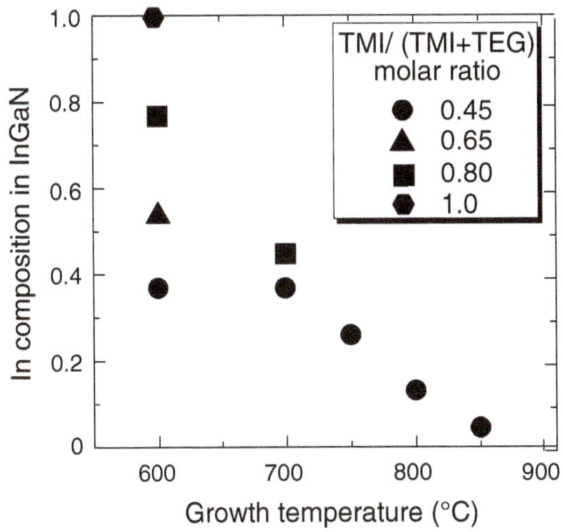

Fig. 4 In composition in InGaN grown at a different temperature with a different TMI/(TMI + TEG) molar ratio

about 0.45 even though the TMI/(TMITEG) molar ratio is high (0.8). In this growth temperature range (≥700 °C), thus, In composition in InGaN is almost governed by growth temperature. This is due to the decomposition of InN component and the consequent evaporation of metallic In from the growing surface. Thus, growth temperature, as well as TMI/(TMI + TEG) ratio, should be carefully managed in the growth of In-rich InGaN.

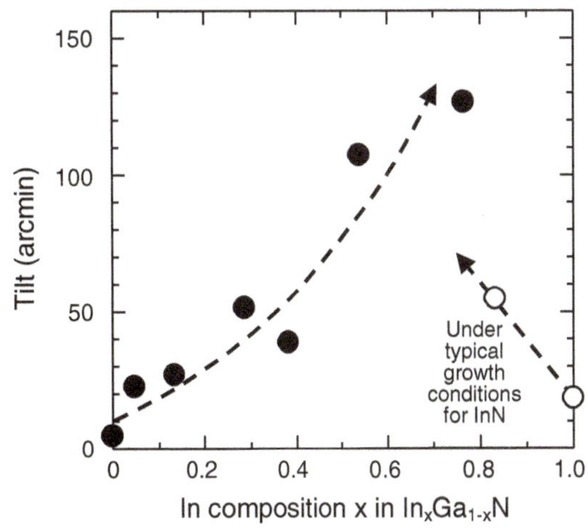

Fig. 5 Tilt fluctuations of InGaN with a different In composition

In order to evaluate crystalline quality of InGaN films, tilt fluctuation is measured using X-ray diffraction method. The results are shown in Fig. 5. The samples with In composition up to 0.4 show relatively good results; tilt fluctuations less than 50 arcmin. On the other hand, the samples with In composition from 0.55 to 0.75 show large tilt fluctuations. In Fig. 5 also shown are the data for InN and InGaN with In content of 0.83. They were grown under the typical growth conditions for InN; at a pressure of 760 Torr and at a growth temperature of 600 °C [16]. Such data are largely deviated from the results for the InGaN with In composition from 0.55 to 0.75. The major difference between the present growth conditions and the typical growth conditions for InN is growth pressure. The result shown in Fig. 5 seems to suggest that InGaN with In composition more than 0.5 should be grown under the growth conditions similar to those for InN.

Mg-doping and p-type conductivity of InGaN

The Mg-doping for InGaN with In contents 0.1–0.4 has been performed using Cp_2Mg as a Mg precursor. The films are grown on GaN/α-Al_2O_3(0001) templates. Figure 6 shows the carrier concentration in Mg-doped InGaN as a function of Cp_2Mg/(TMI + TEG) molar ratio. When Cp_2Mg/(TMI + TEG) molar ratio is 2–5 %, p-type conduction is achieved for InGaN films with In content of 0.25–0.4. These samples are grown at 700–750 °C. Their hole concentrations are in the order of 10^{18} cm^{-3}. In the previous study [17], Cp_2Mg/(TMI + TEG) molar ratio of 4 % or more was needed to get p-type conduction for InGaN with In content of 0.2. In the present case, on the other hand, p-type conduction is achieved for InGaN with In content of 0.24 when Cp_2Mg/(TMI + TEG) molar ratio is only 2 %. Such an effective Mg-doping in this case is believed to be owing to the improved crystalline quality of the present samples. The samples with In content of

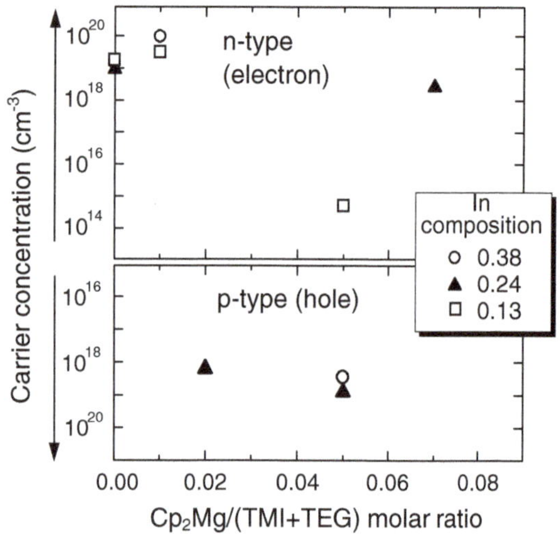

Fig. 6 Electron or hole concentration in Mg-doped InGaN as a function of $Cp_2Mg/(TEG + TMI)$ molar ratio

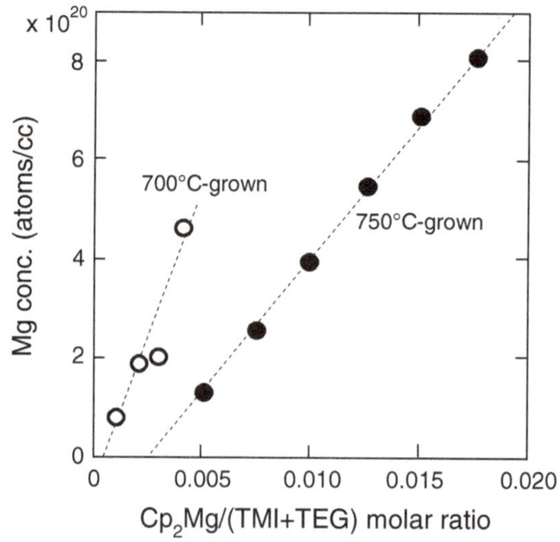

Fig. 7 Relationship between $Cp_2Mg/(TMI + TEG)$ molar ratio and Mg concentration (measured with SIMS) in grown InGaN

0.24–0.38 show n-type conduction again when $Cp_2Mg/(TMI + TEG)$ molar ratio is around 7 % or more. Similar behavior has been observed for Mg-doped GaN [18], indicating that the excess Mg-doping leads to the formation of donor-type defects in the crystals. In this study, p-type conduction has not been obtained for the sample with an In content 0.15, which is grown at 800 °C. One of the possible reasons for this may be the larger activation energy for Mg in a sample with a lower In content [19]. The SIMS analysis has been made for Mg-doped InGaN samples. Figure 7 shows the relationship between $Cp_2Mg/(TMI + TEG)$ molar ratio in the gas phase and Mg concentration in the grown InGaN (measured with SIMS). From this result, one can see that Mg concentration in grown InGaN is decreased as growth temperature increases, even when a constant $Cp_2Mg/(TMI + TEG)$ molar ratio is supplied. From the results in Fig. 7, a very low Mg concentration is expected for a sample grown at 800 °C. This may be also one of the reasons for the unsuccessful p-type conduction of the sample grown at 800 °C with an In content 0.15. In Fig. 7, one can also find that the data lines do not pass the origin. For the samples grown at 750 °C, a $Cp_2Mg/(TMI + TEG)$ molar ratio less than 0.25 % gives no detectable Mg concentration in the grown InGaN. Further investigation will be needed to clarify the Mg-doping behavior in InGaN.

MOVPE growth of InAlN with intermediate In compositions

Growth behavior of InAlN and suppression of adduct formation

It is found that the MOVPE growth behavior of InAlN is more complicated compared with the growth of InGaN. Here, the effects of growth pressure, substrate position and TMA/(TMA + TMI) molar ratio on the growth of InAlN are shown and discussed. Figure 8 shows the growth pressure dependence of the thickness of the material deposited on the substrate. Also shown in this figure is XRD intensity of InAlN peak detected from the deposits. The intensity is normalized by that of the sapphire substrate. As can be seen in Fig. 8, the XRD intensity ratio is the highest at a pressure 730 Torr, while it is very small at a pressure around 300–600 Torr in spite of the large thickness of the deposits. Figure 9 shows the typical surface morphologies (SEM images) of the deposits formed at a different pressure. The surface feature is markedly different depending on pressure. The deposits consist of small grains are obtained at 730 Torr. The thick deposits with micrometer-size grains are obtained at a pressure in the range of 380–530 Torr. When the growth pressure is

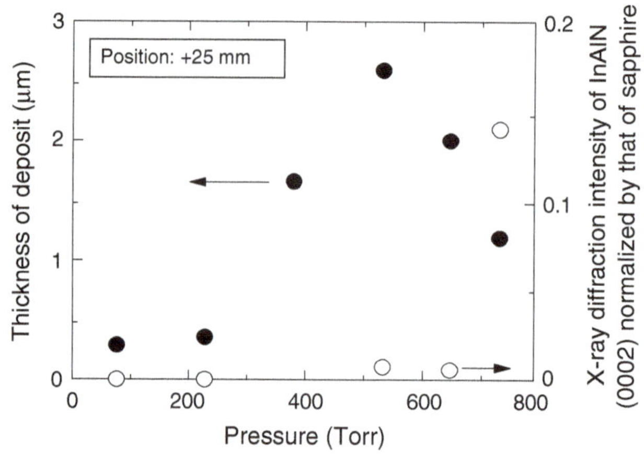

Fig. 8 Growth pressure dependence of the thickness of deposits and XRD intensity of InAlN peak normalized by that of sapphire substrate

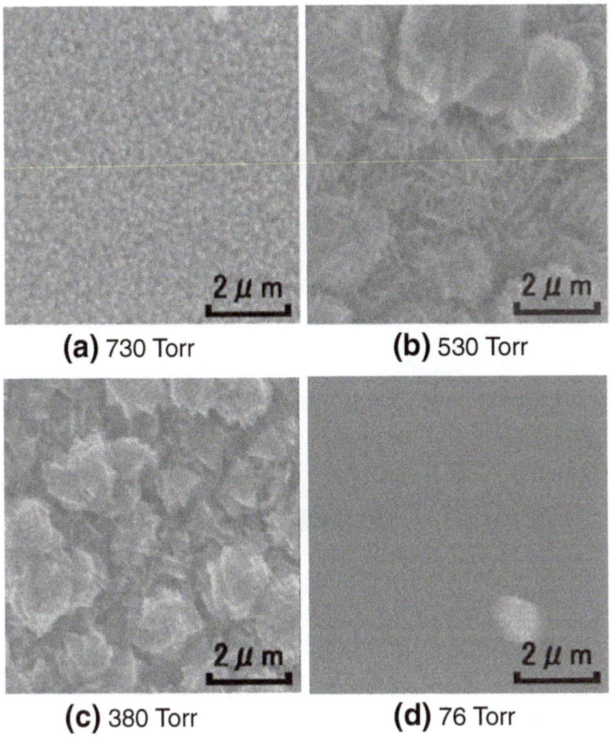

Fig. 9 Surface morphologies (SEM images) of the deposited materials on sapphire substrate at a different pressure (substrate position on the susceptor: +25 mm)

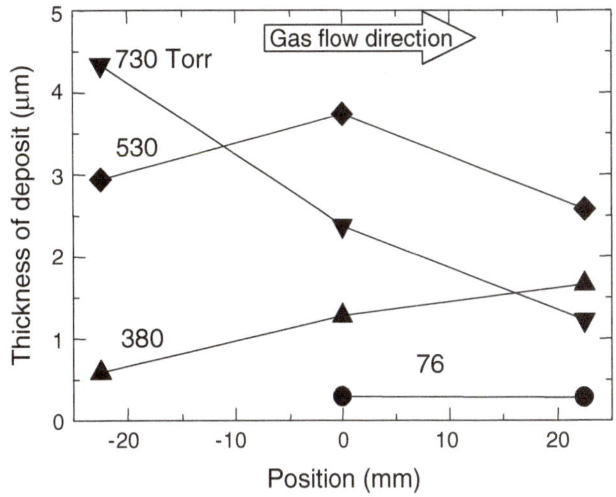

Fig. 10 Substrate position dependence of the thickness of the deposit formed at a different pressure

further reduced to 76 Torr, thin deposits with flat surface are obtained. Accordingly, it is concluded that the deposits with micrometer-size grains formed at a pressure around 380–530 Torr, shown in Fig. 9b and c, scarcely contain InAlN component, indicating that these deposits are composed of adducts through the parasitic reaction between TMA and NH_3 [11].

Figure 10 shows the substrate position dependence of the thickness of the deposited material formed at a different pressure. One can see that the position where the largest thickness is obtained is moved from the upstream side to downstream side as growth pressure is reduced. The amount of the deposits is also decreased with reducing growth pressure. These results are caused by the increase in the gas flow speed and the consequent decrease of parasitic reaction, since the reduction of growth pressure results in increase in gas flow speed under the constant gas supply condition. In the case of the growth at 730 Torr, adducts formation mainly occurs at the upstream side of the susceptor, as shown in Fig. 10. This is due to the considerably low gas flow speed. When we grow InAlN at the downstream side (substrate position +25 mm), therefore, a single-crystalline InAlN can be obtained without obstruction of adducts formation, although the Al content in the grown film is much reduced in this case. The low Al content is due to the TMA consumption by the adduct formation at the upstream side. This is the reason why we could not find adducts formation in the previous study on the InAlN growth at 730 Torr [20]. Figure 11 shows the growth pressure dependence of Al composition in the grown InAlN. At a growth pressure in the range of 300–600 Torr, single-crystalline InAlN films are not obtained due to the adduct formation at the substrate position +25 mm. The Al composition increases with decreasing growth pressure, as seen in the figure. This is due to that the TMA consumption by the adduct formation is reduced with decreasing growth pressure. Figure 12 shows the substrate position dependence of Al composition in InAlN films. At a pressure around 730 Torr and at the substrate position +25 mm, an InAlN with a relatively low Al content is obtained. With decreasing growth pressure Al composition increases and the gradient of Al composition decreases. This is also due to the reduced TMA consumption in the parasitic reaction with decreasing growth pressure. That is, with decreasing

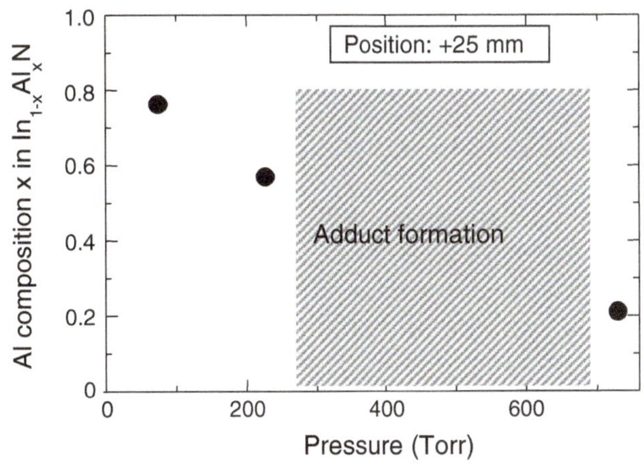

Fig. 11 Growth pressure dependence of Al composition in InAlN

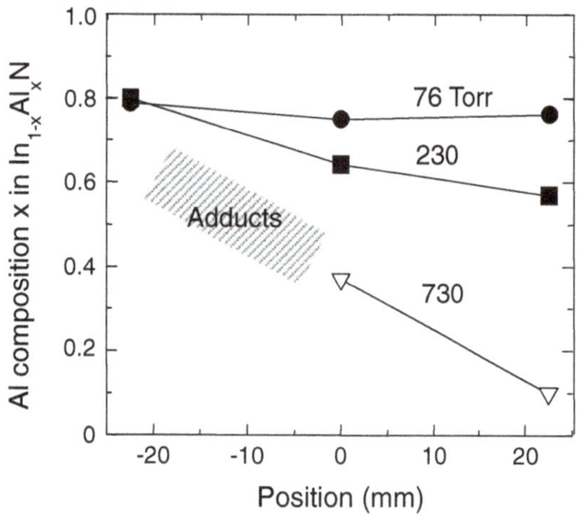

Fig. 12 Substrate position dependence of Al composition in InAlN

growth pressure, more TMA can be delivered to the downstream side. In order to avoid the problems of the adduct formation, we need to select growth pressure less than 300 Torr or at around 730 Torr or more, as shown in Fig. 11. In the case of a pressure less than 300 Torr, however, growth rate of InAlN is very low (less than 0.05 μm/h), as shown in Fig. 8. Therefore, the growth at 730 Torr is the most practical choice to get an InAlN film with a reasonable growth rate (0.5–1 μm/h). In this case, substrate positions on the downstream side should be selected because adduct formation occurs at the upstream side, as can be seen in Fig. 12.

Composition control of MOVPE InAlN

In this study, InAlN films with a thickness 0.5–0.8 μm have been grown on nitrided (0001) sapphire substrates at 600 °C. Due to the reasons described above, a growth pressure of 730 Torr and the substrate position +25 mm are selected. Figure 13 shows the XRD patterns around (0002)-plane of InAlN. By changing TMA/(TMA + TMI) molar ratio, InAlN films with a different In composition are obtained. It is noted that all films prepared here do not show phase separation, although the intensity of XRD peak for the InAlN is decreased and the FWHM of the peaks is increased with increasing Al content. Figure 14 shows the relationship between TMI/(TMA + TMI) molar ratio in the gas phase and Al content in InAlN. InAlN films with Al content from 0 to 0.7 are obtained as seen in Fig. 14. The experimental data are markedly deviated from the line of slope 1. This deviation is due to the TMA consumption by the adduct formation at the upstream side, as discussed above. The carrier concentration of InAlN films grown at 600 °C at the pressure of 730 Torr is in the range of 2×10^{19} cm^{-3} to 5×10^{19} cm^{-3} and does not show a marked dependency on Al composition. On the other hand,

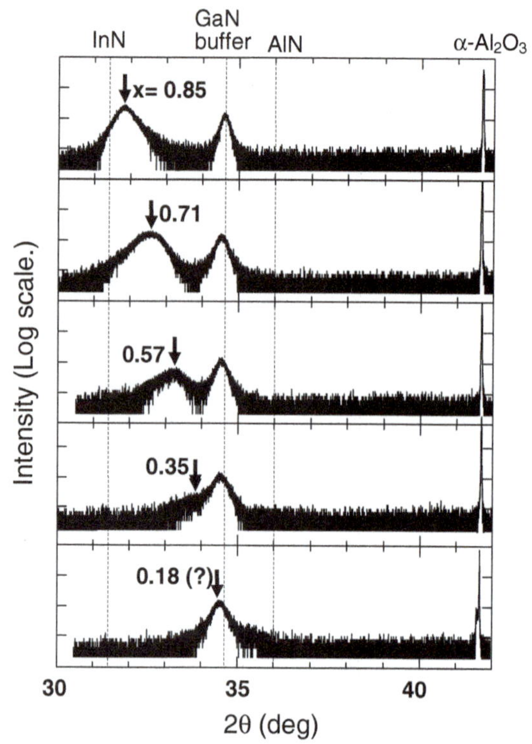

Fig. 13 XRD patterns around (0002)-plane of $In_xAl_{1-x}N$. The samples are grown at a different TMA/(TMA + TMI) molar ratio at the substrate position +25 mm

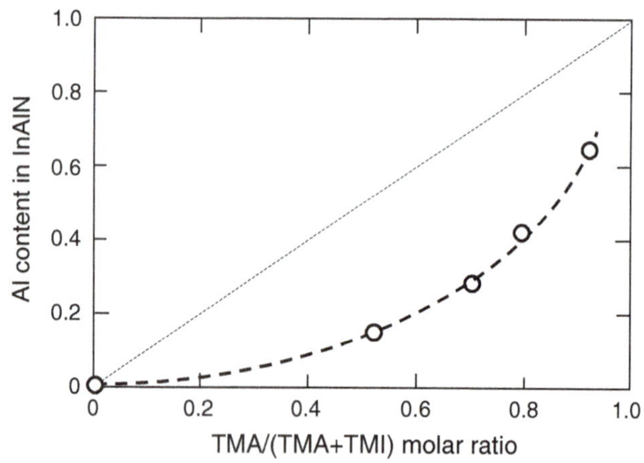

Fig. 14 Relationship between TMA/(TMA + TMI) molar ratio in the gas phase and Al content in grown InAlN. Substrate position +25 mm

the mobility decreases with increasing Al composition; from 200 cm^2/Vs for InN to 10 cm^2/Vs for InAlN with Al content 0.43. The InAlN films with an Al content of 0–0.3 show a PL spectrum even at room temperature [20]. Although the PL intensity is decreased with increasing Al content, the intensity difference between InN and InAlN is less than one order of magnitude even for the film with Al content 0.3.

Preparation and characterization of p-InGaN/n-InAlN hetero-structures

As described above, we have successfully grown both InGaN films with full composition range and InAlN films with In content from 0.3 to 1. In order to demonstrate an ability to prepare these different alloys sequentially, the formation of InAlN/InGaN hetero-structures has been studied. At the substrate position +25 mm, InGaN and InAlN films have been grown sequentially at 150 and 730 Torr, respectively. Figure 15 shows the X-ray diffraction $2\theta/\omega$ profiles for "$In_{0.57}Al_{0.43}N$ on $In_{0.4}Ga_{0.6}N$" structure and "$In_{0.4}Ga_{0.6}N$ on $In_{0.57}Al_{0.43}N$" structure. The insets in the figure show the surface morphologies (SEM images) for such samples. When an $In_{0.57}Al_{0.43}N$ film is grown on an $In_{0.4}Ga_{0.6}N$ film prepared beforehand, the hetero-structure is successfully prepared. On the other hand, an $In_{0.4}Ga_{0.6}N$ film is grown on an $In_{0.57}Al_{0.43}N$ film prepared beforehand, the $In_{0.57}Al_{0.43}N$ film is found to be decomposed during the $In_{0.4}Ga_{0.6}N$ growth. This means that the $In_{0.57}Al_{0.43}N$ is more unstable compared with the $In_{0.4}Ga_{0.6}N$, even though the growth temperature is same for both materials. By using an n-$In_{0.3}Al_{0.7}N$ and p-$In_{0.2}Ga_{0.8}N$ films, a heterojunction has been fabricated on a GaN template for the first time. Figure 16 shows (a) the structure of the fabricated device and (b) the current density–voltage characteristics of the device in the dark and under AM1.5, 100 mW/cm^2 illumination. The thicknesses of the Mg-doped p-InGaN and non-doped n-InAlN layers are 1.2 and 0.1 µm, respectively. The device clearly shows photo-response to the AM1.5, 100 mW/cm^2 illumination. As can be seen in Fig. 5, $V_{oc} = 0.25$ V and $J_{sc} = 0.01$ mA/cm^2 are obtained. Such a low J_{sc} seems to be due to the large bandgap and the low quality of the films. The low quality of the InAlN films seems to be also mainly responsible for the low V_{oc}.

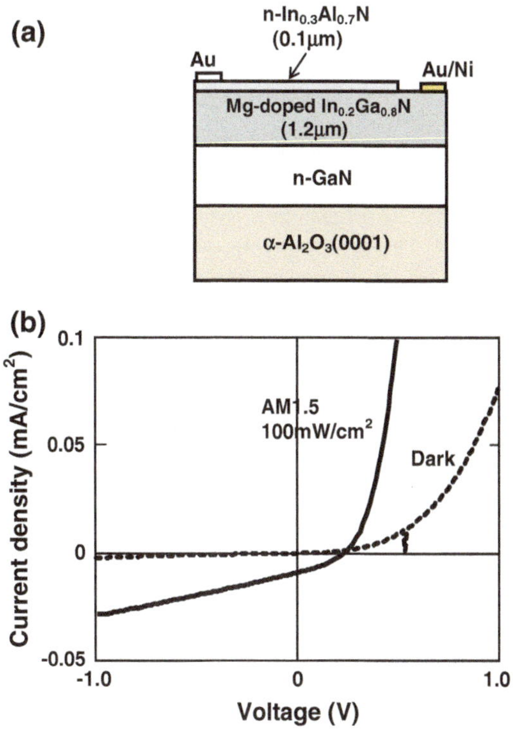

Fig. 16 Schematic drawing of InAlN/InGaN hetero-structure device (**a**) and current density–voltage characteristics of the device in the dark and under AM1.5 illumination (**b**)

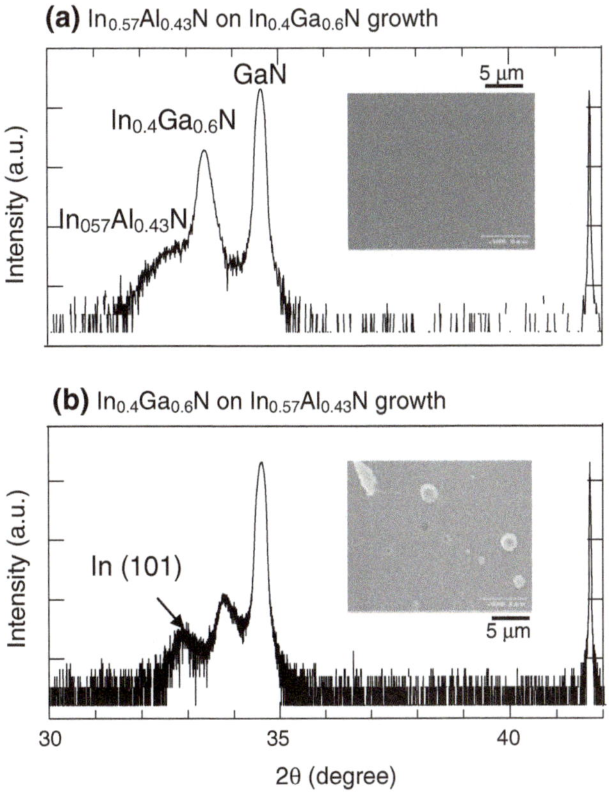

Fig. 15 X-ray diffraction spectra for "$In_{0.57}Al_{0.43}N$ on $In_{0.4}Ga_{0.6}N$" and "$In_{0.4}Ga_{0.6}N$ on $In_{0.57}Al_{0.43}N$" structures. The *insets* show the surface morphologies (SEM images) for such samples

Summary

The InN-based nitride semiconductor alloys such as InGaN and InAlN are one of a few of the promising materials for multi-junction solar cell, because band-gap energy from 0.7 to 2.5 eV can be realized by changing only the composition of these alloys. In order to prepare such alloys, MOVPE growth using a horizontal reactor has been applied in this work. It is shown that InGaN films without phase separation and metallic In segregation can be grown by optimizing gas flow conditions and substrate position on the susceptor. The In composition in InGaN is controlled by growth temperature and TMI/(TMI + TEG) molar ratio, and InGaN films with full composition range are successfully obtained. The Mg-doping behavior and p-type conduction of MOVPE InGaN are also studied using Cp_2Mg as a Mg source. p-type conduction is achieved for InGaN films with In content up to 0.4. A detailed investigation has been also made of the growth behavior of MOVPE InAlN. The InAlN growth using NH_3 and TMA is found to be seriously hindered by the adduct formation based on the parasitic reaction of NH_3 and TMA. By choosing the atmospheric-pressure MOVPE and by optimizing substrate position on the susceptor, a single-crystalline InAlN film with an In content from 0.3 to 1 is successfully grown. Based on these achievements, an n-InAlN/p-InGaN hetero-structure is successfully prepared and its photo-response is confirmed for the first time.

References

1. Yamamoto, A., Islam, Md.R., Kang, T.-T., Hashimoto A.: Recent advances in InN-based solar cells: status and challenges in InGaN and InAlN solar cells. Phys. Stat. Sol. (c) **7**, 1309 (2010)
2. Davydov, V.Yu., Klochikhin, A.A., Emtsev, V.V., Ivanov, S.V., Vekshin, V.V., Bechstedt, F., Furthmulier, J., Harima, H., Mudryi, A. V., Hashimoto, A., Yamamoto, A., Aderhold, J., Graul, J., Haller, E.E.: Band gap of InN and In-rich $In_xGa_{1-x}N$ alloys ($0.36 < x < 1$). Phys. Stat. Sol. (b) **230**, R4 (2002)
3. Bhuiyan, A.G., Sugita, K., Hashimoto, A., Yamamoto, A.: InGaN solar cells: present state of the art and important challenges. IEEE J. Photovolt **2**, 276 (2012)
4. Popovici, G., Morkoc, H.: In: Pearton, S.J. (ed.) GaN and Related Materials II, p. 93. Gordon and Breach Science, Netherlands (2000)
5. Ho, I., Stringfellow, G.B.: Solid phase immiscibility in GaInN. Appl. Phys. Lett. **69**, 2701 (1996)
6. Nagatomo, T., Kuboyama, T., Minamino, H., Omoto, O.: Properties of $Ga_{1-x}In_xN$ films prepared by MOVPE. Jpn. J. Appl. Phys. **28**, L1334 (1989)
7. Yoshimoto, N., Matsuoka, T., Sasaki, T., Katsui, A.: Photoluminescence of InGaN films grown at high temperature by metalorganic vapor phase epitaxy. Appl. Phys. Lett. **59**, 2251 (1991)
8. Iliopoulos, E., Georgakilas, A., Dimakis, E., Adikimenakis, A., Tsagaraki, K., Androulidaki, M., Pelekanos, N. T.: InGaN (0001) alloys grown in the entire composition range by plasma assisted molecular beam epitaxy. Phys. Stat. Sol. (a) **203**, 102 (2006)
9. Terashima, W., Che, S.B., Ishitani, Y., Yoshikawa, A.: Growth and characterization of AlInN ternary alloys in whole composition range and fabrication of InN/AlInN multiple quantum wells by RF molecular beam epitaxy. Jpn. J. Appl. Phys. **45**, L539 (2006)
10. Hums, C., Blasing, J., Dadgar, A., Diez, A., Hempel, T., Christen, J., Krost, A.: Metal-organic vapor phase epitaxy and properties of AlInN in the whole compositional range. Appl. Phys. Lett. **90**, 022105 (2007)
11. Zhao, D.G., Zhu, J.J., Jiang, D.S., Yang, H., Liang, J.W., Li, X.Y., Gong, H.M.: Parasitic reaction and its effect on the growth rate of AlN by metalorganic chemical vapor deposition. J. Cryst. Growth **289**, 72 (2006)
12. Tessarek, C., Figge, S., Aschenbrenner, T., Bley, S., Rosenauer, A., Seyfried, M., Kalden, J., Sebald, K., Gutowski, J., Hommel, D.: Strong phase separation of strained $In_xGa_{1-x}N$ layers due to spinodal and binodal decomposition: formation of stable quantum dots. Phys. Rev. B **83**, 115316 (2011)
13. Tabata, A., Teles, L.K., Scolfaro, L.M.R., Leite, J.R., Kharchenko, A., Frey, T., As, D.J., Schikora, D., Lischka, K., Furthmüller, J., Bechstedt, F.: Phase separation suppression in InGaN epitaxial layers due to biaxial strain. Appl. Phys. Lett. **80**, 769 (2002)
14. Pantha, B.N., Li, J., Lin, J.Y., Jiang, H.X.: Evolution of phase separation in In-rich InGaN alloys. Appl. Phys. Lett. **96**, 232105 (2010)
15. Nola, L., Shen-jie, W., Eun-Hyun, P., Zhe, C.F., Hung-Li, T., Jer-Ren, Y., Ian, F.: Suppression of phase separation in InGaN layers grown on lattice-matched ZnO substrates. J. Cryst. Growth **311**, 4628 (2009)
16. Yamamoto, A., Sugita, K., Hashimoto, A.: Elucidation of factors obstructing quality improvement of MOVPE-grown InN. J. Cryst. Growth **311**, 4636 (2009)
17. Horie, M., Sugita, K., Hashimoto, A., Yamamoto, A.: MOVPE growth and Mg doping of $In_xGa_{1-x}N$ (x-0.4) for solar cell. Sol. Energy Mater. Sol. Cells **93**, 1013 (2009)
18. Tokunaga, H., Waki, I., Yamaguchi, A., Akutsu, N., Matsumoto, K.: Growth condition dependence of Mg-doped GaN film grown by horizontal atmospheric MOCVD system with three layered laminar flow gas injection. J. Cryst. Growth **189/190** 519 (1998)
19. Chang, C., Tang, T., Chang, P., Chen, N., Liang, C.: Magnesium doping of In-rich InGaN. Jpn. J. Appl. Phys. **46**, 2840 (2007)
20. Houchin, Y., Hashimoto, A., Yamamoto, A.: Atmospheric-pressure MOVPE growth of In-rich InAlN. Phys. Stat. Sol. (c) **5**, 1571 (2008)

3

Dislocation reduction in heteroepitaxial $In_xGa_{1-x}N$ using step-graded interlayer for future solar cells

Md. Arafat Hossain · Md. Rafiqul Islam · M. K. Hossain · A. Hashimoto · A. Yamamoto

Abstract The efforts on dislocation reduction have become a potential issue to realize the future high-efficiency solar cells using the InGaN materials. In this work, first, a numerical simulation has been carried out for the reduction of dislocation density in wurtzite InGaN heteroepitaxy using step-graded interlayers. An energy balance model has been developed for evaluating the misfit dislocation (MD) density. The residual strain from previous interlayer has been taken into account with misfit strain in each interlayer. A reaction model for threading dislocations (TDs) has also been developed and solved numerically considering the geometrical parameters. The simulation results confirmed a significant improvement of epilayer quality due to the use of step-graded interlayers. The calculations have been done for 1.5 μm InGaN using 3 step-graded interlayers each containing 10 % composition difference and 0.2 μm thickness. The edge, screw, and mixed type MDs are found to be 4.69×10^{10}, 4.53×10^{9}, and 4.03×10^{11} cm^{-2}, respectively, on the 1/3 < 11–23 > (11–22) slip and similarly evaluated in other possible slips of the $In_{0.4}Ga_{0.6}N$. Significant decreases in MD densities have been evaluated for increasing interlayer up to 4 with 8 % In composition difference. The edge, screw, and mixed MDs are found to be decreased from 3.25×10^{11} to 4.45×10^{10}, 3.2×10^{10} to 3.95×10^{9} and 2.8×10^{12} to 1.77×10^{11} cm^{-2} in 1/3 < 11–23 > (11–22) slip system using four interlayers. Besides, a higher rate of reduction has been reported for the step-graded structure especially for mixed type due to more relative motion and step inclination at each interlayer. The average edge, screw, and mixed type TD densities are found to be 1.48×10^{10}, 3.7×10^{10}, and 1.1×10^{9} cm^{-2}, respectively, at the top surface of the $In_{0.4}Ga_{0.6}N$ epilayer. Finally, experimental work has been done to realize the strain profile using reciprocal space mapping as well as dislocations considering the same films to compare the results. The outcome of the numerical simulation found good agreement with the experimental works as well as with the published experimental results.

Keywords InGaN · Critical thickness · Misfit dislocation · Threading dislocation · Interlayer

Introduction

In the recent years, group III-nitrides and their related alloys have been intensively studied with remarkable breakthrough in the growth of multilayer structures for device applications [1–3]. Among them the ternary alloy InGaN is currently a major topic of research in the field of the modern optoelectronic and microelectronic technologies [4, 5]. Especially, this material is a promising candidate for multi-junction tandem solar cells with high conversion efficiency [6–8]. In order to realize such solar cell, it is essential to grow this alloy with high indium (In) content [6]. As there is no successful technique devised for the bulk growth of InGaN they are usually grown

Md. A. Hossain · Md. R. Islam (✉)
Department of Electrical and Electronic Engineering, Khulna University of Engineering and Technology, Khulna, Bangladesh
e-mail: rafiq043@yahoo.com

M. K. Hossain
Center of Research Excellence in Renewable Energy, King Fahd University of Petroleum and Minerals (KFUPM), Dhahran 31261, Kingdom of Saudi Arabia

A. Hashimoto · A. Yamamoto
Graduate School of Engineering, University of Fukui, Fukui, Japan

heteroepitaxially on GaN template [9]. Therefore, a large lattice (11 % between InN and GaN) as well as thermal expansion coefficient mismatches between the layer and substrate, introduces dislocations which lead to the degradation of the quality of this material and consequently the performance of its constituent devices [10–12]. These dislocations act as scattering centres and midband gap states, which proceed as recombination centres and reduce the minority carrier lifetime [13–15]. As a result, open-circuit voltage (V_{oc}), short-circuit current density (J_{sc}), fill factor, and efficiency of solar cells decrease gradually with the increase of dislocation density [13]. Hence the reduction of dislocation density is a budding issue for InGaN-based solar cells.

The concept of dislocation reduction during the heteroepitaxial growth of InGaN requires an understanding of misfit strain caused by lattice mismatch. Various theoretical as well as experimental evidences confirmed the common mode of strain relaxation as the formation of misfit dislocations (MDs) from different sources [11, 16–18]. The threading dislocations (TDs) which threaded the multilayer may be initiated from MDs or any other sources [16]. Other sources of dislocation such as V-defects and surface inclusions are also reported for single-phase high In-content InGaN on GaN [19]. Due to the adverse effects of dislocations, the researchers are highly interested to find a suitable growth technique which will reduce the dislocations density of future high-performance device fabrication. With this purpose, several efforts have been made to reduce MDs as well as TDs [20]. Ee et al. [21, 22] reported the nano-patterned AGOG (Deposition of Aluminum, Growth of Oxide, and Grain growth) sapphire for the growth of GaN template with reduced TD density in LEDs to enhance the output power. Compositional grading is one of the most effective techniques which can be very effective to reduce both MDs and TDs [23–28]. In a properly designed graded layer, misfit strain as well as dislocation is greatly reduced due to spreading the strain profile throughout the entire layer of thickness instead being concentrated to a single interface. This grading of epitaxial layer may be done in several approaches such as linear or nonlinear grading, step-grading, grading with overshoot interfaces, and reverse grading [23–28]. Most of the experimental and theoretical studies with graded heteroepitaxial layer have been carried out on linear and nonlinear grading [25, 29, 30]. However, the efficient strategy to reduce dislocation density is to initiate annihilation and fusion reaction among them during their glide process. In this process, the inclinations of TDs at layer-substrate or layer–layer interface are responsible for interaction between them and hence result in reduction by reaction [31]. Another approach for TD reduction is the incorporation of Si into the $Al_xGa_{1-x}N$ film that results in a proportional stress transition from compressive to tensile. This dislocation reduction by Si doping also results from the increased TD inclination [32, 33]. Accordingly, linear or nonlinear grading shows poor performance for reducing the TDs in the epilayer due to the less interaction probability.

These objections are a driving force for further investigations of a technique which is best suitable for MDs as well as TDs reduction. In heteroepitaxy growth of InGaN using step-graded interlayer, due to the step increase of In composition in the epitaxial layers, a multiple step inclination of TDs at each interface will occur. This promotes the reaction among the TDs at each interlayer. Few experimental works on different material systems confirmed that this technology will be a promising solution for dislocation reduction during growth of heteroepitaxy [23, 24, 34–36]. Furthermore, the step-wise increase of In-content provides the facility to use of ternary InGaN as a substrate materials for the upper layers with high In-content. As a result, the reduced compressive strain between two successive layers improves the spontaneous emission properties and optical gain with reduced wavelength shift of multiple quantum well (MQW) devices [37–40]. Although available theoretical works have been developed for without grading or linear and nonlinear grading, there is no such work on step-graded interlayers for wurtzite materials. In order to realize future high-performance devices, a detailed analysis of dislocation reduction using step-graded interlayer and further optimization is urgently required.

In this work, an analytical energy balance model for MDs and reaction model for TDs has been developed for the optimized step-graded structure for the fabrication of future solar cells. The reaction model considered possible annihilation and fusion reaction between TDs and TD blocking by MDs. In addition, experimental works have been done according to the modeling and the observed results illustrated. Finally, the results obtained from these simulations have been compared with the available experimental data.

Mathematical modeling

The modeling for misfit and threading dislocations have been developed separately to analyze the different types of dislocations density (edge, screw and mixed) individually without graded layer as well as their effects on interlayers. A flow diagram shown in Fig. 1 presents a brief summary of mathematical modeling done for the proposed step-graded interlayer scheme. Details of this methodology will be discussed in the subsequent sections.

Critical thickness

The most basic form of InGaN heteroepitaxial growth consisting of a layer with constant indium (In) composition is shown in Fig. 2a, which builds up strain with the increase of thickness. Such structure undergoes sudden relaxation with a large network of dislocations forming at the epilayer growth interface [16]. On the other hand, a step-graded interlayer technique is one, where the In composition in InGaN increases step wise up to the epilayer as shown in Fig. 2b. During the growth of each InGaN heteroepitaxy, the lattice as well as thermal expansion coefficient difference with substrate or previous interlayer introduces stain. This strain increases with thickness and after reaching a thickness, namely critical thickness (h_c), builds up enough strain energy to become energetically favorable to relax film through the formation of crystal defects, namely misfit dislocations. For any step-graded structure, multiple values of h_c are calculated for each interlayer and upper epilayer grown stepwise on GaN substrate as shown in Fig. 2b. These values of h_c can be calculated for each layer using the Matthews-Blakeslee's well-known force balance model [16, 36]:

$$h_{ci} = \frac{b_i(1 - v_i \cos^2\theta_i)}{8\pi(1 + v_i)|\varepsilon_{mi}| \sin\theta_i \sin\phi_i} \ln\left(\frac{h_{ci}}{r_{0i}}\right), \tag{1}$$

$$\varepsilon_{mi} = \frac{a_{i-1} - a_i}{a_{i-1}} \tag{2}$$

where the index i in subscript indicates number of interlayer, b is the length of burger vector, v is the Poisson ratio, φ is the angle between the slip plane and normal to the film-substrate interface, θ is the angle between the dislocation line and burger vector, and r_0 is dislocation cut-off parameter. These parameters and the other basic parameters for the $In_xGa_{1-x}N$/GaN heteroepitaxy are listed in Table 1. The in-plane misfit strain between two successive layers will be calculated from Eq. (2). Multiple values of critical thickness will be found according to the parameters for each layer of the step-graded structures.

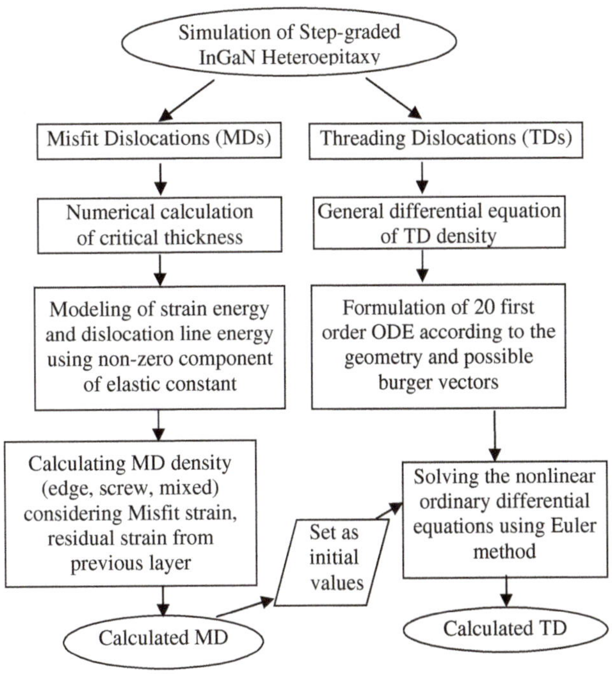

Fig. 1 A flow diagram for mathematical modeling of dislocations in step-graded InGaN heteroepitaxy

Slip systems in InGaN

In wurtzite InGaN, all the slip systems are not active for generation of misfit dislocation like cubic materials such as InGaAs or SiGe [41, 42]. Because all the planes do not experience a shear stress component due the lattice mismatch and should remain inactive in this configuration. Srinivasan et al. [43] estimated theoretically the appropriate slip systems for the MDs in wurtzite InGaN crystal as

Table 1 Typical parameters for $In_xGa_{1-x}N$/GaN heteroepitaxy

Parameters	Values	Parameters	Values
a_{GaN}	3.1892Å [55]	φ	Table 2
a_{InN}	3.5378Å [56]	θ(edge)	90°
c_{GaN}	5.1850Å [55]	θ(screw)	0°
c_{InN}	5.7033Å [56]	θ(mixed)	60°
v_{InN}	0.42 [47]	r_0	$b/2$ [47]
v_{GaN}	0.35 [47]	r_A	500Å [51]
b	Table 2	K_{xy}	[48]

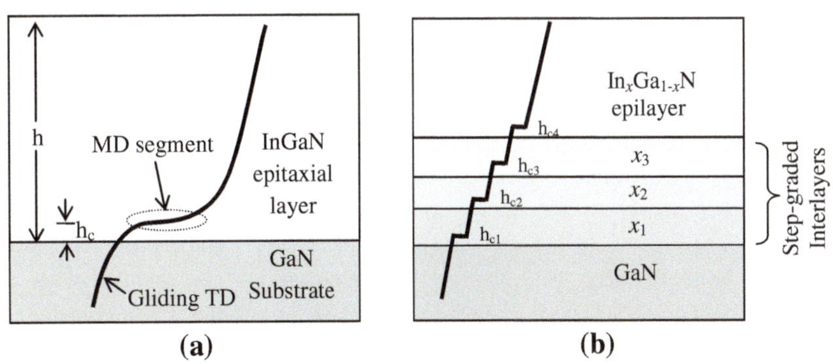

Fig. 2 Growth technique of wurtzite InGaN heteroepitaxy with **a** constant composition layer and **b** step-graded layer

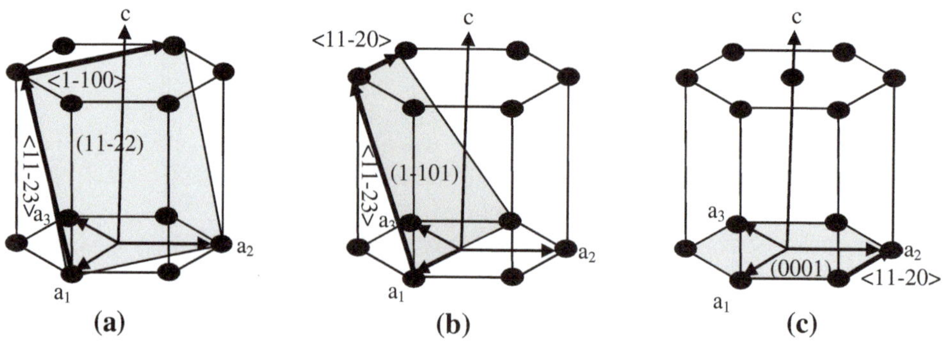

Fig. 3 Slip systems for dislocation in InGaN lattice. **a** 1/3 < 11–23 > (11–22), **b** 1/3 < 11–23 > (1–101), and **c** 1/3 < 11–20 > (0001). Slip planes and directions are as indicated

Table 2 Parameters of favored slip systems in wurtzite InGaN crystal structure showing b and φ

Possible slip	b	φ
1/3 < 11–23 > (11–22)	$\sqrt{a^2(x)+c^2(x)}$	$\arctan\frac{a(x)}{c(x)}$
1/3 < 11–23 > (1–101)	$\sqrt{a^2(x)+c^2(x)}$	$\arctan\frac{\sqrt{3a(x)}}{2c(x)}$
1/3 < 11–20 > (0001)	$\sqrt{3a(x)}$	90°

1/3 < 11–23 > (11–22) and 1/3 < 11–23 > (1–101). However, there are also reports of pure a-type MDs which are theoretically the most efficient MD types for relieving the misfit strain [44]. Therefore, the basal plane slip system 1/3 < 11–20 > (0001) is also included for theoretical consideration. These three slip systems for wurtzite InGaN are shown in Fig. 3a, b, and c. The Table 2 listed the geometrical properties of these favored slip systems.

Misfit dislocations in step-graded InGaN

Initially the strain energy of the epitaxial film increases with thickness, and beyond the critical thickness some of this energy will be accomodated by the formation of misfit dislocation. In our previous works, we have developed an energy balance model for the step-graded structure to calculate the misfit dislocation density [45, 46]. It has been reported that the misfit strain in the heteroepitaxy starts to relax after h_c at each interlayer; the layer will be partially relaxed by MDs. In order to do this, the strain energy of each interlayer and energy accomodated by misfit dislocations have been calcualted. One important term in that calculation was the consideration of residual strain using the nonzero component of elastic constant, after a thickness of h is [46, 47]

$$|\varepsilon_i| = \left| |\varepsilon_{mi}| - \left|\frac{3}{2} b_{ci}\rho_{\mathrm{MDi}}\right| \right| \quad (3)$$

Therefore, the strain energy per unit area of the interface is

$$\frac{\mathrm{dW_i}}{\mathrm{d}A} = \left(c_{11} + c_{12} - \frac{2c_{13}^2}{c_{33}}\right)\left(|\varepsilon_{mi}| - \left|\frac{3}{2} b_{ci}\rho_{\mathrm{MDi}}\right|\right)^2 h_i, \quad (4)$$

where c_{ij} are elastic constant of the film. The energy per unit length of a dislocation lying in the layer-substrate or layer–layer interface of a material with hexagonal symmetry can be written as [7, 8]

$$\frac{\mathrm{dE_{d(i)}}}{\mathrm{d}\ell} = b_{ci}\left(c_{11} + c_{12} - \frac{2c_{13}^2}{c_{33}}\right)\varepsilon_{mi}h_{ci}\left[\ln\left(\frac{h_i}{r_{0i}}\right)/\ln\left(\frac{h_{ci}}{r_{0i}}\right)\right] \quad (5)$$

The energy per unit interface area of an array of evenly distributed dislocations with spacing, l, is then $(1/l)$ $(\mathrm{d}E_{\mathrm{d}(i)}/\mathrm{d}l)$. As all three arrays of non-interacting MDs are assumed to be independent, the total energy per unit area may be expressed as

$$\frac{\mathrm{dE_i}}{\mathrm{d}A} = \frac{3}{\ell}\frac{\mathrm{dE_{d(i)}}}{\mathrm{d}l} = \frac{3}{\ell} b_{ci}\left(c_{11} + c_{12} - \frac{2c_{13}^2}{c_{33}}\right)\varepsilon_{mi}h_{ci}\left[\ln\left(\frac{h_i}{r_{0i}}\right)/\ln\left(\frac{h_{ci}}{r_{0i}}\right)\right]h_i \quad (6)$$

It is assumed that the dislocation spacing l is such that it minimizes the total energy within the ith interlayer. So the misfit dislocation density is found from $\mathrm{d}(E_i + W_i)/\mathrm{d}l = 0$ as

$$\rho_{\mathrm{MDi}} = \frac{1}{\ell_i^2} = \left(\frac{2|\varepsilon_{mi}|}{3b_i \sin\theta\cos\phi}\left(1 - \frac{h_{ci}\ln(h_i/r_{0i})}{h_i\ln(h_{ci}/r_{0i})}\right)\right)^2 \begin{matrix} \text{for } h_i \ominus h_{ci} \\ \text{for } h_i \le h_{ci} \end{matrix} \quad (7)$$

The layer grown upon the partially relaxed layer of thickness h_i will experience a misfit strain, $\varepsilon_{m\,(i+1)}$ reduced by the residual strain ε_i of the previous layer, where the value of ε_i is obtained from Eq. (3) for each interlayer. Therefore, MD density for each interlayer and epilayer will be updated with misfit strain as

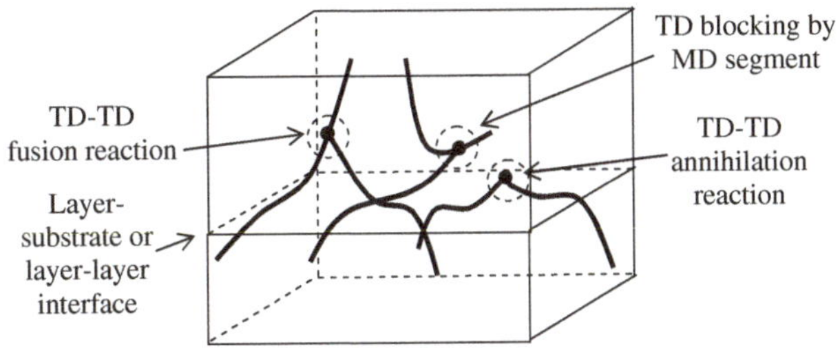

Fig. 4 Threading dislocation reduction mechanism in step-graded heteroepitaxy

$$\left|\varepsilon_{m(i+1)}\right| = \left|\frac{a_{l(i)} - a_{l(i+1)}}{a_{l(i)}}\right| - |\varepsilon_i| \quad (8)$$

Threading dislocations in step-graded InGaN

In the growth of lattice-mismatched heteroepitaxy, TDs are concomitantly generated with MDs. These TDs are non-equilibrium defects that always raise the internal energy of the film/substrate system [48]. Thus, there is a natural driving force to reduce the internal energy and TD density, subsequently. This reduction may occur due to annihilation and fusion reactions between pairs of inclined TDs and blocking by interaction between gliding TD with a MD line on its gliding plane [49]. All of these reactions have been shown schematically in Fig. 4. Annihilation reaction happens when two TD lines with equal and opposite burger vectors ($b_1 = -b_2$) undertake the reaction $b_1 + b_2 = 0$, removing the dislocations entirely. A fusion reaction occurs when two TD (b_1 and b_2) lines becomes one TD with new burger vector, $b_3 = b_1 + b_2$. However, both of these reactions can take place when the distance between interacting dislocations becomes smaller than the characteristic cross-section of a specific reaction: annihilation radius r_A for annihilation reaction and fusion radius r_F for fusion reactions [49, 50]. The general governing equation for TD density reduction of specific type ρ_x can be written for ith interlayer as

$$\left(\frac{d\rho_x}{dh}\right)_{i\text{th layer}} = \left(-\sum_y K_{xy}\rho_x\rho_y + \sum_l\sum_n K_{\text{ln}}\rho_l\rho_n - \sum_m \rho_x\rho_m\right)_{i\text{th layer}} \quad (9)$$

The first summation is for the annihilation and fusion reactions which reduce the TD density of xth type one per reaction. The second summation is for fusion reactions between the other type's l and n that produce the dislocation in the xth type, and the last one is for TD of xth type reduction by MD blocking on its gliding path. The reaction kinetic coefficient K_{xy} presents the degree of possibility to participate in reaction of TDs type x and y in a particular interlayer that depends on the average differential motion between TDs with increasing film thickness. The values of the coefficients for a single layer are obtained using the general expression developed by Romanov et al. [48] as $K_{xy} = 2r_A\left[\sin\psi_x m_x - \sin\psi_y m_y\right]$, where ψ denotes the angle between TD line and surface normal and $\boldsymbol{m}$ indicates the direction of dislocation motion. Using this expression, the coefficient for each interlayer can be calculated for the different values of ψ and $\boldsymbol{m}$ corresponding to their mismatch strain. Though the mismatches strain in without graded layer is higher, the lateral motion of TDs with respect to interlayer thickness is larger for step-graded layer having comparatively lower stepped increase of thickness. Therefore, a multiple inclination of TDs at each interface enables them to increase the probability of interaction in step-graded structure.

In the InGaN heteroepitaxy, there are 20 unique families of dislocations according to their burger vectors and line directions [51]. Among them, numbers 1 to 6 are for edge dislocation whose burger vectors ($+a_1$, $-a_1$, $+a_2$, $-a_2$, $+a_3$ and $-a_3$) lie on the basal plane, numbers 7–18 are for mixed character ($+a_1+c$, $+a_1-c$, $-a_1+c$, $-a_1-c$, $+a_2+c$, $+a_2-c$, $-a_2+c$, $-a_2-c$, $+a_3+c$, $+a_3-c$, $-a_3+c$, and $-a_3-c$), and numbers 19–20 are for screw dislocation whose burger vectors ($+c$ and $-c$) are directed along c axis.

All possible annihilation and fusion reactions between TD–TD and TD-MD have been identified to write a reaction equation for a particular TD. For a specific case, TDs from family 1 can only have annihilation and fusion reactions with TDs from family 9, 10, 11, 12, 15, and 16 that reduce the TDs for the ith interlayer. So we can write

$$\left(\frac{d\rho_1}{dh}\right)_{i\text{th layer}} = \left(-K_{1,9}\rho_1\rho_9 - K_{1,10}\rho_1\rho_{10} - K_{1,11}\rho_1\rho_{11} - K_{1,12}\rho_1\rho_{12} - K_{1,15}\rho_1\rho_{15} - K_{1,16}\rho_1\rho_{16}\right)_{i\text{th layer}} \quad (10)$$

In addition, TDs from family '1' may be generated within the ith interlayer only as the product of fusion reactions between TDs from the following pairs of families 7–8, 7–14, 7–18, 7–20, 8–13, 8–17, 8–19, 13–18, and 14–17. Therefore, for production of TDs from family 1 as a result of fusion, we can write

$$\left(\frac{d\rho_1}{dh}\right)_{i\text{th layer}} = \left(\begin{array}{c} +2K_{7,8}\rho_7\rho_8 + K_{7,14}\rho_7\rho_{14} + K_{7,18}\rho_7\rho_{18} + K_{7,20}\rho_7\rho_{20} + K_{8,13}\rho_8\rho_{13} \\ +K_{8,17}\rho_8\rho_{17} + K_{8,19}\rho_8\rho_{19} + K_{13,18}\rho_{13}\rho_{18} + K_{14,17}\rho_{14}\rho_{17} \end{array}\right)_{i\text{th layer}} \quad (11)$$

Finally, the threading dislocation of family '1' could be blocked by edge type MD in the interlayer [49]. Therefore, TD reduction for type 1 due to this effect can be written as

$$\left(\frac{d\rho_1}{dh}\right)_{ith\ layer} = (-\rho_1\rho_m)_{ith\ layer} \tag{12}$$

Combining, the decrease of TD of type 1 as a result of annihilation (Eq. 10), increases due to fusion reaction (Eq. 11) and blocking by MDs (Eq. 12), we can arrive at the governing equation for the density of TD family '1' in the *i*th interlayer of step-graded InGaN as

$$\left(\frac{d\rho_1}{dh}\right)_{ith\ layer} = \begin{pmatrix} \rho_1(k_{1,9}\rho_9 + K_{1,10}\rho_{10} + K_{1,11}\rho_{11} + K_{1,12}\rho_{12} + K_{1,16}\rho_{16}) \\ +2K_{7,8}\rho_7\rho_8 + K_{7,14}\rho_7\rho_{14} + K_{7,18}\rho_7\rho_{18} + K_{7,20}\rho_7\rho_{20} + K_{8,13}\rho_8\rho_{13} \\ +K_{8,17}\rho_8\rho_{17} + K_{8,19}\rho_8\rho_{19} + K_{13,18}\rho_{13}\rho_{18} + K_{14,17}\rho_{14}\rho_{17} - \rho_1\rho_m \end{pmatrix}_{ith\ layer} \tag{13}$$

Similarly, reaction equations for six different edge TDs, twelve mixed and two screw character TDs, i. e. a total of 20 first-order differential equations have been developed with corresponding burger vector and possible reaction for each interlayer of step-graded InGaN heteroepitaxy.

Simulation results and discussion

Multilayer critical thickness

The simulation of the developed mathematical model and further optimization has been done using available numerical technique. The calculations have been done for a 1.5-μm InGaN step-graded layer using three interlayer each of 0.2 μm with 10 % In composition difference and a final 0.9 μm epilayer with In composition, $x = 0.4$. Multiple critical thicknesses for each interlayer and the upper epilayer have been calculated in different possible slip system for the step-graded InGaN heteroepitaxy. Due to inverse relationship a slow increase in In composition in step-graded structure leads to larger h_c at each interface. Furthermore, the relative lattice mismatch between the adjacent interlayer has been considered according to Eq. (2), which makes the provision of larger values of h_c than that of the without graded layer. The critical thickness at different slip system with increasing In composition is plotted in Fig. 5. From the figure, the critical thicknesses have been found to be 12.4, 13.9, and 3.3 nm in (11–22), (1–101), and (0001) slip, respectively, for 10 % increase of In composition.

Misfit dislocations generation in InGaN

At critical thicknesses the appearance of MDs are, for the first time, energetically favorable and start to relax the misfit strain. However, there is no reason to assume that all MDs will appear suddenly when the film thickness reaches h_c. Instead, after the critical thickness, MD density increases with increasing film thickness h [3]. Different types of MDs are calculated in step-graded $In_{0.4}Ga_{0.6}N$/GaN heteroepitaxy using three interlayers. Figure 6 shows the total MD generation in different slips as a function of layer thickness. All of the slips generate edge, screw, and mixed dislocation depending on the angle between the burger vector and dislocation line direction. Figure 6a presents these three types of MDs separately in the 1/3 < 11–23 > (11–22) slip of the step-graded $In_{0.4}Ga_{0.6}N$. The results in the 1/3 < 11–23 > (1–101) and 1/3 < 11–20 > (0001) slip systems are also shown in Fig. 6b, and c. The mixed MDs are considered as 60° types. There is a significant difference observed among the results where most of the MDs are of mixed types. It can be seen from the Fig. 6a that the total edge, screw, and mixed type MDs are of 4.74×10^{10}, 4.53×10^9, and 4.03×10^{11} cm^{-2}, respectively, generated on the first plane. From the above it is also observed that the MDs of any type generated in 1/3 < 11–20 > (0001) slip are much

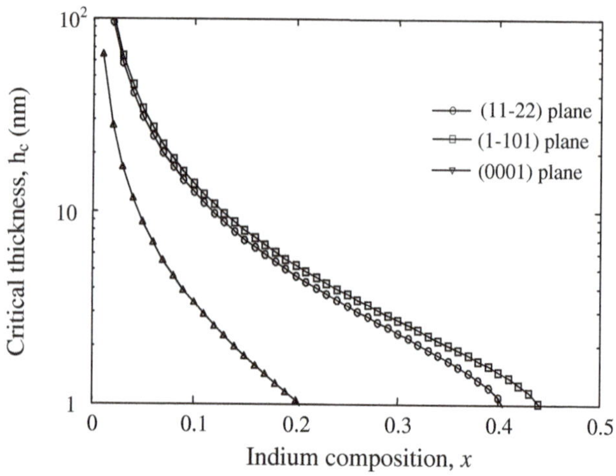

Fig. 5 The critical thickness in different slip systems of $In_xGa_{1-x}N$/GaN heteroepitaxy

Fig. 6 The generation of different types of MDs on 1/3 < 11–23 > (11–22) slip of a step-graded InGaN ($In_{0.4}Ga_{0.6}N/In_{0.3}Ga_{0.7}N/In_{0.2}Ga_{0.8}N/In_{0.1}Ga_{0.9}N/GaN$) heteroepitaxy in different slip systems- **a** 1/3 < 11–23 > (11–22) slip, **b** 1/3 < 11–23 > (1–101) slip, and **c** 1/3 < 11–20 > (0001) slip

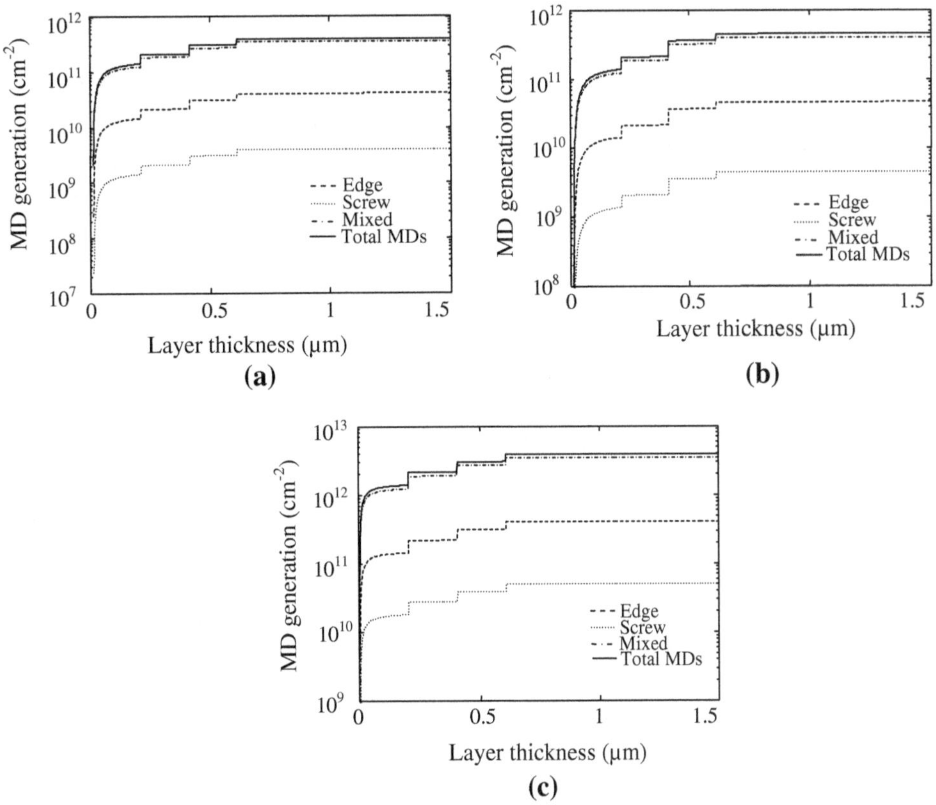

Fig. 7 Total MD generation of **a** edge **b** screw, and **c** mixed type in $In_{0.4}Ga_{0.6}N$(step-graded)/GaN and $In_{0.4}Ga_{0.6}N$(without graded)/GaN heteroepitaxy using three interlayers

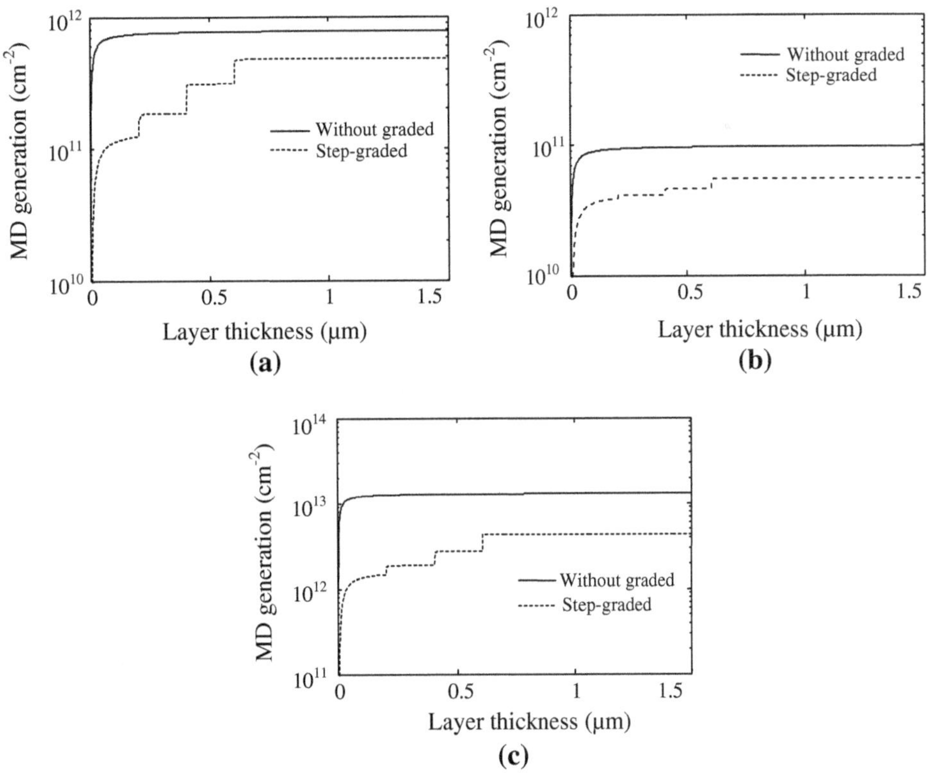

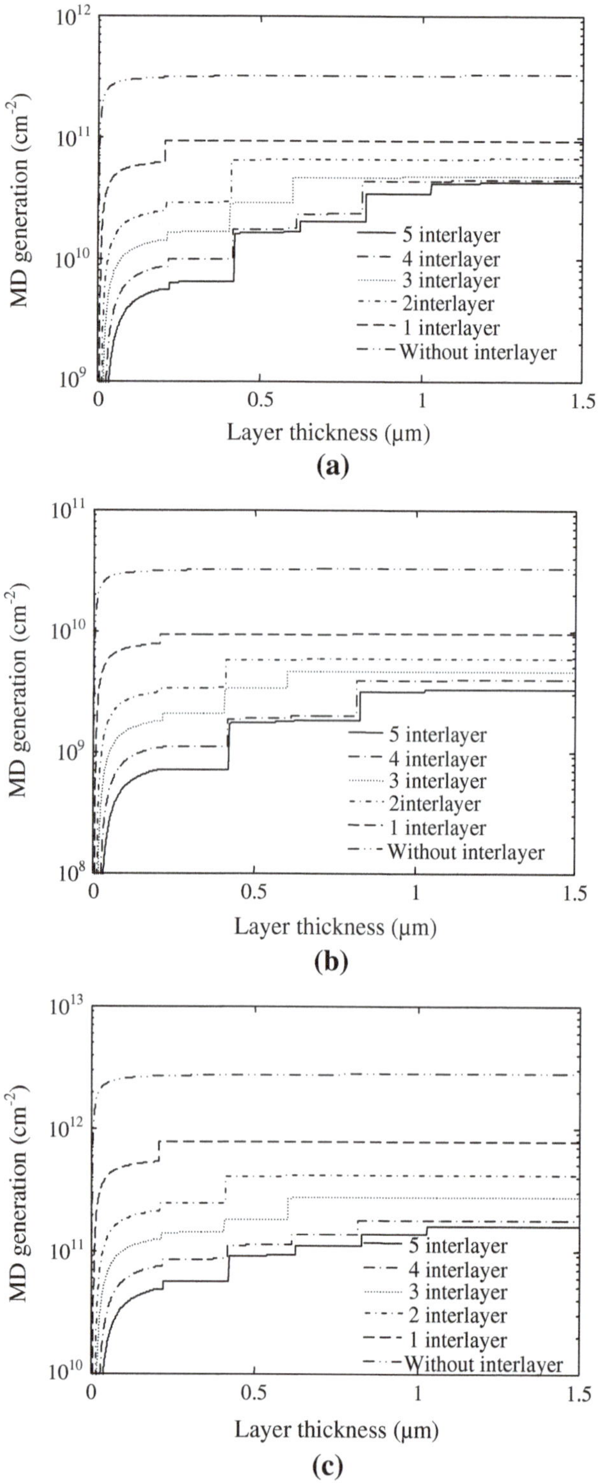

Fig. 8 Effects of interlayer number on **a** edge, **b** screw, and **c** mixed type of dislocations generation in 1/3 < 11–23 > (11–22) slip of step-graded $In_{0.4}Ga_{0.6}N/GaN$ heteroepitaxy

higher than that of the other prismatic slips. This is because the hexagonal c-plane is isotropic and, therefore, burger vectors for all dislocation lines with different values of φ are equivalent. It is also clear from the above figures that the mixed and edge type MDs have higher densities than the screw type ones. This is due to the comparatively lower energy requirement for mixed and edge type MDs than for the screw type ones.

Figure 7 presents the comparison of total generated edge, screw, and mixed type MDs in all possible planes of the step-graded and without graded structure. It is found from Fig. 7a that the total edge type MDs is 4.8×10^{11} and 7.8×10^{11} cm^{-2} for step-graded and without graded structure, respectively, considering 1.5 μm InGaN epilayer thickness. The significant difference of dislocation generation indicates that the step-graded interlayers successfully decrease the dislocation between GaN and InGaN interface. Though the reduction of screw type MDs as indicated in Fig. 7b due to step-graded interlayer is very small, a significant improvement has been observed in case of mixed type MDs as shown in Fig. 7c. The gradual application of misfit strain is primarily responsible for this improvement in dislocation density. With no interlayer the difference in mismatch between epilayer and substrate causes a large misfit and thermal strain between them. The large compressive stress also yields in the InGaN because of large thermal expansion coefficient for GaN (5.5×910^{-6} K^{-1}) than for InN (4.0×10^{-6} K^{-1}), resulting in an introduction of misfit lattice dislocations in the epilayer. On the other hand, with increasing the InGaN interlayer in a step-graded structure composition increases slowly step-wise. Thus the lattice as well as thermal expansion coefficient mismatch of the step-graded interlayer would be gradually change from GaN to $In_{0.4}Ga_{0.6}N$. Furthermore the misfit strain at the new interlayer could be compensated by the residual strain of opposite polarity from the previous layer. Therefore, the stress may be decreased or accommodated within a graded interlayer resulting in lower misfit dislocation.

In order to optimize the interlayer number for a specific structure, we have calculated MD densities at different interlayer numbers. Figure 8a, b and c shows the effect of increasing number of interlayer on edge, screw, and mixed type MDs, respectively, in 1/3 < 11–23 > (11–22) slip system of the $In_{0.4}Ga_{0.6}N/GaN$ heteroepitaxy. We have estimated the densities from 0 to 5 interlayers with the same final In composition ($x = 0.4$). The edge MD densities of 3.25×10^{11}, 9.39×10^{10}, 6.7×10^{10}, 4.74×10^{10}, 4.45×10^{10} and 4.24×10^{10} cm^{-2} have been calculated for without, 1, 2, 3, 4 and 5 interlayer, respectively. Similar improvements have been evaluated for other types of MDs in all slips. From the results it can be seen that a significant improvement of epilayer quality can be achieved with increasing number of interlayer up to 4 where 8 % indium composition differences have been used for each interlayer. The epilayer is further improved for five inter layers but the improvement tendency is very low. In addition, increasing more and more interlayers can

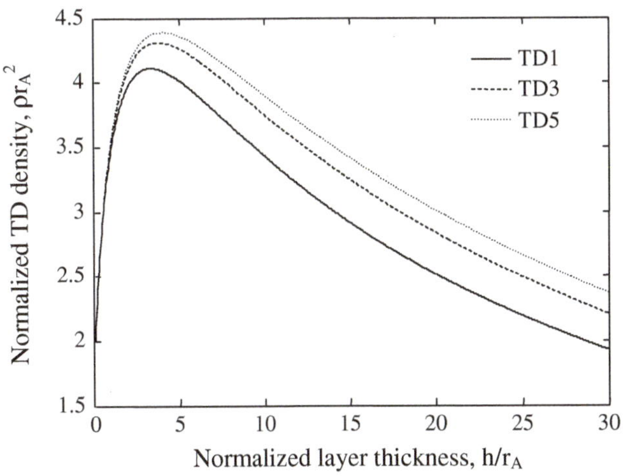

Fig. 9 The variation of edge type TD density with specific number in step-graded $In_{0.4}Ga_{0.6}N$/GaN system

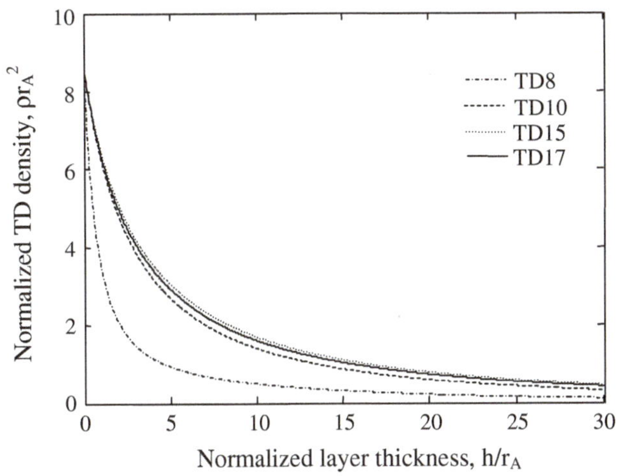

Fig. 10 The variation of mixed type TD density with specific number in step-graded $In_{0.4}Ga_{0.6}N$/GaN system

increase the experimental complexity and will introduce interfacial dislocations each layer. These results are consistent with the outcomes of improved InGaN epilayer quality using step-graded interlayer according to Islam et al. [23], who have shown that inserting more inter layer with 3–5 % indium composition difference did not decrease the dislocations further. Therefore, it is reasonable to conclude that the insertion of interlayer is found effective where the In composition difference between the under layer and epi-layer is around 6.5–8 %.

Threading dislocations in wurtzite InGaN

The threading dislocation densities in the step-graded InGaN heteroepitaxy have been evaluated from the numerical simulation of the reaction model. A total of 20 first orders, nonlinear differential equations according to all possible

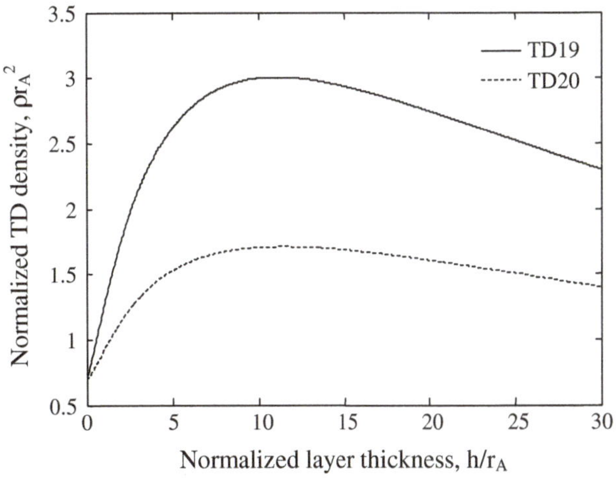

Fig. 11 The variation of screw type TD density with specific number in step-graded $In_{0.4}Ga_{0.6}N$/GaN system

burger vectors have been solved for the same structure of $In_{0.4}Ga_{0.6}N$. To execute the solution, the MDs are considered as the main source of TDs of specific types to set the initial values. The reaction kinetic coefficients between each pair of TDs have been determined for each interlayer considering the annihilation and fusion reaction radius of 500 Å. The solution presents the variation of TD densities with increasing film thickness. Figure 9 shows the density reduction of TD 1, 3, and 5 (edge type) in the step-graded $In_{0.4}Ga_{0.6}N$. It can be reported from the figure that initially the edge type TDs increase rapidly and then begin to decrease with increasing film thickness. The initial increase is due to their lower initial values and their generation from the reaction of other types is higher. Figures 10 and 11 present the variation of mixed and screw type TD densities, respectively, with specific number as a function of film thickness for the step-graded structure. The mixed type TD decreases monotonically with increasing film thickness and these decreases are quicker than any other type of TD. This is because, though the screw dislocations are considered very low initial values, they can be the product of other reaction and allowed to participate in reaction with other TDs. As a result, both the screw-mixed reaction producing vertical edge dislocation and the edge-mixed reaction producing vertical screw dislocations. Therefore, the mixed TD density becomes so low as to prevent the further interaction among TDs, with the end result of saturation in TD density. The nature of screw type TDs are different from that of others. Though both TDs increase with thickness, TD 19 has higher rate than TD 20. The improvement in TD density in the top surface of the epilayer has been understood from the comparison of Fig. 12a, and b. These figures show the decrease of average edge, mixed, and screw TD densities of the step-graded and without graded $In_{0.4}Ga_{0.6}N$ heteroepitaxy, respectively.

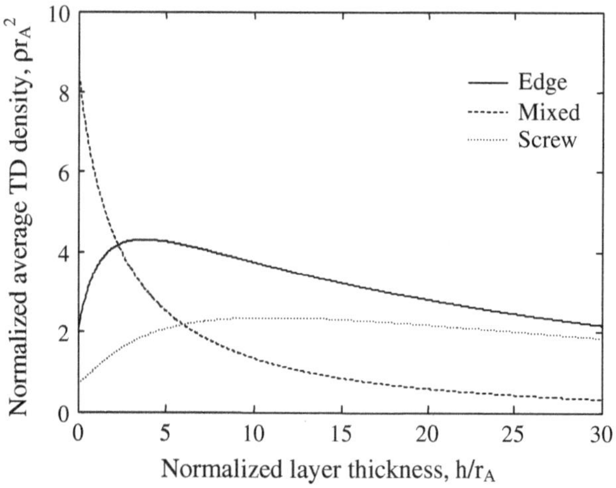

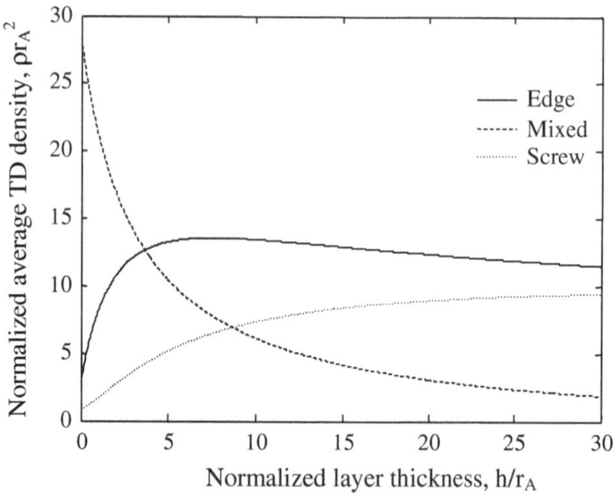

Fig. 12 The average edge, screw and mixed type TDs for **a** $In_{0.4}Ga_{0.6}N$(step-graded)/GaN and **b** $In_{0.4}Ga_{0.6}N$(without graded)/GaN system

The total edge, screw, and mixed type TD densities for the step-graded structures are found to be 8.88×10^{10}, 7.4×10^{10}, and 1.32×10^{10} cm^{-2}, respectively, at the top surface of the epilayer, whereas these values are 4.6×10^{11}, 3.78×10^{11}, and 7.52×10^{10} cm^{-2}, respectively, for the without graded structure.

There are several reasons behind these improvements in epilayer quality of the step-grade structure. Since the dislocations from the substrate are extremely much lower than the dislocation generated from other sources, the MDs are considered as the main source of TDs. In case of without graded structure the most of the MDs are generated near the film–substrate interface due to large mismatch and lower critical thickness. Therefore, the TDs that glided in the upper layers could not be blocked by the MDs. On the other hand, the MDs' generations in the step-graded structure spread throughout the entire layer instead of being concentrated in a single interface. As a result, the TDs that glided out from the first interlayer could be blocked by MDs generated in the second interlayer or TDs from second layer may be blocked by MDs in upper layer and so on. Therefore, the TD reduction in step-graded structure can be enhanced by the MD blocking. Moreover, as the strain inside the thin film drives threading dislocations to incline at each interface and react with each other, the introduction of an intentionally strained layer can be used to facilitate the TDs reduction process. This is the basic of TD reduction mechanism in step-graded layer. Bending of dislocations in without graded structure only occurs very close to the $In_xGa_{1-x}N$/GaN interface, whereas dislocation bending can be noticed at each interface between InGaN interlayers of different composition in step-graded structure [52]. Moreover, due to large misfit dislocation density a strong interaction force among them prohibits the glide motion of the dislocation and further relaxation in without graded layer. Therefore, it requires further dislocation nucleation to relax the strain resulting higher TD densities in epilayer. On the other hand, lower MD densities offers week interaction force with easy gliding of TDs. A well-designed graded layer should relax the misfit strain mostly through the glide motion of existing threading dislocations instead of nucleation of new dislocation. Furthermore, as observed in various experimental works most nucleation sites for TDs are MDs which are extremely lower in step-graded structure than the without graded.

Experimental evidence

Growth procedure

The films are grown on (0001) sapphire substrate using the metal organic vapor phase epitaxy (MOVPE) with a horizontal reactor. The $In_xGa_{1-x}N$ heteroepitaxy is grown on GaN template using two interlayers having intermediate In compositions between epilayer and template. Triethylgallium (TEG), trimethylindium (TMI), and ammonia (NH_3) are used as Gallium (Ga), Indium (In), and Nitrogen (N) sources, respectively. The complete growth procedure of graded layer hetero-structure can be seen from the time chart shown in Fig. 13. At first GaN template of about 2 μm is grown at 1,000 °C. Then step-graded $In_xGa_{1-x}N$ films with different In compositions are grown as interlayers between the GaN template and InGaN epi-layer. The thickness of each interlayer is (150–200 nm) kept more than the critical thickness corresponding to the In composition. Finally, InGaN epi-layer with thickness ~1.5 μm is grown at pressure 150 Torr. The growth temperatures are kept from 750 to 900 °C depending on the optimal growth conditions required for various $In_xGa_{1-x}N$ layers with

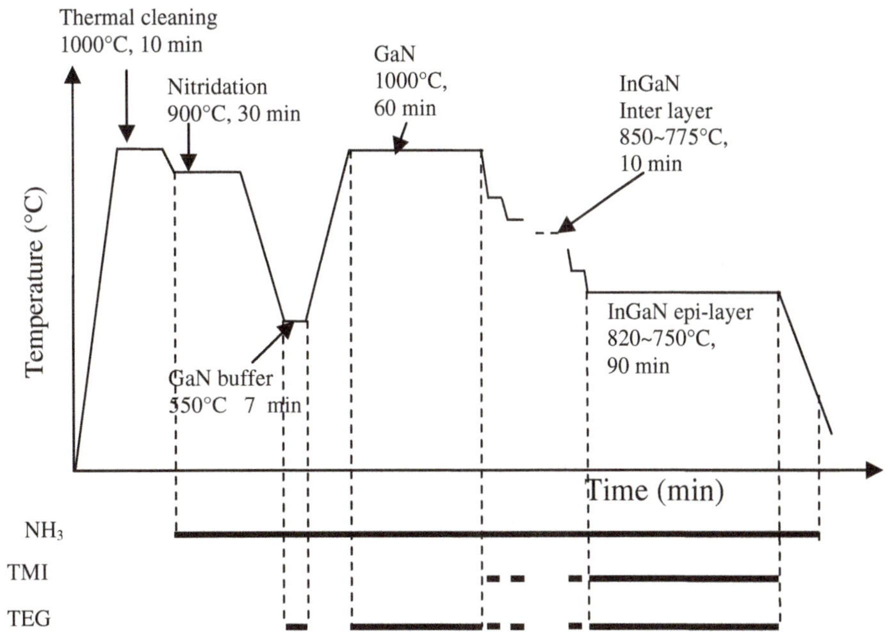

Fig. 13 Time chart for the epitaxial growth of InGaN with step-graded interlayer

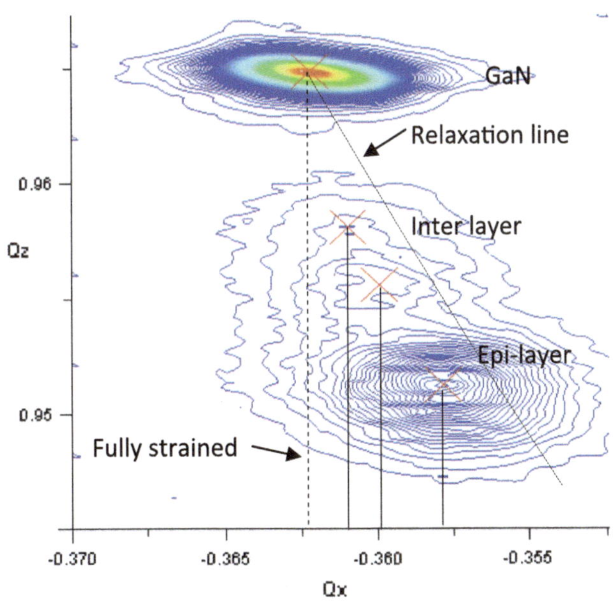

Fig. 14 Reciprocal space maps for the (10–15) reflections of InGaN epi-layer with two inter layers

different In compositions. The In compositions of interlayer and epi-layer are controlled changing the growth temperature [53]. Growth interrupts are employed between the growths of two consecutive interlayers having different In compositions and allowing thermal annealing under NH_3. Also during the growth interrupt, the substrate temperature is adjusted to the optimum value required for the growth of each $In_xGa_{1-x}N$ layer. More details about experimental procedure have been discussed in [23].

Experimental observation of relaxation

Information on the crystal structure and strain of an epitaxial film can be obtained by performing a RSM on an asymmetric reflection using a double-crystal X-ray diffraction system. Additionally, the out-of-plane, c and in-plane, a lattice constant of the hexagonal structure can be measured and a direct comparison of the position of GaN template, InGaN inter layers, and epi-layer diffraction peaks in the RS can be performed. A RSM of the $In_{0.14}Ga_{0.86}N$ epi-layer with two inter layers $In_{0.05}Ga_{0.95}N$ and $In_{0.09}Ga_{0.91}N$, measured around the (10–15) reflection is shown in Fig. 14. The thickness of each inter layer and the epi-layer is around 0.15 and 1.5 μm, respectively. Fully strained and relaxed lines are also shown. The relaxation line has been drawn calculating the lattice constant of InGaN using the relation [54], where the input parameters necessary to complete the calculation are the GaN (c_{GaN}0.51850, a_{GaN}0.31892 nm) [55] and InN (c_{InN}0.57033, a_{InN}0.35378 nm) [56] relaxed lattice constants. The RSM confirms the two inter layers and the epi-layer peaks by the diffraction profile of the InGaN film. It can be seen from the Fig. 14 that the InGaN epi-layers are not completely relaxed, while the interlayers show comparatively more strain. The lattice mismatch of GaN template to 1st inter layer, 1st inter layer to 2nd inter layer, and 2nd inter layer to InGaN epi-layer is 0.33, 0.31, and 0.59 %, respectively, whereas the calculated lattice mismatch of completely relaxed InGaN sample with the same In composition is 1.35 % for without graded layer. Therefore, the step-graded interlayer with the step-by-step relaxation of InGaN epi-layer decreases the lattice mismatch gradually that can reduce the defect density.

Comparison between simulated and experimental results

The result comparison with experimental work is an essential task for validation of any model. Experimental measurements of dislocations have been performed on the sample grown in laboratory. It has been reported that the dislocation density, particularly the edge dislocations, decreases considerably with the insertion of interlayers for higher In composition [23]. An edge dislocation of 3.37×10^{10} cm^{-2} is obtained for a sample with compositionally two step-graded interlayers ($In_{0.06}Ga_{0.94}N$ and $In_{0.13}Ga_{0.87}N$) and 1.5 μm $In_{0.26}Ga_{0.74}N$ final epilayer. On the other hand, using the developed model the edge dislocation density of 2.61×10^{10} cm^{-2} has been evaluated in (11–22) plane of a 1.5-μm step-graded $In_{0.26}Ga_{0.74}N$ heteroepitaxy using two interlayers. In contrast to lattice mismatch calculated from the RSM, we have also estimated the strain relaxation profile for a 1.5-μm $In_{0.14}Ga_{0.86}N$ epilayer using our model with two interlayers having same composition as Fig. 14. It has been found that the mismatches of 0.32, 0.33, and 0.58 % are estimated in first interlayer to GaN substrate, second interlayer to first interlayer, and epilayer to second interlayer, respectively.

Conclusions

From the above realization it can be summarized that a proper management of magnitude and rate of misfit strain is the key requirement to control the misfit and threading dislocations during the growth of InGaN. In this work, the step-graded interlayer technique for wurtzite InGaN heteroepitaxy has been investigated through mathematical modeling and numerical simulation as well as experimental observation for future solar cells.

The multilayer epitaxial structures with 40 % In composition for $In_xGa_{1-x}N$ has been considered using step-graded structure. It has been demonstrated that the insertion of step-graded interlayers significantly reduces the mixed and edge MDs up to a certain number within a specific In composition. It has been found that no significant misfit dislocation density decreases for the heteroepitaxy with In composition difference below 8 %. These results are consistent with the reported results of improved InGaN epilayer quality using step-graded interlayer. In addition, a numerical solution of TDs' reaction models has confirmed their diminishing densities with increasing thickness. The less nucleation sites for TDs and step inclination at each interlayer of the step-graded structure result in significant improvement in the epilayer quality with extremely less TD density, especially of mixed type. The edge and screw dislocations evaluated using this model in (11–22) plane have been found to agree well with the experimentally measured result for InGaN film using step-graded interlayer. The model verification has also been confirmed by comparing the lattice mismatch from the model and measured from RSM.

The above realizations indicate the potentiality of step-graded interlayer technique for InGaN heteroepitaxy. This simulation methodology can be used for other wurtzite heteroepitaxial material to estimate the performance of future high-efficiency solar cells.

References

1. Kong, Y.C., Zheng, Y.D., Zhou, C.H., Deng, Y.Z., Shen, B., Gu, S.L., Zhang, R., Han, P., Jiang, R.L., Shi, Y.: A novel $In_xGa_{1-x}N$/InN heterostructure field effect transistor with extremely high two-dimensional electron-gas sheet. Solid. State. Elec. **49**(2), 199–203 (2005)
2. Nanishi, Y., Saito, Y., Yamaguchi, T.: RF-Molecular beam epitaxy growth and properties of InN and related alloys. Jpn. J. Appl. Phys. **42**(5A), 2549 (2003)
3. Wu, J., Walukiewicz, W., Yu, K.M., Shan, W., Ager III, J.W., Haller, E.E., Lu, H., Schaff, W.J., Metzger, W.K., Kurtz, S.: Superior radiation resistance of $In_{1-x}Ga_xN$ alloys: Full-solar-spectrum photovoltaic material system. J. Appl. Phys. **94**(10), 6477–6482 (2003)
4. Hasan, M.T., Kaysir, M.R., Islam, M.S., Bhuiyan, A.G., Islam, M.R., Hashimoto, A., Yamamoto, A.: 2DEG properties in InGaN/InN/InGaN-based double channel HEMTs. Phys. Stat. Sol. C **7**(7–8), 1997–2000 (2010)
5. Lin, C.F., Zheng, J.H., Yang, Z.J., Dai, J.J., Lin, D.Y., Chang, C.Y., Lai, Z.X., Hong, C.S.: High-efficiency InGaN-based light-emitting diodes with nonporous GaN: Mg structure. Appl. Phys. Lett. **88**(8), 083121–083123 (2006)
6. Yamamoto, A., Islam, M.R., Kang, T.T., Hashimoto, A.: Recent advances in InN-based solar cells: status and challenges in InGaN and InAlN solar cells. Phys. Stat. Sol. C **7**(5), 1309–1316 (2010)
7. Neufeld, C.J., Toledo, N.G., Cruz, S.C., Lza, M., DenBaars, S.P., Mishra, U.K.: High quantum efficiency InGaN/GaN solar cells with 2.95 eV band gap. Appl. Phys. Lett. **93**, 143502–143503 (2008)
8. Horng, R.H., Lin, S.T., Tsai, Y.L., Chu, M.T., Liao, W.Y., Wu, M.H., Lin, R.M., Lu, Y.C.: Improved conversion efficiency of GaN/InGaN thin-film solar cells. IEEE Elec. Device Lett. **30**(7), 724–726 (2009)
9. Liu, Y.: Heteroepitaxial growth of InN and InGaN alloys on GaN(0001) by molecular beam epitaxy. Postgraduate thesis, the University of Hong Kong, Hong Kong (2005)
10. Cho, H.K., Lee, J.Y., Kim, K.S., Yang, G.M.: Formation of misfit dislocations and stacking faults in high indium content $In_xGa_{1-x}N$ layers grown by metal organic chemical vapor deposition. J. Korean Phys. Soc. **39**, S165–S169 (2001)

11. Wang, H., Jiang, D.S., Jahn, U., Zhu, J.J., Zhao, D.G., Liu, Z.S., Zhang, S.M., Qiu, Y.X., Yang, H.: Investigation on the strain relaxation of InGaN layer and its effects on the InGaN structural and optical properties. Phys. B **405**, 4668–4672 (2010)
12. Cho, H.K., Yang, G.M.: Generation of misfit dislocations in high indium content InGaN layer grown on GaN. J. Crystal Growth **243**(1), 124–128 (2002)
13. Song, M., Wu, Z., Fang, Y., Xiang, R., Sun, Y., Wang, H., Yu, C., Xiong, H., Dai, J., Chen, C.: Improved photovoltaic performance of InGaN single junction solar cells by using n-on-p type device structure. J. Optoelectron. Adv. Mater. **12**(7), 1452–1456 (2010)
14. Look, D.C., Sizelove, J.R.: Dislocation scattering in GaN. Phys. Rev. Lett. **82**, 1237–1240 (1999)
15. Sasaoka, C., Sunakawa, H., Kimura, A., Nido, M., Usui, A., Sakai, A.: High-quality InGaN MQW on low-dislocation-density GaN substrate grown by hydride vapor-phase epitaxy. J. Crystal Growth **189**(190), 61–66 (1998)
16. Matthews, J.W., Blakeslee, A.E.: Defects in epitaxial multilayer: 1. Misfit dislocation. J. Crystal Growth **27**, 118–125 (1974)
17. Ayers J. E.: ©2007 by Taylor & Francis Group, LLC, pp. 180–189, ISBN 0-8493-7195-3 (2007)
18. Hanada, T., Shimada, T., Ji, S.Y., Hobo, K., Liu, Y., Matsuoka, T.: Strain relaxation mechanism of InGaN thin film grown on m-GaN. Phys. Stat. Sol. C **8**(2), 444–446 (2011)
19. Gmili, Y.E., Orsal, G., Pantzas, K., Ahaitouf, A., Moudakir, T., Gautier, S., Patriarche, G., Troadec, D., Salvestrini, J.P., Ougazzaden, A.: Characteristics of the surface microstructures in thick InGaN layers on GaN. Opt. Mat. Express **3**(8), 1111–1118 (2013)
20. Kasper, E., Lyutovich, K.: Strain adjustment with thin virtual substrates. Sol. State Elect. **48**(8), 1257–1263 (2004)
21. Ee, Y.K., Biser, J.M., Cao, W., Chan, H.M., Vinci, R.P., Tansu, N.: Metalorganic vapor phase epitaxy of III-Nitride light-emitting diodes on nanopatterned AGOG sapphire substrate by abbreviated growth mode. IEEE J. Sel. Top. Quantum Electr. **15**, 1066–1072 (2009)
22. Ee, Y.K., Li, X.H., Biser, J., Cao, W., Chan, H.M., Vinci, R.P., Tansu, N.: Abbreviated MOVPE nucleation of III-nitride light-emitting diodes on nano-patterned sapphire. J. Crys. Growth **312**, 1311–1315 (2010)
23. Islam, M.R., Ohmura, Y., Hashimoto, A., Yamamoto, A., Kinoshita, K., Koji, Y.: Step-graded interlayers for the improvement of MOVPE $In_xGa_{1-x}N$ ($x \sim 0.4$) epi-layer quality. Phys. Stat. Sol. C **7**(7–8), 2097–2100 (2010)
24. Nakamura, S., Jayavel, P., Koyama, T., Hayakawa, Y.: Investigations on the effect of InSb and InAsSb step-graded buffer layers in $InAs_{0.5}Sb_{0.5}$ epilayers grown on GaAs (001). J. Crystal Growth **300**, 497–502 (2007)
25. Bertoli, B., Sidoti, D., Xhurxhi, S., Kujofsa, T., Cheruku, S., Correa, J.P., Rago, P.B., Suarez, E.N., Jain, F.C., Ayers, J.E.: Equilibrium strain and dislocation density in exponentially graded $Si_{1-x}Ge_x$/Si (001). J. Appl. Phys. **108**(11), 113525-5 (2010)
26. Ocampo, J.F., Suarez, E., Jain, F.C., Ayers, A.E.: Overshoot graded layers for mismatched heteroepitaxial devices. J. Elect. Mat. **37**, 1035 (2008)
27. Ayers, J.E.: Multiple critical layer thicknesses in retro-graded heterostructures. Appl. Phys. Lett. **92**, 102104-3 (2008)
28. Fitzgerald, E.A.: Dislocations in strained-layer epitaxy: theory, experiment, and applications. Mats. Sci. Reports **7**, 87 (1991)
29. Fitzgerald, E.A., Kim, A.Y., Currie, M.T., Langdo, T.A., Taraschi, G., Bulsara, M.T.: Dislocation dynamics in relaxed graded composition semiconductor. Mat. Sci. Engg. **B67**, 53–61 (1999)
30. Bertoli, B., Suarez, N., Ayers, J.E., Jain, F.C.: Misfit dislocation density and strain relaxation in graded semiconductors with arbitrary composition profiles. J. Appl. Phys **106**, 073519-7 (2009)
31. Romanov, A.E., Pompe, W., Mathis, S., Beltz, G.E., Speck, J.S.: Threading dislocation reduction in strained layers. J. Appl. Phys. **85**, 182–192 (1999)
32. Manning, I.C., Weng, X., Acord, J.D., Fanton, M.A., Snyder, D.W., Redwing, J.M.: Tensile stress generation and dislocation reduction in Si-doped $Al_xGa_{1-x}N$ films. J. Appl. Phys. **106**, 023506 (2009)
33. Manning, I.C., Weng, X., Fanton, M.A., Snyder, D.W., Redwing, J.M.: Effects of composition on dislocation microstructure and stress in Si-doped $Al_xGa_{1-x}N$. J. Crys. Growth **312**, 1301–1306 (2010)
34. Mooney, P.M., Jordan-Sweet, J.L., Chu, J.O., LeGoues, F.K.: Evolution of strain relaxation in step-graded SiGe/Si structures. Appl. Phys. Lett. **66**, 3642–3644 (1995)
35. Krishnamoorthy, V., Lin, Y.W., Park, R.M.: Application of 'critical compositional difference concept to growth of low dislocation density ($<10^4/cm^2$) $In_xGa_{1-x}As$ ($x \leq 0.5$) on GaAs. J. Appl. Phys. **72**, 1752–1757 (1992)
36. Macpherson, G., Beanland, R., Goodhew, J.P.: A novel design method for the suppression of edge dislocation formation in step-graded InGaAs/GaAs layers'. Philos. Mag. A **73**(05), 1439–1450 (1996)
37. Zhang, J., Tansu, N.: Improvement in spontaneous emission rates for InGaN quantum wells on ternary InGaN substrate for light-emitting diodes. J. Appl. Phys. **110**, 113110 (2011)
38. Zhang, J., Tansu, N.: Optical gain and laser characteristics of InGaN quantum wells on ternary InGaN substrates. IEEE Photonics J. **5**, 2600111 (2013)
39. Kotani, T., Lugli, P., Hamaguchi, C.: Optical gain analysis of c-InGaN quantum wells on unstrained c-$In_{0.31}Ga_{0.69}N$ templates. Appl. Phys. Lett. **102**, 011128 (2013)
40. Zhao, H., Liu, G., Zhang, J., Poplawsky, J.D., Dierolf, V., Tansu, N.: Approaches for high internal quantum efficiency green InGaN light-emitting diodes with large overlap quantum wells. Opt. Express **19**(S4), A991–A1007 (2011)
41. Tyagi, A., Wu, F., Young, E.C., Chakraborty, A., Ohta, H., Bhat, R., Fujito, K., DenBaars, S.P., Nakamura, S., Speck, J.S.: Partial strain relaxation via misfit dislocation generation at heterointerfaces in (Al, In)GaN epitaxial layers grown on semipolar (11–22) GaN free standing substrates. Appl. Phys. Lett. **95**, 251905 (2009)
42. Young, E.C., Wu, F., Romanov, A.E., Tyagi, A., Gallinat, C.S., DenBaars, S.P., Nakamura, S., Speck, J.S.: Lattice tilt and misfit dislocations in (11–22) semi polar GaN Heteroepitaxy. Appl. Phys. Exp. **3**, 011004 (2010)
43. Srinivasan, S., Geng, L., Ponce, F.A., Narukawa, Y., Tanaka, S.: Slip systems and misfit dislocations in InGaN epilayers. Appl. Phys. Lett. **83**(25), 5187–5189 (2003)
44. Vennegues, P., Bougrioua, Z., Bethoux, J.M., Azize, M., Tottereau, O.: Relaxation mechanisms in metal-organic vapor phase epitaxy grown Al-rich (Al, Ga)N/GaN heterostructures. J. Appl. Phys. **97**(2), 024912–024916 (2005)
45. Hossain, M.A., Hasan, M.M., Islam, M.R.: Strain relaxation via misfit dislocation in step-graded InGaN heteroepitaxial layers grown on semipolar (11–22) and (1–101) GaN. Int. J. Appl. Phys. Math. **02**(01), 049–053 (2012)
46. Hossain, M.A., Islam, M.R.: A theoretical calculation of misfit dislocation and strain relaxation in step-graded $In_xGa_{1-x}N$/GaN layers. Adv. Mat. Res. **403–408**, 456 (2012)
47. Holec, D., Costa, P.M.F.J., Kappers, M.J., Humphreys, C.J.: Critical thickness calculations for InGaN/GaN. J. Crystal Growth **303**, 314–317 (2007)
48. Romanov, A.E., Pompe, W., Beltz, G., Speck, J.S.: Modeling of threading dislocation density reduction in heteroepitaxial layer: Part I. geometry and crystallography. Phys. Stat. Sol. **198**, 599–613 (1996)

49. Romanov, A.E., Pompe, W., Beltz, G., Speck, J.S., Mathis, S.: Threading dislocation reduction in strained layers. J. Appl. Phys. **85**(1), 182–192 (1999)
50. Speck, J.S., Brewer, M.A., Beltz, G., Romanov, A.E., Pompe, W.: Scaling laws for the reduction of threading dislocation densities in homogenous buffer layer. J. Appl. Phys. **80**(7), 3808–3816 (1996)
51. Mathis, S.K., Romanov, A.E., Chen, L.F., Beltz, G., Pompe, W., Speck, J.S.: Modeling of threading dislocation reduction in growing GaN layers. Phys. Stat. Sol. **179**, 125–145 (2000)
52. Haeberlen, M., Zhu, D., McAleese, C., Kappers, M.J., Humphreys, C.J.: Dislocation reduction in MOVPE grown GaN layers on (111) Si using SiNx and AlGaN layers. J. Phys Conf. Ser. **209**, 012017 (2010)
53. Yamamoto, A., Sugita, K., Horie, M., Ohmura, Y., Islam, Hashimoto, A.: Mg-doping and n + -p junction formation in MOVPE-grown $In_xGa_{1-x}N$ ($x \sim 0.4$). 33rd IEEE Photovoltaic Specialist Conference, San Diego (2008)
54. Liou, B.T., Lin, C.Y., Yen, S.H., Kuo, Y.K.: First-principles calculation for bowing parameter of wurtzite $In_xGa_{1-x}N$. Opt. Commun. **249**, 217 (2005)
55. Detchprohm, T., Hiramatsu, K., Itoh, K., Akasaki, I.: Relaxation process of the thermal strain in the GaN/α-Al2O3 heterostructure and determination of the intrinsic lattice constants of GaN free from the strain. Jpn. J. Appl. Phys. **31**, L1454–L1456 (1992)
56. Paszkowicz, W.: X-ray powder diffraction data for indium nitride. Powder Diffr. **14**, 258 (1999)

Strontium and iron-doped barium cobaltite prepared by solution combustion synthesis: exploring a mixed-fuel approach for tailored intermediate temperature solid oxide fuel cell cathode materials

Francesca Deganello · Leonarda F. Liotta · Giuseppe Marcì · Emiliana Fabbri · Enrico Traversa

Abstract $Ba_{0.5}Sr_{0.5}Co_{0.8}Fe_{0.2}O_{3-\delta}$ (BSCF) powders were prepared by solution combustion synthesis using single and double fuels. The effect of the fuel mixture on the main properties of this well-known solid oxide fuel cell cathode material with high oxygen ion and electronic conduction was investigated in detail. Results showed that the fuel mixture significantly affected the area-specific resistance of the BSCF cathode materials, by controlling the oxygen deficiency and stabilizing the Co^{2+} oxidation state. It was demonstrated that high fuel-to-metal cations molar ratios and high reducing power of the combustion fuel mixture are mainly responsible for the decreasing of the area-specific resistance of BSCF cathode materials. Moreover, a new metastable monoclinic phase with $Ba_{0.5}Sr_{0.5}CO_3$ composition was discovered in the as-burned BSCF powders, enlarging the existing information on the BSCF phase formation mechanism.

Keywords Solution combustion synthesis · Intermediate temperature solid oxide fuel cells · Cathode materials · Combustion fuel mixture · BSCF · Perovskite-type compounds

F. Deganello (✉) · L. F. Liotta
Consiglio Nazionale delle Ricerche (CNR), Istituto per lo Studio dei Materiali Nanostrutturati (ISMN), UOS Palermo, Via Ugo La Malfa 153, 90146 Palermo, Italy
e-mail: francesca.deganello@cnr.it

G. Marcì
Dipartimento di Ingegneria Elettrica Elettronica e delle Telecomunicazioni (DIEET), Università di Palermo (UNIPA), 90128 Palermo, Italy

E. Fabbri
National Institute for Materials Science (NIMS), International Research Center for Materials Nanoarchitectonics (MANA), Tsukuba 305-0044, Japan

E. Traversa
International Research Center for Renewable Energy, State Key Laboratory of Multiphase Flow in Power Engineering, Xi'an Jiaotong University, Xi'an 710049, Shaanxi, China

Present Address:
E. Fabbri
Electrochemistry Laboratory, General Energy Research Department, Paul Scherrer Institut, 5232 Villigen, Switzerland

Introduction

The increasing worldwide demand of efficient, low-cost, and eco-friendly energy technologies has been leading in the last years toward an intense research on intermediate temperature (450–700 °C) solid oxide fuel cell (IT-SOFC) devices [1]. In particular, cathode materials to be used in IT-SOFCs should meet some essential requirements, like high oxygen and electronic transport properties (easy oxygen reduction reaction, surface exchange, and diffusion) at intermediate temperatures and high compatibility with the electrolyte material [1, 2]. To meet these requirements, two main strategies have been adopted in the literature: looking for new chemical compositions [3] and developing new preparation routes [4] (or improving the existing preparation and processing procedures [5]) to increase the active area where the cathode reaction can take place.

$Ba_{1-x}Sr_xCo_{1-y}Fe_yO_{3-\delta}$ (BSCF) perovskite-type compounds, already known as oxygen conducting membrane materials [6], have attracted the attention of the scientific community as IT-SOFC cathode materials owing to their very good oxygen conducting properties at intermediate temperatures [7–10]. Nevertheless, their chemical instability [6, 11, 12] and reactivity to CO_2 [13] in the intermediate temperature range have so far limited their industrial applicability. Despite the described limits, BSCF is still a very interesting material to investigate, since the information obtained about the factors governing its high conductivity may help the selection of more stable cathode materials [9, 10, 14–16]. BSCF has been usually prepared by solid-state reaction synthesis [9, 10, 17–19] or by EDTA–citric acid complexing method [5–7, 18, 20, 21], although few other preparation routes have been described as well in the literature, for example the co-precipitation method [22] and the solution combustion synthesis (SCS) [17, 18]. SCS is based on both the concepts of sol–gel and propellant chemistry [23, 24] and it has been successfully used in the literature to obtain powders at lower temperatures, with smaller crystal size and larger surface area or with metastable structures, compared to other preparation methods [23, 25, 26]. The properties of the materials prepared by SCS are essentially governed by the fuel type and by the reducer/oxidizer molar ratio, the so called "Φ" ratio [24, 26, 27]. The effect of the combustion fuel on the properties of heterometallic oxides, and especially of SOFC materials prepared by SCS, has been investigated recently, for both single fuels and fuel mixtures [28–30]. In particular, it has been found that the fuel type affects the microstructural properties of the powders [29–31], the phase formation mechanism [28] and the electrical conducting properties [29, 31, 32] of the final material. Nevertheless, only very general explanations about the effect of the combustion fuel have been so far proposed. Moreover, the use of a mixture of fuels has been rarely reported for the preparation of such materials.

In this paper, fuels such as sucrose, cellulose, citric acid and polyethylene glycol (PEG) were used, for the first time alone or in combination, for the preparation of BSCF powders of $Ba_{0.5}Sr_{0.5}Co_{0.8}Fe_{0.2}O_{3-\delta}$ composition. Moreover, the effect of the fuel mixture composition on the phase formation process, structure, microstructure and conductivity of BSCF prepared by SCS was investigated in detail, going a step further toward the control of the IT-SOFC materials properties.

If the use of citric acid is already well known in the literature [25] the cellulose templating method [33, 34] and sucrose method [35, 36] are less common and they have been not yet used for the preparation of BSCF powders, to the best of our knowledge. Moreover, the experimental conditions used in this work for the synthesis from cellulose and sucrose-containing fuel mixtures were quite different from the literature and need to be commented. Sucrose is usually accompanied by nitric acid at very acid pH, in order to form saccharic acid, which is able to polymerize alone or in the presence of PVA [37, 38]. In this work, sucrose was used without any nitric acid, alone or in combination with PEG, and the resultant pH was around 5.

Experimental procedure

Powder preparation

The same general procedure was used for the preparation of BSCF powders of $Ba_{0.5}Sr_{0.5}Co_{0.8}Fe_{0.2}O_{3-\delta}$ composition, irrespective of the fuel used. $Ba(NO_3)_2$ (99.1 %, VWR Prolabo), $Sr(NO_3)_2$ ($\geq$99 %, Merck), $Co(NO_3)_2{\cdot}6H_2O$ ($\geq$99 %, Fluka), and $Fe(NO_3)_3{\cdot}9H_2O$ ($\geq$98 %, Carlo Erba) were carefully weighted in the required proportions to obtain 3 g of final product (unless a different amount was specified), poured in a 1.2-L stainless steel beaker and dissolved in distilled water at 60 °C under magnetic stirring. Ammonium nitrate and the fuel were then added to the metal precursor solution as oxidant additive and propellant–template–complexant, respectively. Citric acid (anhydrous, Fluka, 99.9 %), cellulose (powder DS-O for thin layer chromatography, Fluka Analytical), sucrose (Eridania), polyethylene glycol (PEG, MW 20000, Fluka Chemika) were used as fuels, alone or in combination. For one of the samples, ammonia solution (NH_4OH 30 vol %) was added to the citric acid fuel as a pH regulator.

Taking into account only the primary fuel, the fuel/metal cation molar ratio was set to 2, unless otherwise specified, while the reducers/oxidizers ratio Φ was set to the stoichiometric value for the specific combustion reaction, following the method described by Jain et al. [24]. The water was left to evaporate at 80 °C (a vertex was inserted in the precursor solution) under constant magnetic stirring until the formation of a homogeneous gel. Initiation of the combustion reaction was performed by putting the stainless steel beaker on a hotplate and setting the temperature to around 250 °C.

The as-burnt powders were fired at higher temperatures (600–950 °C) for the crystallization of the perovskite-type phase and to eliminate any carbon residues derived from the combustion process.

Temperature/time profiles of the combustion process

A K-type thermocouple 1.5 mm in diameter, which covered a temperature range from ambient temperature up to 1,100 °C, was used to measure the temperature variations during the combustion reaction. Concurrently, the signals from the thermocouple were stored and processed using a data acquisition system: a data logger (PICO technology) with a sampling velocity of 20 bit/s and a computer with

Picolog software. The thermocouple was carefully inserted into the combustion medium and the temperature was registered from the time the beaker was put into contact with the hotplate until the end of the combustion process. The duration of the process was calculated from the full width half maximum (FWHM) of the peak profile.

Powder characterization

X-ray diffraction analysis

The same general X-ray diffraction (XRD) measurements were carried out on a SIEMENS D5000 X-ray powder diffractometer equipped with a Kristalloflex 760 X-ray generator and with a curved graphite monochromator that made possible the selection of the Cu Kα radiation (40 kV/30 mA). The 2θ step size was 0.03° for the as-burned powders and the powders fired at 600 °C, whereas a 0.02° step was used for the powders fired at higher temperatures. The integration time was 3 s per step, and the 2θ scan range was from 18° to 90°. The diffraction patterns were analyzed by Rietveld refinement using the GSAS package [37] Chebyschev polynomials and Pearson VII functions were chosen for the background and for the peak profile fitting, respectively. In the structure refinement, lattice constants, zero offset, scale factors, FWHM, Debye–Waller factors and microstrain were considered as variable parameters. In those cases where the XRD patterns were complex and with overlapping peaks, Debye–Waller factors and microstrain were set to a default value (0.025 and 0, respectively). From fitting results, the structural parameters of the investigated compounds and, in particular, the cell edge lengths and the relative phase composition were obtained. When possible, an estimation of the crystal size values was obtained from Scherrer equation, in agreement with the GSAS package procedure. The agreement factors ("*R* values") [38] were generally acceptable and they were reported along with the graphical Rietveld fits showed in the figures. A standard deviation (SD) of ±0.003 Å for the refined cell parameters was estimated in the experimental conditions used in this work.

Simultaneous thermal analysis

Thermo-gravimetric and differential thermal analysis (TG–DTA) measurements were performed using a Perkin Elmer STA 6000 equipment, in the 50–1,000 °C temperature range (heating rate = 10 °C/min) in static air or in nitrogen flow (ca. 20 ml/min), on both as-burned and fired powders.

Nitrogen adsorption

Specific surface area (BET method) [39] and mesopore size distributions (BJH method) [40] were evaluated performing nitrogen adsorption/desorption measurements at −196 °C, using a Sorptomatic 1900 Carlo Erba equipment. All the samples were pre-treated under vacuum at 250 °C for 2 h prior to the measurements.

Temperature-programmed reaction techniques

Reduction properties of the perovskite oxides were studied by temperature-programmed reduction (TPR) measurements. Experiments were carried out with a Micromeritics Autochem 2910 apparatus, equipped with a thermal conductivity detector (TCD) and a mass quadrupole spectrometer (QMS) (Thermostar^{TM}, Balzers). Hydrogen consumption (mmol/g) was calculated through TCD calibration experiments. Before TPR, the samples (0.17 g) were pre-treated in a quartz reactor in O_2 (5 % in He) at 500 °C for 30 min, followed by purging and cooling to room temperature in Ar flow. Then, the flow was switched to the H_2 mixture (5 % in Ar, 50 mL/min) and the temperature was increased with a heating rate of 10 °C/min up to 950 °C. The same instrument was used to study the reactivity towards oxygen of the perovskite oxides, performing temperature-programmed oxidation (TPO) measurements. The fired samples (0.17 g), without any pre-treatment, were flowed under O_2 (5 % in He, 50 mL/min) by increasing linearly the temperature from 25 to 950 °C. The temperature was kept at 950 °C for 2 h under oxygen atmosphere, and then the samples were cooled down to room temperature. Finally, to study the oxygen desorption properties of the perovskite oxides, He-TPD experiments were performed directly on the calcined samples (0.17 g) by flowing He (50 mL/min) from 25 to 950 °C. The temperature was kept at 950 °C for 2 h under He stream, and then cooling down under the same atmosphere. The evolution of gaseous species was monitored by on-line TCD and QMS.

SEM-EDX

Typical micrographs and energy-dispersive X-ray (EDX) analysis of the powder samples and fractured cross section of the co-sintered electrode/electrolyte interfaces were acquired using a FEI Quanta 200 SEM microscope operating in high vacuum at 20 keV, after deposition by sputtering of a thin gold layer.

Electrochemical characterization of symmetric cells

Dense electrolyte pellets, 13 mm in diameter, were obtained by uniaxially pressing $Sm_{0.2}Ce_{0.8}O_2$ (SDC, AGC Seimi Chemical Co., Ltd., Japan) powders at 250 MPa and sintering at 1,500 °C for 5 h. Cathode inks were prepared mixing the BSCF powders with an organic vehicle (α-terpineol as a solvent, di-*n*-butyl phthalate as a plasticizer, and polyvinylbutyral as a binder) until an appropriate viscosity for brushing was achieved. Symmetric cells were prepared by brush painting the cathode slurry to each side of the SDC pellet, drying at 150 °C for 3 h, and firing at 1,000 °C for 2 h.

Electrochemical impedance spectroscopy (EIS) measurements were performed using a multichannel potentiostat (model VMP3, Biologic Science Instruments) in the frequency range between 100 mHz and 500 kHz, with an AC voltage amplitude ranging between 10 and 50 mV, at temperatures ranging between 500 and 800 °C. Silver mesh current collectors and four silver lead wires were pressed against the samples in a quartz reactor by the use of a ceramic screw-and-bolt assembly. The quartz reactor was placed into a temperature-controlled furnace, and dry air was fed through the reactor at a rate of 50 cm^3 min^{-1}. For the measurements at different oxygen partial pressures, the atmosphere in the quartz reactor was changed to different pO_2 values using various O_2–Ar and air–Ar gas mixtures, and monitoring the pO_2 of the outlet gas with a zirconia oxygen sensor.

In the acquired complex impedance plane plots, the intercept with the real axis at high frequencies was considered as the Ohmic resistance (R_{Ohmic}) composed of the electrolyte resistance, possibly electrode-sheet resistance, and lead-contact resistance. The difference between the high frequency and low-frequency intercepts with the real axis, in some cases composed of the sum of two contributions, was considered as the electrode polarization resistance (R_p). The different contributions in the impedance plots were calculated by fitting the experimental data with a Zview software. An equivalent circuits made of $LR_0(R_1Q_1)(R_2Q_2)$ was used to fit the complex impedance plane plots. In the selected equivalent circuits, L represents the set-up inductance, R_0 is the total ohmic resistance, R_1 and R_2 are the resistance values due to the electrode processes, whereas Q_1 and Q_2 are the constant phase elements associated with R_1 and R_2, respectively.

The area-specific resistance (ASR) of the electrode was calculated from the electrode resistance (R_p) by

$$\mathrm{ASR} = \frac{R_p A}{2}$$

where A is the geometrical electrode area and the factor $\frac{1}{2}$ takes into account that symmetrical cells were used.

Results and discussion

Combustion process and formation mechanism of BSCF cathode materials

BSCF powders were prepared by SCS using two single fuels and four fuel mixtures. Fuels such as sucrose, cellulose, citric acid, ammonia and polyethylene glycol were used alone or in combination. Table 1 describes the fuel mixture composition for all the samples studied in this work.

Table 1 Description of the combustion fuel mixtures used for the preparation of BSCF powders

Sample name	Combustion fuel mixture
A1	Cellulose + citric acid (citric acid/cellulose unit molar ratio = 0.5); cellulose unit/metal cations molar ratio = 2; Φ* = 1.1
A2	Citric acid + NH_4OH (until pH = 6); citric acid/metal cations molar ratio = 2; Φ = 1.3
A3	Sucrose; sucrose/metal cations molar ratio = 2; Φ = 1.0
A4	Sucrose + PEG (PEG unit/sucrose molar ratio = 1); sucrose/metal cations molar ratio = 2; Φ = 1.2
A5	Cellulose; cellulose unit/metal cations molar ratio = 4; Φ = 0.8
A6	Sucrose + PEG (PEG unit/sucrose molar ratio = 2.5); sucrose/metal cations molar ratio = 2; Φ = 1.5

* Φ ratios (reducers/oxidizers) are stoichiometric taking into account only the primary fuel. In this table Φ ratios relative to the fuel mixture are also reported

BSCF perovskite-type compounds with cubic structure were obtained from the A1–A6 fuel mixtures by using a solution combustion synthesis followed by a thermal treatment in static air. Figure 1 shows the XRD patterns before (as-burned powders) and after heating to 600, 750 and 950 °C for 5 h of a representative sample, A4, prepared from sucrose and PEG. The XRD patterns of the as-burned samples, regardless of the fuel (single fuel or mixture), showed the formation of two types of Ba–Sr carbonates, one being the already known orthorhombic aragonite-type carbonate (ar-$Ba_{0.5}Sr_{0.5}CO_3$), the other a new metastable δ phase with monoclinic structure (δ-$Ba_{0.5}Sr_{0.5}CO_3$). After heating to 600 °C, δ-$Ba_{0.5}Sr_{0.5}CO_3$ was totally converted into ar-$Ba_{0.5}Sr_{0.5}CO_3$ (Fig. 1). A Fe–Co spinel was also detected in all the powders heated to 600 °C (Fig. 1). After heating to 750 °C, no more carbonates were present in the XRD patterns and two intermediate phases were formed, P1 hexagonal and P′ cubic, of $Ba_{1-x}Sr_xCo_{1-y}Fe_yO_{3-\delta}$ composition, with x (and y) content different from 0.5 (and 0.2) (Fig. 1). Such two phases (P1 and P′) merged into a cubic P phase after thermal treatment at 950 °C. The heating temperature for the formation of a pure BSCF phase reported by other authors in the literature [17, 18, 41] was 50–100 °C lower than the value obtained in this work, though using slightly different experimental conditions.

TG–DTA measurements were performed under static air on both as-burned and calcined BSCF powder samples. The TG–DTA of a representative as-burned sample (A3) is

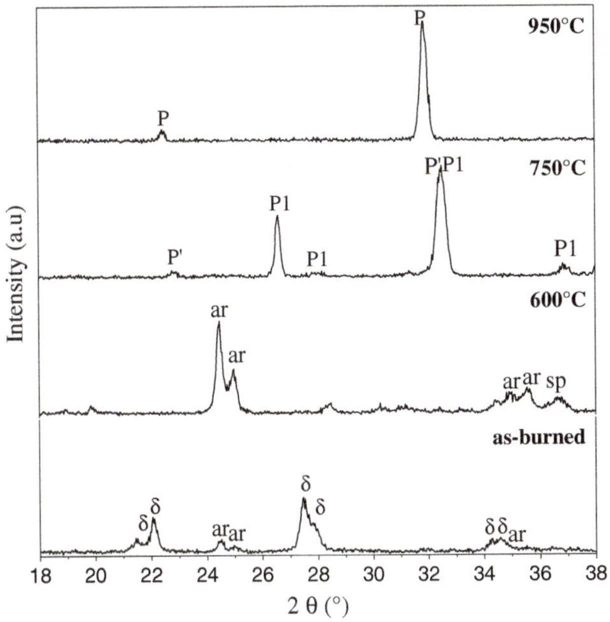

Fig. 1 Evolution of the phase composition with the temperature for a representative BSCF sample (A4). The main peaks of stable aragonite-type $Ba_{0.5}Sr_{0.5}CO_3$ and metastable δ-type $Ba_{0.5}Sr_{0.5}CO_3$ phases are indicated by *ar* and δ, respectively; the desired BSCF cubic phase is indicated by a *P*, while the other intermediate BSCF hexagonal and cubic phases are indicated by *P1* and *P'*, respectively. Peaks belonging to the Fe–Co spinel-type phase are indicated by *sp*

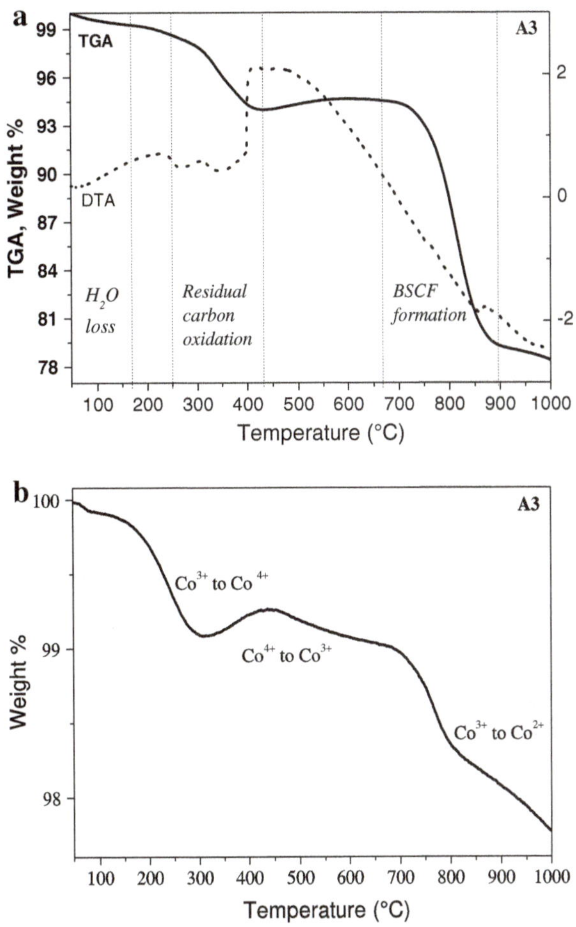

Fig. 2 TG and DTA curves under static air for a representative as-burned sample (**a**); TG curve under static air for the A3 fired sample (**b**)

reported in Fig. 2a. Figure 2b shows TG–DTA profiles of the same sample after calcination at 950 °C/5 h.

In the as-burned sample (Fig. 2a), after the initial water loss, a sharp step ascribable to the residual carbon decomposition was visible at around 400 °C. At around 850 °C a substantial weight loss occurred, which was related to the formation of the final BSCF cubic phase from ar-$Ba_{0.5}Sr_{0.5}CO_3$ and Fe-Co spinel. The BSCF formation was completed at 900 °C. Above that temperature, the weight continued to decrease. The absence of a definite *plateau* in the TGA curves hindered a proper determination of the exact formation temperature and it was ascribed to the peculiar property of the BSCF powders to lose oxygen even under oxidative environment [19, 42]. According to this hypothesis, an evident weight loss was observed in the TGA of A3 after calcination at 950 °C/5 h (Fig. 2b). Such weight loss was around 1.5 wt % in the 200–900 °C temperature range, in agreement with the literature [21]. Following the considerations of Li et al. [43], three different regions were identified in the TGA of A3 sample (Fig. 2b). The oxidation of Co^{3+} and Fe^{3+} to the IV state took place between 200 and 350 °C with a slight weight gain. Above 350 °C the reduction of Co^{4+} and Fe^{4+} to the (III) state was more favored. Then, the temperature-driven reduction of Co^{3+} to Co^{2+} occurred above 800 °C. These redox changes were accompanied by oxygen content variations for charge compensation.

The evolution of the BSCF phase composition with the heating temperature in the 600–950 °C range showed in Fig. 1 was in agreement with the literature [20], whereas the phase composition of the as-burned-powders needs to be commented, since it enlarges the actual picture on the BSCF phase formation mechanism.

Figure 3 shows the XRD pattern and Rietveld refinement for a representative as-burned sample, A2, which showed the largest percentage of the δ-$Ba_{0.5}Sr_{0.5}CO_3$. Table 2 provides the phase composition of such sample together with the structural information about the new metastable phase. A considerable amount of metallic cobalt was also found in the as-burned powders (Table 2). The formation of metallic Co is not surprising since it is known to form, eventually with CoO, in the as-burned powders prepared by SCS, depending on the flame temperature and on the nature of the combustion process [44]. It can be noticed that Fe was not mentioned in Table 2, although it is probably still hidden as amorphous phase, since it appeared as Co–Fe spinel after heating to 600 °C.

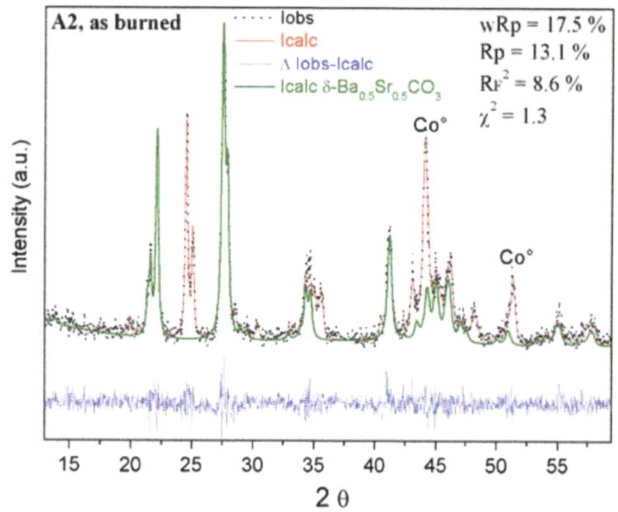

Fig. 3 Graphical results of the Rietveld refinement for a typical as-burned BSCF sample (*A2*), where the new metastable δ-$Ba_{0.5}Sr_{0.5}CO_3$ monoclinic phase represents the main phase; the two main peaks belonging to metallic cobalt are indicated by *Co°*. Rietveld reliability factors are as well reported in the figure

The formation of the new metastable δ-$Ba_{0.5}Sr_{0.5}CO_3$ phase deserves a more detailed discussion. This is very similar to the δ-$BaCO_3$ found in the literature [45, 46], although it has a much smaller lattice volume (145 Å^3) with respect to the δ-$BaCO_3$ phases previously described in the literature (155 and 158 Å^3) (Table 2), in agreement with the smaller ionic radius of Sr^{2+} with respect to that of Ba^{2+}. The XRD pattern of the new δ-$Ba_{0.5}Sr_{0.5}CO_3$ phase identified in this work resembles the "OP1" Ba–Sr carbonate phase observed by Yi et al. [47] on the surface layer of a BSCF membrane at the beginning of the thermal treatment in CO_2 [45], although they indexed it with an orthorhombic Pmma group. In our work, any attempt made to fit the δ-$Ba_{0.5}Sr_{0.5}CO_3$ phase according to an orthorhombic symmetry with the cell parameters suggested by Yi et al. was unsuccessful. Like its δ-$BaCO_3$ analogue, which is known to be stabilized by the incorporation of SO_4^- anions [46] or by an abrupt temperature change [45], it is likely that the new metastable δ-$Ba_{0.5}Sr_{0.5}CO_3$ phase was formed at the beginning of the

Table 2 Phase composition for a typical as-burned A-type sample (A2) and structural results for the new metastable δ-$Ba_{0.5}Sr_{0.5}CO_3$

Phase type	Space group						Structural parameters	wt %
δ-$Ba_{0.5}Sr_{0.5}CO_3$	Monoclinic, P 1 21/m 1 (this work)[a]						$a = 6.811$ Å	55
							$b = 5.178$ Å	
	Atom type	x	y	z	Occupancy	UISO	$c = 4.386$ Å	
	Ba^{+2}	0.233269	0.75	0.918715	0.5	0.03036	$\alpha = 90.0°$	
	Sr^{+2}	0.233269	0.75	0.918715	0.5	0.03036	$\beta = 109.79°$	
	C^{+4}	0.232508	0.25	0.442513	1	0.06314	$\gamma = 90.0°$	
	O^{-2} (1)	0.400891	0.241605	0.812455	0.5	0.03875	$V = 145.5$ Å^3	
	O^{-2} (2)	0.254420	−0.011103	0.434185	0.5	0.01750		
	O^{-2} (3)	0.132924	0.417429	0.335414	0.5	0.00268		
ar–$Ba_{0.5}Sr_{0.5}CO_3$	Orthorhombic, P m c n[b]						$a = 6.266$ Å	24
							$b = 5.207$ Å	
							$c = 8.654$ Å	
							$\alpha = 90.0°$	
							$\beta = 90.0°$	
							$\gamma = 90.0°$	
							$V = 282.4$ Å^3	
Co	Cubic, F m −3 m (ICSD #622439)						$a = 3.563$ Å	21
							$b = 3.563$ Å	
							$c = 3.563$ Å	
							$\alpha = 90.0°$	
							$\beta = 90.0°$	
							$\gamma = 90.0°$	
							$V = 45.2$ Å^3	

A SD of ±0.003 Å for the refined cell parameters was estimated in the experimental conditions used in this work

[a] Atomic positions were adapted for barium carbonate sulfate—Delta ICSD #63257 from Ref. [46]

[b] Atomic positions were taken from [65]

combustion process due to the fast combustion. Figure 4 shows the temperature/time profiles of sample A2, prepared by citric acid-assisted SCS, for 1, 1.5 and 3 g of sample produced. All the fuel mixtures used in this work displayed a single combustion peak similar to those of A2, with a maximum temperature at around 600–700 °C, where the combustion involved the whole mass (visual observation) in few seconds. The temperature/time profiles were reproducible for the same experimental conditions used in the synthesis, including the type of combustion beaker used (geometry and material) and the final mass of the produced samples. Both the duration and the maximum temperature of the combustion increased with the BSCF powder mass produced in the synthesis. Moreover, the amount of δ-$Ba_{0.5}Sr_{0.5}CO_3$ phase, determined by Rietveld analysis of the XRD data, decreased, whereas the ar-$Ba_{0.5}Sr_{0.5}CO_3$ phase and metallic cobalt amounts proportionally increased by increasing the sample mass of the same as-burned sample (table inset of Fig. 4). A larger duration of the combustion process favors the conversion of the metastable δ-$Ba_{0.5}Sr_{0.5}CO_3$ into the more stable ar-$Ba_{0.5}Sr_{0.5}CO_3$.

It should be emphasized that, although the phase composition of the as-burned samples was dependent on the sample weight, such dependence was not observed after heating to 600 °C/5 h, where only the ar-$Ba_{0.5}Sr_{0.5}CO_3$ and Fe–Co spinel were present, irrespective of the amount of powder produced in one batch, in agreement with the existing literature [20].

Sample mass (g)	Maximum T (°C)	Duration (s)	δ-$Ba_{0.5}Sr_{0.5}CO_3$ (wt.%)	ar-$Ba_{0.5}Sr_{0.5}CO_3$ (wt.%)	Metallic Co (wt.%)
1.0	602	23	70	6	14
1.5	648	26	67	16	17
3.0	673	34	55	24	21

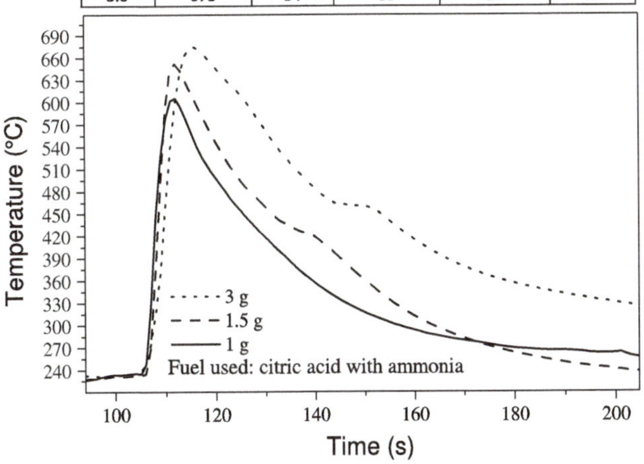

Fig. 4 Temperature–time profiles for the A2 sample, for increasing sample mass of powder prepared in the same batch; the included table shows the corresponding variation of the ar-$Ba_{0.5}Sr_{0.5}CO_3$, δ-$Ba_{0.5}Sr_{0.5}CO_3$, and Co° wt. % with the combustion process duration and combustion profile maximum temperature

Structural parameters and B-site cation stability of BSCF cathode materials

Figure 5 shows the Rietveld refinement of a representative sample, A4, after firing at 950 °C/5 h. A closer look by Rietveld refinement of the XRD patterns of the powders heated to 950 °C revealed the presence of a mixture of BSCF phases with the same cubic symmetry and minor amounts (below 5 wt %) of a rhombohedral phase ($a = b = 5.67$ Å, $c = 35.2$ Å). This phase can be well described with a $Ba_3Co_{10}O_{17}$-like structure oriented parallel to the $(111)_{cubic}$ plane, as already observed by Efimov et al. [48], who stated that the probable composition of this phase is $Ba_{1-x}Sr_xCo_{2-y}Fe_yO_{5-\delta}$. The mixture of cubic phases was observed in all the samples, although some clear differences were present in terms of cell parameter and/or respective weight percentage. The two phases have different cell parameters and oxygen deficiency. The BSCF cubic phase with the smaller cell parameter (phase 1) was the typical BSCF cubic phase (3.986 Å for A4) in agreement with the literature [49]. The other cubic phase (phase 2) had a larger lattice parameter value (3.994 Å for A4), demonstrating a larger oxygen deficiency due to the increased formation of Co^{2+} ions with larger ionic radius. The formation of an oxygen-deficient BSCF with cubic structure can be explained by the fact that BSCF, with respect to other perovskite-type oxides, is able to keep very high oxygen deficiency without forming the ordered brownmillerite structure [50]. From the refined cell parameters and phase composition, the total percentage of Co^{2+} was calculated, taking into account the relationship

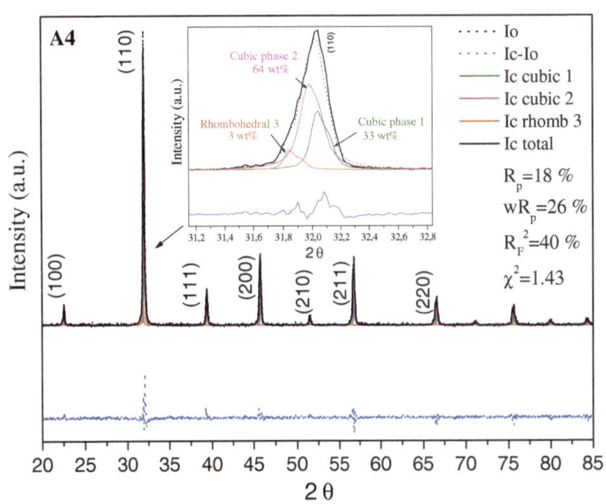

Fig. 5 Graphical results of the Rietveld refinement of the XRD patterns of the A4 sample heated to 950 °C for 5 h. (*h k l*) planes are also indicated for the BSCF cubic phase. Rietveld reliability factors are reported in the figure as well. The *inset* is a magnification of the (110) XRD peak at around 32°2*θ*, which evidences the contribution from the three phases

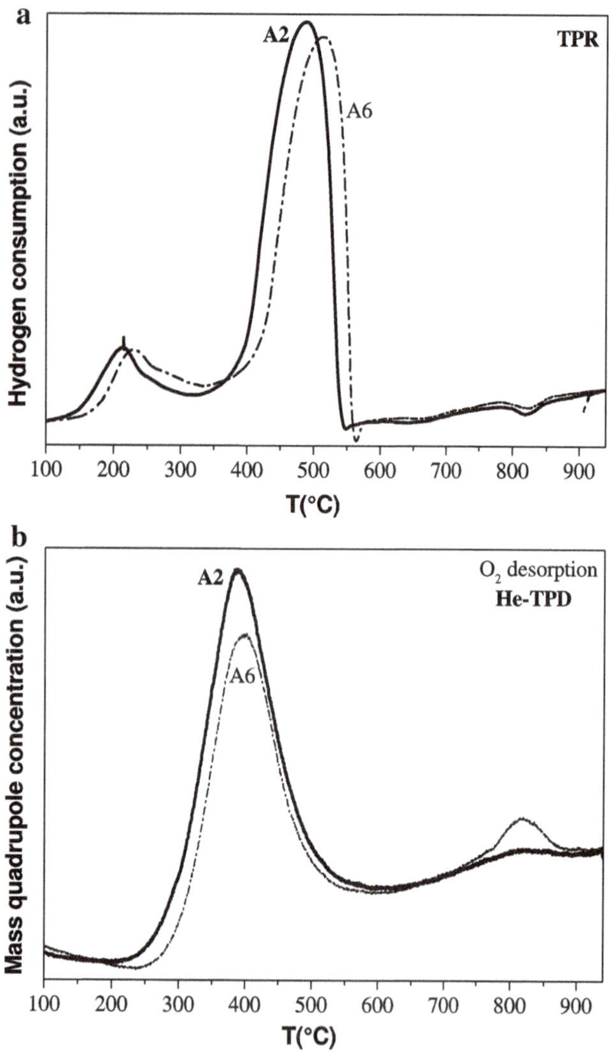

Fig. 6 TPR (**a**) and He-TPD (**b**) profiles for the A2 and A6 samples

found by Kriegel et al. [19] between the BSCF cell parameter and the oxygen deficiency, hypothesizing that BSCF contains more Co^{3+} and/or Co^{2+} than Co^{4+} and more Fe^{3+} than Fe^{4+} [51]. It was found that Co^{2+} changed with the type of fuel mixture, ranging between 5 wt % (A2) and 32 wt% (A6). Figure 6 compares temperature-programmed experiments under reductive (TPR) and inert (He-TPD) environments for the A2 and A6 samples, performed by using the same weight amounts. For both the investigated samples, the shape of TPR curves was very similar: a main reduction peak was evident at around 500 °C, together with another low-intensity peak at around 250 °C (Fig. 6a). The main peak was ascribed to the Co^{3+}, Co^{2+} step reduction to Co^0, in agreement with the hydrogen consumption (5.3–5.5 mmol H_2/g), while the low temperature peak (2.3 mmol H_2/g) was likely related to both the reduction of Fe^{4+} to Fe^{3+} and of Co^{4+} to Co^{3+} [52]. The lower hydrogen consumption of this peak indicates that both samples contained more Co^{3+}/Co^{2+} than Co^{4+}, in agreement with the above discussion. Two additional peaks with very low intensity and total hydrogen consumption of 1.4 mmol H_2/g were also visible above 700 °C, which could be ascribed to the subsequent reduction of Fe^{3+} and Fe^{2+} species into the metallic state [53]. It is important to note that the maximum temperature of the first two peaks, below 600 °C, is distinctly higher for the A6 sample than for the A2 sample (Fig. 6a). A different reduction temperature suggests that the two samples probably contain cobalt ions with a different surrounding and a different oxidation state (Co^{3+}/Co^{2+}). This is supported by the slightly lower hydrogen consumption registered for the main peak of the A6 sample (5.3 mmol H_2/g) with respect to the A2 sample (5.5 mmol H_2/g). The higher Co^{2+} content calculated from XRD results for A6 is in agreement with the higher content of Co^{2+} suggested by TPR results for the same sample.

Figure 6b describes the oxygen desorption from the A2 and A6 samples after heating in static air to 950 °C/5 h. One strong desorption peak was detected at around 400 °C; a second less intense desorption peak was present at around 800 °C. Figure 6b clearly shows that the overall oxygen desorption of sample A6 was smaller than that shown by sample A2, while the desorption peak maximum temperature was almost the same, in agreement with the fact that removal of oxygen was easier and higher for the less oxygen-deficient sample. XRD patterns were registered after the He-TPD experiments, without oxygen, and it was found that the cell parameters of both the cubic phases increased and phase 2 reached a value of 4.005 Å. This phase, whose composition is $Ba_{0.5}Sr_{0.5}Co_{0.8}Fe_{0.2}O_2$ with maximal Co^{2+} content, it is described in the literature as the most deficient cubic phase with disordered oxygen vacancies [50].

TPO experiments, with 5 vol.% O_2, were also performed in order to check for any O_2 consumption or CO_2 evolution. No O_2 consumption was observed, while CO_2 evolution peaks were detected between 500 and 800 °C in both TPO and TPD profiles, probably due to the decomposition of residual carbonate-like species. XRD patterns after TPO were very similar to the XRD patterns of the samples obtained after heating in static air, although the calculated Co^{2+} total amount was slightly larger after TPO, in agreement with the lower O_2 content in the gas (5 vol.% O_2).

Figure 7 shows the TG–DTA curves measured under nitrogen flow for the A1 and A6 samples, after annealing at 950 °C/5 h, where A1 has a smaller total Co^{2+} content (10 wt%) than A6 (32 wt%). At temperatures below 500 °C, where no significant CO_2 evolution was detected by TPD experiments, the weight changes could be easily ascribed to oxygen depletion. The TG–DTA under nitrogen

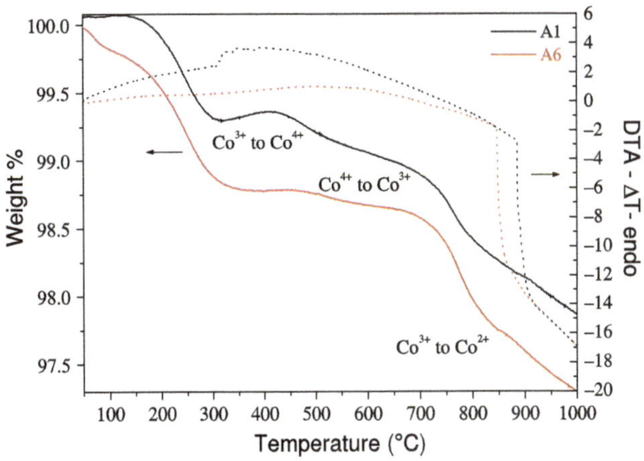

Fig. 7 TG and DTA curves under N_2 flow for A1 and A6 fired samples

were very similar in shape to those obtained in air: the Co^{3+} to Co^{4+} oxidation step and the two Co^{4+} to Co^{3+} and Co^{3+} to Co^{2+} reduction steps were visible, although the weight loss in the 200–500 °C range was generally larger than that for the samples treated in air. It is worth noticing that the Co^{3+} to Co^{4+} oxidation step was much more evident in the A1 sample, with structural parameters and Co^{2+} content close to A2, confirming a larger Co^{3+} percentage, in agreement with the results previously discussed. DTA results evidenced a difference in the temperature for the Co^{3+} to Co^{2+} reduction, which occurred at 890 °C and at 850 °C for the A1 and the A6 sample, respectively, indicating that the Co^{2+} reduced state was better stabilized by using the sucrose–PEG fuel mixture than the cellulose–citric acid one. Both the structural and redox results pointed out that the fuel mixture was able to control the structural parameters of the two BSCF cubic phases through the stabilization of the cobalt (II) oxidation state.

Electrochemical and microstructural properties of BSCF cathode materials

Figure 8a shows the Arrhenius plot of the area-specific resistance (ASR) for the symmetrical cells with BSCF electrodes annealed at 1,000 °C for 2 h on both sides of dense SDC pellets. It is worth noticing that, for a given temperature, the difference in the ASR between the A1 and A6 samples was as large as one order of magnitude, for the same nominal material composition (Fig. 8a). The A6 cathode showed the smallest ASR in the whole tested temperature range (500–800 °C), achieving 0.016 Ω cm^2 at 800 °C and 0.10 Ω cm^2 at 600 °C. These ASR values were comparable with values reported in the literature for BSCF cathodes measured in similar electrochemical conditions [13, 21, 54]. The activation energy values, albeit slightly smaller, were in agreement with literature data [21].

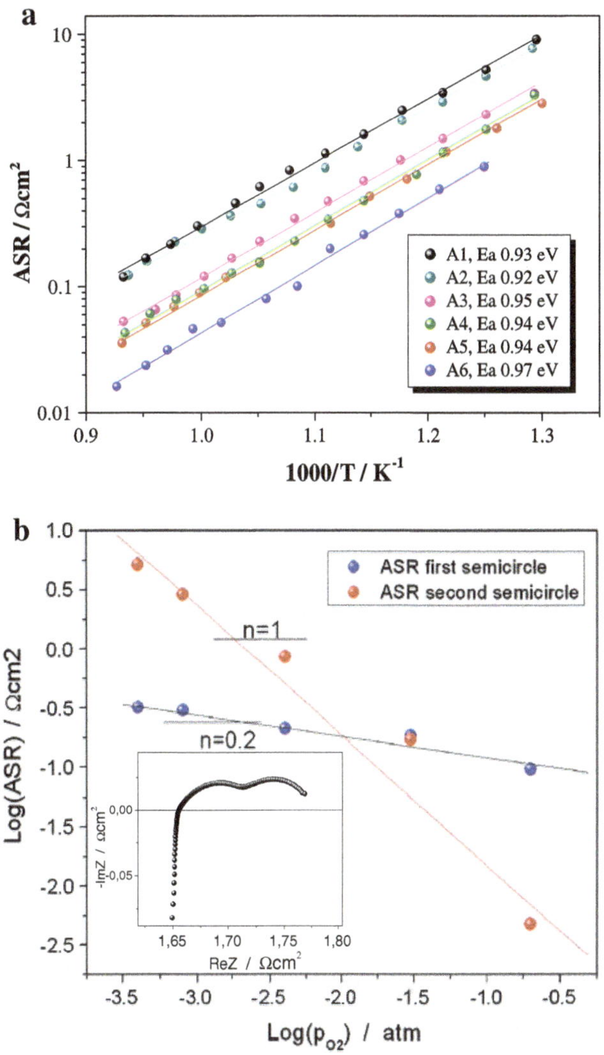

Fig. 8 ASR versus the reciprocal of temperature for BSCF cathodes prepared using different combustion fuel mixtures (**a**); variation with the oxygen partial pressure of the ASR derived from the *first and second semicircles* for the A6 sample, measured at 650 °C. In the *inset* of this figure, a typical EIS spectrum of A6 cathode showing the *two semicircles* is shown (**b**)

A one order of magnitude difference in the ASR values at a given temperature of BSCF with the same chemical composition could be caused by several factors which could be related to differences in microstructure [4, 55–58], oxygen deficiency [2, 9, 10, 58–61] or surface chemical composition (cation enrichment and/or impurity presence) [47]. Other factors could involve a difference in the extent of the adhesion of the powder to the electrolyte, or a difference in the thickness of the cathode layer.

N_2-adsorption experiments conducted on all the investigated BSCF calcined powders, revealed a very low specific surface area (below 5 m^2/g) and porosity (cumulative pore volume around 0.15 cm^3/g and mean pore diameter around 4 nm). Figure 9 shows SEM micrographs (with low

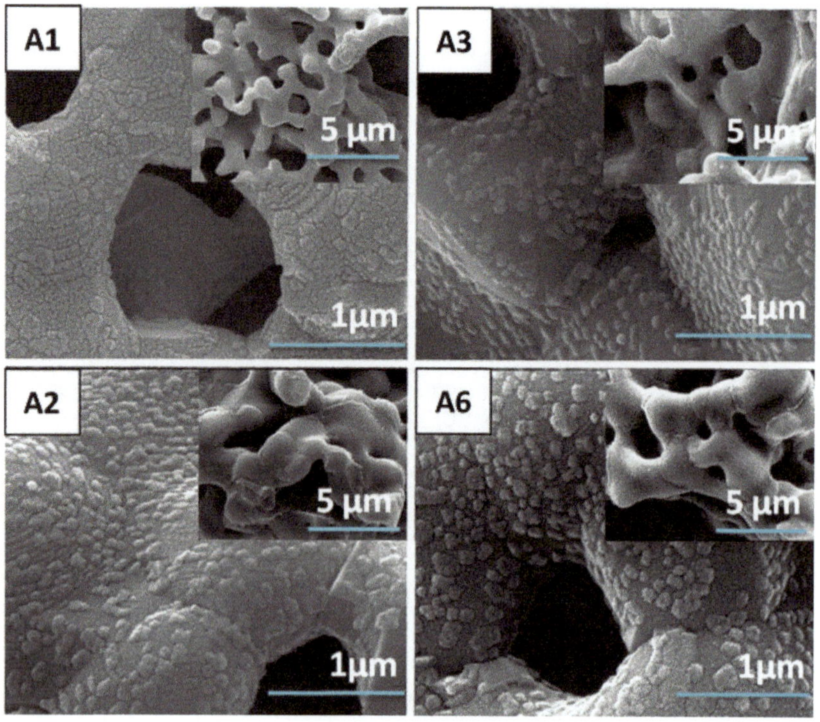

Fig. 9 SEM micrographs of four powders—A1, A2, A3 and A6—at two different magnifications (*lower in the insets*)

magnification images in the insets) for the A1, A2, A3 and A6 powders heated to 950 °C/5 h. Higher porosity and lower mean grain size (around 1 µm) were characteristic of sample A1 (Fig. 9). On the other hand, A2, A3, and A6 were less porous, showed a mean grain size around 3 µm and very small differences could be observed among the three samples (Fig. 9). For example, the A3 sample appeared slightly denser than the A6 sample, both before (Fig. 9) and after (Fig. 10) depositing the powders on the SDC electrolyte. The cathode thickness was very close for all the investigated samples (20 µm) and the cathode–electrolyte adhesion in the half-cells, as observed by SEM, was good (Fig. 10). Peculiar nanoscaled features were well evident in all the investigated BSCF samples (Fig. 9). Some authors ascribed the high oxygen conductivity of BSCF powders to peculiar nanoscaled features on the surface of BSCF powders prepared by coprecipitation [22]. Although this hypothesis cannot be completely ruled out, no evident correlation was found in this work between size/shape/distribution of the nanoscaled features and the observed ASR trend. These features were, for example, almost identical for the A2 and A6 samples (Fig. 8a), which showed very different ASR values at the same temperature (Fig. 8a). On the other hand, A1 and A2, which have very similar ASR values, showed very different morphologies. It can be reasonably postulated that the minor wt% of (111)-oriented rhombohedral phase detected by XRD in all the investigated samples (Fig. 5) was somewhat related to the nanoscaled features observed by SEM.

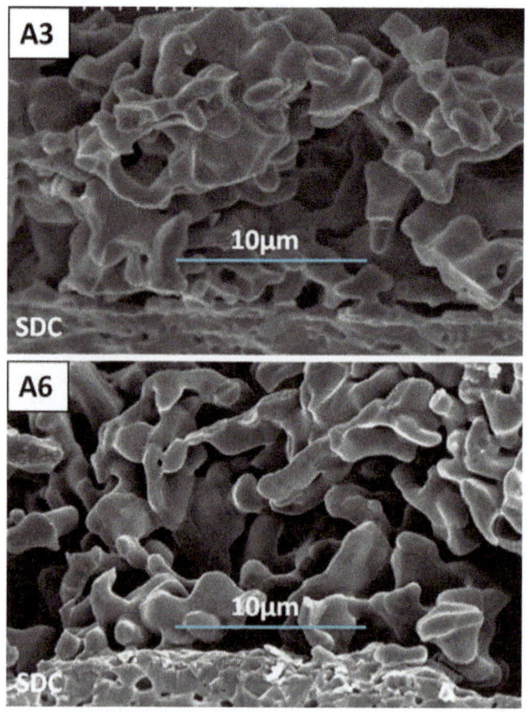

Fig. 10 SEM images of the half cell sections for two representative cathode powders, A3 and A6, after deposition on the SDC pellet

EDX results for three of the six samples are reported in Table 3, as atomic ratios of Ba, Sr and Co with respect to Fe. The EDX results were very similar for all the samples investigated. A slight Sr depletion (ca. 25 at%) was noticeable for all the investigated samples (Table 3) and

Table 3 EDX results (atomic ratios) for some A-type BSCF samples fired at 950 °C/5 h

Sample name	Ba/Fe	Sr/Fe	Co/Fe
Theoretical	2.5	2.5	4.0
A2	2.5	1.9	4.1
A5	2.6	2.1	4.3
A6	2.5	1.7	4.0

The theoretical ratios are reported in the first row of the table. SD ± 0.1

this is in agreement with the observations of Efimov et al. [48] about the chemical composition of the (111) oriented hexagonal phase.

In the present case, it appears that the observed differences in the ASR values of the BSCF cathodes at a given temperature (Fig. 8a) cannot be explained solely by microstructural differences. For example, the very slight microstructural differences between A3 and A6 samples (Figs. 9, 10) cannot explain a difference of half order of magnitude in the ASR (Fig. 8a). On the other hand, the more evident microstructural differences between A1 and A2 samples (Fig. 9), apparently did not affect their electrochemical properties (Fig. 8a).

Microstructural properties are mainly responsible for differences in the electrochemical properties of LSCF cathode materials [62], which was not the case here for BSCF. To understand the difference, one should consider the rate determining steps responsible for the overall oxygen reduction reaction in the case of BSCF and LSCF cathodes. Figure 8b shows the ASR associated to each of the two distinct semicircles observed in the low-frequency region of the Nyquist plots as a function of the oxygen partial pressure, measured at 650 °C. ASR is proportional to $(pO_2)^{-n}$, where the n value gives information about the electrode reaction involved. The n value of 0.2 (Fig. 8b), corresponding to the first semicircle, was indeed indicative of a rate-limiting step related to the charge transfer at the electrode/electrolyte interfaces, while the n value of 1 corresponding to the second semicircle indicated a rate-limiting step due to bulk and surface diffusion of oxygen species. As confirmed in Fig. 8b, the rate determining step of a BSCF cathode is governed by gas diffusion and charge transfer at the electrode/electrolyte interface [63]. On the other hand, in the case of LSCF cathodes the dissociative oxygen adsorption is identified as the main factor governing the oxygen reduction reaction [58, 62, 64]. Therefore, it is reasonable to assume that BSCF electrochemical properties are less affected by the cathode microstructure than LSCF compounds.

Bulk and surface oxygen vacancy diffusion is improved in the presence of disordered oxygen vacancies that enhance the oxygen incorporation in the BSCF [63]. It is worth noticing that the sample A6, with more oxygen

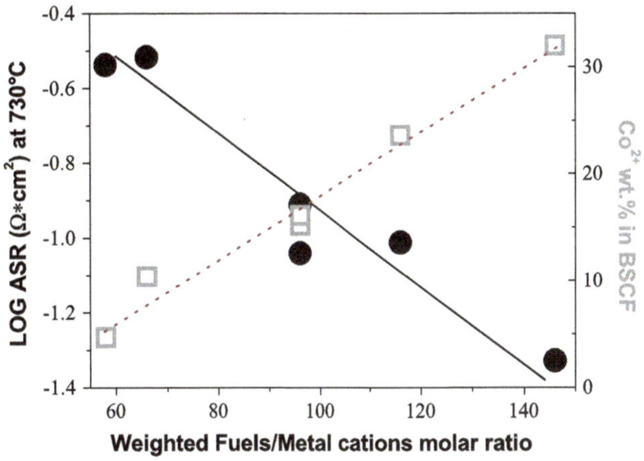

Fig. 11 Correlation between log ASR (at 730 °C), the Co^{2+} in each sample calculated from XRD results and the fuels/metal cation ratio weighted for the reducing power of each fuel in the combustion mixture (WF/*M*)

vacancies and Co^{2+} content, showed a smaller ASR value than the sample A2, at the same temperature (Fig. 8a). The combustion fuel mixture might affect indirectly the ASR values by governing the oxygen deficiency of the BSCF powders. We searched to correlate the measured ASR values with the most important SCS processing parameters, connected with the fuel mixture. In order to consider both the fuel-to-metal cations molar ratio and the reducing power of each different fuel in the combustion mixture, a new parameter was introduced here. The WF/*M* ratio, or the fuel-to-metal cations molar ratio weighted for the reducing power of the specific fuel in the mixture, is equivalent to the fuel-to-metal cations ratio, but it can be used for fuel mixtures, where each fuel has a specific reducing power [24]. Figure 11 shows the relationship of both the log ASR measured at a fixed temperature (730 °C) and the Co^{2+} total content in the sample with the WF/*M* ratio. An inverse correlation was found between the log ASR and the WF/*M* (Fig. 11). On the contrary, the Co^{2+} content increased with increasing WF/*M* (Fig. 11). It is here argued that, during the combustion process and with the other parameters fixed, the larger the reducing power of the fuel mixture, the larger the tendency of the fuel mixture to stabilize the atoms that will form later the BSCF structure in an oxygen vacancy-rich configuration. It can be further argued that, for a given fuel mixture, this tendency is also positively influenced by larger fuel-to-metal cations ratios, where a higher local concentration of fuels with high reducing power exists around the metal cations.

Conclusions

BSCF powders were produced by solution combustion synthesis using various single and double fuels. We showed that the fuel mixture has an explicit influence on the structural parameters, the oxidation state stability at the B-site, the microstructure and the electrochemical properties of the BSCF cathode materials. Moreover, it was observed that such improvement could not be solely ascribed to an improvement in the microstructural features, although a further optimization of the BSCF electrochemical properties by obtaining nanometer-sized powders still remains a challenge due to processing problems. Nonetheless, some important guidelines for the improvement of the electrochemical properties of BSCF cathode powders are proposed in this work. The main conclusion of this work is that BSCF cathode materials with optimized conductivity and high oxygen deficiency can be easily obtained by solution combustion synthesis, making use of high-interacting fuel mixtures having high fuel-to-metal cations molar ratio and high reducing power. In particular, the best electrochemical properties were observed for the A6 sample, where a mixture of sucrose and PEG20000 in a 1:2.5 molar ratio was used. Moreover, a new metastable monoclinic phase with $Ba_{0.5}Sr_{0.5}CO_3$ composition was discovered in the as-burned BSCF powders, extending the existing information on the phase formation mechanism of BSCF compounds. It is expected that most of the information acquired in this work can be successfully used for other chemical compositions. All these results confirmed the importance of the combustion process itself in tailoring the final materials properties, focusing the attention at the fuel mixture–metal cations interaction, contributing to fill an important gap between materials synthesis and materials properties.

Acknowledgments The authors are thankful to Dr. Francesco Giordano (ISMN-CNR UOS Palermo) for technical assistance in performing the XRD measurements and to Dr. Bartolomeo Megna (Laboratorio di Materiali per il Restauro e la Conservazione, Dipartimento di Ingegneria Civile, Ambientale e Aerospaziale, Università di Palermo) for technical assistance in performing the TG–DTA measurements. University of Palermo is also acknowledged for supporting in part this work.

References

1. Brett, D.J.L., Atkinson, A., Brandon, N.P., Skinner, S.J.: Intermediate temperature solid oxide fuel cells. Chem. Soc. Rev. **37**, 1568–78 (2008)
2. Sun, C., Hui, R., Roller, J.: Cathode materials for solid oxide fuel cells: a review. J. Solid State Electrochem. **14**, 1125–1144 (2010)
3. Tarancón, A., Burriel, M., Santiso, J., Skinner, S.J., Kilner, J.A.: Advances in layered oxide cathodes for intermediate temperature solid oxide fuel cells. J. Mater. Chem. **20**, 3799–3813 (2010)
4. Kim, J.H., Park, Y.M., Kim, H.: Nano-structured cathodes based on $La_{0.6}Sr_{0.4}Co_{0.2}Fe_{0.8}O_{3-\delta}$ for solid oxide fuel cells. J. Power Sources **196**, 3544–3547 (2011)
5. Patra, H., Rout, S.K., Pratihar, S.K., Bhattacharya, S.: Effect of process parameters on combined EDTA–citrate synthesis of $Ba_{0.5}Sr_{0.5}Co_{0.8}Fe_{0.2}O_{3-\delta}$ perovskite. Powder Technol. **209**, 98–104 (2011)
6. Liang, F., Jiang, H., Luo, H., Caro, J., Feldhoff, A.: Phase stability and permeation behavior of a dead-end $Ba_{0.5}Sr_{0.5}Co_{0.8}Fe_{0.2}O_{3-\delta}$ tube membrane in high-purity oxygen production. Chem. Mater. **23**, 4765–4772 (2011)
7. Shao, Z., Haile, S.M.: A high-performance cathode for the next generation of solid-oxide fuel cells. Nature **431**, 170–173 (2004)
8. Zhou, W., Ran, R., Shao, Z.: Progress in understanding and development of $Ba_{0.5}Sr_{0.5}Co_{0.8}Fe_{0.2}O_{3-\delta}$-based cathodes for intermediate-temperature solid-oxide fuel cells: A review. J. Power Sources **192**, 231–246 (2009)
9. Yáng, Z., Harvey, A.S., Infortuna, A., Schoonman, J., Gauckler, L.J.: Electrical conductivity and defect chemistry of $Ba_xSr_{1-x}Co_yFe_{1-y}O_{3-\delta}$ perovskites. J. Solid State Electrochem. **15**, 277–284 (2011)
10. Yang, Z., Martynczuk, J., Efimov, K., Harvey, A.S., Infortuna, A., Kocher, P., Gauckler, L.J.: Oxygen-vacancy-related structural phase transition of $Ba_{0.8}Sr_{0.2}Co_{0.8}Fe_{0.2}O_{3-\delta}$. Chem. Mater. **23**, 3169–3175 (2011)
11. Mueller, D.N., De Souza, R.A., Weirich, T.E., Roehrens, D., Mayer, J., Martin, M.: A kinetic study of the decomposition of the cubic perovskite-type oxide $Ba_xSr_{1-x}Co_{0.8}Fe_{0.2}O_{3-\delta}$ (BSCF) ($x = 0.1$ and 0.5). Phys. Chem. Chem. Phys. **12**, 10320–10328 (2010)
12. Efimov, K., Halfer, T., Kuhn, A., Heitjans, P., Caro, J., Feldhoff, A.: Novel cobalt-free oxygen-permeable perovskite-type membrane. Chem. Mater. **22**, 1540–1544 (2010)
13. Yan, A., Yang, M., Hou, Z., Dong, Y., Cheng, M.: Investigation of $Ba_{1-x}Sr_xCo_{0.8}Fe_{0.2}O_{3-\delta}$ as cathodes for low-temperature solid oxide fuel cells both in the absence and presence of CO_2. J. Power Sources **185**, 76–84 (2008)
14. Hung, I.-M., Liang, C.-Y., Ciou, C.-J., Lee, Y.-C.: Conductivity and electrochemical performance of $(Ba_{0.5}Sr_{0.5})_{0.8}La_{0.2}CoO_{3-\delta}$ cathode for intermediate-temperature solid oxide fuel cell. Ceram. Int. **36**, 1937–1943 (2010)
15. Kotomin, E.A., Mastrikov, Y.A., Kuklja, M.M., Merkle, R., Roytburd, A., Maier, J.: First principles calculations of oxygen vacancy formation and migration in mixed conducting $Ba_{0.5}Sr_{0.5}Co_{1-y}Fe_yO_{3-\delta}$ perovskites. Solid State Ionics **188**, 1–5 (2011)
16. Ding, X., Kong, X., Jiang, J., Cui, C., Guo, X.: Characterization and electrochemical performance of $(Ba_{0.6}Sr_{0.4})_{1-x}La_xCo_{0.6}Fe_{0.4}O_{3-\delta}$ ($x = 0$, 0.1) cathode for intermediate temperature solid oxide fuel cells. Mater. Res. Bull. **45**, 1271–1277 (2010)
17. Tan, L., Gu, X., Yang, L., Jin, W., Zhang, L., Xu, N.: Influence of powder synthesis methods on microstructure and oxygen permeation performance of $Ba_{0.5}Sr_{0.5}Co_{0.8}Fe_{0.2}O_{3-\delta}$ perovskite-type membranes. J. Membr. Sci. **212**, 157–165 (2003)
18. Liu, B., Zhang, Y.: $Ba_{0.5}Sr_{0.5}Co_{0.8}Fe_{0.2}O_3$ nanopowders prepared by glycine–nitrate process for solid oxide fuel cell cathode. J. Alloys Compd. **453**, 418–422 (2008)
19. Kriegel, R., Kircheisen, R., Toepfer, J.: Oxygen stoichiometry and expansion behavior of $Ba_{0.5}Sr_{0.5}Co_{0.8}Fe_{0.2}O_{3-\delta}$. Solid State Ionics **181**, 64–70 (2010)
20. Martynczuk, J., Arnold, M., Wang, H., Caro, J., Feldhoff, A.: How $(Ba_{0.5}Sr_{0.5})(Fe_{0.8}Zn_{0.2})O_{3-\delta}$ and $(Ba_{0.5}Sr_{0.5})(Co_{0.8}Fe_{0.2})O_{3-\delta}$ perovskites form via an edta/citric acid complexing method. Adv. Mater. **19**, 2134–2140 (2007)
21. Chen, Z., Ran, R., Zhou, W., Shao, Z., Liu, S.: Assessment of $Ba_{0.5}Sr_{0.5}Co_{1-y}Fe_yO_{3-\delta}$ ($y = 0.0$–1.0) for prospective application as cathode for IT-SOFCs or oxygen permeating membrane. Electrochim. Acta **52**, 7343–7351 (2007)

22. Toprak, M.S., Darab, M., Syvertsen, G.E., Muhammed, M.: Synthesis of nanostructured BSCF by oxalate co-precipitation—As potential cathode material for solid oxide fuels cells. Int. J. Hydrogen En. **35**, 9448–9454 (2010)
23. Specchia, S., Galletti, C., Specchia, V.: Solution combustion synthesis as intriguing technique to quickly produce performing catalysts for specific applications. Stud. Surf. Sci. Catal. **175**, 59–67 (2010)
24. Jain, S.R.: Energetics of propellants, fuels and explosives: A chemical valence approach. Propellants, Explos. Pyrotech. **12**, 188–195 (1987)
25. Deganello, F., Marcì, G., Deganello, G.: Citrate–nitrate auto-combustion synthesis of perovskite-type nanopowders: A systematic approach. J. Eur. Ceram. Soc. **29**, 439–450 (2009)
26. Mukasyan, A.S., Epstein, P., Dinka, P.: Solution combustion synthesis of nanomaterials. Proc. Combust. Inst. **31**, 1789–1795 (2007)
27. Jain, S.R., Adiga, K.C., Verneker, V.R.P.: A new approach to thermochemical calculations of condensed fuel-oxidizer mixtures. Combust. Flame **40**, 71–79 (1981)
28. Öncel, Ç., Gülgün, M.A.: Chemical Synthesis of Mixed-Oxide Powders for Solid Oxide Fuel Cell (SOFC) Electrolyte and Electrodes. In: Sheffield, J.W., Sheffield, Ç. (eds.) Assessment of hydrogen energy for sustainable development, pp. 147–159. Springer, Dordrecht (2007)
29. da Conceição, L., Silva, A.M., Ribeiro, N.F.P., Souza, M.M.V.M.: Combustion synthesis of $La_{0.7}Sr_{0.3}Co_{0.5}Fe_{0.5}O_3$ (LSCF) porous materials for application as cathode in IT-SOFC. Mater. Res. Bull. **46**, 308–314 (2011)
30. da Silva, A.L.A., Castro, G.G.G., Souza, M.M.V.M.: Synthesis of Sr-doped $LaCrO_3$ powders by combustion method. Influence of the fuel agent. J. Therm. Anal. Calorim. **109**, 33–38 (2011)
31. Vijaya Lakshmi, V., Bauri, R., Paul, S.: Effect of fuel type on microstructure and electrical property of combustion synthesized nanocrystalline scandia stabilized zirconia. Mater. Chem. Phys. **126**, 741–746 (2011)
32. Park, K., Nam, S.W.: Thermoelectric properties of $Ca_{2.8}Cu_{0.2}Co_4O_9$ ceramics fabricated by solution combustion method. Curr. Appl. Phys. **11**, 939–944 (2011)
33. Shigapov, A.N., Graham, G.W., McCabe, R.W., Plummer Jr, H.K.: The preparation of high-surface area, thermally-stable, metal-oxide catalysts and supports by a cellulose templating approach. Appl. Catal. A: Gen. **210**, 287–300 (2001)
34. Zhou, W., Shao, Z.P., Ran, R., Jin, W.Q., Xu, N.P.: Functional nano-composite oxides synthesized by environmental-friendly auto-combustion within a micro-bioreactor. Mater. Res. Bull. **43**, 2248–2259 (2008)
35. Das, R.N.: Nanocrystalline ceramics from sucrose process. Mater. Lett. **47**, 344–350 (2001)
36. Alinejada, B., Sarpoolaky, H., Beitollahi, A., Saberi, A., Afshar, S.: Synthesis and characterization of nanocrystalline $MgAl_2O_4$ spinel via sucrose process. Mater. Res. Bull. **43**, 1188–1194 (2008)
37. Larson, A.C., Von Dreele, R.B.: GSAS general structure analysis system. LANSCE, MS-H 805, Los Alamos National Laboratory, Los Alamos, NM 87545, USA (1998)
38. McCusker, L.B., Von Dreele, R.B., Coxd, D.E., Louer, D., Scardi, P.: Rietveld refinement guidelines. J. Appl. Crystallogr. **32**, 36–50 (1999)
39. Brunauer, S., Emmett, P.H., Teller, E.: Adsorption of gases in multimolecular layers. J. Am. Chem. Soc. **60**, 309–319 (1938)
40. Barret, E.P., Joyner, L.G., Halenda, P.P.: The determination of pore volume and area distributions in porous substances. I. Computations from nitrogen isotherms. J. Am. Chem. Soc. **73**, 373–380 (1951)
41. Bo, Y., Wenqiang, Z., Jingming, X., Jing, C.: Microstructure characterization and electrochemical properties of $Ba_{0.5}Sr_{0.5}Co_{0.5}FeO_{3-\delta}$ and its application for anode of SOEC. Int. J. Hydrog. Energy **33**, 6873–6877 (2008)
42. Magnone, E., Miyayama, M., Traversa, E.: Structural properties and electrochemical characteristics of $Ba_{0.5}Sr_{0.5}Co_{1-x}Fe_xO_{3-\delta}$ phases in different atmospheres. J. Electrochem. Soc. **159**, B1059–B1066 (2009)
43. Li, Z., Wei, B., Lü, Z., Huang, X., Su, W.: Structure, electrical and thermal properties of $(Ba_{0.5}Sr_{0.5})_{1-x}Gd_xCo_{0.8}Fe_{0.2}O_{3-\delta}$ perovskite as a solid-oxide fuel cell cathode. Solid State Ionics **207**, 38–43 (2012)
44. Toniolo, J.C., Takimi, A.S., Bergmann, C.P.: Nanostructured cobalt oxides (Co_3O_4 and CoO) and metallic Co powders synthesized by the solution combustion method. Mater. Res. Bull. **45**, 672–676 (2010)
45. Strobel, R., Maciejewski, M., Pratsinis, S.E., Baiker, A.: Unprecedented formation of metastable monoclinic $BaCO_3$ nanoparticles. Thermochim. Acta **445**, 23–26 (2006)
46. Nishino, T.: Characterization of γ-$BaCO_3$. Solid State Ionics **32/33**, 394–397 (1989)
47. Yi, J., Schroeder, M.: High temperature degradation of $Ba_{0.5}Sr_{0.5}Co_{0.8}Fe_{0.2}O_{3-\delta}$ membranes in atmospheres containing concentrated carbon dioxide. J. Membrane Sci. **378**, 163–170 (2011)
48. Efimov, K., Xu, Q., Feldhoff, A.: Transmission electron microscopy study of $Ba_{0.5}Sr_{0.5}Co_{0.8}Fe_{0.2}O_{3-\delta}$ perovskite decomposition at intermediate temperatures. Chem. Mater. **22**, 5866–5875 (2010)
49. Koster, H., Mertins, F.H.B.: Powder diffraction of the cubic perovskite $Ba_{0.5}Sr_{0.5}Co_{0.8}Fe_{0.2}O_{3-\delta}$. Powder Diffr. **18**, 56–59 (2003)
50. Mueller, D.N., De Souza, R.A., Yoo, H.-I., Martin, M.: Phase stability and oxygen nonstoichiometry of highly oxygen-deficient perovskite-type oxides: A case study of (Ba, Sr)(Co, Fe)$O_{3-\delta}$. Chem. Mater. **24**, 269–274 (2012)
51. Harvey, A.S.: Jochen Litterst, F., Yang, Z., Rupp, J.L.M., Infortuna, A., Gauckler, L.J.: Oxidation states of Co and Fe in $Ba_{1-x}Sr_xCo_{1-y}Fe_yO_{3-\delta}$ (x, y = 0.2–0.8) and oxygen desorption in the temperature range 300–1273 K. Phys. Chem. Chem. Phys. **11**, 3090–3098 (2009)
52. Crapanzano, S., Babich, I.V., Lefferts, L.: Selection of mixed conducting oxides for oxidative dehydrogenation of propane with pulse experiments. Appl. Catal. A Gen. **391**, 70–77 (2011)
53. Khan, A., Smirniotis, P.G.: Relationship between temperature-programmed reduction profile and activity of modified ferrite-based catalysts for WGS reaction. J. Mol. Catal. A Chem. **280**, 43–51 (2008)
54. Zhang, Y., Liu, J., Huang, X., Lu, Z., Su, W.: Low temperature solid oxide fuel cell with $Ba_{0.5}Sr_{0.5}Co_{0.8}Fe_{0.2}O_3$ cathode prepared by screen printing. Solid State Ionics **179**, 250–255 (2008)
55. Deganello, F., Esposito, V., Miyayama, M., Traversa, E.: Cathode performance of nanostructured $La_{1-a}Sr_aCo_{1-b}Fe_bO_{3-x}$ on a $Ce_{0.8}Sm_{0.2}O_2$. Electrolyte prepared by citrate–nitrate autocombustion. J. Electrochem. Soc. **154**, A89–A96 (2007)
56. Baqué, L., Caneiro, A., Moreno, M.S., Serquis, A.: High performance nanostructured IT-SOFC cathodes prepared by novel chemical method. Electrochem. Commun. **10**, 1905–1908 (2008)
57. Hieu, N.T., Parka, J., Tae, B.: Synthesis and characterization of nanofiber-structured $Ba_{0.5}Sr_{0.5}Co_{0.8}Fe_{0.2}O_{3-\delta}$ perovskite oxide used as a cathode material for low-temperature solid oxide fuel cells. Mater. Sci. Eng. B **177**, 205–209 (2012)
58. Adler, S.B.: Factors governing oxygen reduction in solid oxide fuel cell cathodes. Chem. Rev. **104**, 4791–4843 (2004)
59. Molenda, J., Świerczek, K., Zając, W.: Functional materials for the IT-SOFC. J. Power Sources **173**, 657–670 (2007)

60. Deganello, F., Liotta, L.F., Longo, A., Casaletto, M.P., Scopelliti, M.: Cerium effect on the phase structure, phase stability and redox properties of Ce-doped strontium ferrates. J. Solid State Chem. **179**, 3406–3419 (2006)
61. Neagu, D., Irvine, J.T.S.: Structure and properties of $La_{0.4}Sr_{0.4}TiO_3$ ceramics for use as anode materials in solid oxide Fuel cells. Chem. Mater. **22**, 5042–5053 (2010)
62. Yin. Q., Yang, Z., Lin, Y.S.: Effects of microstructure on oxygen transport in perovskite-type oxides. J. Mater. Sci. **41**, 4865–4870 (2006)
63. Wang, L., Merkle, R., Maier, J.: Oxygen exchange on mixed conducting oxides: (Ba, Sr)(Co, Fe)$O_{3-\delta}$ SOFC cathode materials. J. Electrochem. Soc. **157**, B1802–B1808 (2010)
64. Zhao, H., Huo, L., Sun, L., Yu, L., Gao, S., Zhao, J.: Preparation, chemical stability and electrochemical properties of LSCF–CBO composite cathodes. J. Mater. Chem. Phys. **88**, 160–166 (2004)
65. Feldhoff, A., Martynczuk, J., Wang, H. (2007) Advanced $Ba_{0.5}Sr_{0.5}Zn_{0.2}Fe_{0.8}O_{3-\delta}$ perovskite-type ceramics as oxygen selective membranes: Evaluation of the synthetic process. Progress in Solid State Chem. 35, 339–353 (2007)

5

Processing and characterizations of a novel proton-conducting $BaCe_{0.35}Zr_{0.50}Y_{0.15}O_{3-\delta}$ electrolyte and its nickel-based anode composite for anode-supported IT-SOFC

Ashok Kumar Baral · Sungmin Choi ·
Byung Kook Kim · Jong-Ho Lee

Abstract This work presents the synthesis and characterizations of a stable and well-sinterable proton conductor $BaCe_{0.35}Zr_{0.50}Y_{0.15}O_{3-\delta}$ in which a transition of conductivity occurred steeply from the order of 10^{-3} to 10^{-2} S cm^{-1} and activation energy changed from 0.35 to 0.22 eV in the temperature range from 350 to 400 °C, due to the dissociation of protons from the dopant-proton defect pairs. The anode composite Ni-$BaCe_{0.35}$ $Zr_{0.50}Y_{0.15}O_{3-\delta}$ (40:60 vol%) prepared by liquid condensation process showed comparatively very high electrical conductivity than the existing solid oxide fuel cell (SOFC) anodes, in the temperature range of 350–800 °C. Fabrication (by screen printing and co-firing processes), performance and post-mortem analysis of anode-supported protonic SOFC cell using these materials are discussed elaborately.

Keywords SOFC · Proton conductor · Electrical conductivity · Cell fabrication

Introduction

The solid oxide fuel cell (SOFC) has attracted a great attention since the decades as an alternate energy source for future generations. Main advantage of SOFC is the fuel flexibility, where a variety of carbohydrates can be used as fuel without requiring an expensive external fuel reformer. However, high operating temperature (i.e., 800–1,000 °C) of SOFC leads to the formation of electrode–electrolyte inter diffusion layers as well as degradation of SOFC components and sealing materials. Over the past few years, proton-conducting ceramics that show conductivity $\sim 10^{-2}$ S cm^{-1} at temperatures below 700 °C have been considered as alternate electrolyte candidates for SOFC, compared to the conventionally used oxygen ion conducting yttrium stabilized zirconia (YSZ). Among the various ceramic proton conductors, perovskite oxides such as yttrium doped barium cerates (BCY) show highest proton conductivity, i.e., $\sim 10^{-2}$ S cm^{-1} at 600 °C with lower activation energy [1]. Similarly, the Y- and Yb-doped barium zirconates (BZY and BZYb) show the conductivity in the order of 10^{-3} S cm^{-1} at 600 °C [2, 3]. Moreover, barium cerates are unstable in CO_2 [4] and moist atmosphere [5] (which are parts of SOFC operating condition), and barium zirconates ($BaZr_{1-x}Y_xO_3$) are highly stable but offer very high grain boundary resistance, causing a reduction in conductivity. Various authors have partially replaced the Ce^{4+} by Zr^{4+} in BCY and found that the Zr rich samples are more stable but conductivity reduces with increase in Zr content [6, 7]. These oxides show pure protonic conduction in wet hydrogen atmosphere, while they exhibit protonic, oxygen ionic and electronic mixed conduction under high oxygen partial pressure at elevated temperatures [6]. Major problem in case the of $BaCe_{1-x-y}Zr_xY_yO_{3-\delta}$ materials is the poor sinterability and they require very high temperature (i.e., above 1,700 °C) for well sintering with pore free microstructure [8–11].

In this work we have developed a stable barium cerate zirconate (BCZY) that exhibits very high proton conductivity at intermediate temperatures and is well sinterable at comparatively lower temperatures. A novel anode

A. K. Baral (✉) · S. Choi · B. K. Kim · J.-H. Lee
High Temperature Energy Materials Center, Korea Institute of Science and Technology, 39-1, Hawolgok-dong, Seongbuk-gu, Seoul, South Korea
e-mail: askbaral@gmail.com

A. K. Baral
Department of Chemistry, University of Calgary, 2500 University Dr NW, Calgary, AB T2N1N4, Canada

composite NiO-BCZY is prepared by liquid condensation method and its sintering and electrical properties are discussed. Fabrication process of anode-supported planar proton-conducting SOFC cell, using BCZY as electrolyte, NiO-BCZY composite as anode and $La_{0.6}Sr_{0.4}Co_{0.2}Fe_{0.8}O_{3-\delta}$ (LSCF) as cathode, is discussed and the performance of the cell is evaluated.

Experimental

The barium cerate zirconate, i.e., $BaCe_{0.35}Zr_{0.50}Y_{0.15}O_{3-\delta}$ (BCZY) was prepared by solid-state reaction method, using the stoichiometry amounts of $BaCO_3$, CeO_2, ZrO_2 and Y_2O_3 with purity above 99.9 %. Phase purity of BCZY and its stability in CO_2 atmosphere were studied by X-ray diffraction technique (using Philips PW 3830 X-ray generator). The particle size distribution was studied by using particle size analyser (LS230, COULTER Co.) after dispersing the powder in ethanol. The powder sample was pressed into pellets (with diameter of 10 mm) by applying a pressure of 100 MPa uni-axially. Sinterability of the material was studied using dilatometer (DIL 402C, NETZSCH, Germany), by heating the pellet from room temperature to 1,500 °C in air at the rate of 3 °C/min. The green pellet was also sintered at 1,600 °C for 30 h and its density was measured by the Archimedes method. Microstructures of as-prepared powder and sintered pellet were studied by scanning electron microscope (SEM) (FEI XL-30, Philips, Netherlands). Platinum paste was painted on both sides of polished sintered pellet (with relative density of 95 %) as electrode and fired at 1,000 °C (1 h) for AC impedance study. The impedance measurements were carried out in wet hydrogen atmosphere (i.e., 3 % H_2O + 4 % H_2 balanced with Ar) by two probe method using Solatron 1260 Impedance/Gain phase analyser. During the measurements, an external field of 100 mV was applied varying the frequency from 0.1 Hz to 10 MHz. The impedance data were recorded in the temperature range of 250–700 °C at an interval of 50 °C, holding the furnace temperature more than 3 h at each step for equilibrium. Impedance plots were fitted with their equivalent circuits using Zview software package to get the resistance and capacitance values.

The composite NiO-BCZY material (with Ni:BCZY volume ratio as 40:60) was prepared by liquid condensation process (LCP) [12, 13]. The granule formed from the LCP was characterized by dilatometer to check the sinterability. DC electrical conductivity of the composite anode sample was studied by linear four probe method in the temperature range of 350–800 °C. The bar-shaped sample was first sintered at 1,500 °C for 3 h and then reduced in H_2 atmosphere at 900 °C for 3 h before using in electrical measurement. The granule was also compressed into disk form for anode substrate.

The electrolyte paste of BCZY was prepared by mixing the stoichiometry amounts of α-terpineol (as solvent), dispersing agent (KD6), binder, plasticizer and BCZY powder in planetary mill. Similarly, cathode pastes of the composite [BCZY—$La_{0.6}Sr_{0.4}Co_{0.2}Fe_{0.8}O_{3-\delta}$ (LSCF)] (with 50:50 volume ratio) and LSCF were also prepared by planetary milling. The commercially available LSCF powder was used to prepare the cathode pastes.

The proton-conducting ceramic fuel cell (i.e., NiO-BCZY/BCZY/BCZY-LSCF/LSCF) was prepared by screen printing and co-firing processes. At first, The NiO-BCZY substrate was pre-sintered at 1,100 °C (3 h) and then the BCZY electrolyte paste was screen printed and co-fired at 1,500 °C (3 h). The cathode pastes of BCZY-LSCF (one layer) and LSCF (two layers) were screen printed and co-fired at 1,050 °C (2 h). Microstructure of the cell and various cell components were studied by SEM before and after the cell test. Performance of the button (2 × 2) cell was studied by current–voltage–power (I–V–P) characteristic and impedance measurement (at OCV) using Impedance analyser (solatron 1260 gain/phase analyser) coupled with potentiostat (Solatron 1287).

Results and discussion

Stability and sinterability of $BaCe_{0.35}Zr_{0.5}Y_{0.15}O_{3-\delta}$ (BCZY)

The powder X-ray diffraction pattern of $BaCe_{0.35}Zr_{0.5}Y_{0.15}O_{3-\delta}$ (Fig. 1a) confirmed the cubic perovskite phase of sample that was calcined in two steps; first at 1,300 °C (5 h) and then at 1,400 °C (15 h). To check the stability of the material against CO_2, the sample was heated at 900 °C for 3 h in CO_2 (which was supplied to the furnace right from the beginning of the heating process). The X-ray diffraction of the BCZY powder heat treated in CO_2 (Fig. 1b) did not show any change in phase, which confirmed that the sample is stable against CO_2. It has been studied that due to basic nature of barium cerates ($BaCeO_3$ or $BaCe_{0.9}Y_{0.1}O_{3-\delta}$), they react with acidic gases such as CO_2 and decompose into $BaCO_3$ and CeO_2 phases [14]. Katahira et al. [6] have studied the chemical stability of Zr-substituted $BaCe_{0.9-x}Z_xY_{0.1}O_{3-\delta}$ by heat treating the samples at 900 °C (2 h) in 100 % CO_2 atmosphere and observed that the material with $x = 0.1$ decomposed to $BaCO_3$ and CeO_2 phases, while the phase of material with $x = 0.4$ remained same after CO_2 treatment, as untreated sample. Authors observed that, Zr act as a phase stabilizer and stability of barium cerate increases with Zr amount. Fabbri et al. [7] have also studied the similar materials and

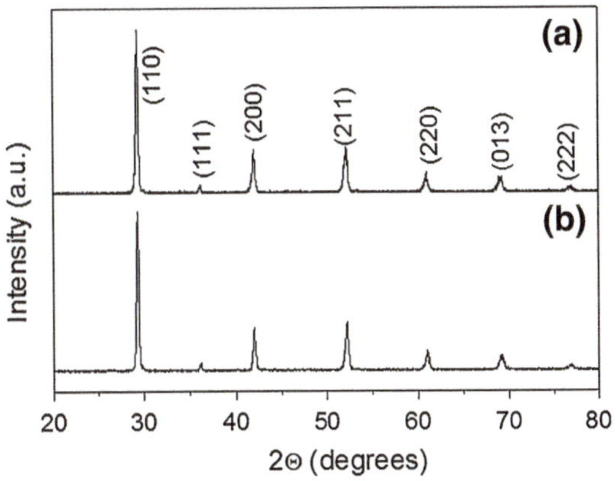

Fig. 1 X-ray diffraction patterns of **a** as-prepared $BaCe_{0.35}Zr_{0.5}Y_{0.15}O_{3-\delta}$ and **b** the sample heated at 900 °C in CO_2 for 3 h

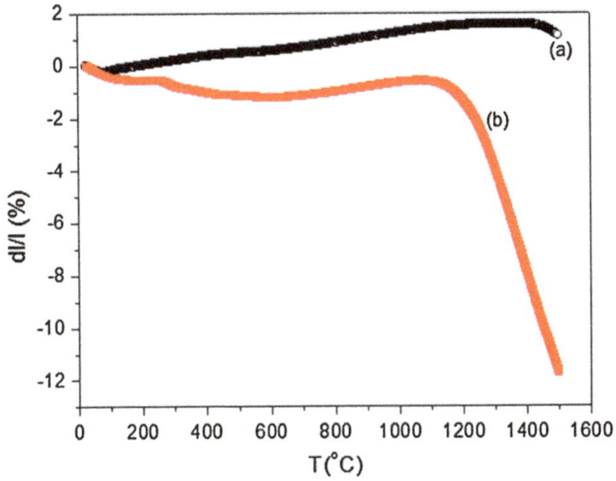

Fig. 2 Shrinkage behavior of pellets made of **a** as-prepared powder and **b** the powder milled for 48 h

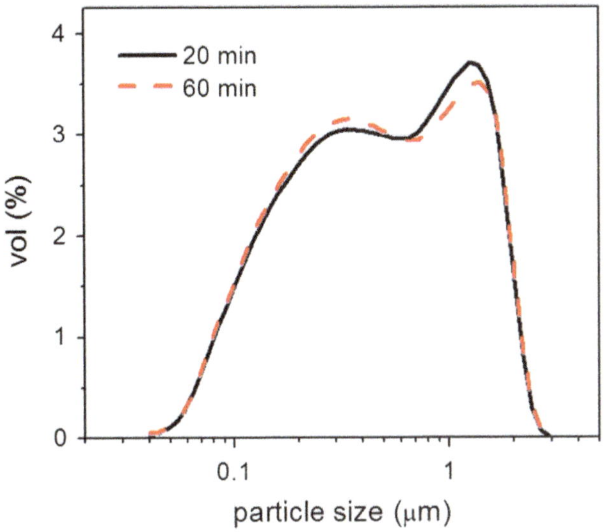

Fig. 3 Particle size distribution curve of as-prepared powder, dispersed in ethanol for 20 and 60 min

found that the chemical stability of samples in CO_2 increases with zirconia content and samples with $x \geq 0.5$ are more stable, as seen in the case of present material.

The powder sample was pressed into pellet and characterized by dilatometry to study its sintering properties. The dilatometry curve (Fig. 2a) did not show any shrinkage of pellet in the temperature range up to 1,500 °C. The pellet prepared in similar condition was sintered at 1,650 °C for 30 h and the relative density of the pellet was about 85 %. The particle size distribution curve (Fig. 3) indicates the presence of secondary agglomerations in addition to the primary particles. The mean particle size D_{10} was about 0.133 and 0.132 micron after dispersing the powder for 20 and 60 min in ethanol, respectively. Similarly, the value of D_{90} was found to be 1.594 and 1.605 micron after dispersing the powder for 20 and 60 min, respectively. The SEM image of as-prepared powder (Fig. 4a) also confirmed the agglomerating nature of particles. Therefore, the as-prepared powder was further ball milled for 48 h and corresponding pellet was also characterized by dilatometry (Fig. 2b), which showed that the shrinkage of pellet is about 12 % at 1,500 °C. Pellet made of this latest ball-milled (for 48 h) powder was sintered at 1,600 °C (30 h) and its relative density was found to be around 95 %. Pellet prepared under this condition were also used for SEM and Impedance studies. The SEM micrograph of sintered pellet (Fig. 4b) also confirmed the dense microstructure of pellet.

Electrical properties of BCZY

Electrical properties of the BCZY sample were studied by AC impedance spectroscopy. Figure 5a, b shows the complex impedance plots of BCZY in wet 4 % H_2 at 300 and 500 °C. At lower temperature range (i.e., up to 400 °C), the impedance plots contain two semicircles, i.e., at higher and intermediate frequencies corresponding to the grain and grain boundary conductivities, respectively. The lower frequency tail/arc in the Fig. 5a corresponds to the electrode polarization. Above 400 °C, the semicircle for grain interior conductivity diminished and the lower frequency tail appeared as a semicircle (Fig. 5b). In that case, the high-frequency intersection of Z-plot with Z' axis is taken as the bulk/grain interior resistance. The Z-plots were fitted with their equivalent circuit (shown in the inset of Fig. 5a) using *Z* view software to get the resistance and capacitance values. At higher temperatures (Fig. 5b), obtained range capacitance values, i.e., $\sim 10^{-8}$ F for high-frequency arc (shown as inset) and $\sim 10^{-6}$ F for lower frequency semicircles indicate that they are corresponding to the grain boundary conduction and electrode polarization, respectively. The Z'' vs. frequency plots (Fig. 6) also

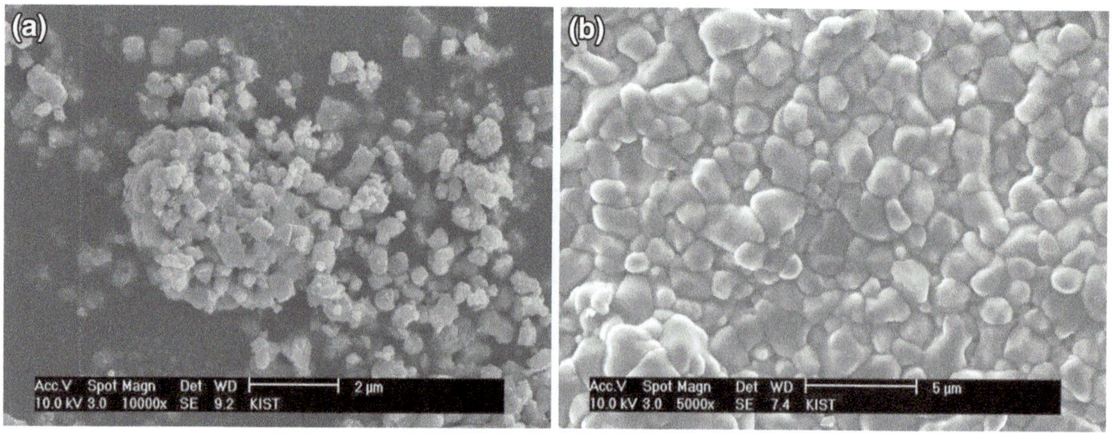

Fig. 4 SEM micrographs of **a** as-prepared powder and **b** pellet (sintered at 1,600 °C for 30 h) that is made of powder milled for 48 h after calcination step

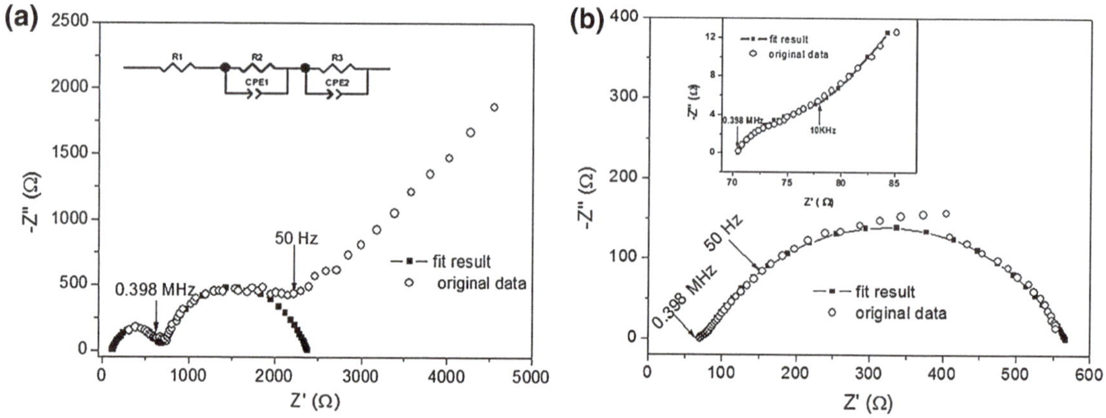

Fig. 5 Complex impedance plots for BCZY sample at **a** 300 °C and **b** 500 °C. *Inset* in **a** shows the equivalent circuit used for fitting, where, R1, R2 and R3 represent the contact, grain and grain boundary resistances, respectively. CPE1, and CPE2 are the constant phase elements in different arcs. *Inset* in **b** shows the presence of high-frequency arc corresponding to grain boundary conductivity

exhibit that, at lower temperatures, there are two peaks at high and intermediate frequencies, which correspond to relaxation of charge carriers at grain and grain boundary, respectively. The tail/peak at lower frequency range is due to the electrode polarization. At elevated temperatures the peak positions shifted to higher frequencies. Comparing to the frequency range of peaks in Fig. 6, it is observed that the peak shown in the plot at 500 °C is due to the electrode polarization which directly corresponds to low-frequency semicircle in the Fig. 5b. This confirmed that the low-frequency semicircle in the Fig. 5b is due to the electrode polarization. Comparing the frequency range in Fig. 5a and inset of Fig. 5b, it can be also clear that, the high-frequency arc shown in the inset of Fig. 5b is due to grain boundary conductivity. Moreover, the total resistance value was obtained by summation of grain and grain boundary resistances. The conductivity of the sample was calculated by using the formula $\sigma = L/(RA)$, where R is the value of

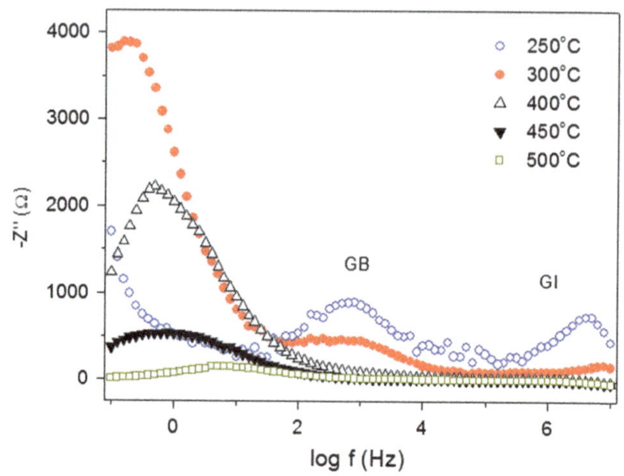

Fig. 6 Variation of Z'' as a function of frequency at various temperatures

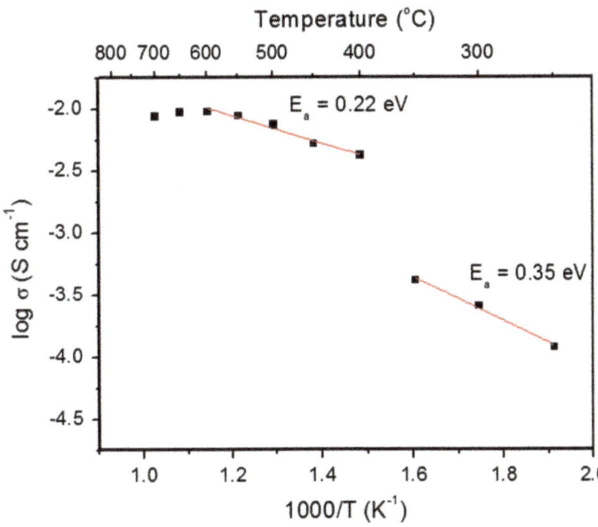

Fig. 7 Arrhenius plot of total conductivity that varies with temperature in BCZY. E_a is the activation energy

total resistance, L is the thickness of pellet and A is the area of cross section.

Figure 7 shows the Arrhenius plot of total conductivity, i.e., log σ vs. $1/T$. The material shows the conductivity $\sim 10^{-3}$ S cm^{-1} in the temperature range of 250–350 °C and $\sim 10^{-2}$ S cm^{-1} above 350 °C. The steep increase in conductivity from the order of 10^{-3} to 10^{-2} S cm^{-1} and decrease in activation energy from 0.35 to 0.22 eV at the temperature range above 350 °C, appears like a super ionic type transition in the material, where state of hydrogen changes from more ordered state to disordered state, as observed in different proton-conducting materials [15]. In the present case, as Y^{3+} replaces Ce^{4+}, the dopant (Y^{3+}) associates an effective -1 charge and protons being positively charged, they make defect association. Protons are said to be trapped by dopants and state of protons become more ordered [16]. The defect pairs (Y-H) act as electric dipoles. When a thermally activated proton jump from one oxygen to another in the octahedra (YO_6), maintaining its defect association with the dopant Y^{3+}, the electric dipole appears to orient about Y^{3+}. In presence of the external electric field, arbitrary oriented dipoles align in the direction of field and a local heat is generated. That is how the electrical energy is dissipated in terms of heat and it is called dielectric loss. Figure 8a, b shows the frequency spectra of dielectric loss tangent (tan δ) (obtained from impedance data) at various temperatures. At lower temperatures (say at 300 °C), tan δ (Fig. 8a) exhibits loss peaks due to the dipolar rotations in the bulk (high-frequency peak) [17] and space charge polarization at grain boundaries (low-frequency peak) [18]. As the temperature increases, diffusion of thermally activated protons from grains to grain boundaries can result in reduction of the space charge width and hence a decrease in space charge polarization at grain boundaries. As a result, the peak due to space charge polarization is disappeared at 350 °C (Figure 8a). However, at 350 °C, the intensity of higher frequency peak increased as the number of dipolar rotation increases due to thermal activation of charge carriers (Fig. 8a). The loss peak due to dipolar rotation is found to disappear at 400 °C (Fig. 8b), and in that case, protons are said to be dissociated from the defect pairs and perform long range order migrations where, the state of protons becomes disordered. The upturn in data points at higher frequency region, at higher temperatures, can be attributed to the conduction loss due to the migration of free charge carriers. This indicates that the transition in conductivity (i.e., about one order) between 350 and 400 °C could be due to the dissociation of protons from defect association with dopants. As the protons are almost free above 400 °C, the activation energy of conductivity is also reduced from 0.35 to 0.22 eV. Moreover, as the BCZY shows very high total conductivity (σ) i.e., around 10^{-2} S cm^{-1} above 400 °C, it increases the possibility to bring the operating temperature of SOFC down to 400 °C. A small decrease in conductivity above 650 °C (Fig. 7) could be due to the dehydration of protons from the sample.

Fabrication of anode-supported planar protonic SOFC cell

The anode composite NiO-BCZY with Ni:BZCY volume ratio of 40:60 was prepared by LCP method [12, 13]. The dilatometry study of the composite was performed to examine its sinterability. Figure 9 shows the shrinkage behavior of both anode composite (NiO-BCZY) and as-prepared BCZY (i.e., used to prepare the anode granules). It is observed that, the anode volume shrunk about 20 % in the temperature range of 50–1,500 °C. The peak around 400 °C is due to removal of phenolite that was used as binder in the processing of granule. The DC electrical conductivity of the bar-shaped anode sample was measured by liner four probe method in the wet H_2 atmosphere. The bar-shaped sample was first reduced at 900 °C in H_2 atmosphere for 3 h before using in conductivity measurement. Figure 10 shows the Arrhenius plot of DC conductivity in the temperature range of 350–800 °C, where the conductivity decreases with increase in temperature showing a metallic behavior, may be due to well dispersion of nickel particle in the sample. The conductivity of Ni-BCZY composite (prepared by LCP) varies between 1,050 and 1,590 S cm^{-1} in the measured temperature range, which is found to be much higher than that of Ni-YSZ based anodes used in conventional SOFCs [19–22] and Ni-BCY based anodes used in protonic SOFCs [23, 24]. This indicates that, as far as electrical conductivity is concerned,

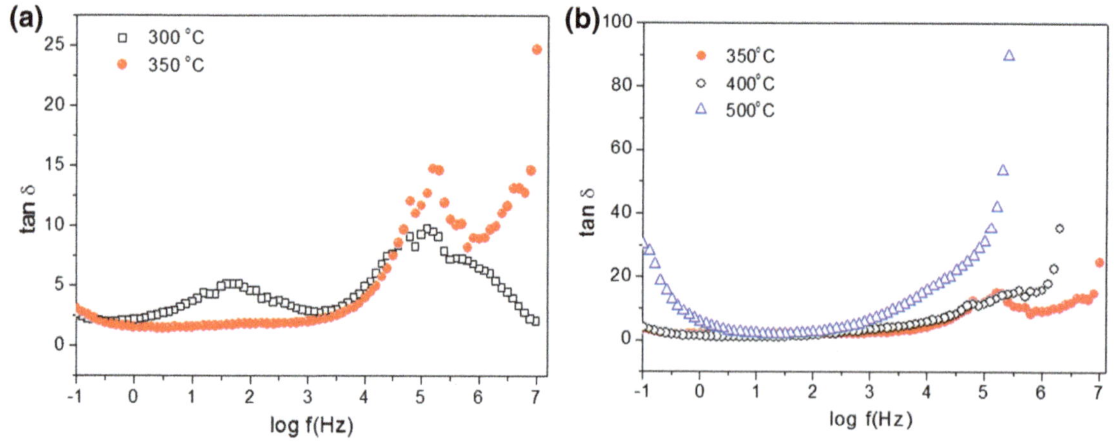

Fig. 8 Spectra of dielectric loss tangent (tan δ) at various temperatures between 300–500 °C

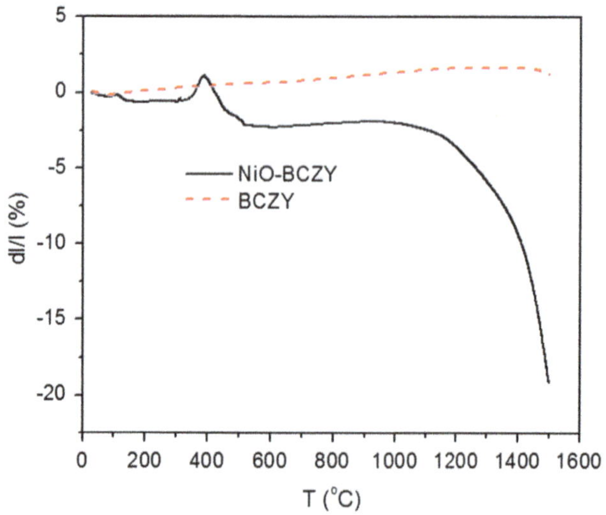

Fig. 9 Shrinkage behavior of as-prepared BCZY and NiO-BCZY composite

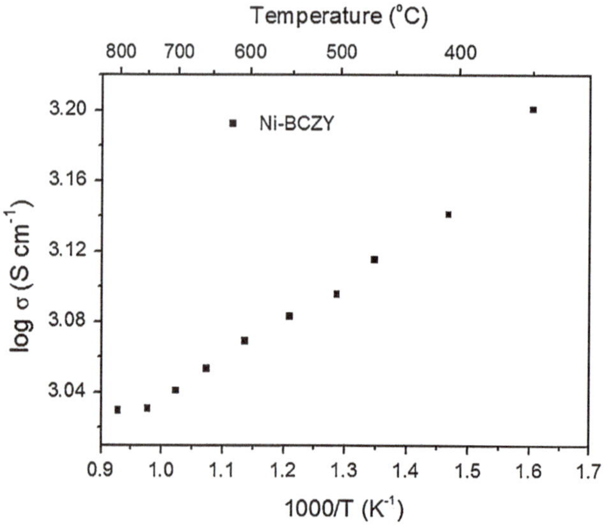

Fig. 10 Arrhenius plot of DC conductivity of Ni-BCZY anode in wet H_2 atmosphere

the Ni-BCZY composite prepared by LCP method could be a suitable anode material for the SOFC to run at lower temperatures.

The anode substrate was prepared by pressing the composite NiO-BCZY granules into disk, with diameter of 45 mm and thickness of 1.5 mm, by using a hot press. The temperature of hot press was maintained at 80 °C and a pressure of 50 MPa was applied. The anode substrate was pre-sintered at 1,100 °C (for 3 h) and then the electrolyte paste of BCZY was screen printed on the anode substrate and co-fired at 1,500 °C for 3 h. The electrolyte (BCZY) paste was screen printed three times, with each time after the screen print, the sample was air dried for 30 min and heat treated at 50 °C for 30 min. The electrolyte (BCZY) paste was prepared by using planetary ball mill, mixing appropriate ratio of BCZY powder, alpha-terpineol as solvent, organic binder and dispersing agent.

Figure 11a shows the cross-sectional image of co-fired anode-electrolyte layers. The electrolyte layer (with thickness ~18 μm) appears to be highly dense and exhibits pore free microstructure. Nickel oxide particles (gray in color in Fig. 11a) are appeared to be well dispersed. However, the anode substrate was found to be little denser than the required anode microstructure. Since the anode was already dense, the anode functional layer was not employed in this cell fabrication. The micrograph of electrolyte surface (Fig. 11b) also indicates that the electrolyte is highly dense and has pore free microstructure. The cell was then cut into 2 × 2 cm^2 dimension by using diamond cutter. The cathode pastes of BCZY-LSFC (50:50 vol %) composite and LSCF were prepared by planetary milling the powders with appropriate ratio of solvent and binder. The BCZY-LSCF composite (one) layer and LSCF (two) layers were screen printed on the electrolyte surface

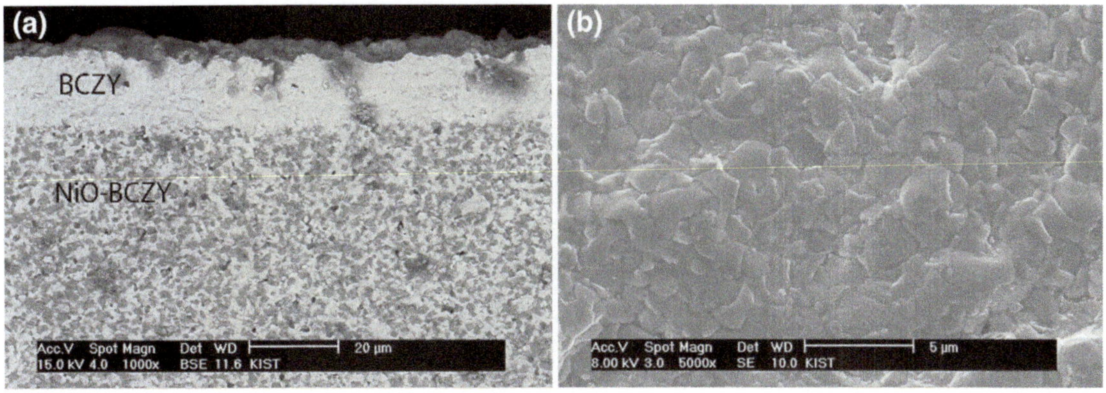

Fig. 11 SEM image of **a** anode-electrolyte cross section and **b** electrolyte surface

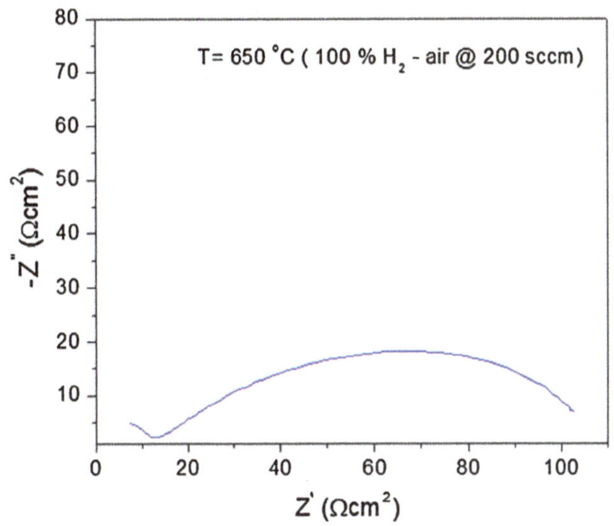

Fig. 12 Impedance plot of cell (with active cathode area of 1 cm^2) recorded at 650 °C (at OCV)

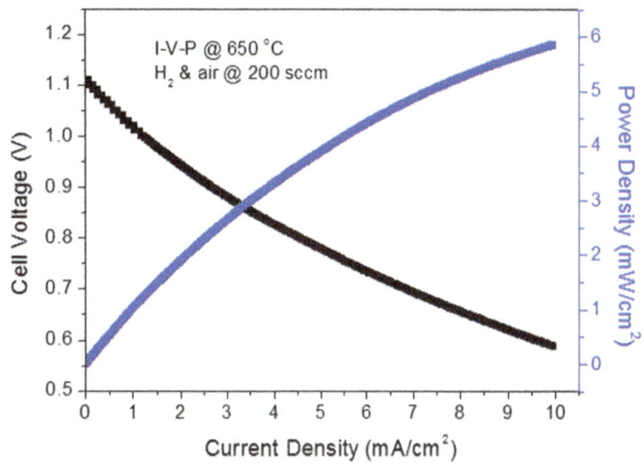

Fig. 13 I–V–P characteristic curves of SOFC cell at 650 °C

in the area of 1 cm^2. The full cell was fired at 1,050 °C for 2 h with continuous supply of air to the furnace at the rate of 200 sccm.

Performance of the proton-conducting SOFC cell NiO-BCZY/BCZY/BCZY-LSCF/LSCF was studied at 650 °C using the impedance gain/phase analyser (Solartron 1560) coupled with potentiostat. During the cell test, anode and cathode sides of the cell were separated by Pyrex glass sealing. Anode was first reduced at 650 °C by supplying the mixed gas of H_2 and N_2 to the anode side step by step in the (H:N) ratio of 10:90; 20:80; 40:60, 80:20; 100:0 (in sccm), and with continuous supply of air (at the rate of 100 sccm) to the cathode side. The open circuit voltage (OCV) of the cell is found to be above 1.1 V at 650 °C (with 100 % H_2 supply), which implied that, the cell was well sealed and there was no electronic leakage through electrolyte. After the anode reduction process, the fuel (H_2) and air was supplied to anode and cathode side, respectively, at the rate of 200 sccm and impedance measurement was carried out at OCV. The impedance plot of the cell at 650 °C is shown in Fig. 12 and it indicates that, the area specific resistance is quite high (i.e., about 100 Ω cm^2). Figure 13 shows the current–voltage–power (I–V–P) characteristic curve of the cell obtained at 650 °C and it exhibits that the power density of the cell is about 6 mW/cm^2. Comparing to the power density (2.35 mW/cm^2 at 900 °C) of protonic SOFC based on Y-doped Barium zirconate (BZY) reported elsewhere [25], the performance of the present anode-supported cell is better but it is still much lower than that observed by Fabbri et al. [26] in case of electrolyte supported protonic SOFC cells.

Figure 14a, b shows the cross-sectional image of cell before and after test. Before the cell test, some cracks were observed in the cathode side as seen in Fig. 14a. After the cell test, the post-mortem analysis exhibits the delamination of cathode near previously existing cracks (Fig. 14b), which might be due to the thermal incompatibility of cathode (LSCF) with BZCY based proton-conducting electrolyte. So the delamination of cathode is expected be the reason for high electrode polarization observed in the

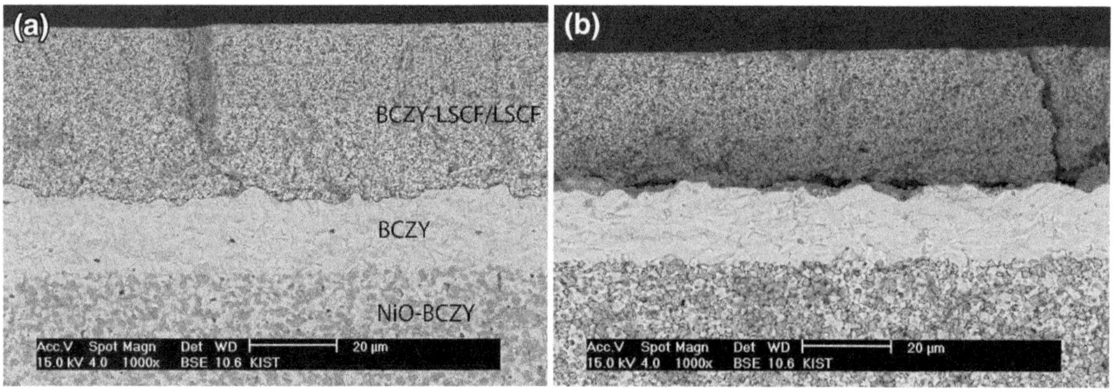

Fig. 14 Cross-sectional images of cell before (**a**) and after test (**b**)

impedance plot (Fig. 12). As a result, a lower power density of the cell is observed.

Conclusions

The proton-conducting solid electrolyte $BaCe_{0.35}Zr_{0.50}Y_{0.15}O_{3\text{-}\delta}$ prepared by solid-state reaction method is highly stable against CO_2. AC electrical study of the sample was thoroughly studied and conductivity of the sample is found to be about 10^{-2} S cm^{-1} above 400 °C with an activation energy of 0.22 eV. The transition in conductivity of the sample occurs between 350 and 400 °C due to dissociation of protons from the dopant-proton defect pairs. High conductivity of electrolyte material extends the possibility of SOFC to run at very lower temperatures (i.e., down to 400 °C). The anode composite Ni-$BaCe_{0.35}Zr_{0.50}Y_{0.15}O_{3\text{-}\delta}$ prepared by LCP method exhibits very high value of conductivity under fuel cell operating condition and at reduced temperatures. The anode-supported planar proton-conducting SOFC was successfully processed by screen printing and co-firing method and tested at 650 °C. OCV of the cell at 650 °C is found to be above 1.1 eV, which is quite promising and indicates no electronic leakage through the electrolyte. The power density of the cell at 650 °C is found to be about 6 mW/cm^2. Despite high conductivity of both electrolyte and anode composite, the performance of the cell is found to be poor, due to the high cathodic polarization as a result of cathode delamination. However, the performance of the cell can be improved with better cathode materials.

Acknowledgments This research was supported by the Fusion Research Program for Green Technologies through the National Research Foundation of Korea (NRF) funded by the Ministry of Education, Science and Technology (2011-0019297) and institutional program of KIST.

References

1. Ma, G., Shimura, T., Iwahara, H.: Ionic conduction and nonstoichiometry in $Ba_xCe_{0.90}Y_{0.10}O_{3\text{-}\alpha}$. Solid State Ion. **110**, 103–110 (1998)
2. Bohn, H.G., Schober, T.: Electrical conductivity of the high-temperature proton conductor $BaZr_{0.9}Y_{0.1}O_{2.95}$. J. Am. Ceram. Soc. **83**, 768–772 (2000)
3. Park, J.-S., Lee, J.-H., Lee, H.-W., Kim, B.-K.: Low temperature sintering of $BaZrO_3$-based proton conductors for intermediate temperature solid oxide fuel cells. Solid State Ion. **181**, 163–167 (2010)
4. Zakowsky, N., Williamson, S., Irvine, J.T.S.: Elaboration of CO_2 tolerance limits of $BaCe_{0.9}Y_{0.1}O_{3\text{-}\delta}$ electrolytes for fuel cells and other applications. Solid State Ion. **176**, 3019–3026 (2005)
5. Bhide, S.V., Virkar, A.V.: Stability of BaCeO3-based proton conductors in water-containing atmospheres. J. Electrochem. Soc. **146**, 2038–2044 (1999)
6. Katahira, K., Kohchi, Y., Shimura, T., Iwahara, H.: Protonic conduction in Zr-substituted $BaCeO_3$. Solid State Ion. **138**, 91–98 (2000)
7. Fabbri, E., D'Epifanio, A., Bartolomeo, E.D., Licoccia, S., Traversa, E.: Tailoring the chemical stability of $Ba(Ce_{0.8-x}Zr_x)Y_{0.2}O_{3\text{-}\delta}$ protonic conductors for intermediate temperature solid oxide fuel cells (IT-SOFCs). Solid State Ion. **179**, 558–564 (2008)
8. Kreuer, K.D.: Proton conducting oxides. Annu. Rev. Mater. Res. **33**, 333–359 (2003)
9. Ricote, S., Bonanos, N., de Lucas, M.C.M., Caboche, G.: Structural and conductivity study of the proton conductor $BaCe_{(0.9-x)}Zr_xY_{0.1}O_{(3\text{-}\delta)}$ at intermediate temperatures. J. Power Sources **193**, 189–193 (2009)
10. Azad, A.K., Irvine, J.T.S.: Synthesis, chemical stability and proton conductivity of the perovksites $Ba(Ce, Zr)_{1-x}$ Scx $O_{3\text{-}\delta}$. Solid State Ion. **178**, 635–640 (2007)
11. Barison, S., Battagliarin, M., Cavallin, T., Doubova, L., Fabrizio, M., Mortalò, C., Boldrini, S., Malavasi, L., Gerbasi, R.: High conductivity and chemical stability of $BaCe_{1-x-y}Zr_xY_yO_{3\text{-}\delta}$

proton conductors prepared by a sol–gel method. J. Mater. Chem. **18**, 5120–5128 (2008)
12. Lee, J.-H., Heo, J.-W., Lee, D.-S., Kim, J., Kim, G.-H., Lee, H.-W., Song, H.S., Moon, J.-H.: The impact of anode microstructure on the power generating characteristics of SOFC. Solid State Ion. **158**, 225–232 (2003)
13. Lee, H.W., Kim, J.S., Lee, J.H., Sung, H.S., You, J.Y., Lee, D.S., Heo, J.W., Shin, H.I.: United States Patent, Patent No. US 7,169,336, B2, (2007)
14. Kreuer, K.D.: On the development of proton conducting materials for technological applications. Solid State Ion. **97**, 1–15 (1997)
15. Pavlenko, N.I., Stasyuk, I.: The superionic phase transition in hydrogen-bonded systems of the $M_3H(XO_4)_2$ class: the effect of nearest-O(2) distortions on the proton dynamics. J. Phys. Condens. Matter **13**, 4081–4095 (2001)
16. Yamazaki, Y., Blanc, F., Okuyama, Y., Buannic, L., Lucio-Vega, J.C., Grey, C.P., Haile, S.M.: Proton trapping in yttrium-doped barium zirconate. Nat. Mater. **12**, 647–651 (2013)
17. Kamishima, O., Abe, Y., Ishii, T., Kawamura, J., Hattori, T.: Dielectric relaxation of protonic defects in hydrated and dehydrated SrZrO3:Yb. Solid State Ion. **177**, 2375–2379 (2006)
18. Kjølseth, C., Fjeld, H., Prytz, Ø., Dahl, P.I., Estournès, C., Haugsrud, R., Norby, T.: Space–charge theory applied to the grain boundary impedance of proton conducting $BaZr_{0.9}Y_{0.1}O_{3-\delta}$. Solid State Ion. **181**, 268–275 (2010)
19. Madsen, L.G., Larsen, P.H., Bonanos, N., Engell, J., Linderoth, S.: Mechanical strength and electrical conductivity of Ni-YSZ cermets fabricated by viscous processing. J. Mater. Sci. **41**, 1097–1107 (2006)
20. Kim, S.-D., Moon, H., Hyun, S.-H., Moon, J., Kim, J., Lee, H.-W.: Performance and durability of Ni-coated YSZ anodes for intermediate temperature solid oxide fuel cells. Solid State Ion. **177**, 931–938 (2006)
21. Fonseca, F.C., de Florio, D.Z., Esposito, V., Traversa, E., Muccillo, E.N.S., Muccilloa, R.: Mixed ionic-electronic YSZ/Ni COMPOSITE for SOFC anodes with high electrical conductivity. J. Electrochem. Soc. **153**, A354–A360 (2006)
22. Zhu, W.Z., Deevi, S.C.: A review on the status of anode materials for solid oxide fuel cells. Mater. Sci. Eng. A **362**, 228–239 (2003)
23. Chevallier, L., Zunic, M., Esposito, V., Bartolomeo, E.D., Traversa, E.: A wet-chemical route for the preparation of Ni–$BaCe_{0.9}Y_{0.1}O_{3-\delta}$ cermet anodes for IT-SOFCs". Solid State Ion. **180**, 715–720 (2009)
24. Zunic, M., Chevallier, L., Radojkovic, A., Brankovic, G., Brankovic, Z., Bartolomeo, E.D.: Influence of the ratio between Ni and $BaCe_{0.9}Y_{0.1}O_{3-\delta}$ on microstructural and electrical properties of proton conducting Ni–$BaCe_{0.9}Y_{0.1}O_{3-\delta}$ anodes. J. Alloys Comp. **509**, 1157–1162 (2011)
25. Shimada, H., Li, X., Hagiwara, A., Iharaa, M.: Proton-conducting solid oxide fuel cells with yttrium-doped barium zirconate for direct methane operation. J. Electrochem. Soc. **160**, F597–F607 (2013)
26. Fabbri, E., Bi, L., Pergolesi, D., Traversa, E.: Towards the next generation of solid oxide fuel cells operating below 600 °C with chemically stable proton-conducting electrolytes. Adv. Mater. **24**, 195–208 (2012)

6

Oxidation pathways towards Si amorphous layers or nanocrystalline powders as Li-ion batteries anodes

Kamal Annou · Mathilde Pelosi · Gregory Gershinsky · Frederic Favier · Yvan Cuminal · Monique Tillard · David Zitoun

Abstract Silicon nanomaterials are obtained by an original approach based on the direct solution phase oxidation of a solid state Zintl phase NaSi used as silicon precursor. Alcohols with different alkyl chains are chosen as oxidizing agents. The materials are characterized by X-ray diffraction, scanning electron microscopy, transmission electron microscopy and Raman spectroscopy. The most relevant parameter lies in the amorphous character of the silicon nanoparticles produced by this route. Amorphous nature of silicon is one of the key features for succeeding in the improvement of anodes for Li-ion batteries. The Si nanostructures have been tested as anodic materials for lithium ion batteries.

Keywords Silicon · Amorphous · Nanoparticles · Nanomaterial · Layers · Li-ion · Battery

K. Annou · M. Pelosi · F. Favier · M. Tillard
Institut Charles Gerhardt, UMR 5253, Université de Montpellier II, Sciences et Techniques du Languedoc, CC015, 2 Place Eugène Bataillon, 34095 Montpellier Cedex 5, France
e-mail: mtillard@univ-montp2.fr

K. Annou · Y. Cuminal
Institut d'Electronique du Sud, UMR5214, Université de Montpellier II, Sciences et Techniques du Languedoc, 2 Place Eugène Bataillon, 34095 Montpellier Cedex 5, France

G. Gershinsky · D. Zitoun (✉)
Department of Chemistry, Bar Ilan Institute of Nanotechnology and Advanced Materials, Bar-Ilan University, 52900 Ramat Gan, Israel
e-mail: david.zitoun@biu.ac.il

Introduction

The success in colloidal bottom-up approach synthesis of Si has been more limited than the textbook examples of ionic semiconductors after some early successes and synthesis pathways investigation still worthwhile [1].

Early examples of gas phase syntheses have been reported for the preparation of well-defined silicon nanocrystals [2] and nanowires [3]. The common precursors used in chemical vapor deposition (CVD) are silanes (like SiH_4 or Si_2H_6) or chlorosilanes, which undergo thermal decomposition or hydrogenolysis, in presence of a metal catalyst. Vapor–liquid–solid mechanism leads to the formation of smooth single crystalline Si nanowires [4, 5]. These routes have been adapted to the synthesis in high boiling point solvents from a very expensive Si_3H_8 precursor to yield amorphous Si particles [6] or crystalline nanowires using a solution–liquid–solid (SLS) mechanism [7]. A successful alternative route to size-controlled Si nanoparticles consists in the thermal decomposition of hydrogen silsesquioxane at high temperature [8].

Examples of solution routes are sparser and involve the reduction of a silicon halide ($SiCl_4$, $RSiCl_3$) by a strong reducing agent [9], butyl-lithium, naphthalides [10–12] or hydrides [13, 14], in presence of a surfactant to limit the nanoparticles growth and coalescence but the surfactant shell impedes their use for electrochemical applications. Alternatively, Si nanowires are electrochemically deposited by step-edge decoration of highly oriented pyrolytic graphite (HOPG) from a platting solution of $SiCl_4$ in aprotic ionic liquid [15].

Therefore, the use of silicon Zintl phases, characterized by the coexistence of Si polyatomic anionic clusters and alkaline or alkaline earth cations, stands as a valuable and simple route towards Si nanostructures. The Kauzlarich group has first developed this pathway by varying the Zintl phase (NaSi or Mg_2Si) dispersed in a solvent, the oxidative agent (such as $SiCl_4$) and the capping agent (Grignard reagent or alkyl lithium) [16–20]. This pioneering work has been followed by in-depth analysis of the quantum size effect on size-selected nanocrystals [21]. Solid state oxidation of Zintl phase also yields embedded Si nanocrystals [22]. Soft oxidation of germanium Zintl phases by alcohols yields crystalline nanoparticles of Ge with potential applications in batteries [23].

The appealing ability to perform a room temperature synthesis of amorphous silicon and the need for better understanding of the reaction pathways, drive us to study the reactivity of silicon anionic clusters (Zintl phases) under oxidation with mild reducing agents.

In this manuscript, we report on the preparation of Si nanoparticles by direct oxidation of Si Zintl anions by anhydrous alcohols. The NaSi Zintl phase serves as Si precursor. Crystallinity, particle size and reaction completion are investigated as a function of the chain-length of the alcohol. The crystallinity has been carefully studied by Raman spectroscopy. The preliminary electrochemical tests as anodic material emphasize the potential application in Li-ion batteries.

Experimental section

Sodium (lingot Na from Merck-Schuchardt, >99 %) is scraped to remove surface impurities and then stored in a glove box filled with dry Ar. Silicon (Si, −325 mesh, 99 %) is purchased from Sigma-Aldrich. Anhydrous alcohols are stored and handled in the glove box: ethanol (Acros Organics, >99 %) 1-butanol (Sigma-Aldrich, >99 %), benzyl alcohol (Acros Organics, 99 %), 1-octanol (Sigma-Aldrich, >99 %).

The NaSi Zintl phase is prepared from stoichiometric proportions of the elements mixed in a weld-sealed stainless steel tube. The Na/Si mixture is heated to 920 K for 2 days during which the reactor is shaken several times to improve homogenization of the melt and then cooled to room temperature at the rate of 40°/h. The air-sensitive products of reactions are handled in the glove box.

Nanoparticles are obtained by direct oxidation of the solid precursor by an alcohol. The NaSi powder (100 mg) is slowly added to 10 mL of anhydrous degassed alcohol under stirring in a glove box or in a Schlenk connected to a vacuum line. The reaction time is optimized for completion of the reaction, and the solution is left to stir for 15–60 min, a time beyond which no major change in the product is observed. At the end of the reaction, the black powder (Si nanoparticles) is separated by centrifugation and washed with anhydrous ethanol. The yield of the reaction is more than 90 %, the only loss coming from the washing steps.

Transmission electron microscopy (TEM) is performed with a JEOL 1200 EX II microscope operating at 100 kV accelerating voltage. The X-ray diffraction patterns are recorded on a Philips analytical X'pert diffractometer equipped with a copper tube and a hybrid monochromator (parabolic multilayer mirror and two-crystal, Cu Kα radiation). The Raman spectroscopy is carried out using a LAbRam Aramis IR^2 spectrometer equipped with a helium neon laser (633 nm). The power of the laser beam is estimated to a maximal value of 6 mW at the sample surface. The use of various filters (named D0.6, D1 and D2) allows a reduction of intensity to 200, 70 and 50 μW, respectively. Scanning electron microscopy (SEM) images are collected on a FEI Magellan equipped with an energy-dispersive X-ray (Oxford 80 mm^2) spectroscopy (EDS) attachment. EDS has been typically collected on images of 25 μm^2 to provide a reliable composition.

Anodes are tested in coin-type cells (2523, NRC, Canada) versus lithium metal (Chemetall Foote Corporation, USA). Electrolyte solution is fluoro-ethylene carbonate (FEC) (Aldrich, <20 ppm water) and dimethyl carbonate (DMC) (Aldrich, <20 ppm water) (1:4 ratio) with 1 M $LiPF_6$ (Aldrich). The cells are assembled in an argon-filled glove box, with a purifying system (MBraun GmbH, Germany), oxygen and water contents below 1 ppm.

Cyclic voltammetry is measured at 30 °C with a Bio-Logic VMP3, multi-channel potentiostat at 1 mV/s between 0.8 and 0.0 V versus Li^+/Li which accounts for a charge and discharge duration of 800 s. The charge capacities, discharge capacities and coulombic efficiencies are calculated in the EC Lab software from the same data.

Results and discussion

Precursor characterization

Si nanoparticles have been prepared from the oxidation of a Zintl phase used as Si precursor (Fig. 1). The NaSi Zintl phase is prepared from the elements by high temperature synthesis. The Rietveld refinement (program Jana) of its experimental XRD powder pattern leads to good agreement factors of $R_p = 2.04$, $R_{wp} = 2.85$ and $\chi^2 = 2.59$ (Fig. 1b). Parameters of the unit cell C2/c; $a = 12.1816$ (5) Å; $b = 6.5541$ (3) Å; $c = 11.1690$ (4) Å and $\beta = 119.010$ (2), values that agree well with those previously reported [24, 25]. The structure of NaSi is built on discrete anionic units, Si_4^{4-}, packed together with sodium cations (Fig. 1a). The NaSi Zintl compound can be formally described as $[Na^+]_4$ $[Si_4^{4-}]$.

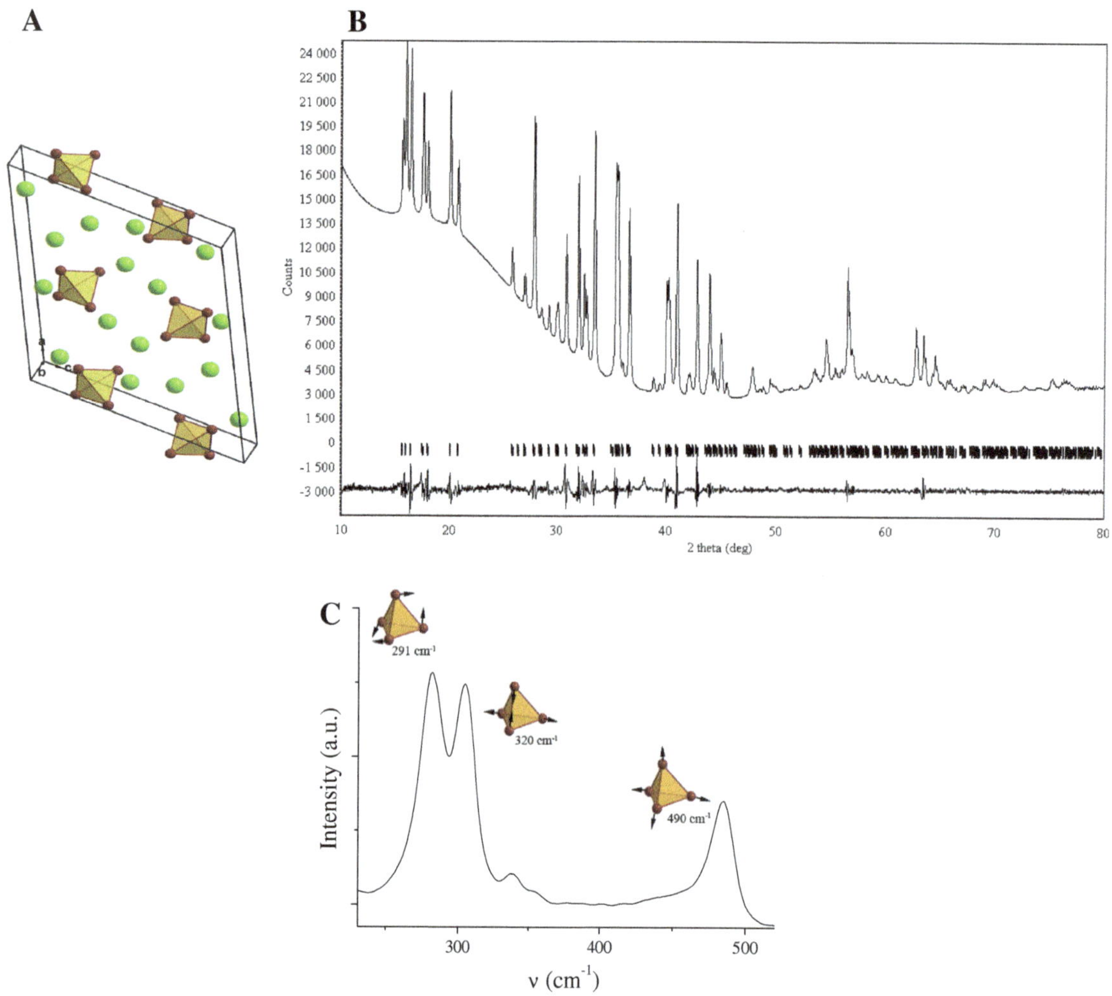

Fig. 1 Representation of the NaSi structure (viewed along *b*-axis) emphasizing the Si_4^{4-} anionic clusters (**a**); X-ray powder diffraction pattern of NaSi, corresponding Bragg positions and Rietveld refinements results (**b**); Raman spectrum of NaSi in the 225–525 cm^{-1} wave number range with the corresponding vibration modes of the Si_4^{4-} clusters (**c**)

Raman spectroscopy is a complementary technique for the full identification of NaSi. The Si_4^{4-} unit (symmetry Td) is characterized by three Raman active vibration modes $A_1(\nu_1)$, $T_2(\nu_3)$ and $E(\nu_2)$, respectively observed at 490, 320 and 291 cm^{-1} (Fig. 1c), consistent with previously reported Raman spectrum [26]. Raman spectroscopy confirms the presence of Si_4^{4-}, in addition to XRD results, and allows the complete and unambiguous identification of the compound NaSi.

Oxidation process

Alcohols are chosen as oxidizing agents for their mild reducing character and their availability. Furthermore, alcohols traces can be easily removed from the surface after synthesis [23]. The alcoholic proton reacts with the Zintl phase to give gaseous H_2 and alkoxide (which protonates upon exposure to air). Indeed, the reaction occurs with gas release and no other by-product has been noticed from ^{1}H-NMR spectrum of the supernatants.

The reaction is carried out using powdered NaSi with anhydrous alcohol as oxidizing agent. Various alcohols are tested including ethanol, 1-butanol, benzyl alcohol and 1-octanol. Powder XRD analyses of products show that NaSi reacts completely during the oxidation reaction (Fig. 2a). In all cases, products of reaction are nanosized and quite amorphous as proved by the wide peaks centered around 29° and 55° angular positions. Sometimes small and narrow peaks are observed that could be indexed within the cubic diamond-type structure of silicon. This crystalline phase is rather an impurity than a product of the reaction and most of the sample can be considered as nanocrystalline or amorphous. Therefore, Raman spectroscopy brings much more information on the nature of the products.

Actually, Raman lines are actually characterized by their shape (width, symmetry) and their position,

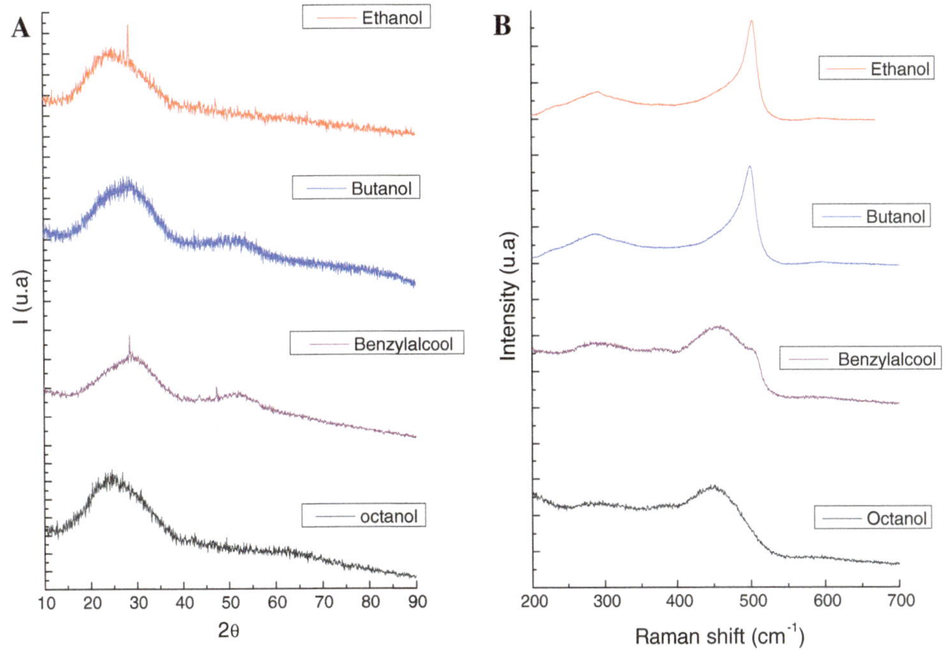

Fig. 2 XRD of Si nanopowders for each oxidizing solvent (**a**); Raman spectra of Si nanopowders for each oxidizing solvent (**b**)

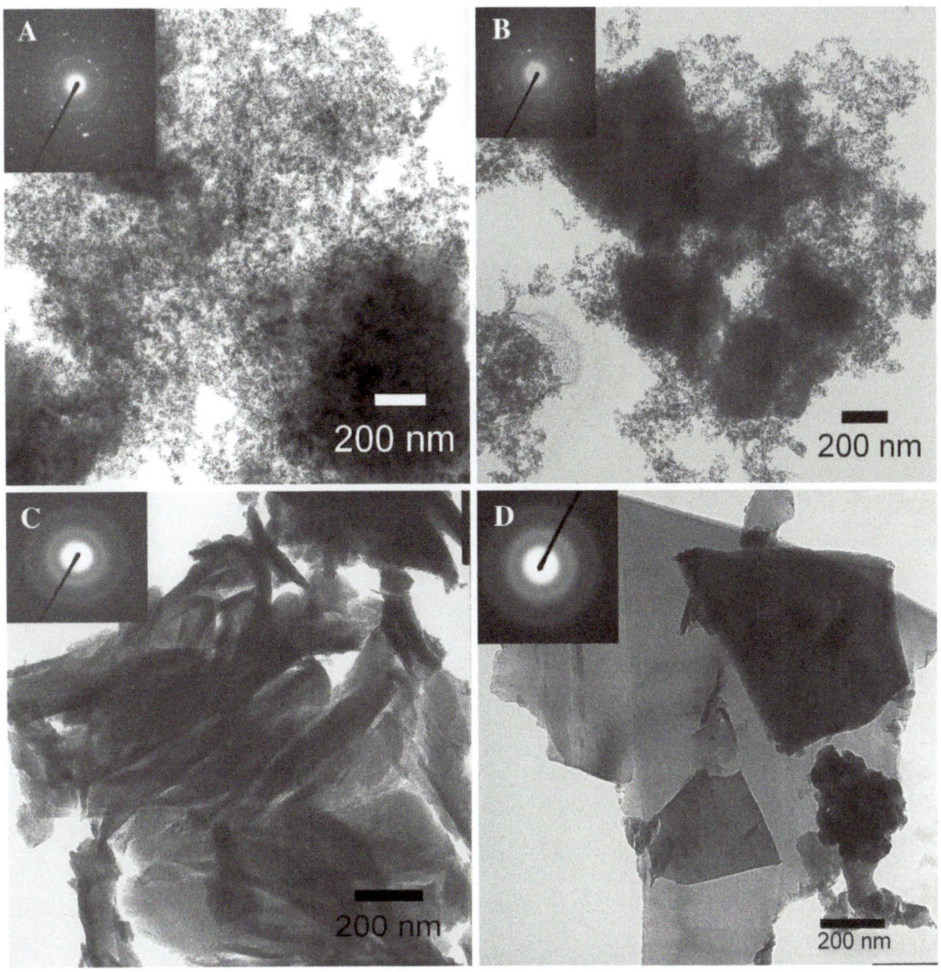

Fig. 3 TEM micrographs of Si nanoparticles from ethanol (**a**); 1-butanol (**b**); benzyl alcohol (**c**); 1-octanol (**d**) (*insets*: SAED)

parameters that are strongly correlated with the structure/morphology (crystalline, amorphous) and with the particle size (micrometric, nanometric). The Raman band position is dependent on the particle size, shifted towards low wave numbers when the average particle size decreases, and broadened when the amorphous character increases [27].

Their spectra, shown in Fig. 2b, present a large and asymmetric band clearly shifted from the position of the Raman line attributed to crystalline silicon (Si–Si stretch at 521 cm^{-1}). The experimental Raman spectrum is fitted using two Gaussian functions centered at 501 and 460 cm^{-1}. The contribution at 501 cm^{-1} is assigned to the nanocrystalline part while that at 460 cm^{-1} corresponds to the amorphous part of the sample (surface and/or amorphous NPs) [28, 29]. The products can be classified into two groups: short chain alcohols (ethanol and 1-butanol) yield mainly nanocrystalline Si with a band at 501 cm^{-1}, while benzyl alcohol and 1-octanol yield mainly amorphous Si with a main band at 460 cm^{-1}. The trend in crystallinity is consistent with a previous investigation of Ge Zintl phase oxidation process [23].

Electron microscopy provides an excellent tool for measuring crystallinity, morphology and elemental composition. For each of the synthesis, the product can be easily dispersed in ethanol without any surfactant or ligand. Selected area electron diffraction (SAED) of the products show two typical patterns. In the case of short chain alcohols (Fig. 3a, b, insets), the nanoparticles display a diffraction pattern where rings could be indexed into the cubic diamond-type structure of Si. On the other hand, for benzyl alcohol and 1-octanol, the electron diffraction is very diffuse and relates to amorphous Si.

Concerning the morphology, using a short linear alcohol (ethanol and 1-butanol), TEM shows chains of particles where the particle size depends on the nature of the oxidizing agent. Nanoparticles of 15 $\pm$ 4 nm in diameter are obtained from ethanol while particles of 11 $\pm$ 4 nm in diameter, with a larger size distribution, are obtained from 1-butanol. In the case of aromatic backbone (benzyl alcohol), TEM shows large submicronic sheets of amorphous materials without any individual particles or networks. The sheets are very thin (<20 nm) since TEM reveals several sheets oriented perpendicular to the grid. Using 1-octanol, the sheets are even larger, beyond micron in size, while the thickness remains nanometric. A reliable measurement of the thickness could not be obtained on this sample since all the sheets are laying on the TEM grid. The control of morphology obtained by varying the solvents can be understood from the difference of reactivity of alcohols towards alkaline elements. It is well established that the hydrogen evolution upon exposure of alkali metals (or Zintl phase) to alcohols depends on the chain length. Reacting with a long chain alcohol is therefore recommended to avoid uncontrollable reactions.

Ethanol and 1-butanol react faster than benzyl alcohol and 1-octanol, yielding nanocrystalline particles. The exothermic character of the reaction causes a local elevation of temperature, a factor that could favor the crystallization of Si.

On the other hand, benzyl alcohol and 1-octanol react slowly towards the Zintl phase and yield an amorphous material with a sheet morphology and dimensions that could be compared with the initial size of the NaSi grains. The soft solvolysis retains the size of the grains but yields a 2D morphology which can be accurately observed on scanning electron micrographs. On Fig. 4, we display the SEM of the different products. The powders present a layered structure on the micron scale for all the products. The platelets are even present in the case of short chain alcohols, for which TEM displays chains of particles. Obviously, the particles are aggregated into a layered material. In the case of benzyl alcohol and 1-octanol, SEM confirms the layered structure of the material with nanoscale thickness and micron scale length.

The formation pathway of the layered materials may arise from the formation of alcohol layers on the surface of the Zintl phase during reaction. Benzyl and octyl moieties are well known to pack as self-assembled monolayers. In particular benzyl alcohol results in layered materials in the case of non-aqueous sol–gel synthesis [30]. Octyl moieties are also long enough to form self-assembled monolayers, which could explain the larger sheets obtained in that case. Moreover, Zintl phase also forms layered compounds and a layered intermediate Na_xSi ($x < 1$) could induce the formation of layered amorphous silicon. To strengthen this assumption, we have measured the Na content in all the samples.

The purity of the material is investigated by energy-dispersive X-ray spectroscopy (EDS) (Fig. 5). All the samples show a high atomic content of Si (above 70 %) with oxygen and sodium as main other elements. The presence of Na is expected since the Zintl phases are also known to result in the formation of clathrates with a significant amount of residual alkaline element [31]. Na amount depends on the alcohol used during the synthesis with a very large discrepancy between the two main groups of oxidizing alcohols. The presence of significant Na content (6.8 % for 1-octanol and 8.0 % for benzyl alcohol) is consistent with the amorphous nature of Si observed for the corresponding samples. Such a high level of impurities can only be accommodated by a high density of defects, i.e. amorphous Si. On the contrary, short chain alcohols yield very low Na content (ethanol 2.4 %, 1-butanol 1.6 %), consistent with the

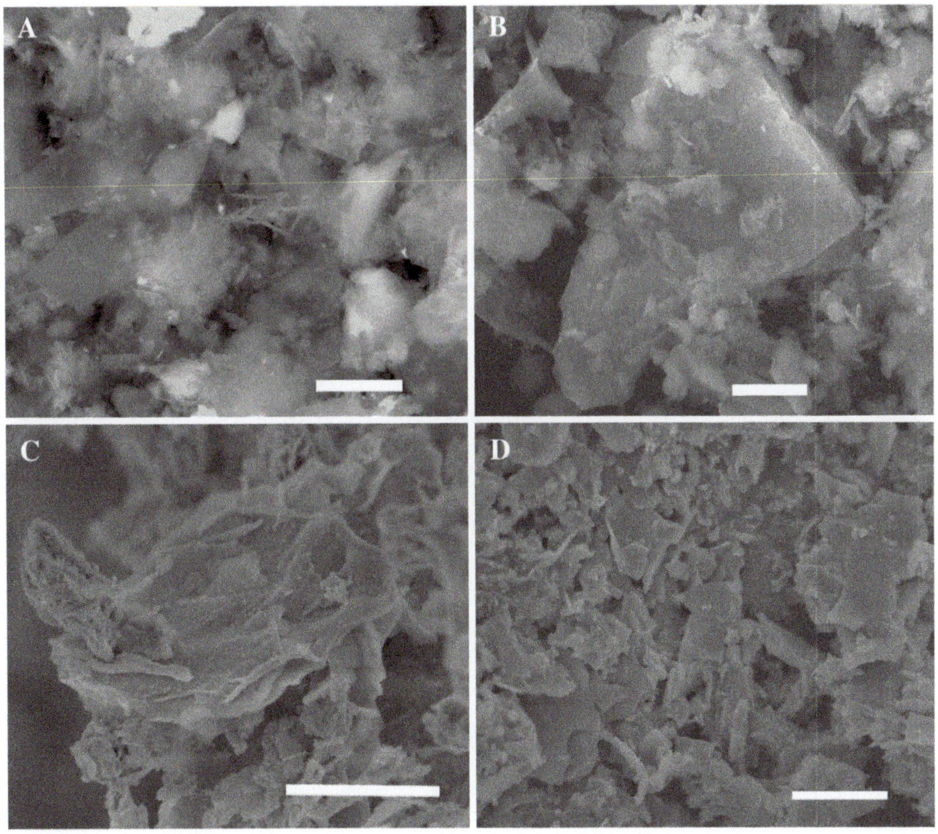

Fig. 4 SEM micrographs of Si nanoparticles from ethanol (**a**); 1-butanol (**b**); benzyl alcohol (**c**); 1-octanol (**d**). (*Scale bar* 2 μm)

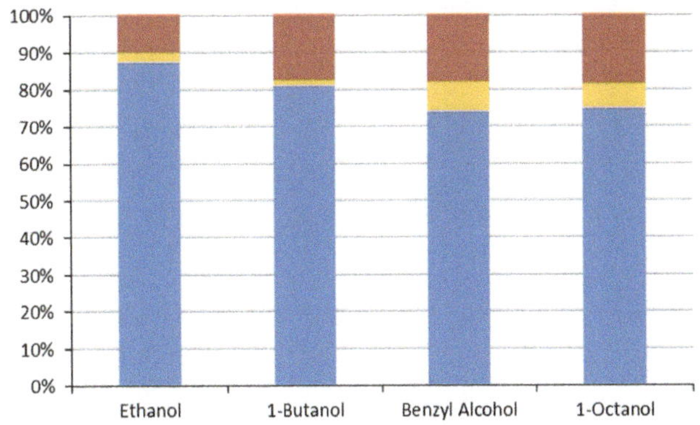

Fig. 5 Elemental atomic composition from EDS on the different samples (Si: *Blue*, Na: *Yellow*, O: *Red*)

nanocrystalline structure where Na is trapped at the developed defective surface. Interestingly, the oxygen content, which can also come from the environment and during post-synthesis storage in air, stays low: 10.2 % only even after exposure to air for several weeks.

The high Si atomic ratio, together with the amorphous or nanocrystalline structure of these materials, makes these products of high interest as anodic materials for Li-ion batteries. The presence of Na impurities is a priori not detrimental since they would be easily replaced by Li during the electrochemical process.

Electrochemical experiments

Si has an outstanding theoretical specific capacity while used as anode material in Li-ion batteries (3,600 vs. 372 $mA.g^{-1}$ for carbonaceous commercial anodes) but bulk Si does not cycle [32]. To address this issue, several routes have been explored which include the use of amorphous [33] or nanostructured [34] silicon thin films, silicon nanowires [5] and, very generally in a broader approach, any nanostructured silicon material [35]. Nanocrystalline silicon has shown remarkable coulombic efficiency when deposited as a thin film on Cu [36]. Nevertheless, the elaboration pathways mentioned above are limited to very low loading of active material [37] and silicon nanopowder has attracted more attention as a readily available material to grow thicker electrodes [38]. One of the most successful routes relies on carboxy methylcellulose (CMC) binder slurry with silicon nanoparticles and carbon black [39–42]. The coulombic efficiency is maintained for hundreds of cycles for a high CMC ratio but inevitably fades [43].

In this work, we have chosen to test a low loading of active material ($\sim$0.1 mg/cm^2) without a specific formulation to test their potential applications as anodic materials. The Si powders prepared in the present work are tested against a pure lithium electrode in coin cells. In the

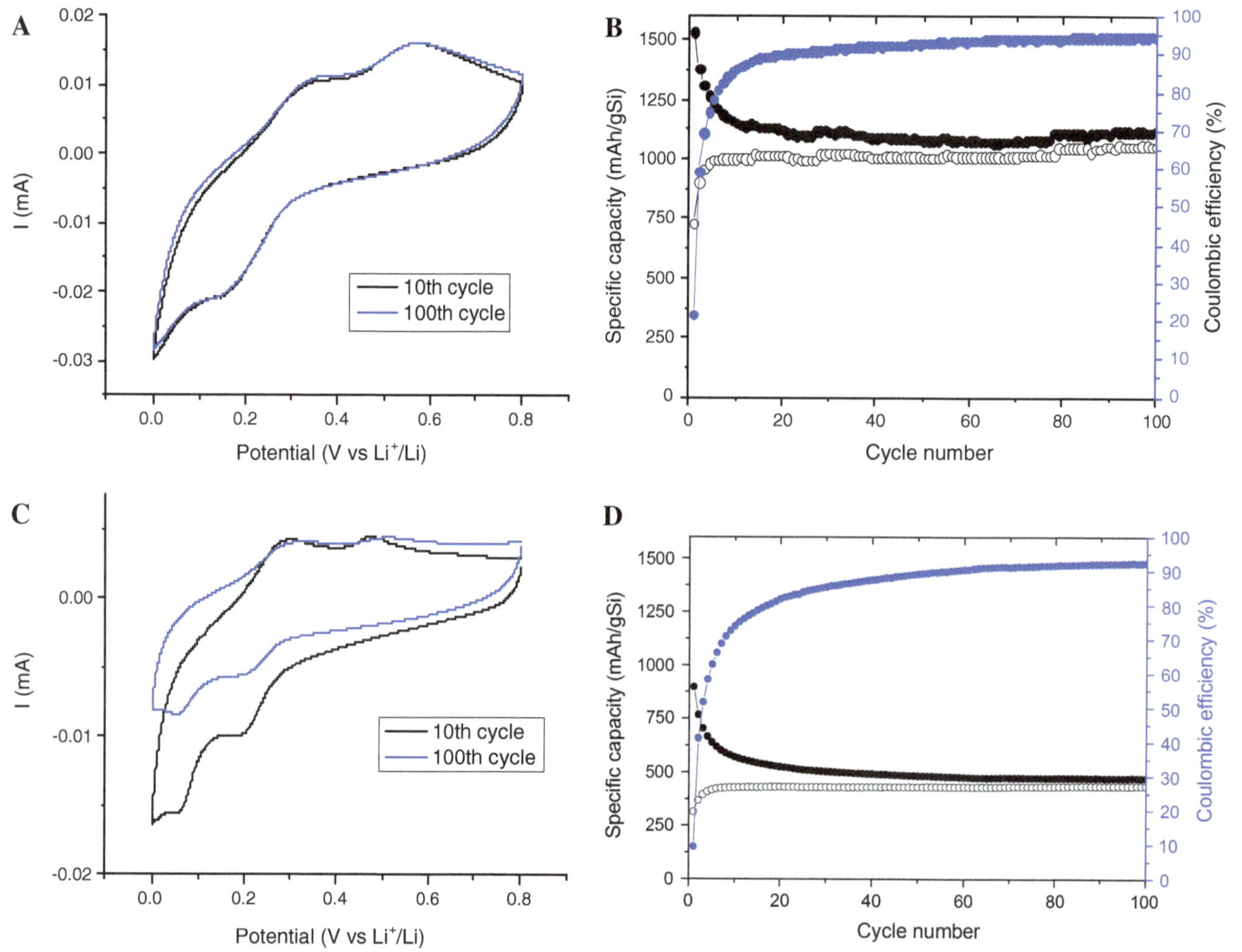

Fig. 6 Electrochemical tests at 3C on amorphous Si (from octanol synthesis: **a**, **b**) and nanocrystalline Si (from butanol synthesis **c**, **d**): cyclic voltammetry of the 10th cycle (*black line*) and 100th cycle (*blue line*) (**a–c**); charge capacity (*black dots*), discharge capacity (*black circles*) and coulombic efficiency (*blue dots*) (**b–d**)

following work on half-cell versus Li metal, we use the term: "lithiation" process to describe the discharge versus Li and "delithiation" for the charge versus Li; contrary to the full cell, where Si is the anodic material. The results are presented for two typical types of nanopowder: an amorphous sample (synthesis from octanol, Fig. 6a, b) and a nanocrystalline one (synthesis from butanol, Fig. 6c, d). For each sample, Fig. 6 presents on the left panel the cyclic voltammetry for the 10th and 100th cycles. On the right panel, Fig. 6 presents the specific charge and discharge capacities for 100 cycles (left *y* axis) together with the coulombic efficiency (right *y* axis). The main point lies in the analysis of the lithiation events and the eventual differences between the samples. For amorphous silicon, we observe two lithiation events at 0.15 and 0.02 V versus Li in discharge and delithiation at 0.36 and 0.57 V versus Li in charge (Fig. 6a). These two events correspond to the transition from Si to two main phases of Li_xSi alloys.

For nanocrystalline Si, Fig. 6c also displays two lithiation plateaus at 0.19 and 0.05 V versus Li and delithiation at 0.30 and 0.47 V versus Li slightly different from the amorphous silicon sample after 10 cycles. However, the weight of the two lithiation events in the overall capacity displays a large discrepancy, which is responsible for the large irreversibility observed for the nanocrystalline sample. On Fig. 6c, the low potential lithiation completely disappears after 100 cycles. The irreversibility of the processes results in the rapid fading of the capacity (Fig. 6d), which stabilizes at 400 mAh/g_{Si} with a coulombic efficiency of 92 % after 100 cycles.

The amorphous silicon sample displays a dominant electrochemical event at higher potential, which is extremely stable from the 10th to the 100th cycle. This reversibility results in the better capacity retention observed for amorphous silicon. The maximum lithium insertion leads to specific discharge capacity of

1,000 mAh.g^{-1} after 3 cycles and the capacity is maintained after 100 cycles. The coulombic efficiency of the amorphous Si sample reaches 95 % after 100 cycles.

The specific capacities measured in this work for amorphous and nanocrystalline Si are in the same range as initial capacities reported in the literature for other kinds of nanoscale Si. These exploratory experiments clearly prove that the electrochemical behavior is related to the crystallinity of Si. Amorphous Si has a clear edge over nanocrystalline Si.

Conclusion

Layers of Si nanomaterials can be prepared by the oxidation of a NaSi Zintl phase by an alcohol. The oxidation consists in the room temperature solvolysis of Zintl phase in anhydrous degassed alcohols. Results clearly emphasize the role of the solvent in the morphology and crystallinity of the materials. While short alcohols yield chains of nanocrystalline Si particles, aromatic and longer chains yield large flakes of amorphous Si as proven by XRD, Raman and SAED. The amorphous Si displays an interesting layered morphology and presents a high content of Na, which could be responsible for the morphology and the amorphous character. All materials display a low level of oxidation, even after storage in ambient conditions.

Preliminary results point out at the major advantage of amorphous Si versus nanocrystalline Si as anodic materials for batteries. The layered materials prepared using this original oxidation reaction are free of organic residues and, when tested in electrochemical cells, exhibit a discharge capacity of 1,100 mAh.g^{-1} after 100 cycles, twice more than the nanocrystalline samples. These results are interesting for Si anodic materials prepared by solution phase chemistry insofar and could be greatly improved by optimization of the electrode preparation.

Acknowledgments This work has been supported by the National Agency for Research (ANR Inxilicium, France) and partially supported by the Israel National Research Center for Electrochemical Propulsion—INREP (Israel Science Foundation Grant No 2797/11).

References

1. Veinot, J.G.C.: Synthesis, surface functionalization, and properties of freestanding silicon nanocrystals. Chem. Comm. **40**, 4160–4168 (2006)
2. Huisken, F., Kohn, B., Paillard, V.: Structured films of light-emitting silicon nanoparticles produced by cluster beam deposition. Appl. Phys. Lett. **74**, 3776–3778 (1999)
3. Morales, A.M., Lieber, C.M.: A laser ablation method for the synthesis of crystalline semiconductor nanowires. Science **279**, 208–211 (1998)
4. Hochbaum, A.I., Chen, R., Delgado, R.D., Liang, W., Garnett, E.C., Najarian, M., Majumdar, A., Yang, P.: Enhanced thermoelectric performance of rough silicon nanowires. Nature **451**, 163–167 (2008)
5. Chan, C.K., Peng, H., Liu, G., McIlwrath, K., Zhang, X.F., Huggins, R.A., Cui, Y.: High-performance lithium battery anodes using silicon nanowires. Nat. Nanotechnol. **3**, 31–35 (2008)
6. Harris, J.T., Hueso, J.L., Korgel, B.A.: Hydrogenated amorphous silicon (a-Si:H) colloids. Chem. Mater. **22**, 6378–6383 (2010)
7. Heitsch, A.T., Fanfair, D.D., Tuan, H.-Y., Korgel, B.A.: Solution–liquid–solid (SLS) growth of silicon nanowires. J. Am. Chem. Soc. **130**, 5436–5437 (2008)
8. Hessel, C.M., Reid, D., Panthani, M.G., Rasch, M.R., Goodfellow, B.W., Wei, J., Fujii, H., Akhavan, V., Korgel, B.A.: Synthesis of ligand-stabilized silicon nanocrystals with size-dependent photoluminescence spanning visible to near-infrared wavelengths. Chem. Mater. **24**, 393–401 (2012)
9. Heath, J.R.: A liquid-solution-phase synthesis of crystalline silicon. Science **258**, 1131–1133 (1992)
10. Zou, J., Baldwin, R.K., Pettigrew, K.A., Kauzlarich, S.M.: Solution synthesis of ultrastable luminescent siloxane-coated silicon nanoparticles. Nano Lett. **4**, 1181–1186 (2004)
11. Baldwin, R.K., Pettigrew, K.A., Ratai, E., Augustine, M.P., Kauzlarich, S.M.: Solution reduction synthesis of surface stabilized silicon nanoparticles. Chem. Commun. **17**, 1822–1823 (2002)
12. Baldwin, R.K., Pettigrew, K.A., Garno, J.C., Power, P.P., Liu, G.Y., Kauzlarich, S.M.: Room temperature solution synthesis of alkyl-capped tetrahedral shaped silicon nanocrystals. J. Am. Chem. Soc. **124**, 1150–1151 (2002)
13. Wilcoxon, J.P., Samara, G.A.: Tailorable, visible light emission from silicon nanocrystals. Appl. Phys. Lett. **74**, 3164–3166 (1999)
14. Tilley, R.D., Warner, J.H., Yamamoto, K., Matsui, I., Fujimori, H.: Micro-emulsion synthesis of monodisperse surface stabilized silicon nanocrystals. Chem. Commun. **14**, 1833–1835 (2005)
15. Fournier, C., Favier, F.: Zn, Ti and Si nanowires by electrodeposition in ionic liquid. Electrochem. Comm. **13**(11), 1252–1255 (2011)
16. Bley, R.A., Kauzlarich, S.M.: Low-temperature solution phase route for the synthesis of silicon nanoclusters. J. Am. Chem. Soc. **118**, 12461–12462 (1996)
17. Mayeri, D., Phillips, B.L., Augustine, M.P., Kauzlarich, S.M.: NMR study of the synthesis of alkyl-terminated silicon nanoparticles from the reaction of SiCl 4 with the Zintl salt, NaSi. Chem. Mat. **13**(3), 765–770 (2001)
18. Lee, S., Cho, W.J., Han, I.K., Choi, W.J., Lee, J.I.: White light emitting silicon nanocrystals as nanophosphor. Phys. Stat. Solid. B **241**, 2767–2770 (2004)
19. Yang, C.S., Bley, R.A., Kauzlarich, S.M., Lee, H.W.H., Delgado, G.R.: Synthesis of alkyl-terminated silicon nanoclusters by a solution route. J. Am. Chem. Soc. **121**(22), 5191–5195 (1999)
20. Pettigrew, K.A., Liu, Q., Power, P.P., Kauzlarich, S.M.: Solution synthesis of alkyl- and alkyl/alkoxy-capped silicon nanoparticles via oxidation of Mg_2Si. Chem. Mat. **15**, 4005–4011 (2003)
21. Dohnalová, K., Poddubny, A.N., Prokofiev, A.A., de Boer, W.D., Umesh, C.P., Paulusse, J.M., Zuilhof, H., Gregorkiewicz, T.: Surface brightens up Si quantum dots: direct bandgap-like size-tunable emission. Light Sci. Appl. **2**, 6378–6383 (2013)
22. McMillan, P.F., Gryko, J., Bull, C., Arledge, R., Kenyon, A.J., Cressey, B.A.: Amorphous and nanocrystalline luminescent Si

and Ge obtained via a solid-state chemical metathesis synthesis route. J. Solid State Chem. **178**, 937–949 (2005)
23. Pelosi, M., Tillard, M., Zitoun, D.: Ge nanoparticles by direct oxidation of Zintl alloys and their electrochemical behavior as anodes of Li-ion batteries. J. Nanopart. Res. **15**, 1872–1883 (2013)
24. Witte, J., von Schnering, H.G., Klemm, W.: Das Verhalten der Alkalimetalle zu Halbmetallen. XI. Die Kristallstruktur von NaSi und NaGe. Z. Anorg. Allg. Chem. **327**, 260–273 (1964)
25. Hohmann, E.: Silicide und Germanide der Alkalimetalle. Z Anorg. Allg. Chem. **257**, 113–126 (1948)
26. Kliche, G., Schwarz, M., von Schnering, H.G.: Raman spectrum of the tetrasilatetrahedrane anion Si_4^{4-}. Angew. Chem. Int. Ed. **26**, 349–351 (1987)
27. Morhange, J.F., Kanellis, G., Balkanski, M.: Raman study of laser annealed silicon. Solid State Commun. **31**(11), 805 (1979)
28. Iqbal, Z., Webb, A.P., Veprek, S.: Polycrystalline silicon films deposited in a glow discharge at temperatures below 250 °C. Appl. Phys. Lett. **36**, 163 (1980)
29. Richter, H., Wang, Z.P., Ley, L.: The one phonon Raman spectrum in microcrystalline silicon. Solid State Commun. **39**, 625–629 (1981)
30. Karmaoui, M., Sá Ferreira, R.A., Mane, A.T., Carlos, L.D., Pinna, N.: Lanthanide-based lamellar nanohybrids: synthesis, structural characterization, and optical properties. Chem. Mater. **18**, 4493–4499 (2006)
31. Ramachandran, G.K., Dong, J.J., Diefenbacher, J., Gryko, J., Marzke, R.F., Sankey, O.F., McMillan, P.F.: Synthesis and X-ray characterization of silicon clathrates. J. Solid State Chem. **145**, 716–730 (1999)
32. Hatchard, T.D., Dahn, J.R.: In situ XRD and electrochemical study of the reaction of lithium with amorphous silicon. J. Electrochem. Soc. **151**, A838–A842 (2004)
33. Jung, H., Park, M., Yoon, Y.-G., Kim, G.-B., Joo, S.-K.: Amorphous silicon anode for lithium-ion rechargeable batteries. J. Power Sources **115**, 346–351 (2003)
34. Kasavajjula, U., Wang, C., Appleby, A.: Nano- and bulk-silicon-based insertion anodes for lithium-ion secondary cells. J. Power Sources **163**, 1003–1039 (2007)
35. Teki, R., Datta, M.K., Krishnan, R., Parker, T.C., Lu, T.-M., Kumta, P.N., Koratkar, N.: Nanostructured silicon anodes for lithium ion rechargeable batteries. Small **5**, 2236–2242 (2009)
36. Graetz, J., Ahn, C.C., Yazami, R., Fultz, B.: Highly reversible lithium storage in nanostructured silicon. Electrochem. Solid-State Lett. **6**, A194–A197 (2003)
37. Wu, H., Chan, G., Choi, J.W., Ryu, I., Yao, Y., McDowell, M.T., Lee, S.W., Jackson, A., Yang, Y., Hu, L., Cui, Y.: Stable cycling of double-walled silicon nanotube battery anodes through solid-electrolyte interphase control. Nat. Nanotechnol. **7**, 310–315 (2012)
38. Martin, C., Crosnier, O., Retoux, R., Bélanger, D., Schleich, D.M., Brousse, T.: Chemical coupling of carbon nanotubes and silicon nanoparticles for improved negative electrode performance in lithium-ion batteries. Adv. Funct. Mater. **21**, 3524–3530 (2011)
39. Li, J., Lewis, R.B., Dahn, J.R.: Sodium carboxymethyl cellulose: a potential binder for si negative electrodes for Li-Ion batteries. Electrochem. Solid-State Lett. **10**, A17–A21 (2007)
40. Lestriez, B., Bahri, S., Sandu, I., Roue, L., Guyomard, D.: On the binding mechanism of CMC in Si negative electrodes for Li-ion batteries. Electrochem. Comm. **9**(12), 2801–2806 (2007)
41. Bridel, J.-S., Azais, T., Morcrette, M., Tarascon, J.-M., Larcher, D.: Key parameters governing the reversibility of Si/Carbon/CMC electrodes for Li-ion batteries. Chem. Mater. **22**(3), 1229–1241 (2010)
42. Guo, J., Wang, C.: A polymer scaffold binder structure for high capacity silicon anode of lithium-ion battery. Chem. Comm. **46**, 1428–1430 (2010)
43. Oumellal, Y., Delpuech, N., Mazouzi, D., Dupré, N., Gaubicher, J., Moreau, P., Soudan, P., Lestriez, B., Guyomard, D.: The failure mechanism of nano-sized Si-based negative electrodes for lithium ion batteries. J. Mater. Chem. **21**(17), 6201–6208 (2011)

7

High-temperature thermoelectric properties of $Cu_{1.97}Ag_{0.03}Se_{1+y}$

Tristan W. Day · Kasper A. Borup · Tiansong Zhang · Fivos Drymiotis · David R. Brown · Xun Shi · Lidong Chen · Bo B. Iversen · G. Jeffrey Snyder

Abstract Copper selenide is of recent interest as a high-performance *p*-type thermoelectric material. Adding Ag to copper selenide increases the dimensionless figure-of-merit zT up to 780 K, with $Cu_{1.97}Ag_{0.03}Se$ reaching a zT of 1.0 at 870 K. The increase compared with Cu_2Se can be explained by analyzing the Hall carrier concentration and effective mass. Addition of Ag reduces the carrier concentration to a nearly optimum value at high temperature. If the Hall carrier concentration were to be further reduced to 6.6×10^{20} cm^{-3} at 750 K, the average zT would be substantially improved for waste heat recovery applications.

Keywords Thermoelectrics · Copper selenide · Carrier concentration optimization · $Cu_{1.97}Ag_{0.03}Se_{1+y}$

Introduction

Materials that convert heat directly into electricity, called thermoelectrics, have been used to construct reliable generators of electrical power for spacecraft [1] and, under an applied current, as solid-state cooling devices [2]. Thermoelectric generators and coolers have the advantages of being silent, having no moving parts, and being free from greenhouse and ozone-depleting gases. The maximum efficiency of a thermoelectric material is governed by the figure-of-merit, zT, which is equal to $S^2T\rho^{-1}\kappa^{-1}$, where S is the Seebeck coefficient, T is the absolute temperature, ρ is the electrical resistivity, and κ is the thermal conductivity. Thermoelectric materials in general must have low electrical resistivity, low thermal conductivity, and a controllable carrier concentration.

A new class of high-performance thermoelectric materials is the superionic coinage-metal chalcogenides. Some examples are $Cu_{2-x}Se$ [3], $AgCrSe_2$ [4], $Ag_{2+x}Se$ [5–8], and Ag_2Te [8]. These materials all undergo a structural phase transition above which the metal ions become mobile, which leads to enhanced phonon scattering and consequently low thermal conductivity. This combined with excellent electrical conductivity makes this class of materials promising for use as thermoelectrics.

$Cu_{2-x}Se$ was found to have a zT of 1.5 at 1,000 K [3]. A related compound is $Cu_{1.97}Ag_{0.03}Se_{1+y}$ [9], which was studied from the late 1960s through the late 1970s as a candidate for use in radioisotope thermal generators (RTGs) [10]. The $y = 0$ compound ($Cu_{1.97}Ag_{0.03}Se$), synthesized by 3 M, reached a zT of 1.1 at 870 K [10], which is comparable to the zT of 1.0 at 870 K achieved in $Cu_{1.97}Ag_{0.03}Se$ in this work. Recently, the "overstoichiometric" composition $Cu_{1.98}Ag_{0.2}Se$ was studied and found to reach a maximum zT of 0.52 at 650 K before the onset of bipolar conduction [11]. While $Cu_{1.97}Ag_{0.03}Se_{1+y}$ was tested under RTG conditions in order to evaluate its performance as an energy converter, no attempt was made to analyze its thermoelectric properties in order to optimize its zT. Furthermore, no such analysis has been done for $Cu_{2-x}Se$. Therefore, we seek to analyze these materials together

T. W. Day · F. Drymiotis · D. R. Brown · G. J. Snyder (✉)
Department of Materials Science, California Institute of Technology, Pasadena, CA 91106, USA
e-mail: jsnyder@caltech.edu

K. A. Borup · B. B. Iversen
Department of Chemistry and iNANO, Center for Materials Crystallography, Aarhus University, Aarhus 8000, Denmark

T. Zhang · X. Shi · L. Chen
CAS Key Laboratory of Energy-conversion Materials, Shanghai Institute of Ceramics, Chinese Academy of Sciences, Shanghai 200050, China

to determine the suitability of a single model for describing them and whether they can achieve greater zT values.

The data will be analyzed using an effective mass approach. The premise of this approach is that at a particular temperature, the intricacies of the band structure are reflected by two parameters, the effective mass and the mobility parameter, with the carrier concentration and chemical potential determined from the Hall effect and the Seebeck effect, respectively.

Experimental

Ingots of Cu_2Se, $Cu_{1.98}Se$, $Cu_{1.97}Ag_{0.03}Se$, and $Cu_{1.97}Ag_{0.03}Se_{1.009}$ were formed by melting Cu (shot, 99.9999 % pure, Alfa Aesar, Puratronic), Ag (shot, 99.9999 % pure, Alfa Aesar, Puratronic), and Se (shot, 99.999 % pure, Alfa Aesar, Puratronic) in quartz ampoules evacuated to less than 6×10^{-5} torr. The ampoules were ramped to 1,373 K at 100 K/h, held at that temperature for 12 h, and then quenched. The ingots were ball-milled to form powders, then re-sealed in quartz ampoules, heated at 1,273 K for 5 days, cooled to 973 K, held at that temperature for 3 days and then quenched. The ingots thus obtained were ball-milled again and then hot-pressed at 973 K and 40 MPa for 6 h [12]. The geometric densities of the hot-pressed pellets were greater than 95 % of their theoretical values.

The thermoelectric properties of the samples were measured with custom-built and commercial apparatus. The thermal diffusivity α was measured with a Netzsch LFA 457 laser flash analysis unit. The total thermal conductivity was calculated from $\kappa = \alpha d C_p$, where d is the geometric density, and C_p is the heat capacity. The Seebeck coefficient S was measured with a custom-built device [13] and with an ULVAC ZEM-3. The resistivity ρ and Hall resistance were measured with a custom-built system [14] using van der Pauw geometry and a 2-T magnetic field to determine the Hall carrier concentration $n_H = 1/eR_H$ and the Hall mobility $\mu_H = R_H/\rho$. C_p was measured on a Netzsch 200F3 differential scanning calorimeter from 317 to 913 K.

Powder X-ray diffraction (PXRD) patterns were collected using a Rigaku SmartLab diffractometer equipped with a Cu K_α source, parallel beam optics, and a Rigaku D/tex detector. A hot-pressed pellet of $Cu_{1.97}Ag_{0.03}Se$ was placed under dynamic vacuum on an Anton-Paar DHS-1100 hot stage with an X-ray transparent graphite dome. PXRD patterns were collected at 300 and 420 K and used for phase identification and refinement of the high-temperature phase using the FullProf suite [15]. Diffractograms were recorded every 20 K from 300 to 500 K in the 2Θ range from 10° to 100°. Between these measurements, the sample was heated at 1 K/min, while fast diffractograms were recorded from 24.5° to 54.5° 2Θ and a scan speed of 12°/min.

Results and discussion

Powder X-ray diffraction shows that $Cu_{1.97}Ag_{0.03}Se_{1+y}$ is not single phase (Fig. 1); rather it is composed of $Cu_{2-x}Ag_xSe$, CuAgSe [16], and at least one more unidentified impurity phase, i.e., some of the Ag enters the Cu_2Se matrix and some forms CuAgSe (which has also been evaluated as a thermoelectric material and found to have low zT values) [17]. In $Cu_{1.97}Ag_{0.03}Se$, the CuAgSe phase is observed to dissolve at about 380 K, slightly before the superionic phase transition just above 400 K. At 420 K, all peaks can be indexed and refined in the high-temperature Cu_2Se structure (antifluorite, space group $Fm\bar{3}m$) except for a few, low-intensity peaks from one or more unidentified impurities. The main phase peaks are satisfactorily described by the antifluorite structure when Cu interstitials are incorporated on the octahedral sites and at trigonal planar sites. The atomic positions were stable when refined. The room-temperature Cu_2Se structure is not known but comparison with phase pure Cu_2Se PXRD patterns [18] reveals another set of peaks not belonging to the impurity at high temperatures, CuAgSe, or pure Cu_2Se. These peaks disappear at the phase transition and can hence either be an impurity, which dissolves, or belong to the main phase if this has a slightly different structure than pure Cu_2Se.

In this study, the compositions Cu_2Se, $Cu_{1.98}Se$, $Cu_{1.97}Ag_{0.03}Se$, and $Cu_{1.97}Ag_{0.03}Se_{1.009}$ were synthesized, and data on $Cu_{1.98}Ag_{0.2}Se$ from a recent publication by Ballikaya et al. [11] are included for a more complete analysis.

The thermoelectric properties of a material depend strongly on the Hall carrier concentration, n_H, whose magnitude can vary with temperature and via chemical doping. In the aforementioned materials, n_H can be decreased by substituting Ag for Cu, or increased by adding Se to create Cu^+ vacancies. Each additional Se^{2-} ion is equivalent to 2 Cu^+ vacancies; each Cu^+ vacancy donates one hole to the valence band. Furthermore, previous work on the band gap of Cu_2Se [19] shows that the valence and conduction bands are separated by a band gap much greater than k_BT, so that only one type of carrier is present. The Hall carrier concentration for all samples is shown in Fig. 2a. Above the phase transition, the Hall carrier concentration of $Cu_{1.97}Ag_{0.03}Se_{1.009}$ is greater than that of $Cu_{1.97}Ag_{0.03}Se$ due to the greater deficiency of metal ions. Likewise, the Hall carrier concentration of $Cu_{1.98}Se$ is greater than that of Cu_2Se. This could be because substitution of Ag for Cu alters the native vacancy concentration

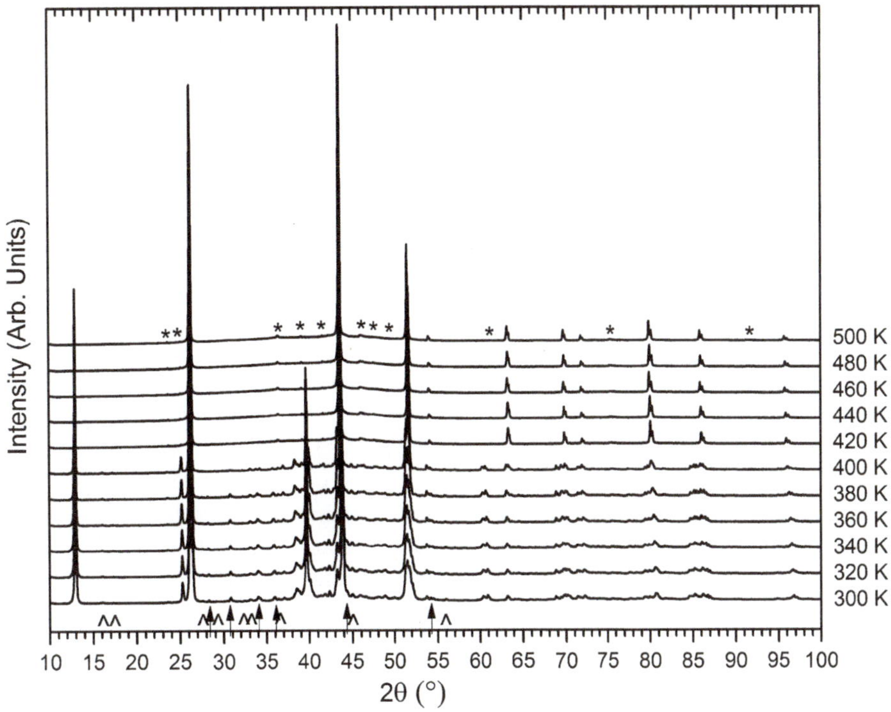

Fig. 1 Temperature resolved PXRD of $Cu_{1.97}Ag_{0.03}Se$ from 300 to 500 K. On the *bottom*, *arrows* mark CuAgSe peaks, while *hat symbol* marks impurities that dissolve at the phase transition. *Asterisk* marks the stable impurity peaks. *Unmarked peaks* on the bottom have corresponding peaks in the low-temperature structure of Cu_2Se

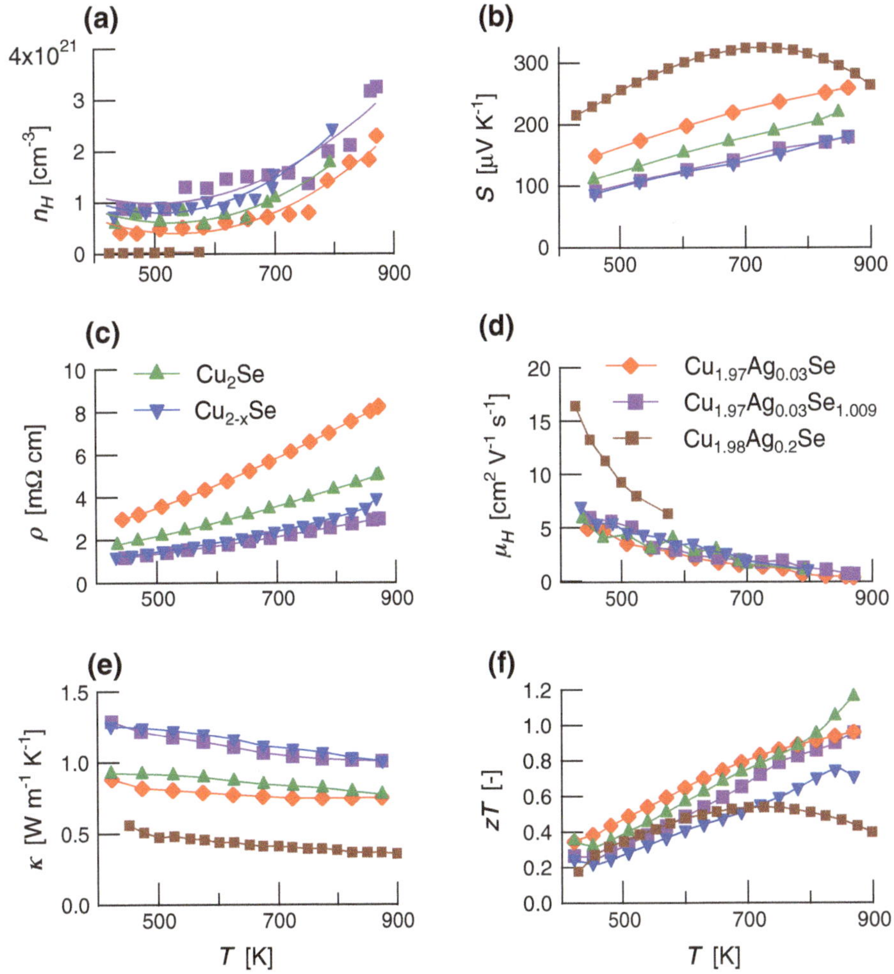

Fig. 2 Transport properties as functions of temperature of $Cu_{2-x}Se$, $Cu_{1.97}Ag_{0.03}Se_{1+y}$ and literature data on $Cu_{1.98}Ag_{0.2}Se$ from Ballikaya et al. [11]. Hall carrier concentration n_H, Seebeck coefficient S, resistivity ρ, Hall mobility μ_H, total thermal conductivity, and figure-of-merit zT are shown in **a**, **b**, **c**, **d**, **e**, and **f**, respectively. The resistivity of $Cu_{1.98}Ag_{0.2}Se$ is not shown because it is several times greater than that of the other compositions. Above the phase transition and up to 780 K, $Cu_{1.97}Ag_{0.03}Se$ achieves higher zT values (**f**) than do Cu_2Se and $Cu_{1.98}Se$ because its Hall carrier concentration is closer to the optimum value (Fig. 3). $Cu_{1.98}Ag_{0.2}Se$ is under-doped compared with both compositions of $Cu_{1.97}Ag_{0.03}Se_{1+y}$, as indicated by its greater Seebeck coefficient (**a**) and lower zT values at all temperatures

of Cu_2Se. The Hall carrier concentration of $Cu_{1.98}Ag_{0.2}Se$ is at least an order of magnitude less ($\sim 10^{19}$ cm^{-3}) than the other compositions ($\sim 10^{20}$–10^{21} cm^{-3}) due to the excess of metal ions [11]. Between 750 and 800 K, the Hall carrier concentrations of $Cu_{2-x}Se$ and $Cu_{1.97}Ag_{0.03}Se_{1+y}$ dramatically increase. The exponential character of this increase seems to indicate bipolar conduction. However, the concomitant decreases in Seebeck and resistivity are not observed. While Hall carrier concentration data for $Cu_{1.98}Ag_{0.2}Se$ were not available above 575 K in Ballikaya et al. [11], the influence of a conduction band separated from the valence band by a band gap of order k_BT is corroborated by the decrease in the Seebeck coefficient of that composition at 725 K, as shown in Fig. 2b.

$Cu_{2\text{-}x}Se$ and $Cu_{1.97}Ag_{0.03}Se_{1+y}$ exhibit the steady increase with temperature of the Seebeck coefficient (Fig. 2b) and of the resistivity ρ (Fig. 2c) expected of a single-carrier semiconductor. $Cu_{1.98}Ag_{0.2}Se$ has greater values of S and ρ than do the other samples in the entire temperature range due to its much lower Hall carrier concentration, and it shows a peak in S at about 725 K.

The Hall mobilities, μ_H, of $Cu_{2-x}Se$, $Cu_{1.97}Ag_{0.03}Se_{1+y}$, and $Cu_{1.98}Ag_{0.2}Se$ above the phase transition are low compared with those of other high-performance p-type thermoelectric materials, such as 40–6 cm^2 V^{-1} s^{-1} in Na-doped PbTe between 600 and 750 K, depending on carrier concentration [20]. The Hall mobility scales with T^{-p} [21]. The average value of p taken from the data shown in Fig. 2d and above the phase transition is about 3.1. Values of p between 1 and 1.5 usually indicate that acoustic phonons limit electron mobility in the material [21], while values greater than 1.5 indicate a temperature-dependent effective mass [22].

The thermal conductivity data are shown in Fig. 2e. The sudden increase in κ around the phase transition temperature is due to the sharp peak in C_p [18]. $Cu_{1.98}Ag_{0.2}Se$ has the lowest thermal conductivity values because it has the lowest lattice thermal conductivity, presumably due to disorder caused by the greater amount of Ag, and because it has the lowest carrier concentration of the compositions studied and therefore the lowest electronic thermal conductivity.

The zT data are shown in Fig. 2f. The zT values of $Cu_{2-x}Se$ and $Cu_{1.97}Ag_{0.03}Se_{1+y}$ all increase continuously from the phase transition temperature to the maximum temperature at which they were measured. The $Cu_{1.97}Ag_{0.03}Se$ sample reaches a zT of 1.0 at 870 K. Cu_2Se reaches a zT of 1.16 at 870 K, but between 450 and 780 K has an average zT of 0.59, whereas $Cu_{1.97}Ag_{0.03}Se$ has an average zT of 0.66 in the same temperature range. Above 780 K, the increasing values of ρ in $Cu_{1.97}Ag_{0.03}Se$ and the decreasing values of κ in Cu_2Se mean that Cu_2Se has a greater zT. $Cu_{1.98}Ag_{0.2}Se$ reaches a peak zT of 0.52 at 650 K and then decreases due to bipolar conduction.

To understand why $Cu_{1.97}Ag_{0.03}Se$ achieves a greater zT up to 780 K than do the other samples, we analyze n_H and the effective mass m^*. The carrier mobility in this model is limited by acoustic phonon scattering, and the effective mass is treated as a constant at each temperature. The results of our analysis are shown in Fig. 3 and Table 1.

We estimated the effective mass m^* at 575 K (the highest temperature for which R_H data were available for $Cu_{1.98}Ag_{0.2}Se$) and 750 K (the lowest temperature at which none of the samples exhibit a sharp increase in Hall carrier concentration) by using m^* as a fitting parameter to fit a theoretical curve to S versus n_H data (Fig. 3a). The theoretical dependence of S on the dimensionless Fermi level η is given by Eq. (1). The following equations are valid if the electron mobility is limited by acoustic phonon scattering, as per the discussion of our Hall mobility data above. k_B is Boltzmann's constant, e is the elementary charge, and $F_j(\eta)$ is the Fermi integral of order j, given by Eq. (2). ε is the electronic energy level normalized by k_BT.

$$S = \frac{k_B}{e}\left(\frac{2F_1(\eta)}{F_0(\eta)} - \eta\right) \tag{1}$$

$$F_j(\eta) = \int_0^\infty \frac{\varepsilon^j d\varepsilon}{1 + \exp(\varepsilon - \eta)} \tag{2}$$

η is set by the Hall carrier concentration and the effective mass m^*. The relationship between these variables is given by Eq. (3).

$$n_H = 4\pi\left(\frac{2m^* k_B T}{h^2}\right)^{3/2} F_{1/2}(\eta) \tag{3}$$

At a fixed temperature, we compute η from S for each sample and then use m^* to compute a theoretical n_H that we fit to the n_H data. The effective mass increases with temperature (Table 1), which we predicted based on the Hall mobility data. The same trends of effective mass, Hall carrier concentration, resistivity, and Seebeck coefficient with temperature were observed by Voskanyan et al. [23], though they estimated different values of the effective mass, e.g., 2.2 m_e at 750 K as opposed to 6.2 m_e at 750 K because they used assumed values of n_H instead of calculating them from R_H. Voskanyan et al. proposed a second valence band as a possible cause of the increasing effective mass. While this may explain of the trend of m^* with T, a two-band model is much more complex than a single-band model, requires more assumptions, and does not guarantee a unique solution. Here, we use a single band in this analysis in order to estimate the maximum achievable zT and optimum Hall carrier concentration in this material. We must emphasize that because the

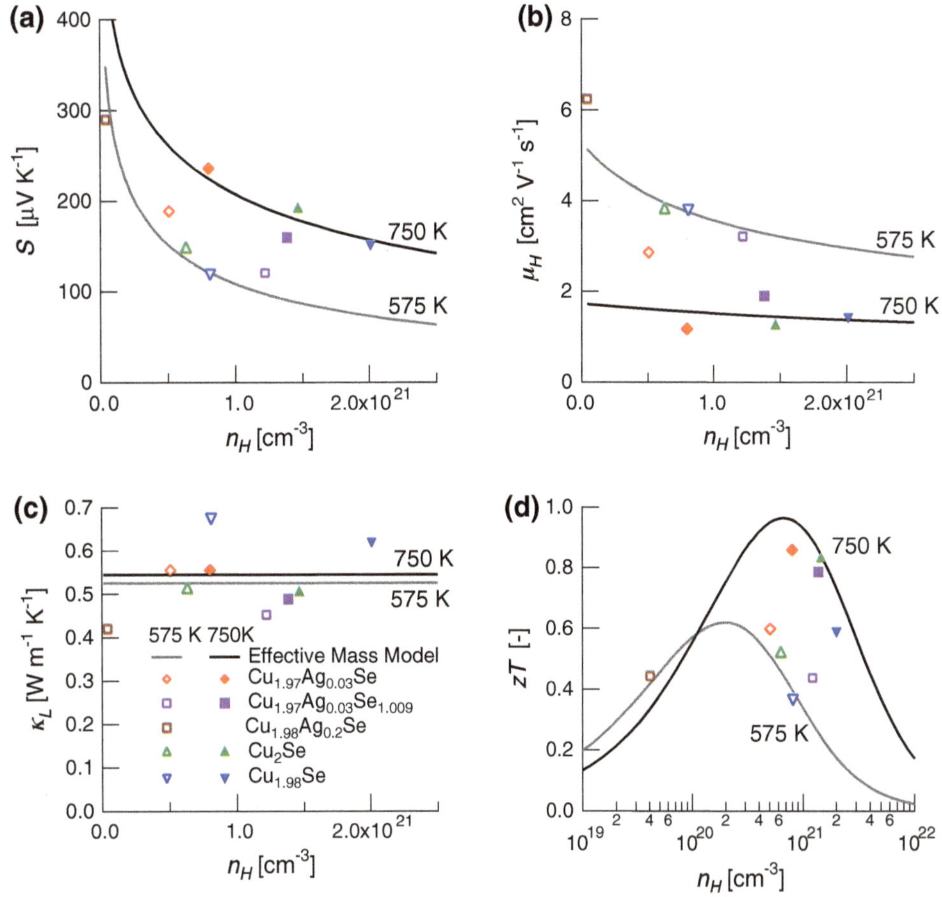

Fig. 3 Analysis of the effective mass and Hall carrier concentration explains and predicts the optimization of $Cu_{1.97}Ag_{0.03}Se$ for thermoelectric use. **a** The Seebeck coefficient as a function of Hall carrier concentration with the effective mass as a fitting parameter. **b** The Hall mobility as a function of Hall carrier concentration with μ_0 as a fitting parameter. The lattice thermal conductivity (**c**) was computed from the resistivity and the Lorenz number *L*. The optimum Hall carrier concentration (**d**) increases with temperature. The Hall carrier concentration of $Cu_{1.97}Ag_{0.03}Se$ also increases with temperature, so it has a Hall carrier concentration close to the optimum value up to 750 K. The *lines* in **c** are average values of κ_L at the indicated temperature

Table 1 Thermoelectric material properties estimated from the effective mass model

	575 K	750 K
m^* (m_e)	3.1	6.2
μ_0 ($cm^2\ V^{-1}\ s^{-1}$)	5.9	1.9
κ_L ($W\ m^{-1}\ K^{-1}$)	0.53	0.54
B (–)	0.28	0.48

effective mass is not constant with temperature, our predictions are valid only at fixed temperatures as a function of Hall carrier concentration.

The estimated values of μ_0 (Table 1) fit to the data shown in Fig. 3b, decrease with temperature, as expected from the raw Hall mobility measurements. The Hall mobility is given by Eq. (4). μ_0 is the mobility of a single electron in the material. The Fermi integral term accounts for energy level degeneracy and scattering. The effective mass m^* and the n_H data are used to compute η, which is then used with the μ_H data to fit μ_0.

$$\mu_H = \mu_0 \frac{F_{-1/2}(\eta)}{2F_0(\eta)} \quad (4)$$

$Cu_{1.97}Ag_{0.03}Se_{1.009}$ has a greater Hall mobility than does $Cu_{1.97}Ag_{0.03}Se$, despite having a greater carrier concentration and more defects.

The thermal conductivity is made up of a lattice contribution κ_L and an electronic contribution κ_E. κ_E is equal to LT/ρ, where L is the Lorenz number, given by Eq. (5). We estimated the Lorenz number of each sample at each temperature studied using the previously determined Fermi level η. The Lorenz numbers of the samples were between 1.5×10^{-8} and $1.9 \times 10^{-8}\ V^2\ K^{-2}$ at 575 and 750 K, respectively.

$$L = \left(\frac{k}{e}\right)^2 \frac{3F_0(\eta)F_2(\eta) - 4F_1(\eta)^2}{F_0(\eta)^2} \quad (5)$$

κ_L of each composition is shown in Fig. 3c, along with the average κ_L at each temperature, the values of which are shown in Table 1. κ_L does not change significantly from 575 to 750 K; therefore, the optimization of zT in this materials system will hinge only on the electrical transport properties. The slight increase in κ_L with temperature in Table 1 is due to uncertainty in the calculated κ_E. Taking the estimates for m^*, μ_0, and κ_L, we can calculate a zT versus Hall carrier concentration curve to determine the maximum zT at a given temperature and the optimum Hall carrier concentration (Fig. 3d). Looking at the zT versus Hall carrier concentration curve for 575 K, it is clear that $Cu_{1.98}Ag_{0.2}Se$ is under-doped, leading to a decrease in zT above 650 K due to bipolar conduction. $Cu_{1.97}Ag_{0.03}Se$ has the Hall carrier concentration closest to the optimum value at every temperature at which we calculated an effective mass, which explains why that composition has the greatest zT of all the compositions included in the analysis.

According to our model, a maximum zT of 1.0 at 750 K is possible in this material system. The dimensionless quality factor B [24–26] (Eq. 6) is a measure of the maximum zT at a given temperature and depends only on material properties and temperature.

$$B = \frac{k_B^{7/2}}{2^{1/2}\pi^{3/2}\hbar^3 e} \frac{\mu_0 (m^*)^{3/2} T^{5/2}}{\kappa_L} \tag{6}$$

The quality factors calculated for this material system increase with temperature (Table 1), meaning the theoretical maximum zT also increases with temperature. According to Fig. 3d, the optimum Hall carrier concentration $n_{H,opt}$ increases with temperature as well, which combined with the increasing trend with temperature of n_H in $Cu_{1.97}Ag_{0.03}Se$ means that that composition has a Hall carrier concentration close to $n_{H,opt}$ at and below 750 K.

Conclusions

Our transport property measurements show that below 780 K $Cu_{1.97}Ag_{0.03}Se_{1+y}$ has superior zT values compared with $Cu_{2-x}Se$ because it has a Hall carrier concentration closer to the optimum value. We have analyzed the Hall carrier concentration and effective mass in these materials and in recently published data on $Cu_{1.98}Ag_{0.2}Se$ for a more complete analysis, and found that these materials together follow the trends expected despite the complexity of the atomic structure and presumed complexity of the electronic structure. This model predicts that a maximum zT of 1.0 at 750 K is possible in this material system.

Each zT versus Hall carrier concentration curve in Fig. 3 indicates that zT of Cu_2Se can be increased simply by reducing the Hall carrier concentration. This means that if Cu_2Se is to be a commercially viable thermoelectric material, the addition of expensive Ag may be unnecessary. Any means of removing charge carriers from the material could improve its thermoelectric properties above the phase transition. Such means could include anion substitution [27, 28] or doping with divalent cations [17]. Furthermore, our analysis is based only on electronic parameters, so separate optimization of the lattice thermal conductivity may improve the zT of this material even further.

Acknowledgments T.W.D. and G.J.S. thank the U.S. Air Force Office of Scientific Research for support. T.W.D. thanks Heng Wang for help with the dimensionless quality factor. T.Z., X.S., and L.C. thank for financial support the National Basic Research Program of China (973-program) under Project No. 2013CB632501 and National Natural Science of Foundation of China (NSFC) under Project No. 51222209. K.A.B. and B.B.I. thank the Danish National Research Foundation (DNRF93).

References

1. LaLonde, A.D., et al.: Lead telluride alloy thermoelectrics. Mater. Today **14**(11), 526–532 (2011)
2. Bell, L.E.: Cooling, heating, generating power, and recovering waste heat with thermoelectric systems. Science **321**, 1457–1461 (2008)
3. Liu, H., et al.: Copper ion liquid-like thermoelectrics. Nat. Mater. **11**(5), 422–425 (2012)
4. Gascoin, F., Maignan, A.: Order-disorder transition in $AgCrSe_2$: a new route to efficient thermoelectrics. Chem. Mater. **23**(10), 2510–2513 (2011)
5. Ferhat, M., Nagao, J.: Thermoelectric and transport properties of beta-Ag_2Se compounds. J. Appl. Phys. **88**(2), 813–816 (2000)
6. Aliev, F., Jafarov, M., Eminova, V.: Thermoelectric figure of merit of Ag_2Se with Ag and Se excess. Atomic Struct. Non-electron. Prop. Semicond. **43**(8), 1013–1015 (2009)
7. Day, T. et al.: Evaluating the Potential for High Thermoelectric Efficiency of Silver Selenide. J. Mater. Chem. C. **1**, 7568–7573 (2012)
8. Miyatani, S.: Ionic conduction in β-Ag$_2$Te and β-Ag$_2$Se. J. Phys. Soc. Jpn. **14**(8), 996–1002 (1959)
9. Hampl, E.F.: U. S. Patent 3,853,632. Minnesota Mining and Manufacturing Company, USA (1974)
10. Brown, D.R., et al.: Chemical stability of $(Ag,Cu)_2Se$: a historical overview. J. Electron. Mater. **42**(7), 2014–2019 (2013)
11. Ballikaya, S. et al.: Thermoelectric properties of Ag-doped Cu_2Se and Cu_2Te. J. Mater. Chem. A. **1**, 12478–12484 (2013)
12. LaLonde, A.D., Ikeda, T., Snyder, G.J.: Rapid consolidation of powdered materials by induction hot pressing. Rev. Sci. Instrum. **82**(2), 025104 (2011)
13. Iwanaga, S., et al.: A high temperature apparatus for measurement of the Seebeck coefficient. Rev. Sci. Instrum. **82**(6), 063905 (2011)
14. Borup, K.A., et al.: Measurement of the electrical resistivity and Hall coefficient at high temperatures. Rev. Sci. Instrum. **83**(12), 123902 (2012)
15. Rodríguez-Carvajal, J.: Recent advances in magnetic-structure determination by neutron powder diffraction. Phys. B **192**(1–2), 55–69 (1993)
16. Miyatani, S.: Electronic and ionic conduction in (AgxCu1-x)2Se. J. Phys. Soc. Jpn. **34**(2), 423–431 (1973)
17. Ishiwata, S., et al.: Extremely high electron mobility in a phonon-glass semimetal. Nat. Mater. **12**(6), 512–517 (2013)
18. Brown, D.R., et al.: Phase transition enhanced thermoelectric figure-of-merit in copper chalcogenides. APL Mater. **1** (2013). Art. No. 052107
19. Sorokin, G.P., Papshev, Y.M., Oush, P.T.: Photoconductivity of Cu_2S, Cu_2Se, and Cu_2Te. Sov. Phys. Solid State **7**(7), 1810–1811 (1966)

20. Pei, Y., et al.: High thermoelectric figure of merit in heavy hole dominated PbTe. Energy Environ. Sci. **4**(6), 2085–2089 (2011)
21. Fistul, V.I.: Heavily Doped Semiconductors. Plenum Press, NY (1969)
22. Ravich, Y.I., Efimova, B.A., Smirnov, I.A.: Semiconducting Lead Chalcogenides. Plenum, New York (1970)
23. Voskanyan, A.A., et al.: Electrical properties of copper selenide. Sov. Phys. Semicond. **12**(11), 1251–1253 (1978)
24. Chasmar, R.P., Stratton, R.: The thermoelectric figure of merit and its relation to thermoelectric generators. J. Electron. Control. **7**, 52–72 (1959)
25. Pei, Y., Wang, H., Snyder, G.J.: Band engineering of thermoelectric materials. Adv. Mater. **24**(46), 6125–6135 (2012)
26. Wang, H., et al.: Material design considerations based on thermoelectric quality factor. In: Koumoto, K., Mori, T. (eds.) Thermoelectric Nanomaterials, pp. 3–32. Springer, Heidelberg (2013)
27. Drymiotis, F., et al.: Enhanced thermoelectric performance in the very low thermal conductivity Ag2Se0.5Te0.5. Appl. Phys. Lett. **103** (2013). Art. No. 143906
28. Mansour, B., El Akkad, F., Hendeya, T.: Electrical and thermoelectric properties of some copper chalcogenides. Phys. Status Solidi A **62**(2), 495–501 (1980)

8

Modified graphite and graphene electrodes for high-performance lithium ion hybrid capacitors

Ji Hoon Lee · Weon Ho Shin · Soo Yeon Lim · Byung Gon Kim · Jang Wook Choi

Abstract Lithium ion capacitors (LICs) have recently received considerable attention as a new class of energy storage system because they possess the combined advantages of lithium ion batteries and supercapacitors. LICs typically consist of activated carbon cathodes and pre-lithiated graphite anodes. Despite the promising electrochemical performance, most LICs still hold room for further improvement in terms of power density, which is largely related to the limited (de)intercalation kinetics of graphite. In an attempt to address these limited kinetics, we have developed a simple treatment to modify the morphology and surface characteristics of graphite engaging hydrogen peroxide. The treatment increases the exposed edge planes and generates more stable solid-electrolyte-interphase layers, which facilitate substantially improved power and cycling performance of the graphite anodes. Especially when integrated with a urea-reduced graphene cathode, the modified graphite-based LIC exhibits significantly higher energy and power densities compared to those of the pristine graphite-based and other reported counterparts.

Keywords Lithium ion capacitor · Graphene · Graphite · Enolization · Hybrid capacitor

J. H. Lee · W. H. Shin · S. Y. Lim · B. G. Kim ·
J. W. Choi (✉)
Graduate School of EEWS (Energy, Environment, Water, and Sustainability) and Center for Nature-inspired Technology (CNiT), KAIST Institute NanoCentury, Korea Advanced Institute of Science and Technology (KAIST), 291 Daehakro, Yuseong-gu, Daejeon 305-701, Republic of Korea
e-mail: jangwookchoi@kaist.ac.kr

Introduction

Energy storage systems (ESSs) are expected to play an important role in resolving various energy and environmental issues because ESSs are essential means for efficient electricity distribution as well as use of renewable energy resources [1–8]. Along this direction, supercapacitors have fulfilled critical functions such as 'energy smoothing' and 'momentary energy loading' in grid-scale ESSs by utilizing their high rate capabilities and long cycle lives [7–14]. However, most supercapacitors still suffer from limited energy densities originating from the intrinsic charge/discharge mechanism based on adsorption/desorption of carrier ions, thus restricting the range of applications. As an effort to overcome this limitation, a new class of ESSs, so-called lithium ion capacitors (LICs), has been recently developed and has shown promising electrochemical performance [15–30]. As the name indicates, LIC is a combined system between Li ion batteries (LIBs) and supercapacitors. In this hybrid system, the cathode works based on the adsorption/desorption of ions as in supercapacitors, whereas the anode operates as a Li ion source by engaging the insertion/desertion processes of Li ions as in LIBs. Typical LICs exhibit energy densities around 30 Wh/kg, which is more than three times larger than those of electrical double layer capacitors [16, 26].

Although a variety of carbonaceous materials have been considered for LIC anodes [23, 28, 31–37], pre-lithiated graphite has been the most widely used one, which is usually due to its unique voltage characteristic that charge and discharge take place at a low constant voltage at 0.1 V (vs. Li/Li^+) over nearly the entire capacity. This voltage character enables high operating voltage ($\sim$4 V) and stable solid-electrolyte-interphase (SEI) formation throughout cycling, as thoroughly discussed in previous literature [15].

This low constant voltage renders graphite clearly more desirable for high energy density and long cycle life as compared to other carbonaceous materials including hard carbon (HC), graphene, and carbon nanotubes.

Despite these well-established advantages, the graphite anode suffers from relatively sluggish kinetics for (de)insertion of Li ions during charge and discharge, restricting the rate capability of the entire cell. In particular, it has been known that the lithiation process (charge) is most rate-limiting [18, 19, 28, 38] among a series of continuous electrochemical events. The graphite could be replaced by other carbonaceous materials such as HC that can afford to facilitate rapid kinetics for Li (de)insertion. However, the improved rate performance, in turn, deteriorates the specific capacity and thus the cell energy density, indicating a trade-off between different carbonaceous materials [23, 28, 30]. For the same purpose of increasing the rate performance, ball-milling processes were also employed for graphite during the electrode fabrication but leading to only limited success [18].

Herein, we develop a simple hydrogen peroxide treatment for graphite LIC anodes, which remarkably improves the rate capabilities of the cells without sacrificing other advantageous properties of graphite, such as high specific capacity and long cycle life. The enhanced rate performance was associated with more desirable particle morphology and SEI formation for efficient Li ion diffusion in the graphite anode. Moreover, in an attempt to increase the cell energy density further, the most commonly used activated carbon cathode was replaced by a functionalized graphene cathode, thus addressing both power and energy density issues simultaneously.

Experimental section

Preparation of modified graphite

The graphite anode was modified to improve rate performance. Hereafter, we denote the modified graphite as MG. For preparation of MG, 1 g of pristine graphite (PG, Alfa Aesar. Synthetic, APC 7-11, micron, 99 %) was first introduced in 100 mL of 5 wt% hydrogen peroxide (H_2O_2, Sigma Aldrich) aqueous solution, and the solution was then heated to 70 °C for 1 h. Next, the suspension was sonicated at a power of 200 W for 1 h. Once the sonication was completed, MG was filtrated through alumina membrane filters (pore size 0.2 μm, Whatman) and was then washed with deionized (DI) water and ethanol several times.

Preparation of urea-reduced graphene oxide (URGO)

The functionalized graphene was used as a LIC cathode to increase the energy density. For this purpose, reduced graphene oxide (GO) was prepared following the procedure reported previously [15]. Briefly, GO suspension was first prepared from graphite (Alfa Aesar) by a modified Hummer's method and was then reduced by a urea treatment. 1 g of graphite and 0.5 g of sodium nitrate ($NaNO_3$, Sigma Aldrich) were introduced into 23 mL of sulfuric acid (H_2SO_4, Sigma Aldrich) and stirred vigorously at 0 °C for 3 h. Then, 3.5 g of potassium permanganate ($KMnO_4$, Sigma Aldrich) was gradually added at 30 °C until the color of the suspension turned dark gray. 46 mL of DI water was added very slowly until the temperature reached 98 °C. Lastly, warm DI water (60 °C, 140 mL) and H_2O_2 (35 wt%, 5 mL, Sigma Aldrich) were added, and the suspension was maintained at 60 °C for 3 h. The final bright brown solution was filtered and washed with DI water several times. 200 mg of GO powder was added to 200 mL of DI water. Finally, the GO solution was reduced using urea ($H_2N–CO–NH_2$, Sigma Aldrich). 6 g of urea was added to 200 mL of the GO solution with a concentration of 1 mg/mL, and the mixed solution was refluxed at 98 °C for 35 h. After the reduction process, the color of the solution was black. The urea-reduced GO (URGO) solution was then filtered and washed with DI water and ethanol several times to remove any residual impurities.

Characterization of the graphite anode

The surface functional groups of both MG and PG were characterized by X-ray photoelectron spectroscopy (XPS, Thermo VG scientific, Sigma Probe). A high-resolution dispersive Raman microscope (Raman, Horiba Jobin–Yvon) was used to characterize the degree of defects in the graphitic array. The chemical compositions of both types of graphite were investigated by element analysis (EA, Flash 2000 series). Finally, field emission-scanning electron microscope (FE-SEM, Sitrion) was used to observe the morphologies of both types of graphite.

Electrochemical test

In full-cell configurations, both types of graphite were used as anode materials, whereas AC (Kurary, YP-80F) and URGO were used as cathode materials. The anodes were prepared by first preparing slurries. The active materials, super-P (Timcal), and poly (vinylidene difluoride) (PVDF, Sigma Aldrich) were dispersed in 1-methyl-2-pyrrolidine (NMP, Sigma Aldrich) in a weight ratio of 8:1:1. The slurries were then cast onto copper foil. The AC cathode was prepared through the same procedure except that AC and aluminum foil were used as the active material and current collector, respectively. After the casting, all of the electrodes were dried in a vacuum oven at 70 °C for 12 h. From control experiments that dictate available capacities for further charging to 4 V after the pre-lithiation, the mass ratio of both electrodes was determined to be cathode : anode = 2:1. The actual mass loadings of the active materials were 2 and 1 mg/cm^2 for the cathode and anode, respectively. A mixture of ethylene carbonate and diethylene carbonate (1:1 = *v*:*v*) containing 1 M lithium hexafluorophosphate ($LiPF_6$) and 3 wt% vinylene carbonate (PANAX E-TEC, Korea) was used as electrolyte. Polypropylene membranes (Celgard 2400) were used as separators. The cell assembly was performed in an argon-filled glove box.

Prior to full-cell characterization, the graphite anodes underwent pre-lithiation by employing a half-cell configuration in which lithium metal (Hoshen) was used as both the counter and reference electrodes. The same electrolyte and separators were used as in the full-cells. For the pre-lithiation, the cells went through two full galvanostatic cycles in the voltage range of 0.005–1.5 V (vs. Li/Li^+) followed by a charging process down to 0.05 V (vs. Li/Li^+) at a 0.1 C (=0.037 A/g) rate. After the pre-lithiation, LIC full-cells were fabricated by assembling the cathodes and the pre-lithiated graphite anodes in 2,032 coin cells. Electrochemical performance of LIC full-cells was evaluated by using a WBCS 3000 battery cycler (Wonatech, Korea) at 25 °C. For the charge, the constant current constant voltage (CCCV) mode was applied by holding the cell voltage at 4 V for 10 min before the discharge. For the discharge, the constant current mode was employed. Specific capacity (mAh/g), energy density (Wh/kg), and power density (W/kg) were calculated based on the combined weight of the active materials from both sides of the electrodes. As in the previous reports [15], the energy density was calculated by obtaining the integrated area of the galvanostatic discharge curve (voltage vs. capacity). The power density was calculated by dividing the obtained energy density by discharge time.

Results and discussion

The SEM images taken before and after the hydrogen peroxide treatment are presented in Fig. 1. While both PG and MG show similar morphologies in low magnification images (Fig. 1a, c), they show clearly distinctive morphologies in a higher magnification (Fig. 1b, d). PG shows relatively smooth surfaces over most particles (Fig. 1b), whereas MG shows sharp edges (Fig. 1d and Fig. S1 in Electronic Supplementary Material) as in other similar cases that underwent mild oxidation treatments [39, 40]. It is anticipated that the exposure of the sharp edges is associated with the hydrogen peroxide treatment that rearranges the inter-particle structures while etching the edge carbon atoms. The oxidation takes place mainly at the edges because a majority of dangling bonds to be oxidized are located at the edges [41, 42]. The edge-oriented etching of carbon atoms in MG was also supported by Raman spectroscopy results (Fig. S2 in Electronic Supplementary Material). The intensity ratio between D band ($\sim$1,350 cm^{-1}) and G band ($\sim$1,580 cm^{-1}) generally reflects the degree of defects in the graphitic array. As compared with PG, MG exhibits a more intensified D band, suggesting that a larger number of defects were generated in MG, which is indeed commensurate with carbon etching at the edges during the hydrogen peroxide treatment.

The effect of the hydrogen peroxide was first examined in lithium half-cell measurements. As displayed in Fig. 2a, both types of graphite show similar CV curves in the entire potential range of 0.005–2.5 V (vs. Li/Li^+) when measured at 0.03 mV/s. The lithiation and delithiation take place at $\sim$0.01 and 0.2 V (vs. Li/Li^+), respectively, which are consistent with previous reports [19]. The reductive peaks at $\sim$0.8 V (vs. Li/Li^+) corresponding to SEI formation were also observed for both types of graphite. However, unlike PG, MG exhibits a small broad peak around 1.7 V (vs. Li/Li^+) (inset of Fig. 2a), which is ascribed to functional groups at the edges of graphite that enable an enolization process with Li ions. This phenomenon seemingly results from the fact that the surface functional groups generated during the hydrogen peroxide treatment are sequentially oxidized to carbonyl containing groups such as carbonate, carboxyl, carbonyl, and lactone groups [41–43], and these groups react with Li ions based on $>C{=}O + Li^+ + e^- \leftrightarrow >C{-}O{-}Li$. Similar enolization processes have been reported in other previous reports [15, 44–46].

The enolized functional groups on the surfaces of graphite have a significant effect on the cycle lives of graphite anodes. For the testing of cycle life, both cells were tested at C/10 (=0.037 A/g) for the first five cycles to generate stable SEI layers and then at C/3.3 (=0.111 A/g) for subsequent cycles. As displayed in Fig. 2b, while both samples show similar initial specific capacities, the MG anode exhibits much better cycling performance compared to that of the PG anode. After 30 cycles, the MG anode retains 102 % of the original capacity, whereas the PG anode retains only 54 % of the original capacity. The improved cycle life of the MG anode results from formation of stable SEI layers [41, 42, 47–50] because it has been known that lithium carboxylic salts (enolized functional groups in the current study) can be chemically bonded to the outer SEI layers generated from electrolyte reduction which has turned out to be more stable during repeated cycles than those formed directly on the graphite surfaces. In detail, the chemically bonded SEI layers have been previously found to (1) adhere robustly to the edge planes owing to the role of the enolized surface functional groups

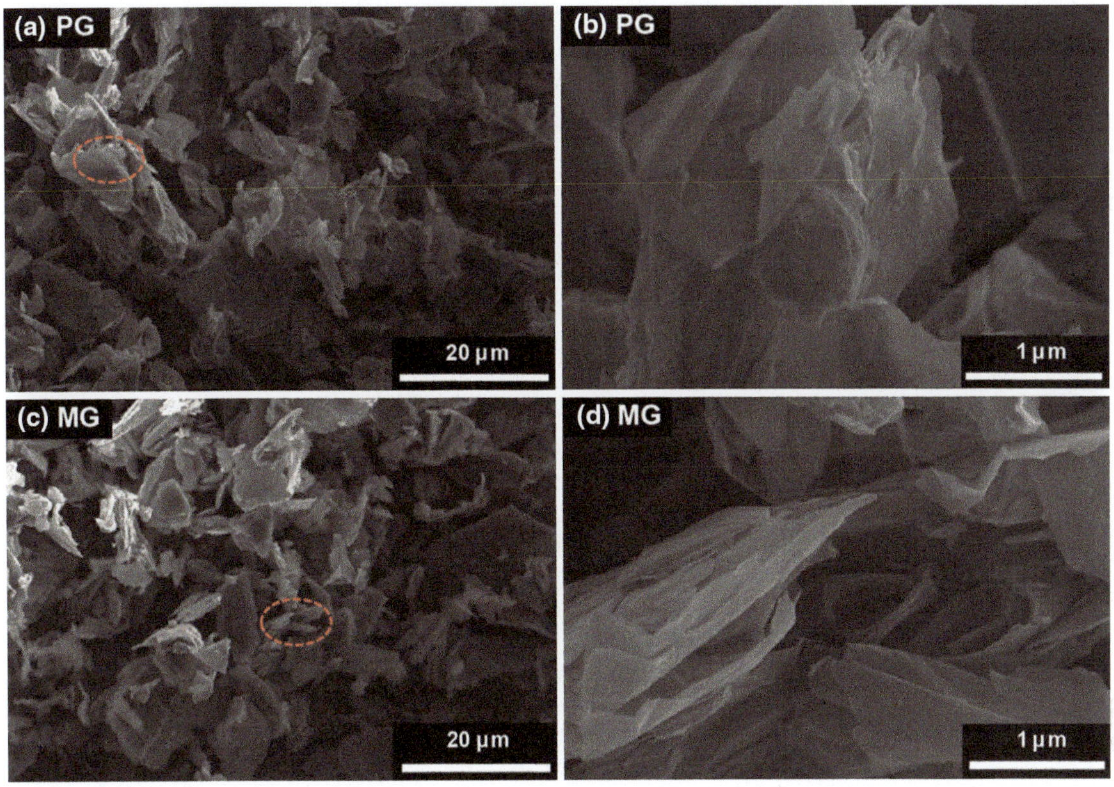

Fig. 1 **a**, **b** Low and high magnification SEM images of PG. **c**, **d** SEM images of MG with the same magnifications as in (**a**) and (**b**)

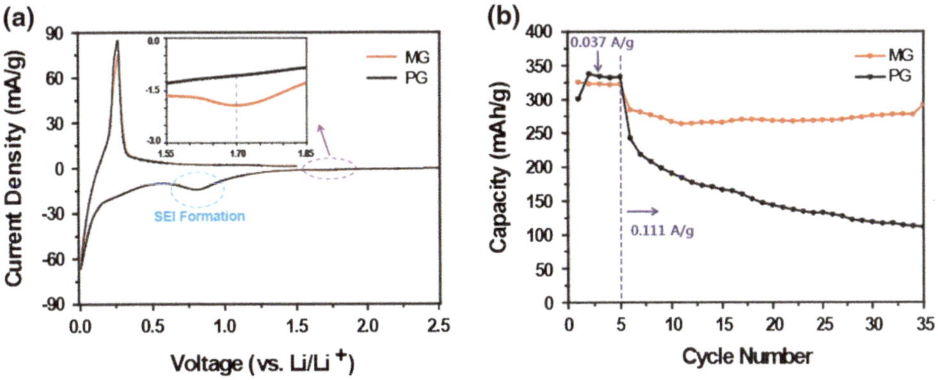

Fig. 2 **a** CV data for MG and PG as anodes in lithium half-cell configurations. (*Inset*) magnified data in the voltage range of 1.55–1.85 V (vs. Li/Li^+) indicative of lithium enolization. **b** Cycling performance of MG and PG anodes in lithium half-cell configurations. The current density is 0.037 A/g for the first five cycles and 0.111 A/g for the subsequent cycles. The mass loadings of both active electrodes are 1 mg/cm^2

serving as glue or bonding agents for Li_2CO_3 based SEI layers [41], (2) block the co-intercalation of bulky counter ions such as PF_6^- [51], and (3) thus prevent the exfoliation of exposed edge planes. It is also anticipated that the well-developed edge planes of the MG play a role in persistent Li (de)intercalation and therefore the better cycle life. Although XPS analyses were performed from the expectation of distinctive chemical characteristic of MG, O1s spectra appear quite similar for both cases, which is likely due to the limited number of the functional groups formed at the edges of MG. This observation is also consistent with the small reductive CV peak in Fig. 2a corresponding to the enolization. Thus, the amount of chemically bonded SEI layers is predicted to be limited. However, the modified SEI layers can still play a critical role for efficient Li diffusion because Li (de)intercalation takes place mainly through the edge planes where the modified SEI layers are in direct contact.

EA and XPS were performed to clarify the relationship between the surface functional groups and element

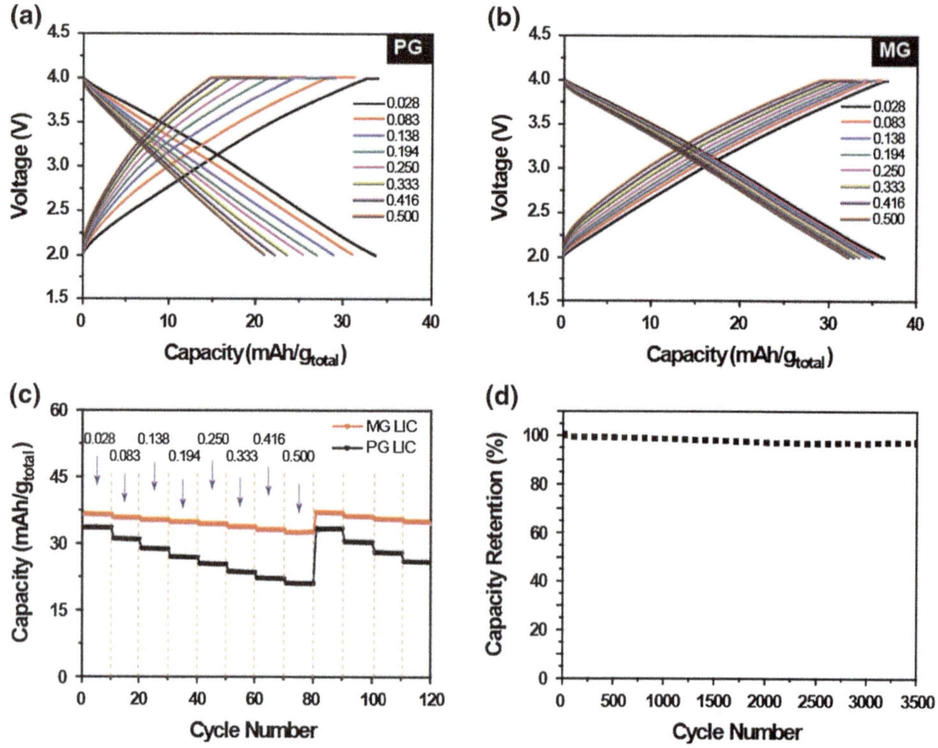

Fig. 3 The galvanostatic charge/discharge profiles of **a** PG and **b** MG LICs measured at various current densities from 0.028 to 0.5 A/g_{total}. For each cycle, the CC and CCCV modes were applied for the discharge and charge, respectively. **c** Rate capabilities of both LICs at various current densities in the range of 0.028–0.5 A/g_{total}. The shown capacities are discharge capacities. **d** Cycle life of MG LIC measured at a current density of 0.138 A/g_{total} (=4 C). All of the data in this figure are based on the weights of both active electrodes (AC cathode: 2 mg/cm^2, MG/PG anode: 1 mg/cm^2)

contents. (Table 1 in Electronic Supplementary Material) In the bulk EA analysis, as expected, carbon signal shows up dominantly for both types of graphite. In the XPS analysis, however, the detected surface oxygen amount was different (6.94:4.48 wt% = MG:PG), supporting that the carbonyl containing functional groups were generated during the hydrogen peroxide treatment (Fig. S3 in Electronic Supplementary Material).

After pre-lithiation processes for the graphite anodes, the full-cells consisting of the graphite anodes and the AC cathodes were evaluated under galvanostatic testing. From Fig. 3a and b, the following points are noticeable in LIC operations: (1) The MG LIC shows far better rate performance. As the current density increases, the specific capacity of the PG LIC drops more rapidly compared to that of the MG LIC. This distinctive trend is summarized in Fig. 3c. While the MG LIC shows a slightly higher capacity at the lowest rate of 0.028 A/g_{total}, the capacity difference between both LICs becomes more significant with increased current rate. When the current density increases by 18 times from 0.028 to 0.5 A/g_{total}, the MG LIC retains 88.9 % of the original capacity (36 → 32 mAh/g_{total}). However, for the same increase in the current density, the PG LIC retains only 62.3 % (33.6 → 21 mAh/g_{total}). (2) In both cases of LICs, it was seen that the charging processes are more kinetically limiting than the discharging processes, as the charging capacities are smaller at all of the current rates, and this gap becomes more apparent at increased current rates. This observation reflects that lithiation processes of graphite are more kinetically sluggish than delithiation processes. Remarkably, after the hydrogen peroxide treatment, the charging kinetics becomes markedly improved, diminishing the capacity differences between charging and discharging processes, which suggests that the exposed edge planes and modified SEI layers make Li ion diffusion into graphite more efficient [38, 41, 42, 49]. This trend is also consistent with the half-cell results in Figs. 2b and S4 (Electronic Supplementary Material). (3) The increased capacity of the MG LIC at the lowest current rate is likely to be associated with modified staging processes after the hydrogen peroxide treatment. As displayed in Fig. 4, while both types of graphite anodes exhibit almost same capacities, the MG anode shows a longer discharge stage I at ~0.1 V (vs. Li/Li$^+$) and a shorter discharge stage II at ~0.14 V (vs. Li/Li$^+$) as compared to those of the PG anode [52]. It is anticipated that the extended stage I at lower potential contributes to the cell capacity increase at the lowest current rate. In detail, when the LIC is under discharge condition, the potential of the AC cathode swings down while that of graphite anode gradually increases until their potential gap touches the lower cut-off voltage of 2 V [15]. Therefore, the longer discharge stage I at the lower anode potential enables the cathode potential to move more down so that the total LIC capacity can be increased.

Furthermore, the MG LIC exhibits stable cycling over a large number of cycles. As shown in Fig. 3d, when measured at 4C (=0.138 A/g_{total}), the MG LIC loses only 3 %

of its original capacity after 3,500 cycles, indicating that the modified SEI layers are stable over repeated cycles and therefore the LIC cells satisfy the one of the most important conditions in LIC operation. The cycling performance of the PG LIC is also compared in Fig. S5 (Electronic Supplementary Material). When measured at 2C (=0.083 A/g_{total}), the PG LIC shows continuous capacity fading in the first 50 cycles (33 → 28 mAh/g_{total}), but thereafter its capacity becomes saturated at 30 mAh/g_{total}, which might be related to an SEI stabilization process.

In an effort to further improve the overall cell capacity at each current rate, we also integrated functionalized graphene cathode in replacement of the AC cathode. In a previous study [15], we have demonstrated that the LIC cathode consisting of URGO delivers substantially higher capacities compared to those of the electrodes based on AC or hydrazine-reduced graphene oxide. Thus, in the current investigation, we combined the URGO cathode with the MG anode, and its rate capability is comparatively presented in Fig. 5a. The URGO-MG cell shows increased capacities at both high and low current rates. For example, the URGO-MG cell exhibits 23 and 11 % higher capacities at 0.62 and 13.8 C, respectively, compared to the AC-MG cell, implying that the new electrodes on both the anode and cathode sides improve both energy and power densities of overall LICs simultaneously.

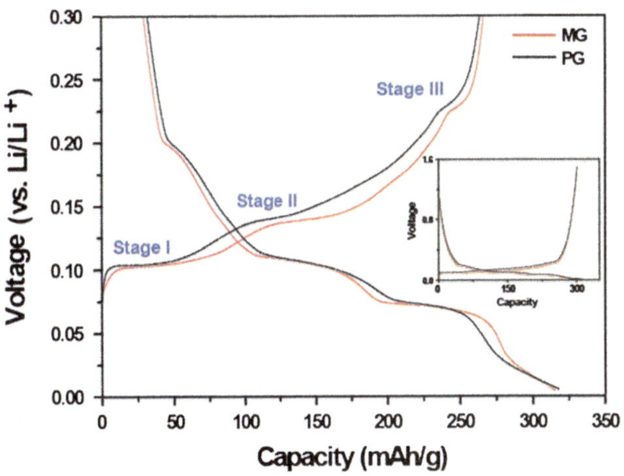

Fig. 4 The galvanostatic charge/discharge curves of MG and PG measured at a current density of 0.037 A/g in the voltage range of 0.005–1.5 V (vs. Li/Li^+). The *inset* shows the charge/discharge curves in the entire potential range

Based on these electrochemical results, we have constituted a Ragone plot (Fig. 5b) that also includes the data from other representative LIC cells in previous reports, such as AC-HC [28], AC-soft carbon (SC) [23], and AC-lithium titanate ($Li_4Ti_5O_{12}$; LTO) [27], for comparison. From this plot, the following points can be addressed: (1) As described in Fig. 3a–c, the AC-MG cell shows higher energy densities at given power densities compared to those of the AC-PG cell. For instance, while the AC-PG LIC shows an energy density of 61 Wh/kg_{total} at a power density of 1,452 W/kg_{total}, the AC-MG LIC delivers a 1.55 times higher energy density of 95 Wh/kg_{total} at a similar power density of 1,489 W/kg_{total}. (2) The URGO-MG LIC shows even higher power and energy densities. The URGO-MG LIC provides 12.7–25.3 and 33.2–77.5 % increased energy densities than those of AC-MG and AC-PG LICs, respectively, in the power density range covered in this study. Apparently, the energy density of the URGO-MG LIC is superior to those of other reported LICs (1.4–7.2 times) in the given power density range: AC/LTO (15–35 Wh/kg_{total}) [26], AC/SC (50–100 Wh/kg_{total}) [22], and AC/HC (65–78 Wh/kg_{total}) [27]. (3) The energy density of the URGO-MG LIC reaches 135 Wh/kg_{total}. Although other cell components than electrodes should be counted for fair comparison with other established systems, the energy density of 135 Wh/kg_{total} based on the electrode mass only must be still at least close to those of the conventional secondary batteries such as lead-acid battery (30–40 Wh/kg_{total}), nickel–cadmium battery (50–60 Wh/kg_{total}), and nickel–metal hydride battery (60–100 Wh/kg_{total}) [5, 12].

Conclusion

In conclusion, in an attempt to address the limited rate performance of conventional LICs, the graphite anode, the very

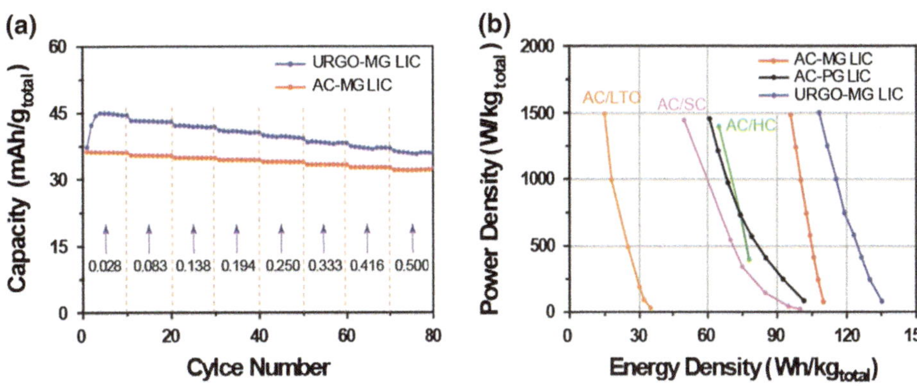

Fig. 5 **a** Rate capabilities of URGO-MG and AC-MG LICs at various current densities from 0.028 to 0.5 A/g_{total}. **b** A Ragone plot (energy density vs. power density) for the LICs in this study alongside others reported recently. All of the data in this figure are based on the weights of both active electrodes (Cathode: 2 mg/cm^2, Anode: 1 mg/cm^2)

limiting cell component for rate performance, was modified by a simple hydrogen peroxide treatment. After the treatment, the MG anode is endowed with chemically bonded SEI layers and exposed edge planes both of which facilitate more efficient Li diffusion, especially during charging (lithiation) processes. In both Li half-cell and LIC full-cell tests, the enhanced Li diffusion is reflected by the improved rate capability and stable capacity retention over a large number of cycles. Moreover, when paired up with the established functionalized graphene cathode, the so-called URGO, the LIC cell exhibits excellent energy and power densities, which are also superior to those of previously reported analogous systems. In particular, the energy density of the URGO-MG LIC approaches those of the existing battery systems. Overall, these new anode and cathode materials, respectively, address the causes of the limited power and energy densities in previous LICs. The current investigation delivers a lesson that even well-known electrode materials can still be modified further to improve key electrochemical properties of energy storage cells.

Acknowledgments This work was financially supported by the National Research Foundation of Korea (NRF) grant funded by the Korea government (MEST) (NRF-2012-M1A2A2026587).

References

1. Yang, Z.G., Zhang, J.L., Kintner-Meyer, M.C.W., Lu, X.C., Choi, D.W., Lemmon, J.P., Liu, J.: Electrochemical energy storage for green grid. Chem. Rev. **111**, 3577–3613 (2011)
2. Etacheri, V., Marom, R., Elazari, R., Salitra, G., Aurbach, D.: Challenges in the development of advanced Li-ion batteries: a review. Energy Environ. Sci. **4**, 3243–3262 (2011)
3. Hall, P.J., Mirzaeian, M., Fletcher, S.I., Sillars, F.B., Rennie, A.J.R., Shitta-Bey, G.O., Wilson, G., Cruden, A., Carter, R.: Energy storage in electrochemical capacitors: designing functional materials to improve performance. Energy Environ. Sci. **3**, 1238–1251 (2010)
4. Armand, M., Tarascon, J.M.: Building better batteries. Nature **451**, 652–657 (2008)
5. Tarascon, J.M., Armand, M.: Issues and challenges facing rechargeable lithium batteries. Nature **414**, 359–367 (2001)
6. Ibrahim, H., Ilinca, A., Perron, J.: Energy storage systems: characteristics and comparisons. Renew. Sustain. Energy Rev. **12**, 1221–1250 (2008)
7. Choi, B.G., Yang, M., Hong, W.H., Choi, J.W., Huh, Y.S.: 3D macroporous graphene frameworks for supercapacitors with high energy and power densities. ACS Nano **6**, 4020–4028 (2012)
8. Jeong, H.M., Lee, J.W., Shin, W.H., Choi, Y.J., Shin, H.J., Kang, J.K., Choi, J.W.: Nitrogen-doped graphene for high-performance ultracapacitors and the importance of nitrogen-doped sites at basal planes. Nano Lett. **11**, 2472–2477 (2011)
9. Zhang, J.T., Zhao, X.S.: On the configuration of supercapacitors for maximizing electrochemical performance. ChemSusChem. **5**, 818–841 (2012)
10. Sharma, P., Bhatti, T.S.: A review on electrochemical double-layer capacitors. Energy Convers. Manag. **51**, 2901–2912 (2010)
11. Stoller, M.D., Park, S.J., Zhu, Y.W., An, J.H., Ruoff, R.S.: Graphene-based ultracapacitors. Nano Lett. **8**, 3498–3502 (2008)
12. Simon, P., Gogotsi, Y.: Materials for electrochemical capacitors. Nat. Mater. **7**, 845–854 (2008)
13. Burke, A.: Ultracapacitors: why, how, and where is the technology. J. Power Sources **91**, 37–50 (2000)
14. Lee, J.H., Park, N., Kim, B.G., Jung, D.S., Im, K., Hur, J., Choi, J.W.: Restacking-inhibited 3D reduced graphene oxide for high performance supercapacitor electrodes. ACS Nano (2013).
15. Lee, J.H., Shin, W.H., Ryou, M.-H., Jin, J.K., Kim, J.-H., Choi, J.W.: Functionalized graphene for high performance lithium ion capacitors. ChemSusChem. **5**, 2328–2333 (2012)
16. Park, M.S., Lim, Y.G., Kim, J.H., Kim, Y.J., Cho, J., Kim, J.S.: A novel lithium-doping approach for an advanced lithium ion capacitor. Adv. Energy Mater. **1**, 1002–1006 (2011)
17. Cericola, D., Kotz, R.: Hybridization of rechargeable batteries and electrochemical capacitors: principles and limits. Electrochim. Acta **72**, 1–17 (2012)
18. Sivakkumar, S.R., Milev, A.S., Pandolfo, A.G.: Effect of ball-milling on the rate and cycle-life performance of graphite as negative electrodes in lithium-ion capacitors. Electrochim. Acta **56**, 9700–9706 (2011)
19. Sivakkumar, S.R., Nerkar, J.Y., Pandolfo, A.G.: Rate capability of graphite materials as negative electrodes in lithium-ion capacitors. Electrochim. Acta **55**, 3330–3335 (2010)
20. Sivakkumar, S.R., Pandolfo, A.G.: Evaluation of lithium-ion capacitors assembled with pre-lithiated graphite anode and activated carbon cathode. Electrochim. Acta **65**, 280–287 (2012)
21. Naoi, K.: 'Nanohybrid capacitor': the next generation electrochemical capacitors. Fuel Cells **10**, 825–833 (2010)
22. Amatucci, G.G., Badway, F., Du Pasquier, A., Zheng, T.: An asymmetric hybrid nonaqueous energy storage cell. J. Electrochem. Soc. **148**, A930–A939 (2001)
23. Schroeder, M., Winter, M., Passerini, S., Balducci, A.: On the use of soft carbon and propylene carbonate-based electrolytes in lithium-ion capacitors. J. Electrochem. Soc. **159**, A1240–A1245 (2012)
24. Aravindan, V., Chuiling, W., Madhavi, S.: High power lithium-ion hybrid electrochemical capacitors using spinel LiCrTiO4 as insertion electrode. J. Mater. Chem. **22**, 16026–16031 (2012)
25. Cericola, D., Novak, P., Wokaun, A., Kotz, R.: Hybridization of electrochemical capacitors and rechargeable batteries: an experimental analysis of the different possible approaches utilizing activated carbon, Li4Ti5O12 and LiMn2O4. J. Power Sources **196**, 10305–10313 (2011)
26. Du Pasquier, A., Plitz, I., Menocal, S., Amatucci, G.: A comparative study of Li-ion battery, supercapacitor and nonaqueous asymmetric hybrid devices for automotive applications. J. Power Sources **115**, 171–178 (2003)
27. Jung, H.G., Venugopal, N., Scrosati, B., Sun, Y.K.: A high energy and power density hybrid supercapacitor based on an advanced carbon-coated Li4Ti5O12 electrode. J. Power Sources **221**, 266–271 (2013)
28. Kim, J.H., Kim, J.S., Lim, Y.G., Lee, J.G., Kim, Y.J.: Effect of carbon types on the electrochemical properties of negative electrodes for Li-ion capacitors. J. Power Sources **196**, 10490–10495 (2011)
29. Konno, H., Kasashima, T., Azumi, K.: Application of Si–C–O glass-like compounds as negative electrode materials for lithium hybrid capacitors. J. Power Sources **191**, 623–627 (2009)

30. Shin, W.H., Jeong, H.M., Kim, B.G., Kang, J.K., Choi, J.W.: Nitrogen-doped multiwall carbon nanotubes for lithium storage with extremely high capacity. Nano Lett. **12**, 2283–2288 (2012)
31. Wu, Z.S., Ren, W.C., Wen, L., Gao, L.B., Zhao, J.P., Chen, Z.P., Zhou, G.M., Li, F., Cheng, H.M.: Graphene anchored with Co3O4 nanoparticles as anode of lithium ion batteries with enhanced reversible capacity and cyclic performance. ACS Nano **4**, 3187–3194 (2010)
32. Wu, Z.S., Ren, W.C., Xu, L., Li, F., Cheng, H.M.: Doped graphene sheets as anode materials with superhigh rate and large capacity for lithium ion batteries. ACS Nano **5**, 5463–5471 (2011)
33. Endo, M., Kim, C., Nishimura, K., Fujino, T., Miyashita, K.: Recent development of carbon materials for Li ion batteries. Carbon **38**, 183–197 (2000)
34. Wang, G.X., Shen, X.P., Yao, J., Park, J.: Graphene nanosheets for enhanced lithium storage in lithium ion batteries. Carbon **47**, 2049–2053 (2009)
35. Wang, Q., Li, H., Chen, L.Q., Huang, X.J.: Monodispersed hard carbon spherules with uniform nanopores. Carbon **39**, 2211–2214 (2001)
36. Liu, X.M., Huang, Z.D., Oh, S.W., Zhang, B., Ma, P.C., Yuen, M.M.F., Kim, J.K.: Carbon nanotube (CNT)-based composites as electrode material for rechargeable Li-ion batteries: a review. Compos. Sci. Technol. **72**, 121–144 (2011)
37. Buiel, E., Dahn, J.R.: Li-insertion in hard carbon anode materials for Li-ion batteries. Electrochim. Acta **45**, 121–130 (1999)
38. Zaghib, K., Song, X., Guerfi, A., Kostecki, R., Kinoshita, K.: Effect of particle morphology on lithium intercalation rates in natural graphite. J. Power Sources **124**, 505–512 (2003)
39. Buqa, H., Golob, P., Winter, M., Besenhard, J.O.: Modified carbons for improved anodes in lithium ion cells. J. Power Sources **97–8**, 122–125 (2001)
40. Zhao, H.P., Ren, J.G., He, X.M., Li, J.J., Jiang, C.Y., Wan, C.R.: Purification and carbon-film-coating of natural graphite as anode materials for Li-ion batteries. Electrochim. Acta **52**, 6006–6011 (2007)
41. EinEli, Y., Koch, V.R.: Chemical oxidation: a route to enhanced capacity in Li-ion graphite anodes. J. Electrochem. Soc. **144**, 2968–2973 (1997)
42. Peled, E., Menachem, C., BarTow, D., Melman, A.: Improved graphite anode for lithium-ion batteries: chemically bonded solid electrolyte interface and nanochannel formation. J. Electrochem. Soc. **143**, L4–L7 (1996)
43. Nakajima, T., Gupta, V., Ohzawa, Y., Koh, M., Singh, R.N., Tressaud, A., Durand, E.: Electrochemical behavior of plasma-fluorinated graphite for lithium ion batteries. J. Power Sources **104**, 108–114 (2002)
44. Chen, H., Armand, M., Demailly, G., Dolhem, F., Poizot, P., Tarascon, J.M.: From biomass to a renewable LixC6O6 organic electrode for sustainable Li-ion batteries. ChemSusChem **1**, 348–355 (2008)
45. Xu, W., Read, A., Koech, P.K., Hu, D.H., Wang, C.M., Xiao, J., Padmaperuma, A.B., Graff, G.L., Liu, J., Zhang, J.G.: Factors affecting the battery performance of anthraquinone-based organic cathode materials. J. Mater. Chem. **22**, 4032–4039 (2012)
46. Song, Z.P., Xu, T., Gordin, M.L., Jiang, Y.B., Bae, I.T., Xiao, Q.F., Zhan, H., Liu, J., Wang, D.H.: Polymer-graphene nanocomposites as ultrafast-charge and discharge cathodes for rechargeable lithium batteries. Nano Lett. **12**, 2205–2211 (2012)
47. Kumar, T.P., Stephan, A.M., Thayananth, P., Subramanian, V., Gopukumar, S., Renganathan, N.G., Raghavan, M., Muniyandi, N.: Thermally oxidized graphites as anodes for lithium-ion cells. J. Power Sources **97–8**, 118–121 (2001)
48. Menachem, C., Peled, E., Burstein, L., Rosenberg, Y.: Characterization of modified NG7 graphite as an improved anode for lithium-ion batteries. J. Power Sources **68**, 277–282 (1997)
49. Menachem, C., Golodnitsky, D., Peled, E.: Effect of mild oxidation of natural graphite (NG7) on anode-electrolyte thermal reactions. J Solid State Electr. **5**, 81–87 (2001)
50. Menachem, C., Wang, Y., Flowers, J., Peled, E., Greenbaum, S.G.: Characterization of lithiated natural graphite before and after mild oxidation. J. Power Sources **76**, 180–185 (1998)
51. Besenhard, J.O., Winter, M., Yang, J., Biberacher, W.: Filming mechanism of lithium-carbon anodes in organic and inorganic electrolytes. J. Power Sources **54**, 228–231 (1995)
52. Winter, M., Besenhard, J.O., Spahr, M.E., Novak, P.: Insertion electrode materials for rechargeable lithium batteries. Adv. Mater. **10**, 725–763 (1998)

9

Processing of pristine carbon nanotube supported platinum as catalyst for PEM fuel cell

Chanchal Gupta · Priyanka H. Maheshwari · S. Sasikala · R. B. Mathur

Abstract Pristine multiwalled carbon nanotube supported platinum catalyst (Pt/CNT) has been developed by reduction of Pt salt on CNTs in ethylene glycol solution. The reactions were carried out under different pH conditions followed by detailed physical and electrochemical characterizations. The I–V performance of unit Polymer electrolyte membrane fuel cell shows a peak Power density of 325 $mWcm^{-2}$ with catalyst prepared in alkaline medium, an increase of >250 % as compared to 87 $mWcm^{-2}$ obtained, while employing catalyst prepared in acidic medium and tested under similar conditions. This is attributed not only to the uniform distribution of Pt nanoparticles with small particle size in range of 2–3 nm, but also to the metallic or zero oxidation state in which most of the Pt is present when reduction is carried out in alkaline medium.

Keywords Carbon · Nanotubes · Catalyst · Fuel cell · Mid-wave potential · Polarization

Introduction

Fuel cells are considered as the clean power delivery devices for the future that produces electricity by the electrochemical redox reactions [1–3]. It is assembled as a stack of identical repeating unit cells comprising membrane electrode assembly (MEA) in which hydrogen is oxidized at the anode and oxygen is reduced at the MEA cathode, all compressed by bi-polar plates that introduce gaseous reactants and coolants to the MEA and harvest the electric current. The cathode oxygen reduction reaction (ORR) and anode hydrogen oxidation reaction (HOR) both occur on the surfaces of Pt-based catalysts with pure water and heat as the only by-products. The catalyst performance is, therefore, very decisive to the fuel cell functioning and much effort has been guided towards its optimization [4–6]. Cost and durability are other major concerns in the commercialization of fuel cells. Platinum is not only costly, but may either be contaminated by impurities from reactant gases and/or may agglomerate at fuel cell operating temperatures. Hence, Pt on conductive supports has been proposed [7, 8]. The most commonly and commercially used support is the high surface area carbon black (Vulcan XC-72), which in principle meets the criterion of performance and cost. But in terms of durability, its performance is unsatisfactory under high temperature and humidity, low pH and strong oxidizing and reducing atmospheres [8, 9]. Its amorphous nature may cause the platinum to lose its catalytic activity due to sintering, or migration on the carbon support and subsequent dissolution into the electrolyte. A number of other catalyst supports have, therefore, been investigated [10, 11]. In the above scenario, carbon nanotubes (CNTs) prove to be a much promising alternate as catalyst support. Synthesis methods have improved over the years to produce economically viable, size-controlled CNTs in macroscopic quantities for commercial applications. CNTs would impact future science and technology since it is a highly potential material with unique structure, high crystallinity, conductivity, high accessible surface area and resistance to corrosion [12]. For prolong fuel cell operation or at high current densities, fuel starvation increases with water retention [13–17]. Teflon

C. Gupta · P. H. Maheshwari (✉) · R. B. Mathur
Physics and Engineering of Carbon, CSIR-National Physical Laboratory, New Delhi, India
e-mail: hedap@mail.nplindia.ernet.in

S. Sasikala
CSIR-Central Electrochemical Research Institute, Chennai, India

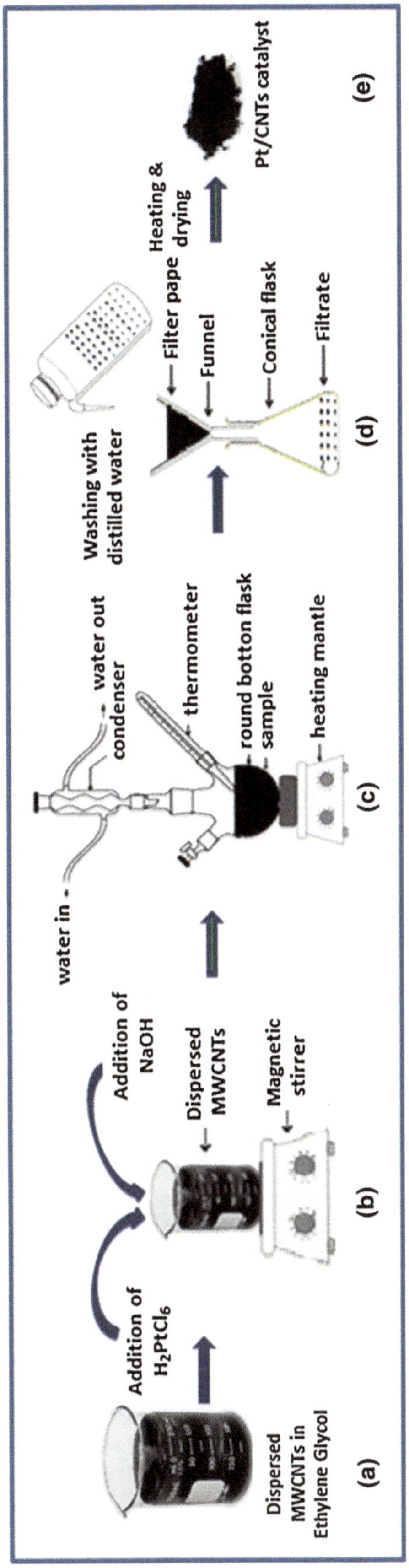

◀ **Fig. 1** Schematic of the process of catalyst development showing **a** dispersion of MWCNTs in EG, **b** addition of Platinic acid and pH adjustment, **c** refluxing, **d** filtration and washing, and **e** prepared catalyst

addition to overcome the problem decreases the conductivity. CNTs, however, have an inherent hydrophobic nature that can significantly reduce the mass transport problems.

Although research is being carried out to investigate the role of CNTs as catalyst support, there is always a scope for further improvement. Various treatments including sonochemical [18], electrochemical [19], microwave heating [20–22], and reflux heating [23] have been proposed to incorporate Pt nanoparticles onto the CNT surface. Lee et al. [24] reviewed the recent developments in the synthesis of the Pt electrocatalyst supported on CNTs for PEM fuel cell. Studies with various types of solvents [25], different reducing agents [25, 26], and varying amount of Pt [18, 26] are present in the open literature.

The present study demonstrates the development of Pt/CNT by reduction of Chloroplatinic acid on MWCNTs under different pH conditions, thus providing a detailed account of how the varying conditions of pH during the reduction process affect the properties of the synthesized nanocomposites and their corresponding behavior as a catalyst for PEM fuel cell. The effect of pH has been studied previously with high Pt loading (nearly 40 %) on CNTs. It was observed that Pt loading decreased with increasing alkalinity of the solution [23, 27]. Such high loadings are not recommended with CNTs [28, 29]. The authors have, therefore, reduced the loading to 20 % to overcome the above problem. Moreover, pristine MWCNTs have been used that not only reduces the synthesis step, but also prevents depletion of CNTs due to chemical modification. Ethylene glycol is used as a dispersing medium for CNTs which also generates reducing species in situ at high temperatures. Refluxing not only maintains the water content and pH value of the solution, but also allows uniform heating that leads to the formation of homogeneous nucleation centers [23] which act as sites for platinum deposition [30, 31]. The study is supported by exhaustive characterization. X-ray diffraction (XRD), Thermo gravimetric analysis (TGA), Transmission electron microscopy (TEM) and X-ray photoelectron spectroscopy (XPS) have been carried to determine platinum loading, structure, particle size, and oxidation states of platinum nanoparticles. Raman analysis has been carried out to evaluate the presence of defects and the interaction between Pt nanoparticles (NPs) and hexagonal carbon sheet. Electrochemical characterizations like cyclic voltammetry (CV), Linear scan voltammetry (LSV) for HOR and ORR, respectively, and unit PEM fuel cell

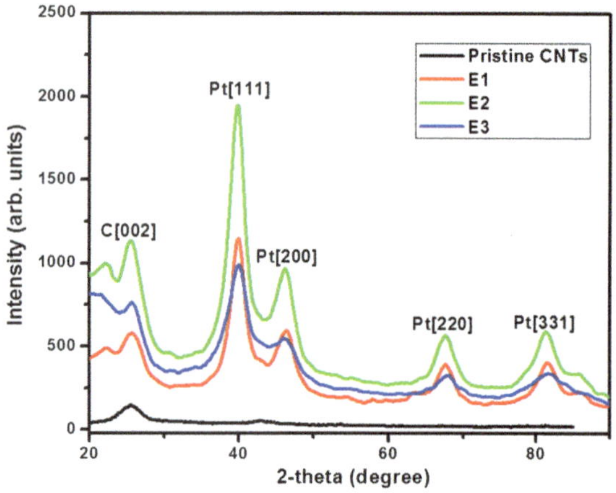

Fig. 2 X-ray diffraction curves of pristine CNTs and sample E1, E2, and E3

performance have been carried out to elucidate the catalytic performance of the prepared samples.

Materials and methods

Development of Pt/CNTs

Carbon nanotube supported platinum (Pt/CNTs) samples were prepared using commercial grade Nanocyl 7000 multi-walled carbon nanotubes (MWCNTs) with diameter in the range of 20–30 nm and aspect ratio >1,000. These were uniformly dispersed in ethylene glycol (EG, 99 % pure obtained from Merck Ltd.) by ultra-sonication. Solution of 0.01 M H_2PtCl_6 (obtained from ACROS Organics GR grade 99 % pure) in isopropyl alcohol was added to the above solution drop by drop with constant magnetic stirring such that the ratio of Pt: CNT is 1: 4. The pH of the above solution was measured and was found to be less than 2. To prepare samples in different pH mediums 0.1 M NaOH was added to adjust the pH to 2, 7, and 11. These were further refluxed at 140 °C for 3 h in air. The process of reduction immobilizes Pt on CNTs forming Pt/CNT nanocomposites. The solution was filtered followed by washing with copious amount of de-ionized water. The filtrate was further dried to obtain powdered catalyst. The schematic of the above process is shown in Fig. 1. Dilute solution of $NaBH_4$ was added to the residue (of the filtrate) to detect the presence of platinum. The samples prepared by refluxing the solution in acidic (pH 2), neutral (pH 7), and alkaline mediums (pH 11) have been designated as E1, E2, and E3, respectively, in the following text.

Characterization techniques

The structural details were obtained by X-ray diffraction examination. XRD of the catalyst samples was performed on Rikagu powder X-ray diffractometer model: XRG 2KW using Cu Kα radiation. The mean crystallite size and lattice parameters were calculated from line broadening and *d*-spacing using Debye–Scherrer formula. The morphological studies of the prepared Pt/CNT nanocomposites were carried out with the help of TEM using Tecnai G2 F30 S-Twin instrument. For TEM analysis, samples were sonochemically dispersed in ethanol and the solution was drop casted on copper grid. Raman spectroscopy was carried out using a Renishaw InVia Reflex Micro Raman Spectrometer equipped with CCD detector at room temperature and in air. Green laser (excitation line 514 nm) was used to excite the samples. Five scans per sample were recorded wherein the samples were exposed to the laser power of 25 mW for 10 s. XPS was carried out to determine the chemical states of platinum and carbon in Pt/CNTs samples using instrument PHI Model 5000 Series. TGA was carried out on TGA/DSC 1600 by Mettler Toledo. The experiments were carried out in air from room temperature to 1,000 °C at the rate of 10 °C min^{-1}.

Electrochemical characterization

The electrochemical measurements were performed using Biologic instrument, VSP model with EC-Lab software. Conventional three-electrode system was used for CV measurements with saturated calomel electrode (SCE), Pt and test samples as reference, counter, and working electrodes, respectively. 0.5 M $HClO_4$ was used as electrolyte. Kinetic activity was measured ex situ by LSV in rotating disk electrode (RDE) apparatus with 5 mV/s sweep rate and 1,600 rpm speed of the rotating disk.

For fuel cell performance, Toray carbon paper samples of size 25 cm^2 were teflonized and gas diffusion layer (GDL) was prepared by coating 1.5 mg/cm^2 of carbon black (Vulcan XC-500) followed by sintering at 350 °C for 30 min. The prepared Pt/CNT samples (E1, E2 and E3) were coated onto the GDL layer such that the amount of Pt was maintained at 0.3 mg/cm^2. Nafion Membrane-1135 (obtained from DuPont) was sandwiched in between the two electrodes (as prepared above) and hot pressed at 120–130 °C for 3 min to make the membrane electrode assembly (MEA). Fuel cell was evaluated at 60 °C and 100 % humidified conditions. Measurements of cell potential with varying current densities were conducted using electronic load procured from Bitrode Instruments, US (model: LCN4-25-24/LCN 50-24). The voltage corresponding to the applied current was measured at an interval of 30 s. Hydrogen and oxygen streams were supplied to the anode and cathode at atmospheric pressures, respectively.

Table 1 Miller indices, *d*-spacing and crystallite size for the given diffraction angle for samples E1, E2 and E3

Sample name	Miller indices	2θ (°)	FWHM (°)	*d*-spacing (nm)	Crystallite size (nm)	Average crystallite size of Pt (nm)
E1	C(002)	25.67	4.04	3.46	2.01	
	Pt(111)	40.07	2.24	2.24	3.76	3.59
	Pt(200)	46.55	3.28	1.95	2.63	
	Pt(220)	67.65	2.66	1.38	3.59	
	Pt(311)	81.70	2.39	1.17	4.39	
E2	C(002)	25.73	3.76	3.46	2.18	
	Pt(111)	39.99	2.45	2.25	3.44	3.46
	Pt(200)	46.35	3.49	1.96	2.47	
	Pt(220)	67.72	2.53	1.38	3.78	
	Pt(311)	81.39	2.55	1.18	4.11	
E3	C(002)	25.86	3.86	3.44	2.11	
	Pt(111)	40.07	3.48	2.24	2.42	2.97
	Pt(200)	46.06	3.56	1.96	2.43	
	Pt(220)	67.88	2.28	1.38	4.22	
	Pt(311)	81.92	3.74	1.18	2.81	

Results and discussion

Figure 2 shows the XRD curves for pristine CNTs and Pt/CNT samples. Diffraction peak at 25.7° is attributed to graphite crystallographic planes (002) of CNTs. Pt/CNT samples show diffraction peaks at nearly 39.9°, 46.3° and 67.7°, corresponding to face-centered cubic structure of platinum. No peaks of Co or Ni are obtained which suggests that these catalyst impurities are very less. Table 1 gives a detail account of the miller indices, *d*-spacing and crystallite size for the given diffraction angle for samples E1, E2 and E3. The FWHM values of the samples increase with increase in pH of the synthesizing medium and correspondingly the average crystallite size decreases. The crystallite size was found to be minimum (~3 nm) for sample E3.

This can be explained by the following mechanism of the reduction of Chloroplatinic acid as a function of pH. In acidic solution with low pH it is present as $[PtCl_6]^{2-}$ and $[PtCl_{6-x}(H_2O)_x]^{(2-x)-}$ [32].

$$[PtCl_6]^{2-} + 2\,H^+ + xH_2O \leftrightarrow [PtCl_{6-x}(H_2O)_x]^{(2-x)-} + xCl^- + 2\,H^+ \quad (1)$$

With addition of external OH^- ions in form of NaOH, there may be replacement of H_2O or Cl^- from the aqua species $[PtCl_{6-x}(H_2O)_x]^{(2-x)}$ formed in step (1).

$$[PtCl_{6-x}(H_2O)_x]^{(2-x)-} + 2\,H^+ + yOH^- \leftrightarrow \left[PtCl_{6-x}(OH)_y(H_2O)_{x-y}\right]^{(2-x+y)-} + 2H^+ \quad (2)$$

$$[PtCl_{6-x}(H_2O)_x]^{(2-x)-} + 2\,H^+ + zOH^- \leftrightarrow [PtCl_{6-x-z}(OH)_z(H_2O)_x]^{(2-x)-} + zCl^- + 2H^+ \quad (3)$$

Chen et al. [33] from Extended X-ray absorption fine structure studies have confirmed that Pt–O bond length (2.03 Å) is smaller than Pt–Cl bond length (2.33 Å). Hence, due to steric contraction effect and electrostatic charge effect, the species formed in alkaline medium are smaller in size than complexes formed in acidic medium [26]. This implies that the size of platinum species deposited on the carbon support greatly depends on the pH of the environment.

The results were further confirmed by the TEM studies. Figure 3 shows the TEM images of samples E1 and E3. TEM micrographs reveal that Pt NPs are attached to the outer walls of CNTs. For sample E1, agglomeration of platinum particles is observed at some places as shown in Fig. 3a–c and the Pt particle size distribution shows maximum particles in the range of 4–5 nm (Fig. 3d). However, for sample E3, homogeneous distribution of Pt NPs with particle size in the range of 2–3 nm has been observed (Fig. 3e–h). The particle sizes determined by XRD are in close agreement with observed values from TEM. Figure 3b, d shows the presence of closed ends of CNT which further implies that during the synthesis process the CNTs remained highly intact and Pt NPs are deposited only on their surface.

A small size with a uniform distribution of the catalyst NPs is an important parameter affecting the fuel cell performance. However, the oxidation state is another important factor which decides the activity of the catalyst. It has been observed that platinum in metallic or zero oxidation state shows higher catalytic activity than its corresponding oxidized species [34]. XPS was used as a quantitative technique to measure the elemental composition and their oxidation states present in the samples so as to illustrate the

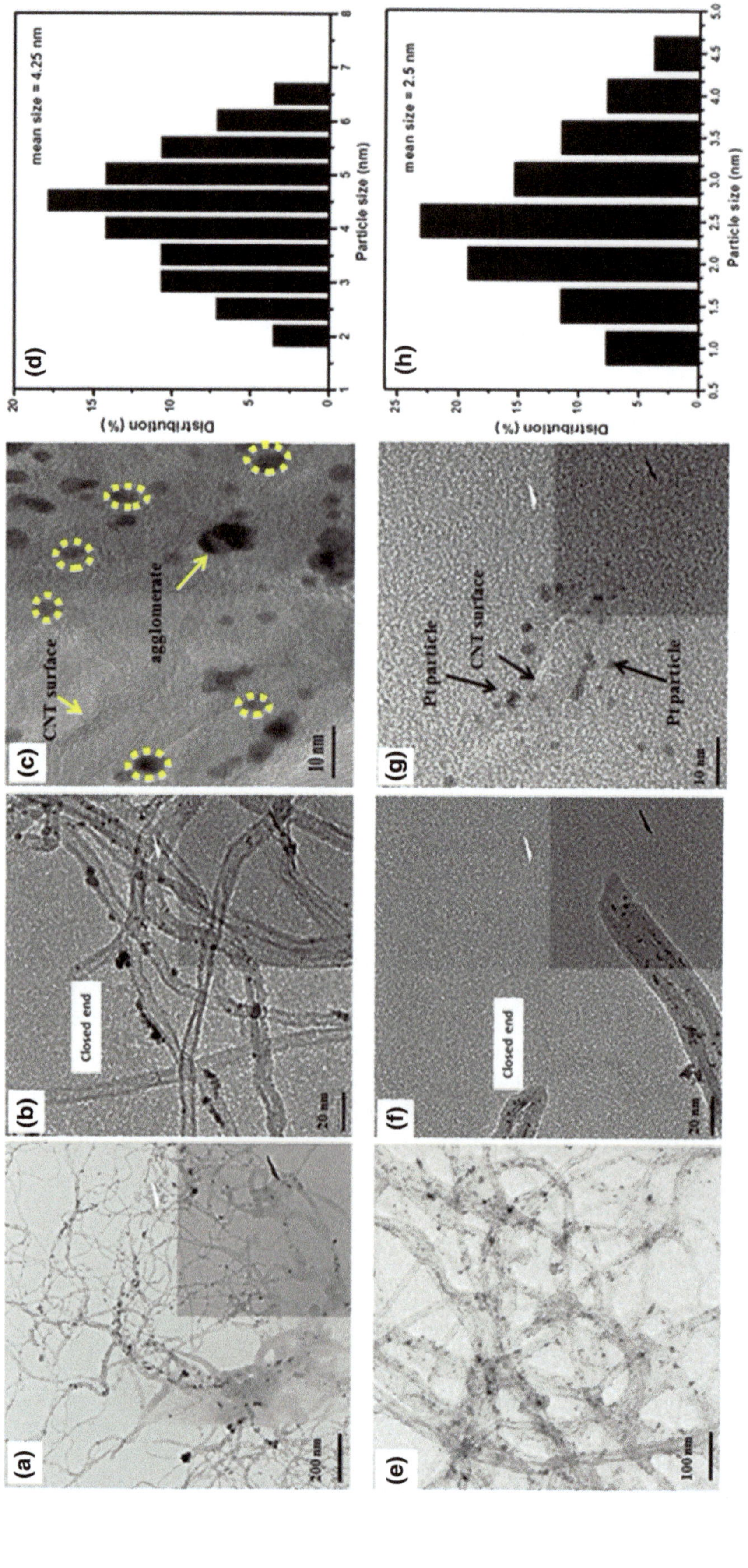

Fig. 3 TEM micrographs of sample E1 showing non-uniform distribution and agglomeration of catalyst particles (**a**–**c**) and sample E3 showing catalyst distribution (**d**); closed ends of CNTs after processing (**e**); and formed catalyst nanoparticles size of 2–3 nm (**f**)

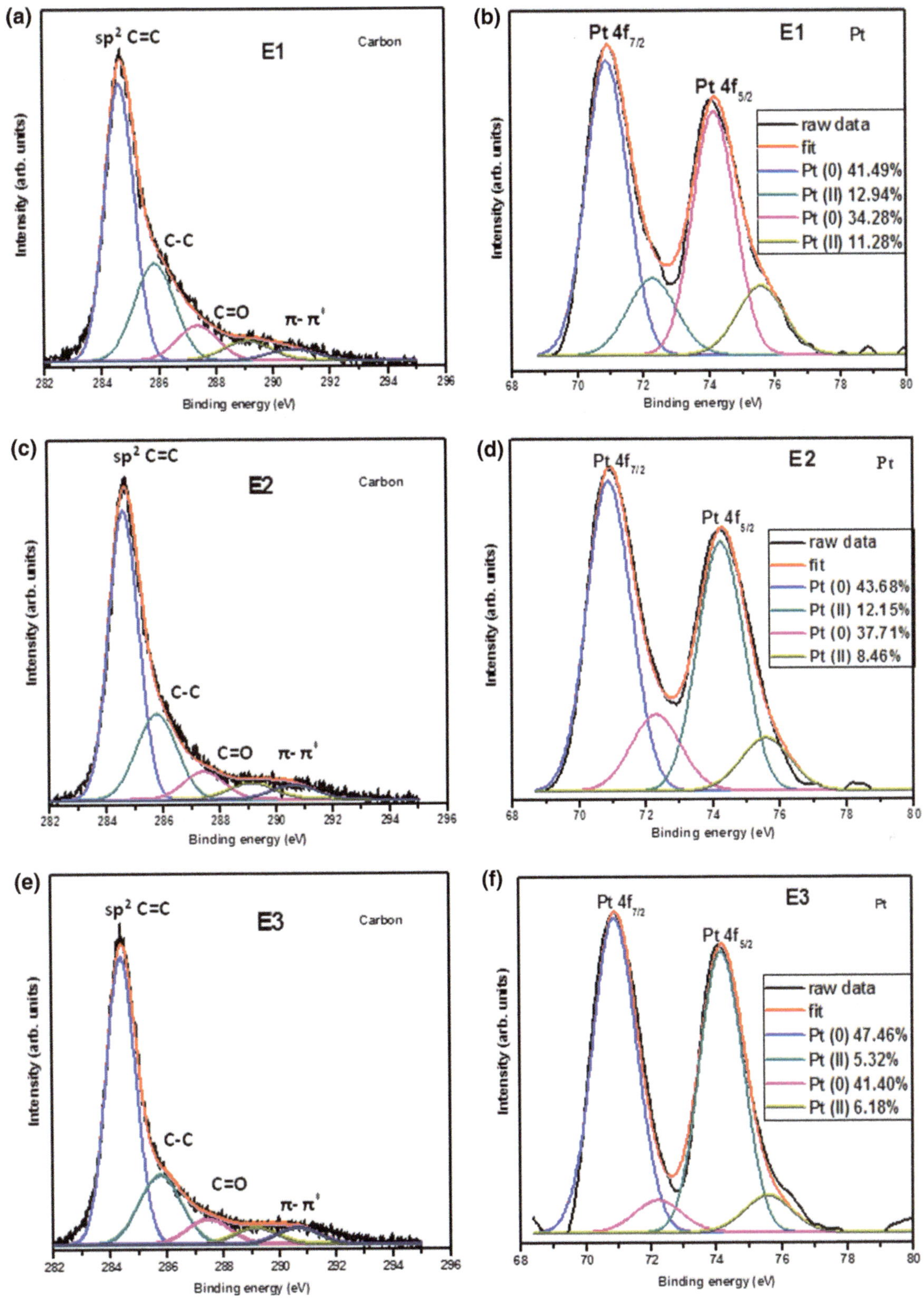

Fig. 4 XPS curves of catalyst samples E1, E2 and E3 for carbon 1 s (**a**, **c**, and **e**) and Platinum 4f peaks (**b**, **d**, and **f**)

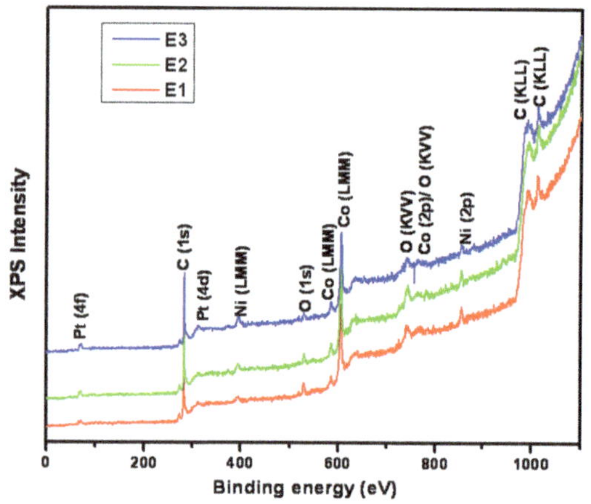

Fig. 5 The XPS data for the catalyst samples E1, E2, and E3

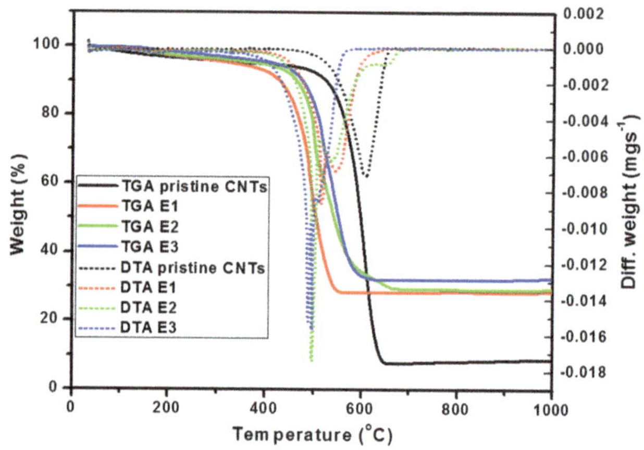

Fig. 7 TGA and DTA curves of pristine CNTs and catalyst samples E1, E2, and E3

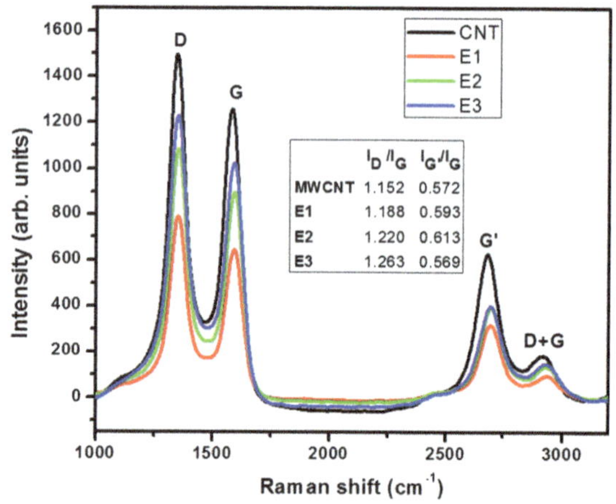

Fig. 6 Raman spectra for pristine CNTs and catalyst samples E1, E2, E3 along with the calculated values of I_D/I_G and $I_{G'}/I_G$

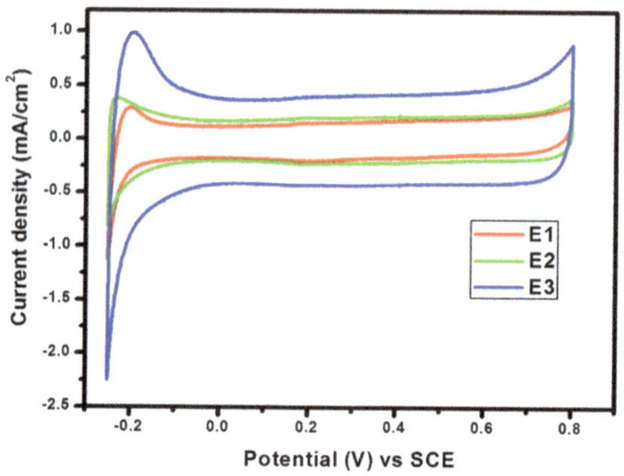

Fig. 8 Cyclic voltammetry curves for catalyst samples E1, E2, and E3

chemical changes occurring due to different pH environments. The deconvoluted XPS curves (by fitting to pure Gaussian peaks) of samples E1, E2 and E3 for carbon (C1 s) and Platinum (Pt4f) are shown in Fig. 4. The peak of elemental carbon is obtained at 284.6 eV. Deconvoluting the carbon 1 s spectrum shows additional shoulder peaks at 285.8, 287.4, 289.1, and 290.7 eVs which corresponds to oxidized or functionalized carbon species.

Although the percentage of graphitic carbon in all samples is quite high, but there is visible contribution of other species also. The presence of sp^3 hybridized carbon depicts the defects which were further confirmed from Raman analysis. As it is clear from the curves, the percentage of functionalized carbon decreases with increasing pH of the synthesizing medium. This is probably because acidic environment contributes more towards the formation of oxidized/functionalized carbon species.

XPS spectrum in Fig. 4b, d, f reveals two Pt peaks, $4f_{7/2}$ and $4f_{5/2}$ at ~70.90 and 74.14 eV, respectively, corresponding to Pt in metallic state [Pt(0)]. Deconvoluting the curves, however, presents us with shoulder peaks with a peak shift of nearly 1.4 eV (for $4f_{7/2}$) and 2.5 eV (for $4f_{5/2}$) corresponding to Pt–OH and Pt–O species, respectively. As it is clear from the curves, the percentage of Pt(0) increases and oxidized Pt decreases with increasing pH of the medium in which the catalyst is prepared. Thus, while in sample E1, 24.2 % Pt is present in oxidized state; it reduces to nearly 11.4 % in sample E3. This result further indicates that alkaline medium facilitates reduction of platinum. Thus, in all three Pt/CNT samples, the metallic state of Pt is predominant, but was found to be highest in E3, i.e., 88.6 %.

Pt/CNT-catalyzed cathode ORR may not be a complete four-electron reaction. There may be slight contribution from two-electron process also [35–37] and hydrogen peroxide formed in the process negatively affects the performance and stability of cathode and electrolyte membrane [38]. The presence of Pt–OH due to incomplete reduction may initiate the two-electron ORR process thus affecting the fuel cell performance.

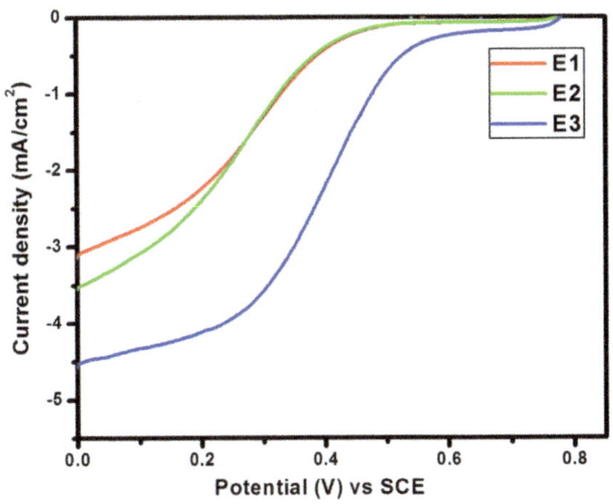

Fig. 9 Linear scan voltammograms for catalyst samples E1, E2, and E3

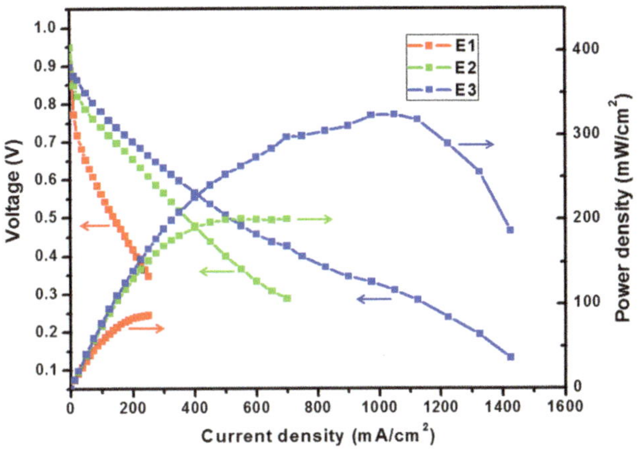

Fig. 10 Comparative current–voltage performance of PEM fuel cell using samples E1, E2, and E3 as catalyst

The XPS data for the entire range have been given in Fig. 5 that shows the presence of C, O, Pt, Ni and Co.

Figure 6 shows the Raman spectra of the different nanocomposite samples. Three bands: D (defect/disorder induced) band, G (graphite like) band, and G′ (second order harmonic to D) band have been identified. Also present is the small D + G band, which requires a defect for its activation [39]. The intensities of the bands were determined by the area under the spectral curve. The intensity of the G band (I_G) has been used as a reference in determining the relative intensities of the D band (I_D) and G′ band ($I_{G'}$).

The relative intensity of D band (ratio I_D/I_G) for pristine CNT sample is nearly 1.15, which indicates a large no. of defects. This is probably due to the small size of the nanotubes with closed ends (as also shown in the TEM images in Fig. 3b, e. The ends contain sp^3 carbon due to pentagon–heptagon pairs and, therefore, contribute to the D band intensity. The incorporation of Pt NPs on the tubes further results in generating stresses in the hexagonal carbon layer that is reflected by the increase in the relative intensity of the D band.

The G′ band is attributed to the interlayer coupling. Its relative intensity increases very sightly in nanocomposite samples as compared to pristine MWCNTs, which indicates that the Pt nanoparticles are attached only on the outer wall of the nanotubes. This probably affects the coupling of only a few surface layers leaving the bulk unchanged.

Figure 7 shows the TGA curves of the Pt/CNT samples. TGA for pristine MWCNT without Platinum addition shows a residue of about 8.5 % which is probably due to the presence of oxides of metal (Co, Ni) used during the CNT synthesis process. For the Pt/CNT samples E1, E2, and E3, the residual content is nearly 31.9, 29.4, and 28.8 %, respectively, indicating that almost entire platinum (i.e., ~20 %) is reduced on the CNT surface. During synthesis of Pt/CNTs, it was observed that the residue left after filtration was colorless. Further addition of $NaBH_4$ also did not lead to any change in color, thus indicating almost complete reduction of platinum on CNT surface. There is an increase in the residual content of the three samples, which should be ideally 28.5 %. This can be explained as follows. Pt will not undergo oxidation when heated in air; however, the oxidized species of Pt (as demonstrated by the XPS data) may result in oxide formation and lead to increase in the residual weight. As the oxidized Pt decreases with increasing pH of the medium in which the catalyst is prepared, similar trend is observed in the residue.

The DTG curves show decrease in the thermal stability of the prepared catalyst samples as compared to pristine MWCNTs. However, the decomposition temperature is much higher than the PEMFC operating temperature.

Figure 8 shows the cyclic voltammetry curves for Pt/CNTs samples prepared in different pH medium. CV measurements of Pt/CNTs samples were conducted in 0.5 M $HClO_4$ solution over a scan range of −0.25 to 0.8 V and exhibit typical characteristic of crystalline Pt electrodes in the hydrogen and oxygen adsorption–desorption regions. The ECSA was calculated from the integrated charge associated with the hydrogen adsorption region which is based on a monolayer hydrogen adsorption on platinum and gives a measure of the HOR. ECSA for E1, E2 and E3 was found to be 102.7, 157.1 and 214.3 m^2/g, respectively. The increase in the ESCA can be correlated to the decrease in the size of the deposited Pt NPs that leads to its effective utilization.

The ORR kinetics of Pt/CNTs samples were investigated by LSV curves as shown in Fig. 9. From the LSV curves, the onset potential for the ORR was measured which gives potential at which the ORR initiates and was found to be highest for sample E3. High mid-wave

Table 2 Electrochemical kinetic parameters for the unit PEM fuel cells using catalyst samples E1, E2 and E3

	E_0 (V)	$j_{0.9}$ (A/cm^2)	b (V/decade)	α
E1	0.903	2.11	0.0954 ± 0.0038	0.1346
E2	0.948	3.41	0.0595 ± 0.0038	0.2155
E3	0.940	4.83	0.0549 ± 0.0065	0.2337

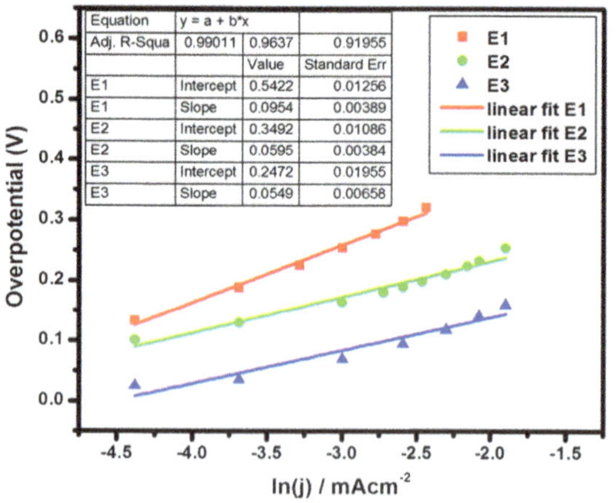

Fig. 11 Tafel plots of the electrochemical reactions with catalyst samples E1, E2, and E3. The tafel slops have been obtained by liner fitting

potential for E3 further reflects low binding energy of the catalyst surface with the reactants which in turn implies fast catalytic activity as the catalyst surface is quickly available for reducing more oxygen.

The half-cell results were further reflected in the performance of the unit PEM fuel cell. Figure 10 shows the comparative fuel cell performance with samples E1, E2, and E3 employed as catalyst. From the curves it is clear that there is a marked difference in the fuel cell performance that increases with increasing pH of the medium in which the catalyst is prepared. Maximum power density, i.e., 325 mWcm^{-2} has been achieved with sample E3 (employed as catalyst), much higher than that observed for catalyst samples E1 (87 mWcm^{-2}) and E2 (201 mWcm^{-2}). The obtained power density is much higher as compared to that of commercial catalyst as reported elsewhere [26, 40, 41]. Since the entire cell parameters are kept constant except the catalyst, the results have been explained by detailed analysis of the activation region of the polarization curves [42].

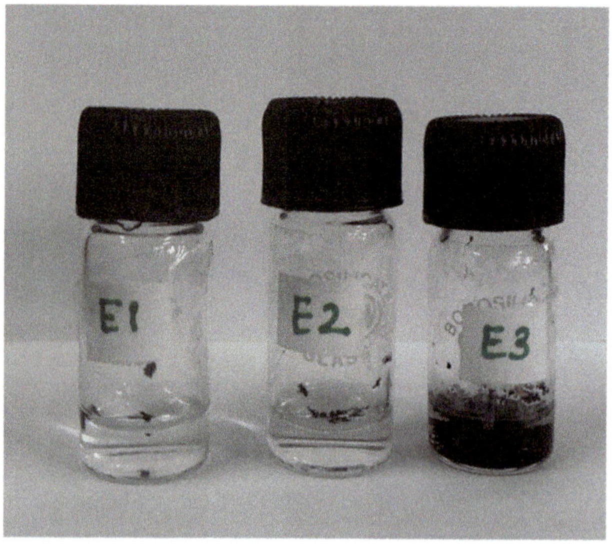

Fig. 13 Photograph with samples E1, E2 and E3 dispersed in water and ultrasonicated for 1 min, showing hydrophilicity of sample E3

The cell potential (E) versus current density (j) data for different catalyst systems can be analyzed by fitting the data to equation:

$$E = E_{eq} - \Delta V_{act} - \Delta V_{ohmic} - \Delta V_{diff} \quad (4)$$

where E_{eq} is the equilibrium voltage and ΔV_{act} is the activation overpotential, ΔV_{ohmic} is the over potential due to ohmic resistances in the cell; and ΔV_{diff} is the concentration polarization due to poor mass transport. We are mainly concerned with ΔV_{act} due to slow electrode kinetics given by

$$\Delta V_{act} = (RT/\alpha nF)\ln(j/j_0) = a\log j - a\log j_0 \quad (5)$$

The above Eq. (5) is linear of the form $y = ax + b$, where a (= RT/αnF) is the Tafel slope, j is the current

Fig. 12 Plausible reaction mechanism when the reduction is carried out in alkaline mediums

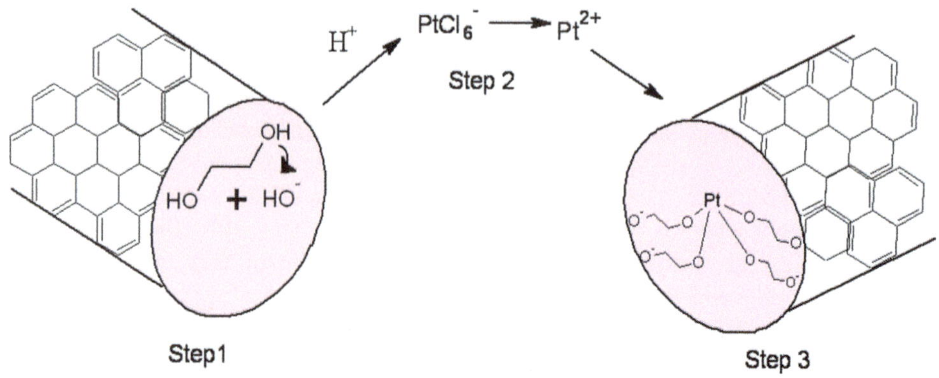

density, j_o is the exchange current density, n is the number of electrons transferred per mole of reactant, F is Faraday's constant, and α the charge transfer coefficient represents the fraction of the overpotential assisting the reaction. Thus, slow reaction kinetics lead to an offset in open circuit potential E_0 by an amount

$$E_0 - E_{eq} = a\log j_0 \quad (6)$$

Thus Eq. (4) can be written as:

$$E = E_0 - a\log j - \Delta V_{ohmic} - \Delta V_{diff} \quad (7)$$

The electrode kinetic parameters for the cells are summarized in Table 2, while the tafel plots are shown in Fig. 11. The value of $j_{0.9}$ represents the current density at the cell potential of 0.9 V, known as kinetic current density, while α is calculated from the slope of the tafel plots. The reason for presenting j at a particular cell potential (0.9 V) instead of j_0 is because even a slight variation in tafel slope can cause anomalous variation in the values of j_0 [43, 44]. The values of charge transfer coefficient (α) and $j_{0.9}$ increase for samples E1 to E3, whereas the tafel slopes of the polarization curves decrease indicating feasibility of the reaction. Similar values of the kinetic parameters have also been reported elsewhere [42, 45, 46]. The kinetics of the reaction can be explained by the probable mechanism. Addition of OH- ions (increasing basicity) along with EG has the tendency to catalyze the entire reduction reaction as shown in Fig. 12. This is because OH^- ions has the tendency to release protons from EG (step 1).

The $OH–(CH_2)_2–O^-$ ion thus produced will bind on the CNT surface for stabilizing its charge. This will render the CNT surface hydrophilic. To confirm the above hypothesis 2 mg of each sample (E1, E2, and E3) was dispersed in 5 ml of water and ultrasonicated for 1 min. Figure 13 shows that sample E3 dissolves in water, while E1, and E2 do not. These additional OH^- ions will also be available to reduce Pt which can bind with EG (forming square planar complexes) which is already on CNT surface (step 3). Thus, the reaction will become much more feasible as compared to when it is carried out in acidic medium (where it will be difficult to release proton). Further coordinate bonds will be stronger leading to stability of the catalyst formed.

Conclusions

Multiwalled carbon nanotube supported platinum nanoparticles prepared by reflux heating are smaller in size and well dispersed. During synthesis, high viscosity of ethylene glycol can effectively stabilize the diffusion of Pt nanoparticles. The catalytic activity of CNT-supported Pt nanoparticles greatly depends on the particle size and distribution and its oxidation state which in turn is largely affected by the pH of the synthesizing medium. The I–V performance of unit PEM fuel cell shows a peak power density of 325 mW/cm^2 with catalyst prepared in alkaline medium, an increase of nearly 61 and 250 % as compared to that obtained while employing catalyst prepared in neutral and acidic mediums, respectively, and tested under similar conditions.

Acknowledgments The authors are grateful to Prof. R.C. Budhani, Director, NPL, New Delhi for his support, encouragement and permission to publish the results. Thanks are also due to Dr. Vidyanand Singh and Dr. Govind for carrying out the TEM and XPS studies. One of the authors CG wants to thank CSIR, India for financial support. These studies have been carried out under the CSIR project entitled "Development of Advanced Materials for Next Generation Energy Efficient Devices (D-NEED)".

References

1. Haile, S.M.: Fuel cell materials and components. Acta. Mater. **5**, 5981–6000 (2003)
2. Park, S., Vohs, J.M., Gorte, R.J.: Direct oxidation of hydrocarbons in a solid-oxide fuel cell. Lett. Nat. **404**, 265–267 (2000)
3. Steele, B.C.H., Heinzel, A.: Materials for fuel-cell technologies. Nature **414**, 345–352 (2001)
4. Hall, P.J., Bain, E.J.: Energy-storage technologies and electricity generation. Energy Policy **36**, 4352–4355 (2008)
5. Wind, J., Späh, R., Kaiser, W., Böhm, G.: Metallic bipolar plates for PEM fuel cells. J. Power Sour. **105**, 256–260 (2002)
6. Mathur, R.B., Maheshwari, P.H., Dhami, T.L., Tandon, R.P.: Characteristics of the carbon paper heat-treated to different temperatures and its influence on the performance of PEM fuel cell. Electrochim. Acta **52**, 4809–4817 (2007)
7. Zhang, B., Chen, L.J., Ge, K.Y., Guo, Y.C., Peng, B.X.: Preparation of multiwall carbon nanotubes-supported high loading platinum for vehicular PEMFC application. Chin. Chem. Lett. **16**, 1531–1534 (2005)
8. Zhang, W., Sherrell, P., Minett, A.I., Razal, J.M., Chen, J.: Carbon nanotube architectures as catalyst supports for proton exchange membrane fuel cells. Energy Environ. Sci. **3**, 1286–1293 (2010)
9. Shao, Y., Yin, G., Gao, Y., Shi, P.: Durability study of Pt/C and Pt/CNTs catalysts under simulated PEM fuel cell conditions. J. Electrochem. Soc. **153**, A1093–A1097 (2006)
10. Maruyama, J., Abe, I.: Enhancement effect of an adsorbed organic acid on oxygen reduction at various types of activated carbon loaded with platinum. J. Power Sour. **148**, 1–8 (2005)
11. Shao, Y., Liu, J., Wang, Y., Lin, Y.: Novel catalyst support materials for PEM fuel cells: current status and future prospects. J. Mater. Chem. **19**, 46–59 (2009)
12. Shao, Y., Yin, G., Zhang, J., Gao, Y.: Comparative investigation of the resistance to electrochemical oxidation of carbon black and carbon nanotubes in aqueous sulfuric acid solution. Electrochim. Acta **51**, 5853–5857 (2006)

13. Steiner, N.Y., Mocotéguy, P.H., Candusso, D., Hissel, D.: A review on polymer electrolyte membrane fuel cell catalyst degradation and starvation issues: causes, consequences and diagnostic for mitigation. J. Power Sour. **194**, 130–145 (2009)
14. Zhang, J., Wu, J., Zhang, H.: PEM fuel cell testing and diagnosis, pp. 301–305. Elsevier, Newnes (2013)
15. Chen, J., Siegel, J.B., Matsuura, T., Stefanopoulou, A.G.: Carbon corrosion in PEM fuel cell dead-ended anode operations. J. Electrochem. Soc. **158**, B1164–B1174 (2011)
16. Dyer C.K., Moseley P.T., Ogumi Z., Rand D.A.J., Scrosati B.: Encyclopedia of Electrochemical Power Sources, pp. 852–854. Elsevier, Newnes Technology & Engineering (2013)
17. Jung, J., Park, B., Kim, J.: Durability test with fuel starvation using a Pt/CNF catalyst in PEMFC. Nanoscale Res. Lett. **7**, 1–8 (2012)
18. Xing, Y.: Synthesis and electrochemical characterization of uniformly-dispersed high loading Pt Nanoparticles on sonochemically-treated carbon nanotubes. J. Phys. Chem. B **108**, 19255–19259 (2004)
19. Guo, D.J., Li, H.L.: High dispersion and electrocatalytic properties of platinum on functional multi-walled carbon nanotubes. Electroanalysis **17**, 869–872 (2005)
20. Zhang, W., Chen, J., Swiegers, G.F., Ma, Z.F., Wallace, G.G.: Microwave-assisted synthesis of Pt/CNT nanocomposite electrocatalysts for PEM fuel cells. Nanoscale **2**, 282–286 (2010)
21. Chen, W.X., Lee, J.Y., Liu, Z.L.: Microwave-assisted synthesis of carbon supported Pt nanoparticles for fuel cell application. Chem. Commun. **21**, 2588–2589 (2002)
22. Chen, W., Zhao, J., Lee, J.Y., Liu, Z.: Microwave heated polyol synthesis of carbon nanotubes supported Pt nanoparticles for methanol electro-oxidation. J. Mater. Chem. Phys. **91**, 124–129 (2005)
23. Knupp, S.L., Li, W., Paschos, O., Murray, T.M., Snyder, J., Haldar, P.: The effect of experimental parameters on the synthesis of carbon nanotube/nanofiber supported platinum by polyol processing techniques. Carbon **46**, 1276–1284 (2008)
24. Lee, K., Zhang, J., Wang, H., Wilkinson, D.P.: Progress in the synthesis of carbon nanotube- and nanofiber supported Pt electrocatalysts for PEM fuel cell catalysis. J. Appl. Electrochem. **36**, 507–522 (2006)
25. Kim, P., Joo, J.B., Kim, W., Kim, J., Song, I.K., Yi, J.: $NaBH_4$-assisted ethylene glycol reduction for preparation of carbon-supported Pt catalyst for methanol electro-oxidation. J. Power Sour. **160**, 987–990 (2006)
26. Fang, B., Chaudhari, N.K., Kim, M.S., Kim, J.H., Yu, J.S.: Homogeneous deposition of platinum nanoparticles on carbon black for proton exchange membrane fuel cell. J. Am. Chem. Soc. **131**, 15330–15338 (2009)
27. Stevens, D.A., Dahn, J.R.: Electrochemical characterization of the active surface in carbon-supported platinum electrocatalysts for PEM fuel cells. J. Electrochem. Soc. **150**, A770–A775 (2003)
28. Oh, H.S., Oh, J.G., Kim, H.: Modification of polyol process for synthesis of highly platinum loaded platinum–carbon catalysts for fuel cells. J. Power Sour. **183**, 600–603 (2008)
29. Oh, H.S., Oh, J.G., Hong, Y.G., Kim, H.: Investigation of carbon-supported Pt nanocatalyst preparation by the polyol process for fuel cell applications. Electrochem. Acta **52**, 7278 (2007)
30. Bittencourt, C., Hecq, M., Felten, A., Pireaux, J.J., Ghijsen, J., Felicissimo, M.P., Rudolf, P., Kee, W.D., Tendeloo, G.V.: Platinum–carbon nanotube interaction. Chem. Phys. Lett. **462**, 260–264 (2008)
31. Hull, R.V., Li, L., Xing, Y., Chusuei, C.C.: Pt nanoparticle binding on functionalized multiwalled carbon nanotubes. Chem. Mater. **18**, 1780–1788 (2006)
32. Spieker, W.A., Liu, J., Miller, T.L., Kropf, A.J., Regalbuto, J.R.: An EXAFS study of the co-ordination chemistry of hydrogen hexachloroplatinate (IV) Speciation in aqueous solution. App. Catal. A Gen. **232**, 219–235 (2002)
33. Chen, X., Chu, W., Wang, L., Wu, Z.: Geometry of Pt(IV) in H_2PtCl_6 aqueous solution: an X-ray absorption spectroscopic investigation. J. Mol. Struct. **920**, 40–44 (2009)
34. Yoshida, H., Yazawa, Y., Hattori, T.: Effects of support and additive on oxidation state and activity of Pt catalyst in propane combustion. Catal. Today **87**, 19–28 (2003)
35. Perez, J., Gonzalez, E.R., Ticianelli, E.A.: Oxygen electrocatalysis on thin porous coating rotating platinum electrodes. Electrochim. Acta **44**, 1329–1339 (1998)
36. Lima, F.H.B., Ticianelli, E.A.: Oxygen electrocatalysis on ultra-thin porous coating rotating ring/disk platinum and platinum–cobalt electrodes in alkaline media. Electrochim. Acta **49**, 4091–4099 (2004)
37. Rizo, R., Herrero, E., Feliu, J.M.: Oxygen reduction reaction on stepped platinum surfaces in alkaline media. Phys. Chem. Chem. Phys. **15**, 15416–15425 (2013)
38. Wu, J., ZiYuan, X., Martin, J.J., Wang, H., Zhang, J., Shen, J., Wu, S., Merida, W.: A review of PEM fuel cell durability: degradation mechanisms and mitigation strategies. J. Power Sour. **184**, 104–111 (2008)
39. Saito, R., Hofmann, M., Dresselhaus, G., Jorio, A., Dresselhaus, M.S.: Raman spectroscopy of graphene and carbon nanotubes. Adv. Phys. **60**, 413–550 (2011)
40. Andersen, S.M., Borghei, M., Lund, P., Elina, Y.R., Pasanen, A., Kauppinen, E., Ruiz, V., Kauranen, P., Skou, E.M.: Durability of carbon nanofiber (CNF) & carbon nanotube (CNT) as catalyst support for proton exchange membrane fuel cells. Solid State Ion. **231**, 94–101 (2013)
41. Gosh, A., Basu, S., Verma, A.: Graphene and functionalized graphene supported platinum catalyst for PEMFC. Fuel Cells **13**, 355–363 (2013)
42. Maheshwari, P.H., Mathur, R.B.: Enhanced performance of PEM fuel cell using MWCNT reinforced carbon paper. RSC Adv. **4**, 22324–22333 (2014)
43. Selvarani, G., Sahu, A.K., Choudhury, N.A., Sridhar, P., Pitchumani, S., Shukla, A.K.: A phenyl-sulfonic acid anchored carbon-supported platinum catalyst for polymer electrolyte fuel cell electrodes. Electrochim. Acta **52**, 4871–4877 (2007)
44. Suárez-Alcántara, K., Rodríguez-Castellanos, A., Dante, R., Solorza-Feria, O.: $Ru_xCr_ySe_z$ electrocatalyst for oxygen reduction in a polymer electrolyte membrane fuel cell. J. Power Sour. **157**, 114–120 (2006)
45. O'Hayre, R.P.: Fuel cell fundamentals. In: Cha, S.W., Colella, W., Prinz, F.B. (eds.) Fuel cells mass transport, pp. 137–168. Wiley, New York (2006)
46. Barbir, F.: PEM fuel cells: theory and practice, pp. 33–72. Elsevier, Burlington (2005)

10

Hydrogen selectivity and carbon behavior during gasoline steam reforming over nano-Al_2O_3 catalysts

Dalia R. Abd El-Hafiz · Mohamed A. Ebiad · Radwa A. El-salamony

Abstract In this work, the influence of metal type (Ni and Co) and preparation method (co-precipitation and impregnation) on catalytic gasoline steam reforming (CGSR) using nano-Al_2O_3 support were studied at different reaction temperature (500–800 °C). The structure and surface properties of fresh and spent catalysts were tested by different characterization tools such as TGA, TPR, XRD, HRTEM, BET surface areas and Raman spectroscopy. Results show that product distribution is dependent on both preparation method and metal type. Co/Al_2O_3 is the most selective catalyst for hydrogen production from GSR reaction; H_2 selectivity reached 70 % over this catalyst at 800 °C. While, when considering hydrogen production and carbon deposition in the meantime, Ni/Al_2O_3 is the best one because of the generation of non-deactivating carbon on its surface. CO conversion represented as $CO_2/(CO + CO_2)$ is an indication of the WGS reaction, over all of the prepared catalysts, WGS reaction was preferred at low temperature. On the other hand, we can show that Ni catalysts favor WGS reaction more than that of co catalyst. Moreover, a higher thermal stability and graphitization degree of deposit carbon were obtained on Ni–Al_2O_3 and Co/Al_2O_3.

Keywords Gasoline reforming · Carbon deposition · Hydrogen production · Nano-Al_2O_3 catalysts

Introduction

Nowadays, the automotive industry pays great attention to catalytic reforming of gasoline (CRG) to produce hydrogen, owing to the existing infrastructure and high power density of gasoline. This hydrogen can be transformed cleanly and efficiently into electricity using fuel cells. CRG has two advantages when used as fuel: first, the higher energy efficiency with respect to the use of internal combustion engine. Second, it has an environmental impact by eliminating the pollutant emissions such as NO_x produced from internal combustion engines.

The main catalytic reactions for hydrogen production from petroleum fuels [1, 2] are steam reforming, partial oxidation, and auto-thermal reforming. Among those methods, catalytic steam reforming (CSRG) shows the highest conversion to H_2 [3, 4]. In order to achieve high H_2 selectivity on a metallic catalyst for the CSRG, the catalysts should have a high C–C bond breaking rate and a low methanation reaction rate. Mono-metallic noble catalysts, such as Pt, depending on the support could demonstrate high H_2 selectivity. However, the high cost of Pt makes it economically infeasible to use [5]. The commercial catalysts used in steam reforming are Ni/Al_2O_3 and Ni/MgO [6, 7]. According to the literature [8–10] Ni is a potential catalyst for the steam reforming reactions because it has high rate of C–C bond breakage, good water gas shift activity, and moderate methanation reaction capacity.

High specific surface area, homogeneous dispersion of metals, design of meso-pore structure, control of metal–support interaction, and also preparation of catalysts in nano-form are other important factors showing better catalytic activity and stability in steam reforming reaction [11]. Nano-sized catalysts are prepared either by impregnation or co-precipitation of metal with the support [12].

However, coking is a major problem associated with reforming of heavy hydrocarbon fuels such as gasoline over different catalysts. It is well known that the origin of carbon deposit could be different using different catalysts

D. R. Abd El-Hafiz (✉) · M. A. Ebiad · R. A. El-salamony
Egyptian Petroleum Research Institute, Cairo 11727, Egypt
e-mail: dalia_epri@yahoo.com

[13]. There are several studies that reported the better stability and suppression of carbon deposition when using cobalt, compared with nickel [14, 15].

Therefore, the aim of the present work was to study the performance of nano-sized Ni– and Co–Al_2O_3 prepared by different methods as catalysts in gasoline steam reforming reaction. The detailed study of the amount and type of carbon deposition will be carried out.

Experimental

Catalyst preparation

Impregnation method

First, nano-sized porous gamma-alumina (γ-Al_2O_3) was prepared by control precipitation method [16] in which two solutions of ammonium bicarbonate [(NH_4HCO_3, 98 %) (Merck)] and (0.066 M) aluminium nitrate [($Al(NO_3)_3{\cdot}9H_2O$, 95 %) (Merck)] were added dropwise to deionized water. The pH of $Al(OH)_3$ gel was adjusted at ~10 using NaOH solution. The precipitate was aged at 70 °C for 3 h, filtered, and washed with deionized water, dried, and then calcined in air at 550 °C for 5 h to produce γ-Al_2O3 powders. Co/Al_2O_3 and Ni/Al_2O_3 samples were prepared by impregnating γ-Al_2O_3 support with aqueous solutions of $Co(NO_3)_2{\cdot}6H_2O$ or $Ni(NO_3)_2{\cdot}6H_2O$, respectively, to produce 10 % metal loading. All samples were calcined at 700 °C for 5 h.

Co-precipitation method

Likewise of γ-Al_2O_3 preparation method the Ni–Al_2O_3 sample was prepared by a control precipitation method using a mixture of $Al(NO_3)_3{\cdot}9H_2O$ and (0.066 M) $Ni(NO_3)_2{\cdot}6H_2O$. The catalysts Co/Al_2O_3, Ni/Al_2O_3, and Ni–Al_2O_3 are typical as Co/Al, Ni/Al and Ni–Al respectively.

Characterization methods

Thermal gravimetric analysis (TGA) was carried out using SETARAM Labsys TG-DSC16 equipment in the temperature range from room temperature up to 1,000 °C under N_2 flow in case in fresh catalysts and O_2 in case of spent one to follow the thermal stability of the prepared catalyst.

X-ray diffraction analysis (XRD) was carried out by Shimadzu XD-1 diffractometer using Cu-target and Ni-filtered radiation, at step size of 0.02° and scan step time 0.4 s. The phase identification was made by comparing to the Joint Committee on Powder Diffraction Standards (JCPDS). The crystallite size of the prepared samples was determined from the XRD peaks using the Scherrer equation.

Transmission electron microscopy (TEM) and energy dispersive X-ray (EDS) spectroscopy take place using a JEOL 2100F TEM at an accelerating voltage of 200 kV. The samples were sonicated for 30 min in ethanol and then a drop of this solution placed on a carbon coated Cu grid.

The textural properties were determined from the N_2 adsorption–desorption isotherms measured at liquid nitrogen temperature (−196 °C) using NOVA2000 gas sorption analyzer (Quantachrome Corporation) system. All of the prepared samples were degassed at 200 °C for 17 h in nitrogen atmosphere before measurement to ensure a dry clean surface. The adsorption isotherm was constructed as the volume adsorbed (Vcm^3/g) versus the equilibrium relative pressure P/P_o.

Temperature programmed reduction (TPR) measurements were carried out to investigate the redox properties (the ease of reducibility of metal oxide) over the resultant materials. The experiments were performed in automatic equipment (Chem BET 3000, Quantachrome). Typically, 100 mg of pre-calcined sample was loaded into a quartz reactor and pretreated by heating under inert atmosphere (20 ml/min nitrogen) at 200 °C for 3 h prior to running the TPR experiment and then cooled down to room temperature in N_2. At that point, the sample was submitted to a constant rate of heat treatment (10 °C/min up to 1,000 °C) in a gas flow (80 ml/min) of the mixture hydrogen/nitrogen (5/95 Vol%) as a reducing gas. A thermal conductivity detector (TCD) was employed to monitor the amount of hydrogen consumption.

Raman spectra were carried out at room temperature, using HR UV 800 confocal scanning spectrometer (Horiba Jobin Yvon) equipped with a Peltier-cooled charge-coupled device. The Raman scattering was excited using a 632.81-nm excitation wavelength supplied by an internal He–Ne laser through an Olympus high-stability BXFM confocal microscope. Patterns were recorded in the 50–1,000 cm^{-1} Raman shift range with a spectral resolution of 0.5 cm^{-1}. Lab SPEC v. 5 software was used for data acquisition and processing.

Catalytic activity

Gasoline steam reforming was carried out at atmospheric pressure in a continuous fixed bed, vertical tubular reactor [17]. 1 g of catalyst diluted with quartz particles (1:1) was used for catalytic tests. Before each run, the catalyst was reduced in hydrogen flow (20 cm^3/min) at 800 °C for 2 h. Water/gasoline mixture (gasoline/H_2O = 1:8 mol/mol) was fed into the evaporator (at 150 °C) by a dosing pump (Model RP-G6; FMI, USA) with a flow rate 0.2 ml/min. Nitrogen carrier gas was fed into the evaporator with a flow rate 40 cm^3/min to uniformly carry the steam, and the steam was then fed to the reactor. The reaction temperature varied from 500 to 800 °C.

The products were analyzed by gas chromatography (GC) (Agilent 6890 plus HP, Varian Natural Gas Analyzer type C model CP-3800). Selectivity values were calculated as the mole percentage of products obtained, excluding water; selectivity of product = (mol of product/total mol of H_2 and carbon-containing products in the outlet) × 100.

Results and discussion

Catalyst characterization

Thermal gravimetric analysis

The thermal behavior of the prepared samples was investigated using TGA and DTA from room temperature up to 1,000 °C (Fig. 1). The curve shows an endothermic process and an exothermic one in parallel with two weight loss steps; an endothermic peak from 30 to 400 °C is due to loss of physically adsorbed water and the hydroxyl of the catalyst's surface. A high temperature exothermic peak may be related to the crystallization and/or the weight loss due to components' loss of the catalysts caused by certain reactions occurring at the high temperature [18]. From the whole weight loss process of catalysts, it can be seen that the profiles of Ni/Al_2O_3 and Co/Al_2O_3 are almost the same and that their weight loss is higher than that of $Ni–Al_2O_3$.

X-ray diffraction analysis

XRD patterns of the prepared catalysts are shown in Fig. 2; generally, the broad features of all diffractro-grams were ascribed to the prepared nano-structured solids. For $Ni–Al_2O_3$ catalyst, only high intensity and broad diffraction peaks at $2\theta = 37.7°$, 44.56° and 65.96° of $NiAl_2O_4$ (JCPDS 00-001-1299) are observed. No Bragg reflections were detected for NiO, which suggests that the co-precipitation method leads to the incorporation of small particle size NiO in Al_2O_3 structure to form the $NiAl_2O_4$ spinel phase [19].

However, XRD pattern of Ni/Al_2O_3 catalyst prepared by impregnation method shows peaks attributed to NiO phase at 2θ 43.29° and 62.77° (JCPDS 00-001-1258) in addition to that for Al_2O_3, which ascribed to the inhomogeneous distribution of the Ni species [19]. Figure 2b shows an enlarged pattern (from 2θ 60° to 70°) to show the difference between Al_2O_3 and $NiAl_2O_4$ peaks, which were difficult to be distinguished due to its broadening. Moreover, the particle size of $NiAl_2O_4$ is smaller for $Ni–Al_2O_3$ (7.2 nm) catalyst than for Ni/Al_2O_3 (8.66 nm) as shown in Table 1.

It is noteworthy that XRD pattern of Co/Al_2O_3 shows different peaks, peaks at 2θ 31.4°, 37° and 59.53° (JCPDS 00-001-1152) were ascribed to cubic Co_3O_4 crystallites [20], in addition to Al_2O_3 peaks. The increase in intensity and sharpness of Al_2O_3 peak at 37° may be due the overlap with the peak assigned to the cubic Co_3O_4 crystallites. No diffraction peak of CoO, cobalt aluminate or other Co species was found as reported by Batista et al. [21]. The average crystal sizes of metals on catalyst surfaces were 17.9 nm for NiO and 26.32 nm for Co_3O_4 in Ni/Al_2O_3 and Co/Al_2O_3, respectively.

Table 1 Textural properties of the prepared samples

Catalyst	BET surface area (m^2/g)	Pore volume (cm^3/g)	Crystal size (metal oxide) (nm)	Crystal size (support) (nm)
Ni–Al	249	0.57	–	7.2
Ni/Al	91	0.26	17.9	8.7
Co/Al	173	0.67	26.3	9.7

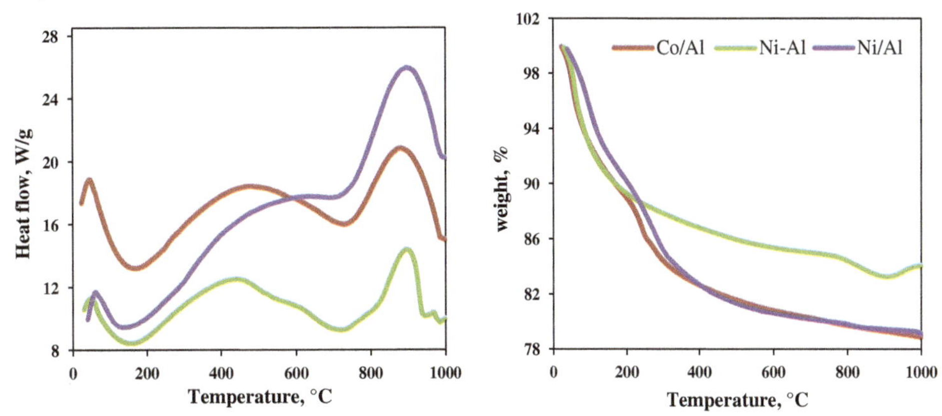

Fig. 1 TGA/DTA profiles of the prepared samples

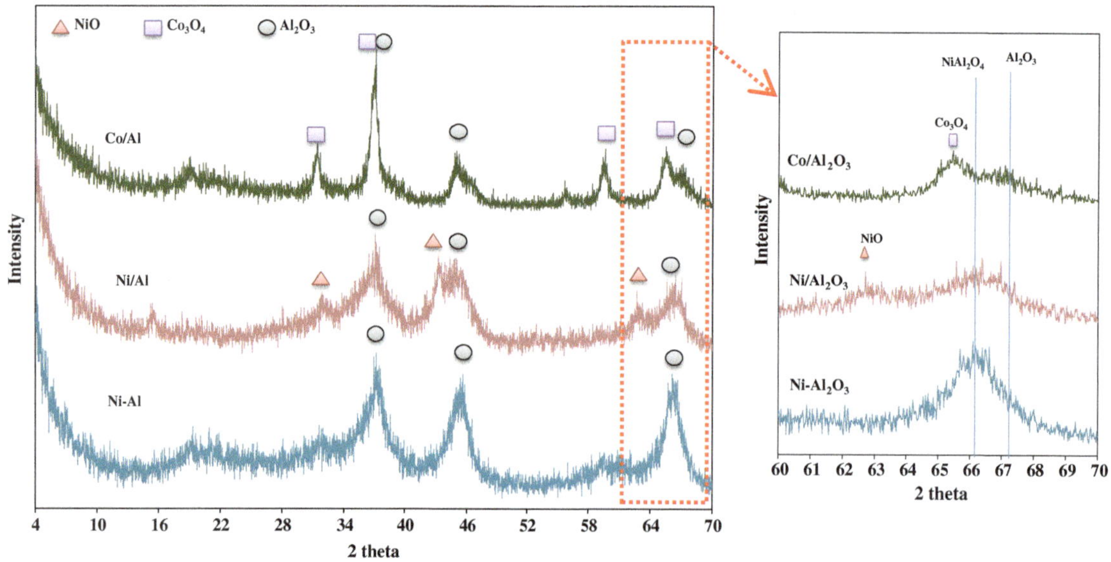

Fig. 2 XRD pattern for the prepared catalyst

High-resolution transmission electron microscopy

The HRTEM photographs obtained from the prepared samples are shown in Fig. 3. Analysis of the photographs for Ni–Al revealed particles with defined dimensions with a slight contraction of the crystal lattice due to partial substitution of Al by Ni, having a grain size ranging from 2 to 4 nm. Although no separate NiO phases were discernible by the XRD of Ni–Al, STEM/EDS elemental mapping (not shown) identified nickel clusters, which indicate that the co-precipitation method leads to incorporation of Ni inside the Al_2O_3 matrix. In Ni/Al and Co/Al systems, two phases are observed, corresponding to the presence of the Al_2O_3 in needle structure, and metal oxide in crystalline structure (indicated by red circle) as expected by XRD data. In the two later catalysts, the mean particle size is estimated to be in the range 10–14 and 18–22 nm, respectively.

Surface areas

The nitrogen adsorption desorption isotherms of the prepared samples are shown in Fig. 4. Ni–Al sample exhibited IV-type isotherm with H2-type hysteresis loop, typical characteristics of mesoporous materials [22] as shown by BJH pore size distributions, and T-plot represented in Fig. 5, which shows the meso-pores material with narrow pore size distributions centered at 4 nm. On the other hand, Ni/Al and Co/Al samples show type II isotherm with H3 hysteresis loop which attributed to slit-shaped pores or plate-like particles with spaces between the parallel plates.

The values of the specific BET surface area and total pore volumes of the prepared samples are listed in Table 1. Compared with the Ni/Al_2O_3 prepared by impregnation method, $Ni–Al_2O_3$ catalyst shows the much higher specific surface area and pore volume, this because in co-precipitation method, Ni species can be uniformly dispersed. Moreover, the inhomogenous distribution of nickel species in Ni/Al catalyst led to blocking of some pores and decrease specific surface area. On the other hand, the impregnation by Co has little effect on the loss of surface area of Al_2O_3 (173 m^2/g), this may be attributed the good dispersion of Co metal on alumina surface.

Temperature programmed reduction (TPR)

Continuing with the idea revealed by Chen et al. [23], that the strong metal–support interaction can effectively suppress the formation of deposit carbon, the TPR results obtain information about the effect of metal type and preparation method on metal–support interaction. The TPR profile for $Ni–Al_2O_3$ catalyst (Fig. 6), shows only one broad peaks with a maximum at about 850 °C, which mainly due to the reduction of NiO strongly interacted with the support by the formation of amorphous nickel aluminate spinel structure ($NiAl_2O_4$) as shown from XRD data [19]. This strong metal–support interaction can restrict the agglomeration of active metal and avoid metal sintering during the reaction.

The TPR profile of Ni/Al_2O_3 catalyst (Fig. 6) shows three distinct regions of temperature, the weakest peak centered at about 304 °C assigned to the reduction of free NiO. The second broad peak at 415 °C was an indication of surface NiO species weakly interacting with the support [24, 25]. The third peak, at about 712 °C, related to the

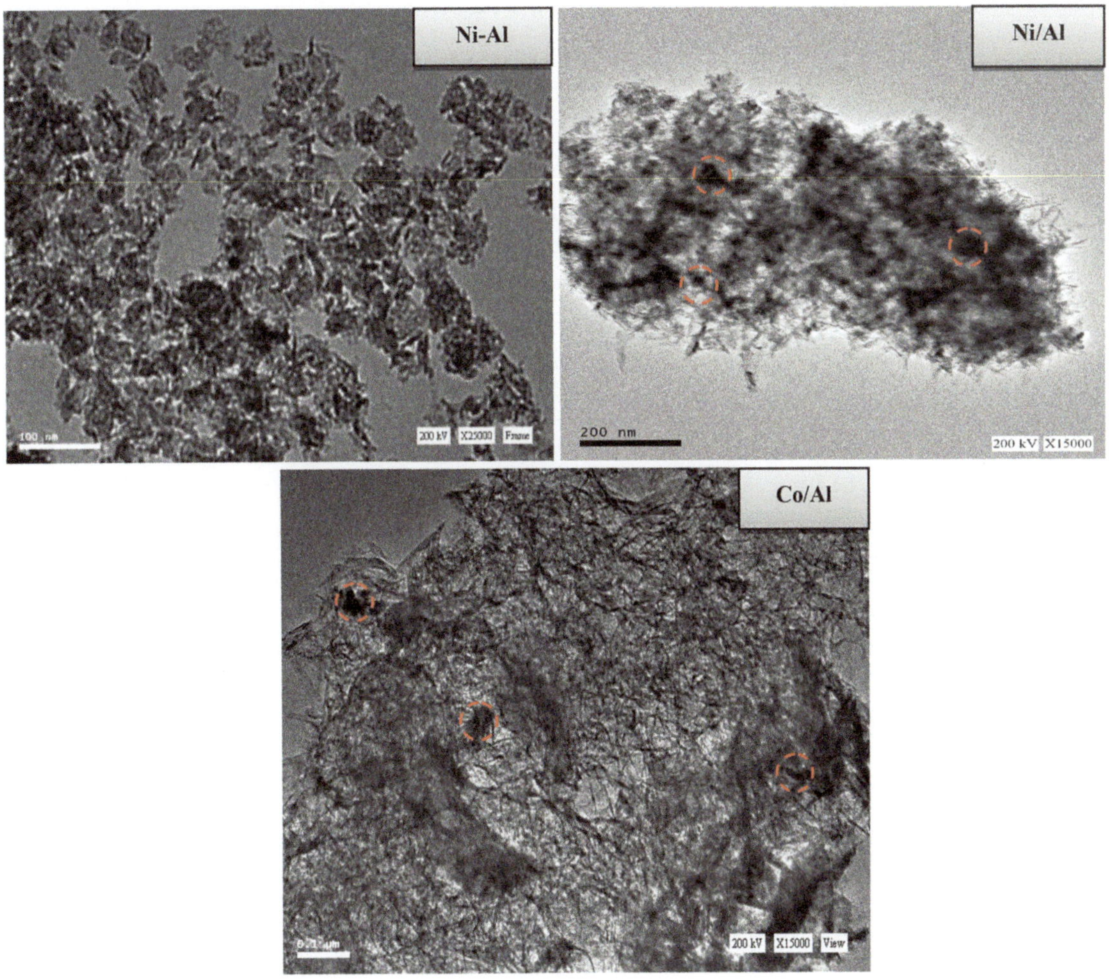

Fig. 3 TEM image of the prepared catalysts

reduction of the stoichiometric and non-stoichiometric nickel aluminate ($NiAl_2O_4$), which are more difficult to reduce [26]. Those different reduction temperatures were related to the inhomogeneous distribution of the Ni species in the Al_2O_3 support as confirmed by XRD data. So, the catalyst prepared by co-precipitation method shows good Ni dispersion and difficult reduction degree than that prepared by impregnation method.

On the other hand, the TPR profile for Co/Al_2O_3 catalysts shows two H_2 consumption peaks, reflecting a two-step reduction process ($Co_3O_4 \rightarrow CoO \rightarrow Co^0$), as previously described by Jacobs et al. [27]. The reduction of Co_3O_4 to CoO is achieved at 440 °C, and the reduction of CoO to Co^0 occurs around 700 °C (Fig. 2a) [28]. Another small peak at about 840 °C may be due to the small amount of Co species that incorporated (strongly interacted) in the Al_2O_3 structure.

Catalytic activity

Gasoline conversions and product selectivities for steam reforming on the three alumina catalysts are summarized in Table 2. The gasoline used in this work was desulfurized gasoline ($\sim$5 ppm sulfur). The main products of gasoline steam reforming are H_2, CH_4, CO, CO_2, C_2H_4, C_2H_6, C_3H_8, C_4H_{10}, C_5H_{12}, and other heavy ends. Several reactions can occur during gasoline steam reforming depending on the reaction conditions [29, 30]; those reactions could be steam reforming reaction (reaction 1), water gas shift reaction (reaction 2), Methanation of CO and CO_2 (reactions 3, 4) and steam reforming of methane (reverse of reaction 3).

$$\text{Steam reforming (SR)}: \; C_nH_m + n\,H_2O \rightarrow n\,CO + (n + m/2)\,H_2 \quad (1)$$

$$\text{Water gas shift (WGS)}: \; CO + H_2O \rightarrow CO_2 + H_2 \quad (2)$$

$$\text{Methanation of CO}: \; CO + 3H_2 \leftrightarrow CH_4 + H_2O \quad (3)$$

$$\text{Methanation of } CO_2: \; CO_2 + 4H_2 \rightarrow CH_4 + 2H_2O \quad (4)$$

Figure 7 shows that the higher carbon product (heavy ends) overall reaction temperature is lower on Co/Al_2O_3

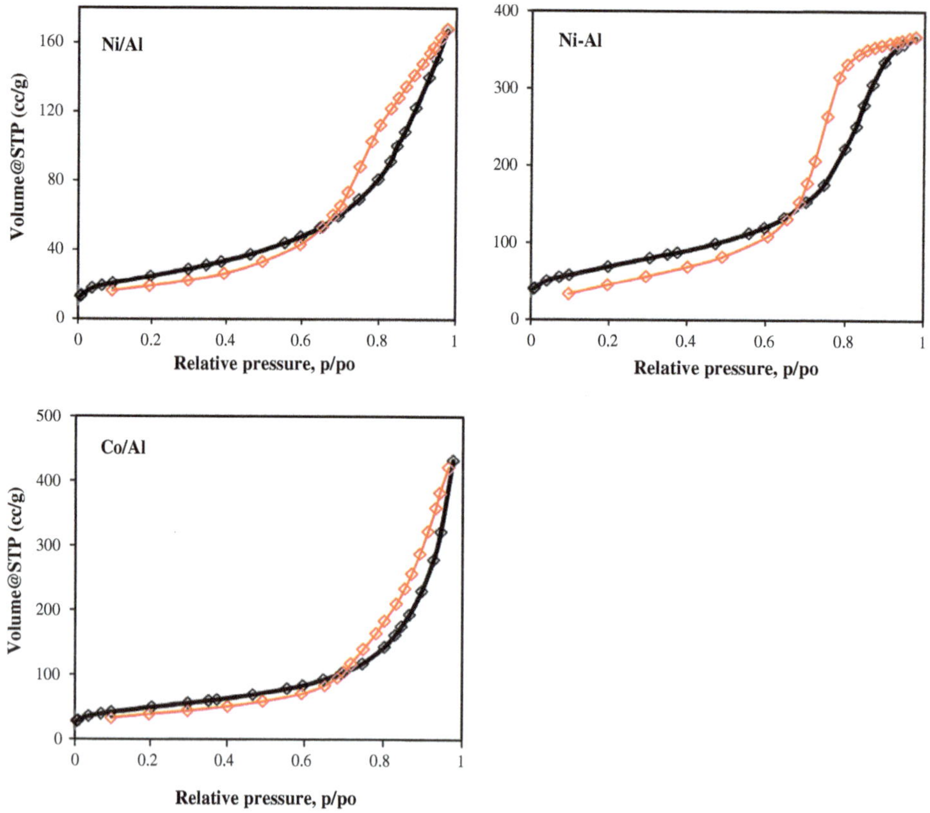

Fig. 4 Nitrogen adsorption–desorption isotherms of the prepared samples

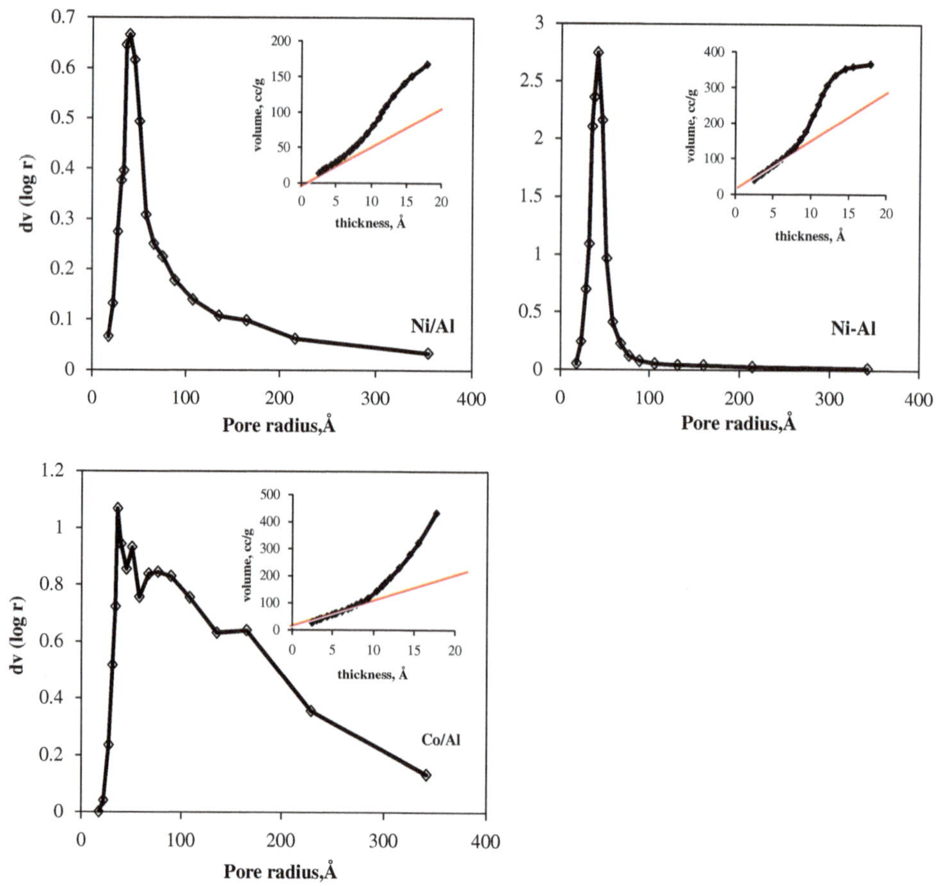

Fig. 5 Pore size distributions and T-plot of the prepared samples

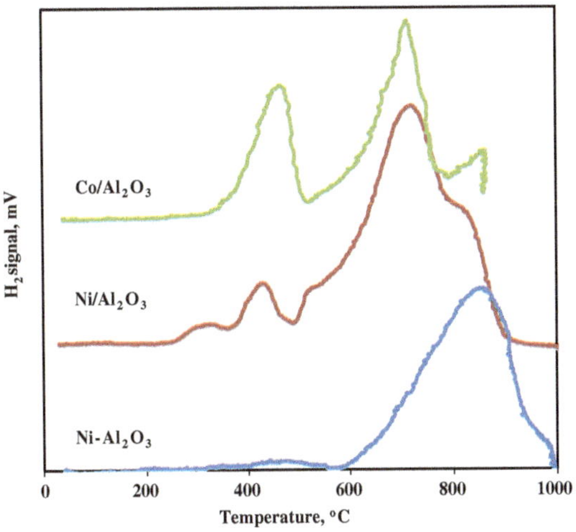

Fig. 6 TPR pattern of the prepared catalysts

than on Ni–Al$_2$O$_3$ than Ni/Al$_2$O$_3$, which indicates the higher catalytic activity of Co/Al$_2$O$_3$ catalyst in gasoline reforming at lower reaction temperature. The higher catalytic activity of Ni–Al$_2$O$_3$ than that of Ni/Al$_2$O$_3$ may be attributed to the highly homogeneous dispersion and small particle sizes of Ni species in case of catalyst prepared by co-precipitation method, which enhances the reaction efficiency in gasoline reforming.

The product distribution as a function of temperature on all catalysts is compared in Fig. 8. Generally, hydrogen and methane are the major products, while the CO and CO_2 exhibited the lowest selectivity.

The selectivity to H_2 increases as the temperature increases over all catalyst to reach maximum value of 69.88 % at 800 °C over a Co/Al catalyst. Therefore, it is effective to enhance the selectivity to H_2 using Co instead

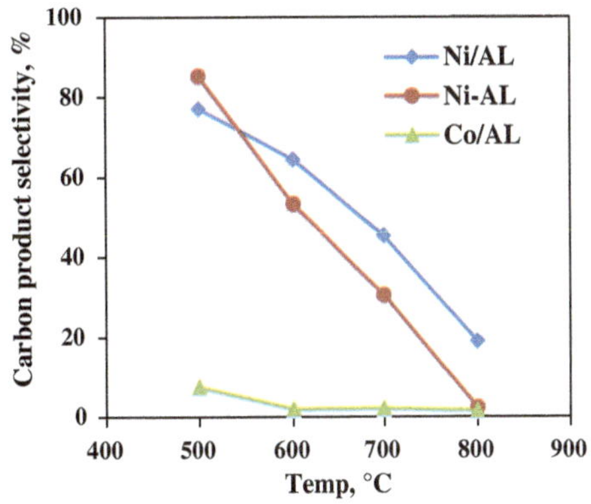

Fig. 7 Selectivity to carbon produced

of Ni as catalyst for gasoline steam reforming. Moreover, the Ni–Al$_2$O$_3$ catalyst is more selective to H_2 than Ni/Al$_2$O$_3$ catalyst owing to the good Ni desperation resulted from co-precipitation method as shown by XRD data. So, we can clearly find that the product distribution of steam reforming of gasoline is dependent on both preparation method and metal type.

Figure 9 shows the value of $H_2 + CO$ over the prepared catalysts at all temperature ranges; from this figure we can show that the steam reforming reaction is favored over a Co/Al2O3 catalyst than the other catalysts.

Comparing the CH_4 and the CO_2 formation during GSR reaction over different catalysts we can conclude that the reaction mechanism are different over Ni catalysts than that over Co one. From Fig. 10, Co/Al is the most selective catalyst to methane production at lower reaction temperature, while, as the temperature increase the selectivity to methane decrease to reach the lower value 10.15 % at 800 °C. This conduct with low amount

Table 2 Product distribution from GSR

Temp (°C)	Product selectivity (mol %)											
	Ni–Al$_2$O$_3$				Ni/Al$_2$O$_3$				Co/Al$_2$O$_3$			
	500	600	700	800	500	600	700	800	500	600	700	800
Hydrogen	0.82	15.24	20.28	49.75	10.82	13.11	19.22	36.48	23.64	29.58	44.40	69.88
Methane	8.16	29.54	48.55	45.25	9.16	17.03	30.97	37.09	65.44	67.29	39.85	10.15
CO	3.82	1.79	0.71	2.53	1.85	0.84	1.87	6.18	2.72	1.25	13.03	18.14
CO_2	2.35	0.57	0.30	0.50	1.38	4.86	2.89	1.57	1.14	0.33	0.97	0.59
Ethane	2.78	7.68	4.56	0.43	3.07	3.99	2.65	0.86	0.03	0.01	0.00	0.01
Ethylene	3.79	14.90	18.49	0.88	3.74	11.06	22.69	9.75	0.01	0.01	0.00	0.01
Propane	4.29	12.28	4.12	0.08	5.04	9.15	9.45	1.26	0.02	0.01	0.00	0.02
i-Butane	3.15	0.19	0.02	0.05	2.92	0.24	0.34	1.09	0.04	0.01	0.01	0.00
n-Butane	1.39	6.78	1.41	0.04	5.63	7.93	5.97	1.47	0.14	0.06	0.08	0.04
i-pentane	24.98	3.21	0.29	0.05	0.65	12.27	1.47	1.11	0.09	0.03	0.02	0.12
n-Pentane	8.40	1.34	0.13	0.02	15.79	4.71	0.62	0.59	0.06	0.02	0.01	0.06
Hexane and heavy end	36.07	6.48	1.15	0.42	39.97	14.80	1.85	2.54	6.67	1.41	1.62	0.98

Fig. 8 Product distribution of gasoline steam reforming over the prepared catalysts

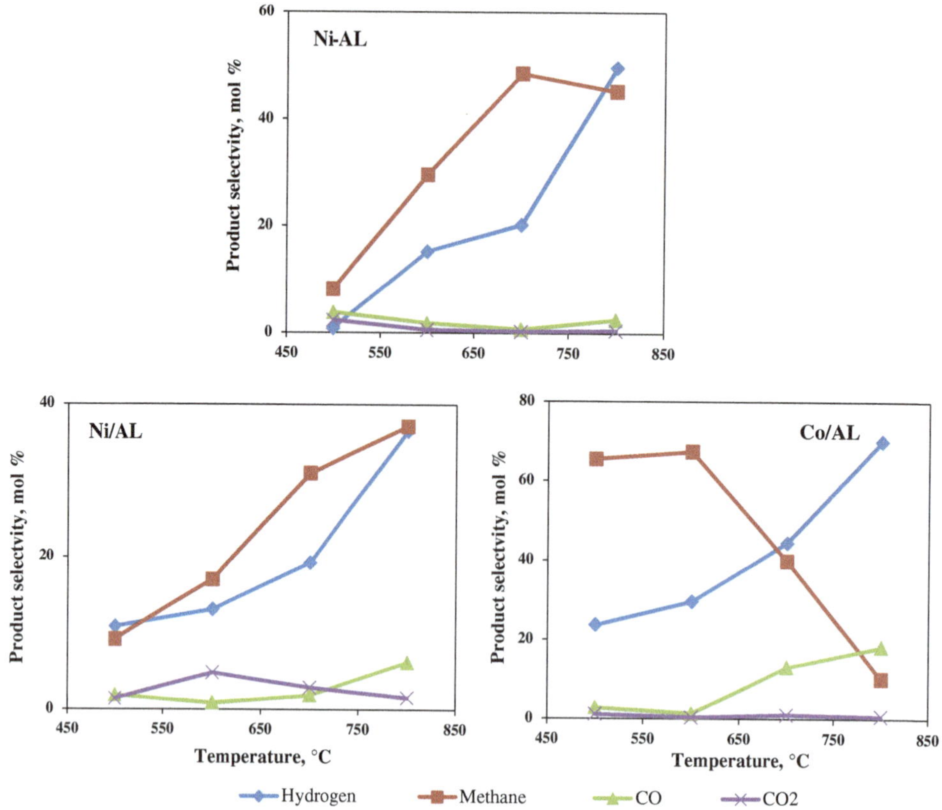

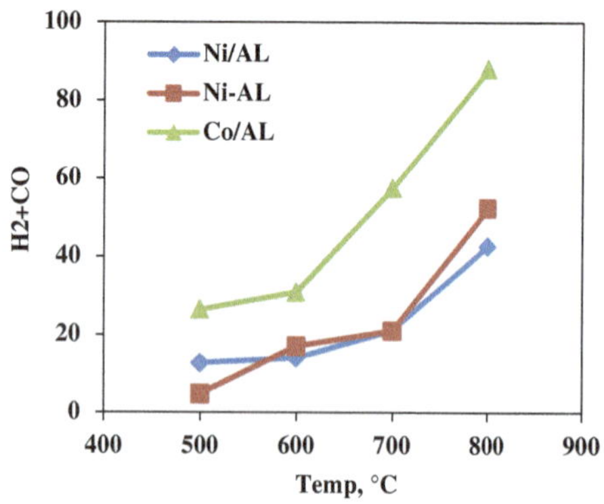

Fig. 9 $H_2 + CO$ over the prepared catalysts

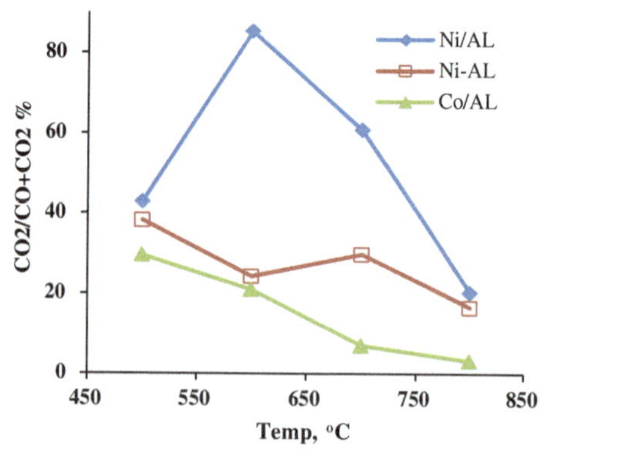

Fig. 11 Effect of reaction temperature on CO conversion (or WGSR)

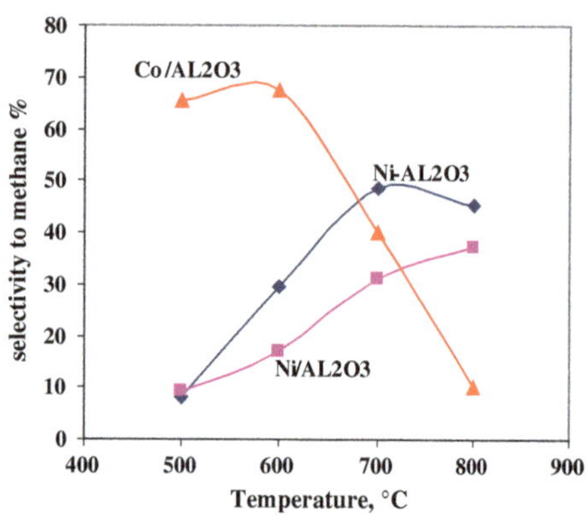

Fig. 10 Methane selectivity over the prepared catalysts

of CO_2 at higher temperature reflecting the CO_2 reforming of methane can take place over this catalyst (reaction 5) [31].

$$CH_4 + CO_2 \rightarrow 2CO + 2H_2 \qquad (5)$$

Figure 11 shows the $CO_2/(CO + CO_2)$ ratio, which represents the CO conversion for the WGS reaction. All of the prepared catalysts favor WGS reaction at low reaction temperature. On the other hand, we can show that Ni catalysts favor WGS reaction more than that of Co catalyst. However, Ni/Al_2O_3 is a good GSR catalyst not only to produce H_2 by SR but also to decrease the CO ratio by WGS reaction. This also explains why Ni/Al_2O_3 catalyst

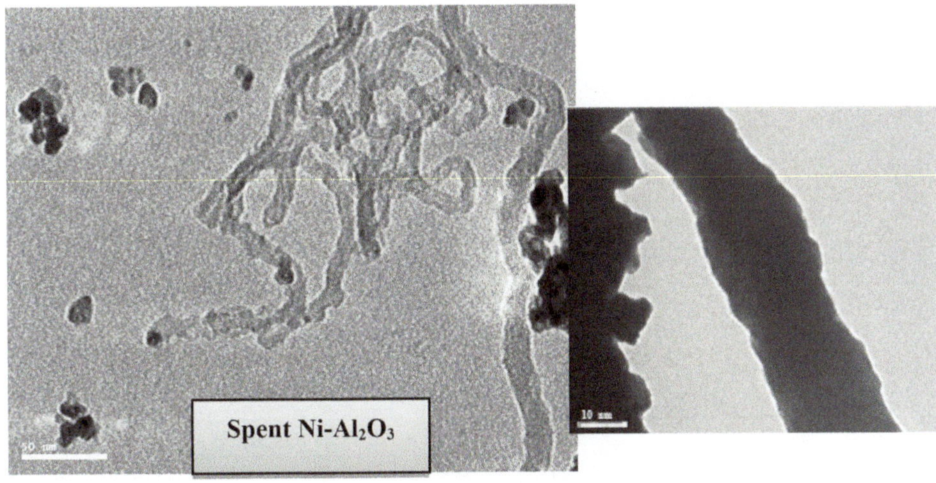

Fig. 12 TEM image of spent Ni–Al_2O_3 catalyst

shows high hydrogen yield at a lower temperature than Ni–Al_2O_3 catalyst.

From the practical application point of view, the use of gasoline as fuel in fuel cells needs high H_2 yield with very low amount of CO (<2 ppm). So, Co–Al_2O_3 catalyst can be used as a primary catalyst to produce high H_2 yield, while Ni/Al_2O_3 catalyst can be used as a secondary catalyst for conversion of CO to CO_2 and production of more hydrogen with low CO by WGS reaction.

Characterization of spent catalysts

Generally, the stability of any catalysts depends on the degree of metal sintering at high temperature and/or the carbon deposited behavior, which can block the active center leading to the decrease of catalytic activity. So, in this section we studied the characterization of the three catalysts after 10 h of GSR reaction.

High-resolution transmission electron microscopy

The three catalysts were analyzed by TEM to gain a deeper insight into the type of carbon deposited on the surface. TEM images for Ni/Al_2O_3, Ni–Al_2O_3 and Co/Al_2O_3 catalysts after 10 h of gasoline steam reforming reaction are presented in Fig. 12. The catalysts show coke formation with different morphologies. The image of Ni–Al_2O_3 reveals that the carbons are not an internal empty tube but opaque nano-rod. However, TEM images of the spent Ni/Al_2O_3 catalysts reveal the formation of amorphous carbon in addition to CNTs with Ni particles at the top of the tubes (indicated by the red circle in Fig. 13). In the case of Co/Al_2O_3 catalyst (Fig. 14), Web-like micrometer long CNTs partially covering the catalyst particle surface can be clearly observed. The figure also reveals the formation of carbon nano-fiber beside CNTs as confirmed by high oxidation temperature in thermal analysis data.

X-ray diffraction analysis

Figure 15 shows the XRD pattern of spent catalysts after gasoline steam reforming. Ni–Al_2O_3 catalyst shows strong peaks at 2 = 26.5°, 42.9° and 53.2° representing the (002), (100) and (004) graphitic basal plane reflection in the form of CNTs [ref. JCPDS card (00-058-1638)]; the peak assigned $NiAl_2O_4$ was also observed. It is worth noting that, although the catalyst was reduced before reaction there is not any peak assigned to the presence of Ni^0 metal was observed due to the absence of free NiO in the fresh catalyst and the difficult reducibility of $NiAl_2O_4$ spinel under the reaction condition. Moreover, the graphite carbon phase is prominent compared to the peaks of Al_2O_3 support.

For Ni/A_2O_3, $NiAl_2O_4$ was observed in addition to Ni^0 (at 2θ reflections of 44.5°, 51.7°) and the peak attributed to NiO phase completely disappeared. The sharpness and high intensity of Ni^0 peaks may be indicating that the Ni metal sintering is unavoidable for this catalyst under the reaction conditions.

On the other hand, the small intensity of all peaks in Co/Al_2O_3 catalyst may be attributed to the large amount of carbon deposited on the metal site and/or the pillaring formation.

Thermo-gravimetric analysis

Table 3 shows the amount of carbon deposited per gram catalysts. Ni/Al_2O_3 catalyst seems to show a better resistance to coke formation, or promoting the gasification of the carbon formed on the catalyst surface, followed by Ni–Al_2O_3 and Co/Al_2O_3 catalysts. This result is greatly consistent with the catalytic activity toward CO_2 production, indicating that the current reactions (carbon gasification) primarily depend on the surface metal and also on the preparation methods.

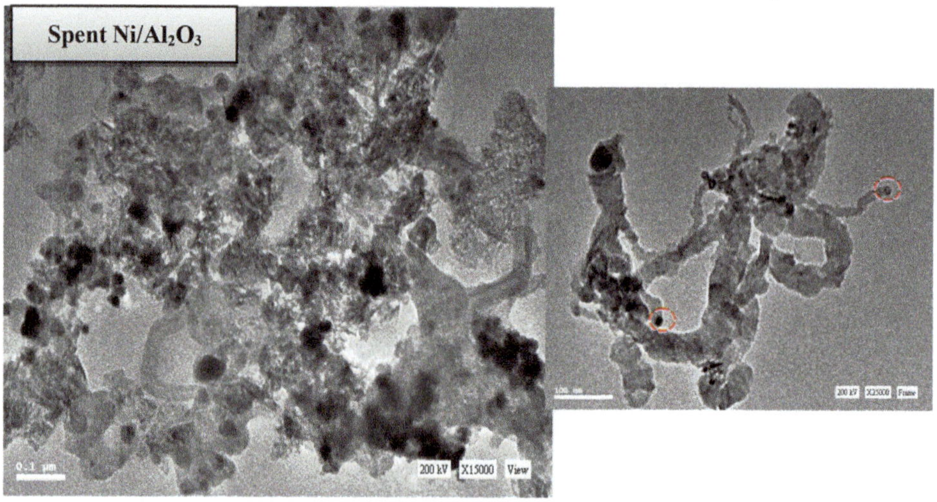

Fig. 13 TEM image of spent Ni/Al_2O_3 catalyst

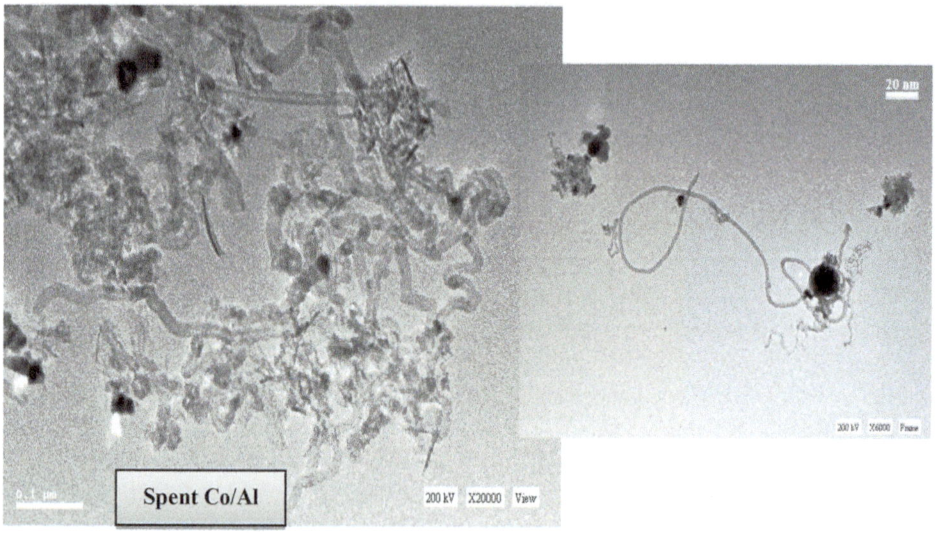

Fig. 14 TEM image of spent Co/Al_2O_3 catalyst

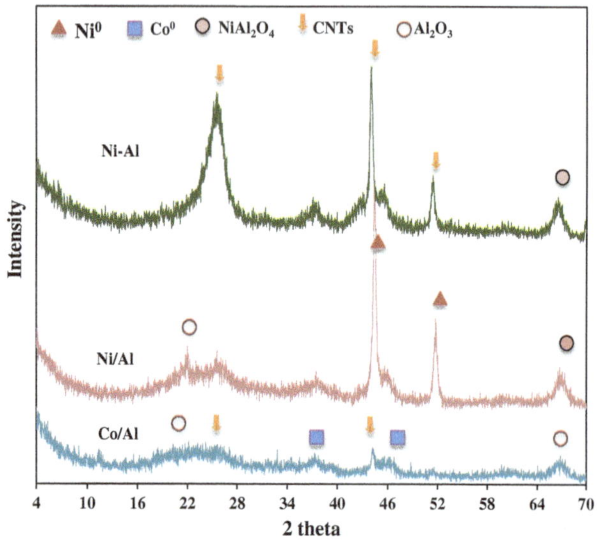

Fig. 15 XRD pattern for the spent catalysts

Table 3 The amount of carbon formed (g/g catalysts)

Catalysts	Weight loss (g/g catalysts)
Ni/Al_2O_3	0.37
$Ni–Al_2O_3$	0.76
Co/Al_2O_3	0.85

As shown in Fig. 16, the three catalysts showed different behavior toward weight loss. A single-step degradation was observed over a Ni–Al2O3 catalyst at temperature above 600 °C indicating the deposition of carbon in the graphitic form [32, 33], while in case of Ni/Al_2O_3, most of carbon formed in the form of mono atomic carbon and filamentous coke (lower than 550 °C). The mono atomic carbon can be oxidized easy at low temperature, while the filamentous carbon is more stable and can be oxidized at higher temperatures [34]. This indicates that the

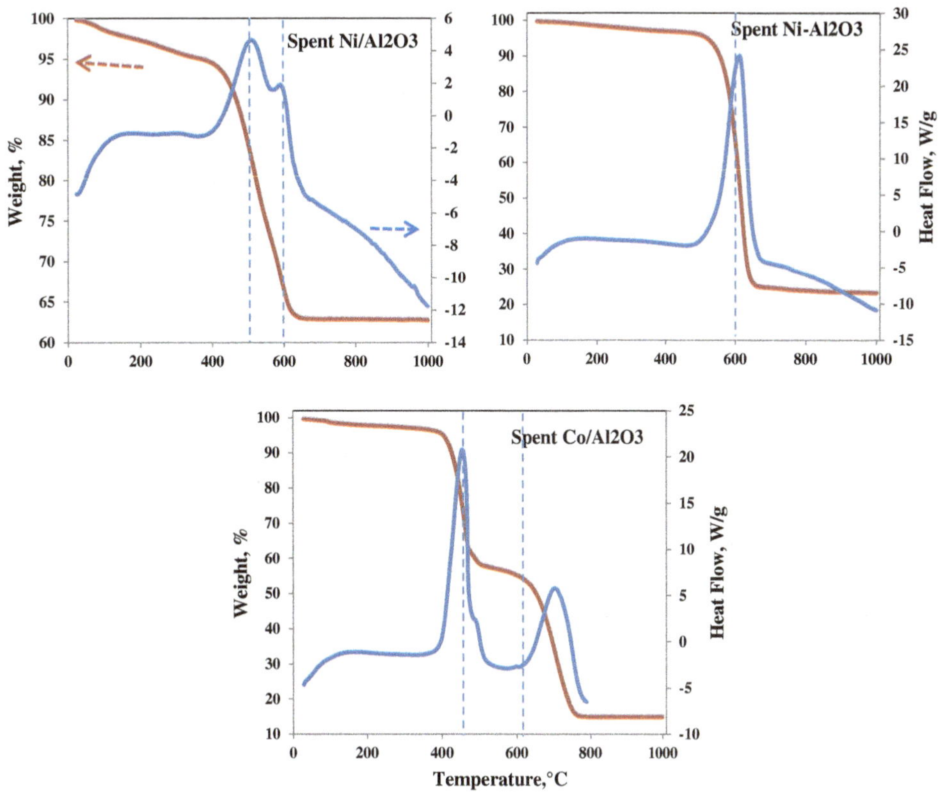

Fig. 16 TGA profile for nano-Al_2O_3 catalyst after 10 h gasoline steam reforming

preparation of Ni–Al_2O_3 by co-precipitation method enhances the graphitization degree of the deposit carbon.

However, in Co/Al_2O_3 catalyst, about 57.74 % of deposit carbon was in the form of mono atomic carbon and filamentous coke (lower than 550 °C). And 42.26 % of deposit carbon was carbon in the high graphitic form (higher than 600 °C), the higher oxidation temperature of this peak (700 °C) is due to the formation of carbon nano-fibers beside the CNTs [35], which is in a good agreement with the TEM data.

Raman of spent catalysts

Raman spectra for the three spent catalysts are shown in Fig. 17. All spectra had two bands: the first band is between 1,200 and 1,450 cm^{-1} (D-band) which is assigned to the disorder vibration of C–C bond [36], and the second band is between 1,500 and 1,700 cm^{-1} (G-band) assigned to the carbon material with sp^2 orbital structure [37]. The G-band is a graphitic band produced by the high degree of symmetry and order of these graphitic carbon materials; it provides information about the electronic properties of the filamentous carbon and is used to identify well-ordered CNTs [38, 39]. These findings suggest that the carbon species found on the surface of the catalysts tested in gasoline steam reforming consisted of deactivating (encapsulated carbon) and non-deactivating carbon (whisker carbon) [36, 37], whose origins are diverse [40].

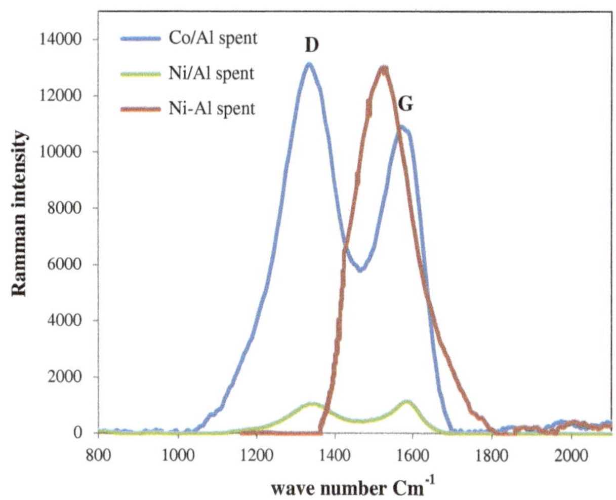

Fig. 17 Raman spectra of the spent catalysts

From Fig. 17 we can note that the intensities of the two bands were very high for Co/Al_2O_3 catalysts compared to Ni/Al_2O_3 one and that the intensity of the two bands was almost symmetric in Ni/Al_2O_3, which means that in case Ni/Al_2O_3 the carbon filaments was coexistent with amorphous carbon. In contrast,the high intensity of the D-band in Co/Al_2O_3 indicates that, this catalyst favors the formation of polymeric encapsulated carbon [26], which leads to rapid deactivation. Moreover, the spectra of Ni–Al_2O_3 catalyst shows that, the intensity of G-band (between 1,400 and 1,700 cm^{-1}) is very high with respect to that of D-band (which may be neglected) indicting the deposition of carbon in the graphitic form as suggested by TGA and TEM data.

Conclusion

In this work, we prepare three different nano-alumina catalysts; the first two catalysts are nickel alumina catalysts prepared by two different methods (co-precipitation and impregnation method); the third catalyst was Co/Al_2O_3 prepared by impregnation method to study the effect of metal type on the performance of the prepared catalysts. It was found that Co/Al_2O_3 shows the most reactivity toward hydrogen production from GSR reaction (highest activity and highest H_2/CO ratio). Moreover, Ni–Al_2O_3 catalyst exhibited promising properties in gasoline steam reforming reaction than Ni/Al_2O_3. This may be attributed to small particle sizes, high surface area, and strong metal–support interaction of the catalyst prepared by the co-precipitation method.

On the other hand, the characteristic properties of the three catalysts after 10 h of GSR shows that graphitic carbon was formed on the surface of the Ni–Al_2O_3 catalyst, and the amorphous one covered most of the surface of the Ni/Al_2O_3 catalyst. Although the formation of $NiAl_2O_4$ plays an important role in decreasing the sintering process, it increases the graphitization degree of carbon deposit. However, over Co/Al_2O_3 catalyst the major amount of carbon was in the form of mono atomic carbon and filamentous coke and the small amount was graphitic carbon in the form of carbon nano-fibers beside the CNTs.

References

1. Cheekatamarla, P.K., Lane, A.M.: Efficient bimetallic catalysts for hydrogen generation from diesel fuel. Int. J. Hydrog. Energy **30**, 1277 (2005)
2. Hoang, D.L., Chan, S.H.: Modeling of a catalytic autothermal methane reformer for fuel cell application. Appl. Catal. A **268**, 207 (2004)
3. Simson, A., Farrauto, R., Castaldi, M.: Steam reforming of ethanol/gasoline mixtures: Deactivation, regeneration and stable performance. Appl. Catal. B **106**, 295 (2011)
4. Villegas, L., Guilhaume, N., Provendier, H., Daniel, C., Masset, F., Mirodatos, C.: A combined thermodynamic/experimental study for the optimisation of hydrogen production by catalytic reforming of isooctane. Appl. Catal. A **281**, 75 (2005)
5. Roy, B., Martinez, U., Loganathan, K., Datye, A.K., Leclerc, C.A.: Effect of preparation methods on the performance of Ni/Al2O3 catalysts for aqueous-phase reforming of ethanol: part I-catalytic activity. Int. J. Hydrog. Energy **37**, 8143–8153 (2012)
6. Gaur, S., Haynes, D.J., Spivey, J.J.: Rh, Ni, and Ca substituted pyrochlore catalysts for dry reforming of methane. Appl. Catal. A **403**, 142 (2011)
7. Gamba, O., Moreno, S., Molina, R.: Catalytic performance of NiePr supported on delaminated clay in the dry reforming of methane. Int. J. Hydrog. Energy **36**, 1540 (2011)
8. Sinfelt, J.H., Yates, D.J.C.: Catalytic hydrogenolysis of ethane over the noble metals of Group VIII. J. Catal. **8**, 82–90 (1967)
9. Grenoble, D.C., Estadt, M.M., Ollis, D.F.: The chemistry and catalysis of the water gas shift reaction: 1. The kinetics over supported metal catalysts. J. Catal. **67**, 90–102 (1981)
10. Vannice, M.A.: The catalytic synthesis of hydrocarbons from H2CO mixtures over the Group VIII metals: V. The catalytic behavior of silica-supported metals. J. Catal. **50**, 228 (1977)
11. Therdthianwong, S., Therdthianwong, A., Siangchin, C., Yongprapat, S.: Synthesis gas production from dry reforming of methane over Ni/Al2O3 stabilized by ZrO2. Int. J. Hydrog. Energy **33**, 991 (2008)
12. Garbarino, G., Campodonico, S., Perez, A.R., Carnasciali, M.M., Riani, P., Finocchio, E., Busca, G.: Spectroscopic characterization of Ni/Al2O3 catalytic materials for the steam reforming of renewables. Appl. Catal. A **452**, 163 (2013)
13. Takanabe, K., Aika, K., Inazu, K., Baba, T., Seshan, K., Lefferts, L.: Steam reforming of acetic acid as a biomass derived oxygenate: bi-functional pathway for hydrogen formation over Pt/ZrO2 catalysts. J. Catal. **243**, 263 (2006)
14. Delacruz, V.M.G., Guez, R.P., Ternero, F., Holgado, J.P., Caballero, A.: Preparation of hydrogen and carbon nanotubes over cobalt-containing catalysts via catalytic decomposition of ethanol. J. Phys. Chem. C **116**, 2919 (2012)
15. Alonso, D.S.J., Juan, J.J., Gomez, M.J.I., Martınez, M.C.R.: Ni, Co and bimetallic Ni–Co catalysts for the dry reforming of methane. Appl. Catal. A **371**, 54 (2009)
16. Parida, K.M., Pradhan, A.C., Das, J., Sahu, N.: Synthesis and characterization of nano-sized porous gamma-alumina by control precipitation method. Mater. Chem. Phys. **113**, 244 (2009)
17. Ebiad, M.A., Abd El-Hafiz, D.R., Elsalamony, R.A., Mohamed, L.S.: Ni supported high surface area CeO2–ZrO2 catalysts for hydrogen production. RSC Adv. **2**, 8145 (2012)
18. Xi-kun, G., Ping-ping, X., Shu-dong, L.: Preparation of Cu/La2O3-ZrO2-Al2O3 catalyst and its catalytic properties for selective reduction of NO. J. Fuel Chem. Technol. **36**, 732 (2008)
19. Hao, Z., Zhu, Q., Jiang, Z., Hou, B., Li, H.: Characterization of aerogel Ni/Al2O3 catalysts and investigation on their stability for CH4-CO2 reforming in a fluidized bed. Fuel Process. Technol. **90**, 113 (2009)
20. Padilla, R., Benito, M., Rodrıguez, L., Serrano, A., Munoz, G., Daza, L.: Nickel and cobalt as active phase on supported zirconia catalysts for bio-ethanol reforming: Influence of the reaction mechanism on catalysts performance. Int. J. Hydrog. Energy **35**, 8921 (2010)
21. Batista, M.S., Santos, R.K.S., Assaf, E.M., Assaf, J.M., Ticianelli, E.A.: High efficiency steam reforming of ethanol by cobalt-based catalysts. J. Power Sour. **134**, 27 (2004)
22. Seo, J.G., Youn, M.H., Song, I.K.: Hydrogen production by steam reforming of liquefied natural gas (LNG) over nickel catalyst supported on mesoporous alumina prepared by a non-ionic surfactant-templating method. Int. J. Hydrog. Energy **34**, 1809 (2009)

23. Chen, J.X., Wang, R.J., Zhang, J.Y., He, F., Han, S.: Effects of preparation methods on properties of Ni/CeO2-Al2O3 catalysts for methane reforming with carbon dioxide. J. Mol. Catal. A Chem. **235**, 302 (2005)
24. Kwak, S.B., Kim, J., Kang, M.: Hydrogen production from ethanol steam reforming over coreeshell structured NixOy, FexOy, and CoxOy-Pd catalysts. Int. J. Hydrog. Energy **35**, 11843 (2010)
25. Zhou, Y., Guo, Q., Zhang, M., Yagi, J., Hatakeyama, L.H., Chen, J., Sakurai, M., Kameyama, H.: A novel catalyst with plate-type anodic alumina supports, Ni/NiAl2O4/g–Al2O3/alloy, for steam reforming of methane. Appl. Catal. A **347**, 200 (2008)
26. de Sousa, F.F., de Sousa, H.S.A., Oliveira, A.C., Junior, M.C.C., Ayala, A.P., Barros, E.B., Viana, B.C., Filho, J.M., Oliveira, A.C.: Nanostructured Ni-containing spinel oxides for the dry reforming of methane: effect of the presence of cobalt and nickel on the deactivation behaviour of catalysts. Int. J. Hydrog. Energy **37**, 3201 (2012)
27. Jacobs, G., Chaney, J.A., Patterson, P.M., Das, T.K., Davis, B.H.: Fischer–Tropsch synthesis: study of the promotion of Re on the reduction property of Co/Al2O3 catalysts by in situ EXAFS/XANES of Co K and Re LIII edges and XPS. Appl. Catal. A **264**, 203 (2004)
28. Jacobs, G., Davis, B.H.: Low temperature water–gas shift: applications of a modified SSITKA–DRIFTS method under conditions of H2 co-feeding over metal/ceria and related oxides. Appl. Catal. A **333**, 192 (2007)
29. Akande, A.J., Idem, R.O., Dalai, A.K.: Synthesis, characterization and performance evaluation Ni/Al2O3 catalysts for reforming of crude ethanol for hydrogen production. Appl. Catal. A **287**, 159 (2005)
30. Rabe, S., Vogel, F., Truong, T.B., Shimazu, T., Wakasugi, T., Aoki, H., Sobukawa, H.: Catalytic reforming of gasoline to hydrogen: kinetic investigation of deactivation processes. Int. J. Hydrog. Energy **34**, 8023 (2009)
31. Ferencz, Zs, Baán, K., Oszkó, A., Kónya, Z., Kecskés, T., Erdohelyi, A.: Dry reforming of CH4 on Rh doped Co/Al2O3 catalysts. Catal. Today **228**, 123–130 (2014)
32. Zhang, L., Li, W., Liu, J., Guo, C., Wang, Y., Zhang, J.: Ethanol steam reforming reactions over Al2O3SiO2-supported Ni–La catalysts. Fuel **88**, 511 (2009)
33. Freni, S., Carvallaro, S., Mondello, N., Spadaro, L., Frusteri, F.: Production of hydrogen for MC fuel cell by steam reforming of ethanol over MgO supported Ni and Co catalysts. Catal. Commun. **4**, 259 (2003)
34. Wang, P., Tanabe, E., Ito, K., Jia, J., Morioka, H., Shishido, T., et al.: Filamentous carbon prepared by the catalytic pyrolysis of CH4 on Ni/SiO2. Appl. Catal. A **231**, 35 (2002)
35. Awadallah, A.E., Aboul-Enein, A.A., Aboul-Gheit, A.K.: Various nickel doping in commercial NieMo/Al2O3 as catalysts for natural gas decomposition to COx-free hydrogen production. Renew. Energy **57**, 671 (2013)
36. Pinheiro, A.L., Pinheiro, A.N., Valentini, A., Filho, J.M., Sousa, F.F., Sousa, J.R., Rocha, M.G.C., Bargiela, P., Oliveira, A.C.: Analysis of coke deposition and study of the structural features of MAl2O4 catalysts for the dry reforming of methane. Catal. Commun. **11**, 11 (2009)
37. Damyanova, S., Pawelec, B., Arishtirova, K., Martinez Huerta, M.V., Fierro, J.L.G.: The effect of CeO2 on the surface and catalytic properties of Pt/CeO2–ZrO2 catalysts for methane dry reforming. Appl. Catal. B **89**, 149 (2009)
38. Dresselhaus, M.S., Dresselhaus, G., Jorio, A., Souza Filho, A.G., Saito, R.: Raman spectroscopy on isolated single wall carbon nanotubes. Carbon **40**, 2043 (2002)
39. Pan, C., Bao, Q.: Well-aligned carbon nanotubes from ethanol flames. J. Mater. Sci. Lett. **21**, 1927 (2002)
40. Amin, A.M., Croiset, E., Epling, W.: Review of methane catalytic cracking for hydrogen production vol 3. Int. J. Hydrog. Energy **36**, 2904 (2011)

11

Novel and efficient texturing approach for large-scale industrial production line of large-area monocrystalline silicon solar cell

U. Gangopadhyay · S. Jana · S. Das · P. Ghosh

Abstract Texturing monocrystalline silicon (c-Si) for solar cell fabrication is still a key issue due to consumption of a large amount of costly isopropyl alcohol (IPA) in conventional NaOH/KOH solution. Controlling IPA amount in NaOH/KOH-texturing bath for longer life of the solution in industrial batch production is a difficult task. This article reports the use of a chemical bath with a composition of NaOH–NaOCl (sodium hydroxide–sodium hypochlorite) for texturing monocrystalline silicon surface for solar cell fabrication. We report such a novel concept of texturing of the c-Si surface without using IPA and discarding the saw damage removal step. The optimized etching condition with NaOH–NaOCl-containing solution does not have any negative effect on monocrystalline silicon solar cell efficiency. Easy control of the process due to the self-controlled reaction is one of the major advantages and there is no need of sophisticated temperature-controlling system during texturing in the industrial batch process. Moreover, FTIR analysis shows the formation of Si–Cl bond during texturing, which improves the quality of diffused junction due to the gettering effect of Cl. Solar cell of conversion efficiency 14.5–15.8 % on CZ–Si substrate of area 125 mm × 125 mm was successfully fabricated using a novel texturing approach in industrial production line with yield >85 %.

Keywords Texturization · NaOH–NaOCl solution · Monocrystalline silicon · Inverted pyramids

U. Gangopadhyay (✉) · S. Jana · S. Das · P. Ghosh
Meghnad Saha Institute of Technology, Nazirabad,
Kolkata 700107, India
e-mail: utpal_ganguly@yahoo.com

Introduction

Texturing of monocrystalline silicon is necessary for fabrication of better-quality solar cells. Therefore, industrial production of large-area monocrystalline silicon for solar cell fabrication needs effective and economically viable technique for texturing of monocrystalline silicon (c-Si). Conventionally, anisotropic chemical etching of silicon is performed with a mixture of alkaline solution of sodium hydroxide (NaOH) or potassium hydroxide (KOH), and isopropyl alcohol (IPA) is widely used for texturing of monocrystalline silicon wafers [1–3]. In a typical texturing condition, the IPA concentration is higher than the NaOH (or KOH) concentration. IPA increases the wettability of silicon surface, but increasing IPA concentration beyond a limit decreases the rate of silicon removal very strongly [1]. Therefore, excessive IPA addition to the texturing solution leads to the detrimental effect of no texturization even after long-time treatment. Moreover, IPA is a volatile compound and so it has to be added frequently to the texturization bath during the batch process to compensate for the evaporated amount, leading to a higher consumption of IPA. Controlling IPA amount in NaOH/KOH-texturing bath for longer life of the solution in the industrial batch production is a difficult task [4]. IPA is also expensive. All these together are the constraints of NaOH/KOH-texturing processes during monocrystalline silicon solar cell fabrication in large-scale industrial production line. However, due to the high cost of IPA and constraint on controlling its excess amount, there is always a search for an alternate chemical composition for texturing without using IPA. So, elimination of IPA during texturing of the commercial solar cell fabrication is the key issue of the overall texturing cost reduction approach, as well as maintenance of the continuous flow of texturing batch process in industrial process line. To overcome these constraints, use of sodium carbonate (Na_2CO_3) or potassium carbonate (K_2CO_3) has been reported by some researchers [5, 6]. This article reports the use of a chemical bath with a composition of NaOH–NaOCl (sodium hydroxide–sodium hypochlorite) for texturing monocrystalline silicon surface without using IPA for solar cell fabrication. Moreover, this proposed novel concept for texturing of silicon surface also discards the saw damage removal step in the industrial production line. The optimized etching condition with NaOH–NaOCl-containing solution does not have any negative effect on monocrystalline silicon solar cell efficiency. Self-controlling reaction is one of the major advantages of the NaOH–NaOCl texturization process, which leads to much less complicacies of using a temperature control system in the industrial batch process. The role of NaOH–NaOCl solution for cleaning organic contaminations while texturing c-Si and comparison of this process with conventional NaOH-IPA approach have been reported in this paper.

Experiment

Texturing

In our experiment, we have used *CZ* p-type, monocrystalline silicon (c-Si) with resistivity of about 0.5–2.0 Ωcm for texturing. In our texturization experiment, we varied the three different concentrations of NaOH solution (i.e., 10, 20 and 40 %) for a fixed 12 % NaOCl solution each time. The ratio of the amounts of NaOH to NaOCl also varied along with the time of etching. All the etched wafers were studied under optical microscope and etching thickness was studied with a standard gravimetric method. Scanning electron microscopy (SEM) and FTIR were studied for selected monocrystalline wafers. Evaluation of surface reflectance after texturing was also completed. We used all these measurement tools in the case of conventional NaOH-IPA texturing and compared the results of the two different approaches.

Solar cell fabrication

During the solar cell fabrication experiment, textured *CZ* p-type monocrystalline silicon (c-Si) wafers were diffused by pentavalent impurity (phosphorus) in an open-tube furnace using conventional $POCl_3$ diffusion source at 850 °C for 7 min pre-deposition followed by 16 min drive-in. PSG removal was done by short time dipping in BHF solution followed by DI-water rinsing and drying. The sheet resistance of n^+ emitter layer was 55–60 Ω/sq. All textured c-Si wafers were oxidized at 850 °C for 20 min. After edge isolation and oxide layer removal, about 70 nm layer of SiNx (refractive index ~2.0) was deposited on the front side by means of PECVD at 450 °C for ARC. The front and back metallization of the diffused silicon wafers was carried out using standard Ag paste (product no. 3349, Ferro Electronic Materials) and Al paste (product no. Fx53-038, Ferro Electronic Materials) for screen-printed metallization technique followed by baking and co-firing at the temperature of 750 °C in a conveyor belt furnace (Seirra-Therm, USA).

The illuminated current–voltage (LIV) characteristics, dark current–voltage (DIV) characteristics of NaOCl–NaOH-textured and conventional NaOH-IPA-textured c-Si solar cells were determined.

Results and discussion

For etching performance comparison with two different approaches, (a) proposed NaOH–NaOCl solution and (b) conventional NaOH-IPA solution, we used contaminated c-Si wafers. Prior to the texturization experiment, FTIR analysis of the c-Si samples was done. From the FTIR analysis of the samples, it was observed that at the start all silicon wafers had significant contamination [7, 8] all over the surface as shown in Fig. 1. Three different values of concentration of NaOH such as 10, 20 and 40 % were taken and the ratio of NaOH:NaOCl was considered as 1:2 for the first four experiments. The etching time durations for those experiments were set between 15 and 20 min. In the next set of experiment, mainly the ratio of NaOH:NaOCl was varied in the textured solution and only three different time settings were considered. The details of the experimental conditions along with average reflectivity of the different etched monocrystalline silicon(c-Si)

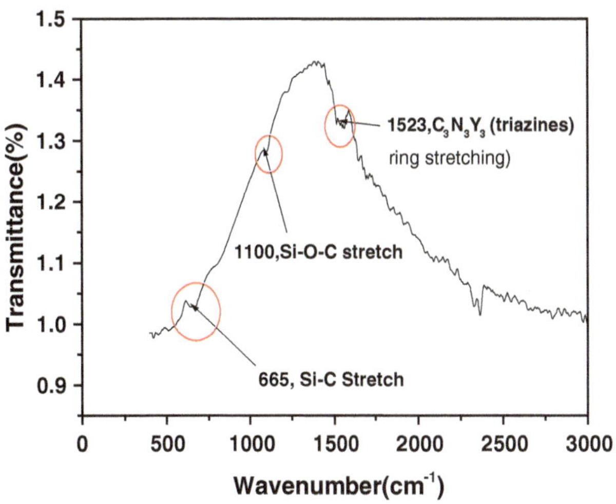

Fig. 1 FTIR analysis of the starting c-Si wafer before the texturization experiment

Table 1 Different conditions of NaOH–NaOCl texturing

Sample ID	Conc. of NaOH (%)	Conc. of NaOCl (%)	NaOH:NaOCl	Time of etching (min)	Average reflectivity (%)
S1	10	12	1:2	20	19.19
S2	20	12	1:2	20	18.31
S3	40	12	1:2	15	18.37
S4	40	12	1:2	20	19.56
S5	40	12	1:3	30	19.28
S6	40	12	1:4	10	17.49
S7	40	12	1:4	24	21.03
S8	40	12	1:5	10	17.12
S9	40	12	1:5	24	20.85
S10	40	12	1:6	10	15.47

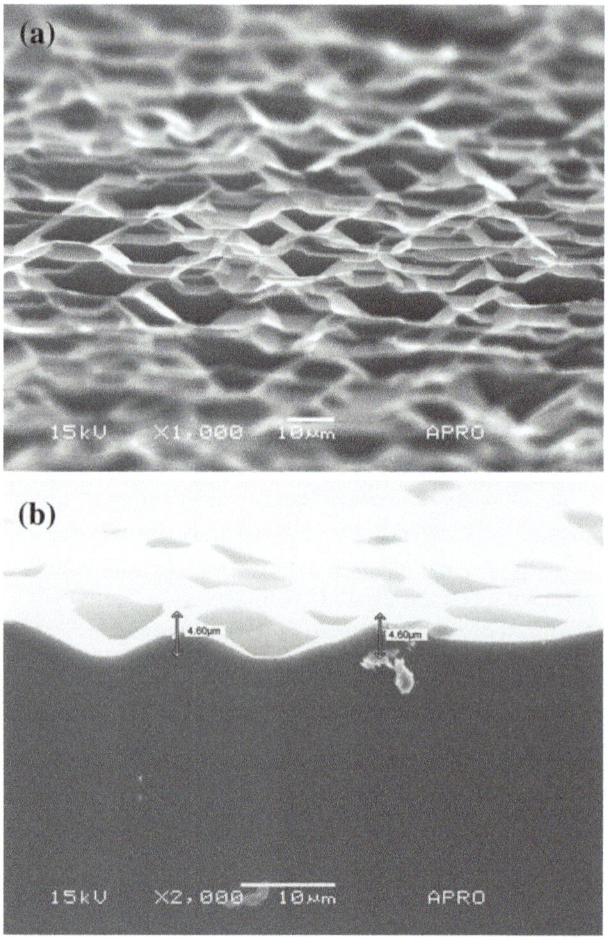

Fig. 2 SEM picture of NaOH–NaOCl-textured c-Si wafer: **a** surface, **b** cross-sectional edge view where NaOH (40 %): NaOCl = 1:2, textured time 20 min

Fig. 3 Surface morphology of a typical NaOH–NaOCl-textured c-Si with higher NaOH and NaOCl ratio (for typical 1:4)

surfaces are listed in Table 1. It was observed that the average reflectivity of the textured sample S2 was lowest compared to samples S1 and S3. This might be due to the different etching mechanism of NaOH–NaOCl solution for different combination ratios. As NaOCl is a strong oxidizing agent [9], it oxidizes the silicon layer first on dipping in NaOH–NaOCl solution. This was followed by etching due to the presence of NaOH in the solution. Moreover, as the silicon layer was chemically oxidized, the probability of growth of non-uniform oxide on the silicon layer was significantly large. Anisotropy of NaOH solution is highest at its 20 % concentration [10]. The chance of inverted pyramidal structure formation through the automatically unintentionally formed oxide mask on the c-Si surface is more than that in the other two cases, i.e., 10 and 40 % in NaOH:NaOCl = 1:2 ratio (sample S1 and S2). So, sample S2 was more textured than S1 and S3. This was probably due to the lowest average reflectance of sample S2 out of the three samples (S1, S2 and S3). The calculated standard deviation for reflectivity of the c-Si samples under the experiment is 1.7 using standard deviation formulas. Figure 2 shows the SEM micrograph of a typical NaOH–NaOCl-textured surface with etching condition as mentioned in Table 1 for sample 2. Inverted pyramidal structures were formed at some places on the surface of c-Si as shown in Fig. 2a. The etch depth was around 4 µm as

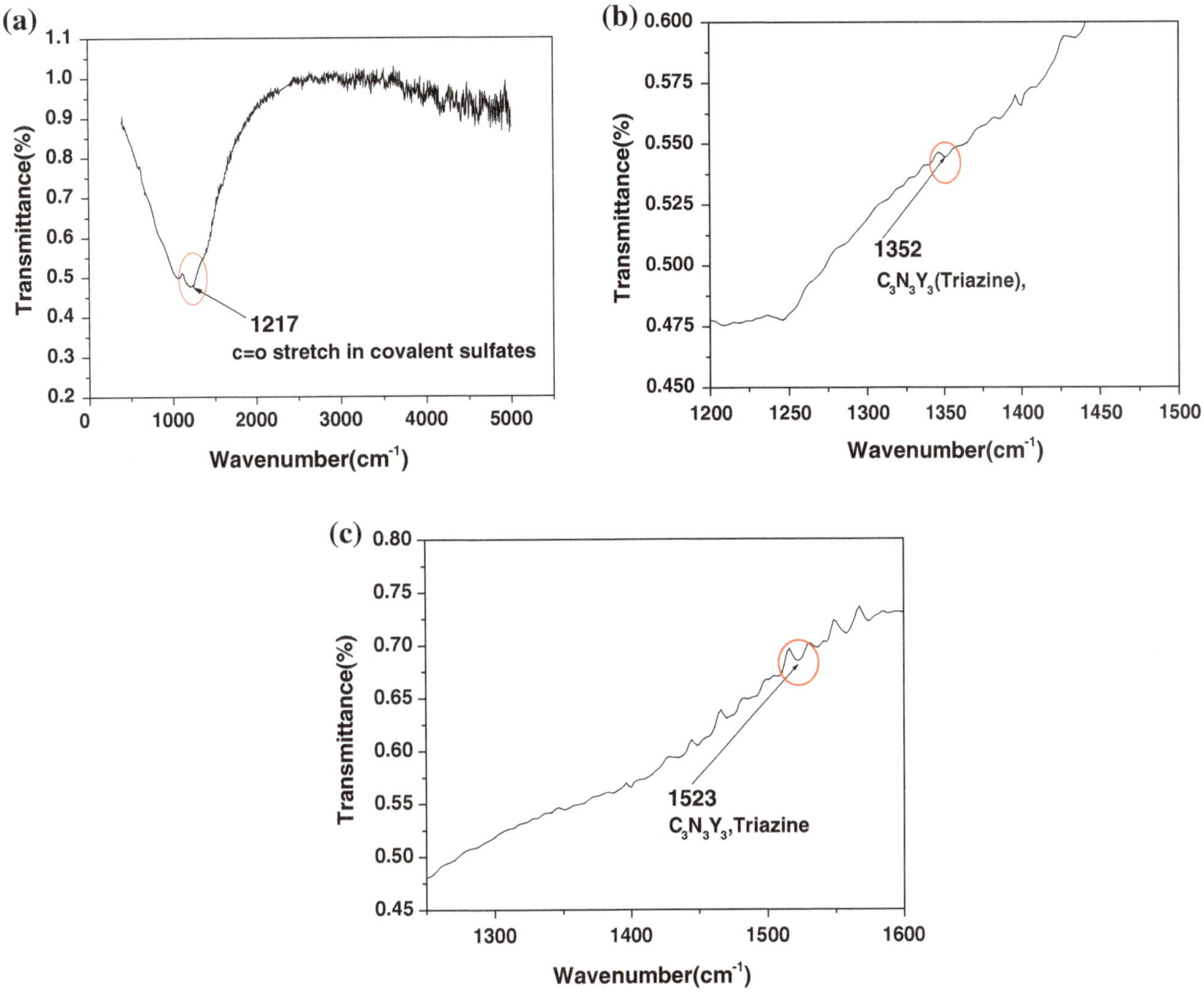

Fig. 4 FTIR analysis of NaOH–NaOCl c-Si surface after short time texturing (10 min) of s samples S6, S8 and S10, respectively

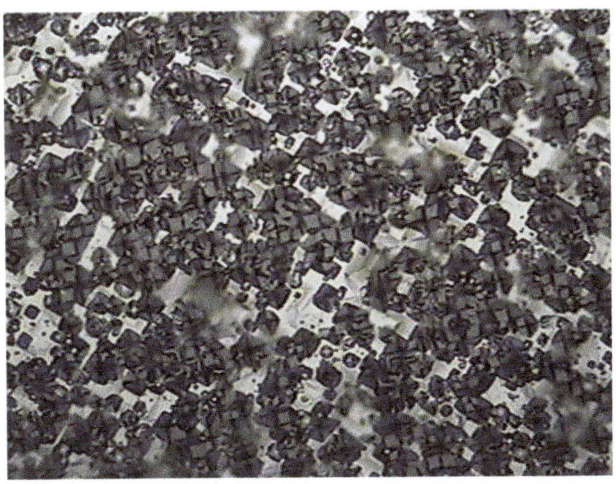

Fig. 5 Optical micrograph of a conventional NaOH-IPA-textured c-Si surface

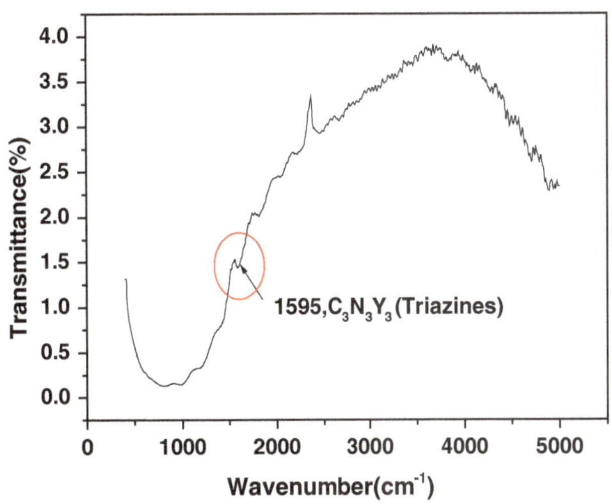

Fig. 6 FTIR analysis of NaOH-IPA-textured c-Si surface

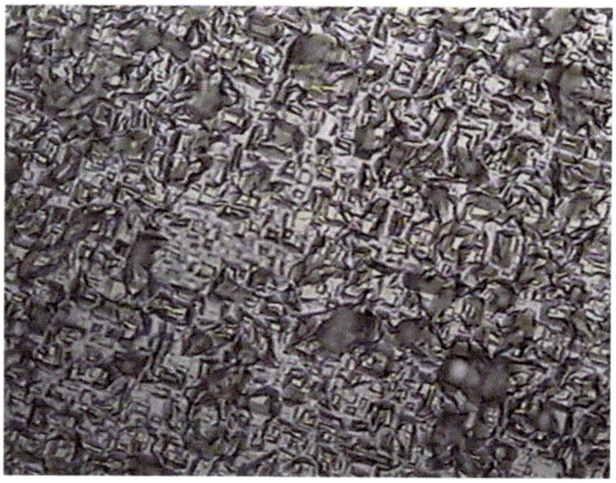

Fig. 7 Optical micrograph of a typical NaOH–NaOCl-textured c-Si surface (condition of texturization: concentration of NaOH: 20 %, concentration of NaOCl: 12 %, NaOH:NaOCl = 1:2 and time = 20 min.)

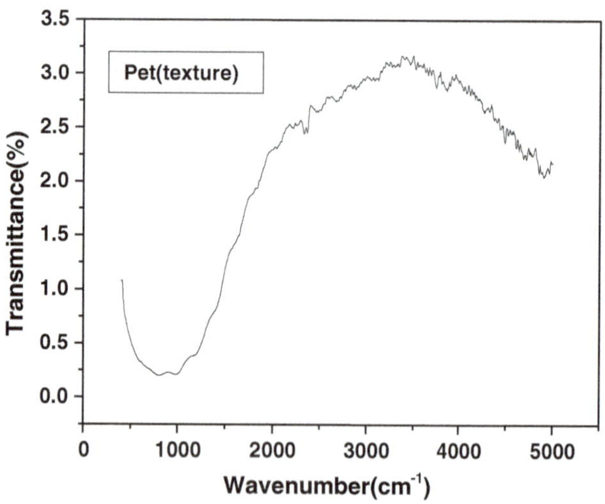

Fig. 8 FTIR analysis of a typical NaOH–NaOCl-textured c-Si surface with texturing time >10 min

indicated in Fig. 2b. Moreover, it was also observed that the average reflectance (wavelength range 300–1,100 nm) of textured c-Si wafers depended on the three major experimental parameters, such as concentration of NaOH, ratios of NaOH:NaOCl and times of etching for a fixed NaOCl concentration as mentioned in Table 1. In this case, NaOCl plays a dominating role due to its higher amount compared to NaOH, as a result of which the etch rate is strongly reduced in the ⟨100⟩ direction and increased in the ⟨111⟩ direction, resulting in a much flatter surface. Higher reflectance compared to 20 % NaOH in a 1:2 ratio was due to the flatter surface nature of the samples S5, S7, S9 and S10, as shown in Table 1. Figure 3 shows the nature of surface structure formed with a higher amount of NaOCl in the NaOH–NaOCl solution. However, for lower time (~10 min), the samples S6, S8 and S10 showed low average reflectance. This was due to the very less etching rate of the silicon surface after the addition of NaOCl to the etching solution, resulting in lesser amount of contaminant removal from the surfaces of S6, S8 and S10 samples, as indicated in FTIR analysis curves (Fig. 4a–c). Therefore, we did not use the samples S6, S8 and S10 despite their lower reflectance values, because the surface contaminants were still present even after texturization. As done earlier, we thoroughly analyzed the NaOH-IPA-textured c-Si samples via optical microscopic study and FTIR analysis, respectively. It was observed from optical microscopic study throughout the entire c-Si surface that the textured c-Si surface was not at all satisfactory even after 2 min saw damage removal followed by 45 min of NaOH-IPA texturization as shown in Fig. 5. FTIR analysis of the NaOH-IPA-textured c-Si sample showed some impurity still present on the silicon surface as indicated in Fig. 6, hindering the initiation of the texturing process. In the case of surface texturization with our proposed NaOH–NaOCl approach, we discarded the saw damage removal step. From the optical microscopic study of the surface throughout the NaOH–NaOCl-textured surface, it was observed that the surface of c-Si was uniformly textured with the NaOH–NaOCl-texturing approach as shown in Fig. 7. The FTIR observation of NaOH–NaOCl-textured surface indicated the complete removal of organic contaminants from the surface of c-Si during texturing in

Fig. 9 SEM micrographs of textured c-Si surface using **a** conventional NaOH-IPA solution and the **b** proposed NaOH–NaOCl solution, respectively, for texturing

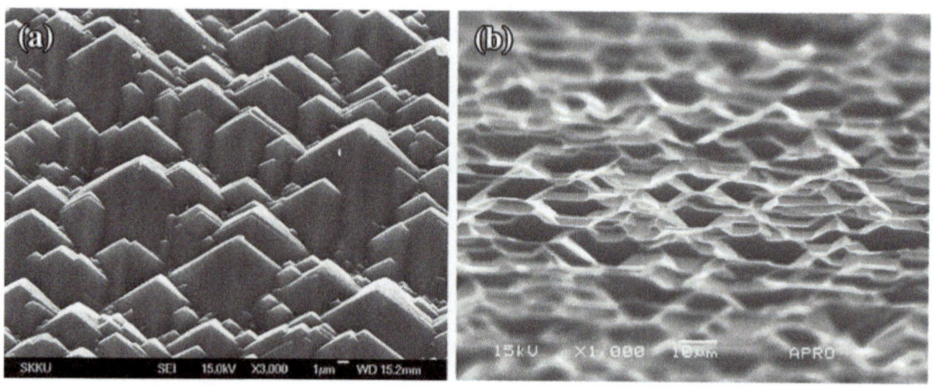

NaOH–NaOCl bath as shown in Fig. 8. This was due to the NaOCl present in the texturization solution, which oxidized all organic contaminants from the c-Si surface. The oxidized contaminants thus formed were finally etched out by NaOH present in the textured solution. This chemical process might help in the initiation of the texturization process. From the SEM analysis as showed in Fig. 9, we observed that the nature of the NaOH–NaOCl-textured surface was entirely differently structured than that observed in the case of conventional NaOH-IPA-textured surface. Inverted pyramidal structures might have originated on the c-Si surface during the NaOH–NaOCl texturization sequence. Formation of inverted pyramidal structures on the c-Si surface in this case was likely observed in the case of selective texturization through an oxide mask [11]. The formation of inverted pyramid structures in the case of the proposed NaOH–NaOCl etching might be due to the combined effect of in situ non-uniform oxidation via etching due to the presence of constituent chemicals NaOCl and NaOH in the texturing solution. The oxidizing chemical NaOCl present in the textured solution may help in the formation of thin, non-uniform oxide growth on the c-Si surface, and OH^- in the NaOH solution might help in c-Si surface etching. Moreover, the chance for formation of more inverted pyramids through an unintentional opening window of oxide is more due to the highest anisotropy of the 20 % concentrated NaOH solution.

Figure 10 shows the reflectivity of c-Si wafers textured using conventional NaOH-IPA and new NaOH–NaOCl-texturing approaches. It was observed that the overall reflectance (wavelength range 300–1,100 nm) of c-Si wafer textured using conventional NaOH-IPA approach was lower than that using the proposed NaOH–NaOCl

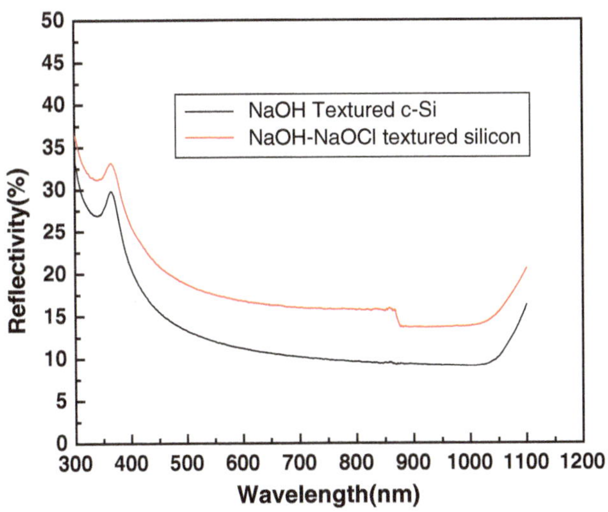

Fig. 10 Reflectance of NaOH-IPA-textured and NaOH–NaOCl-textured c-Si surfaces

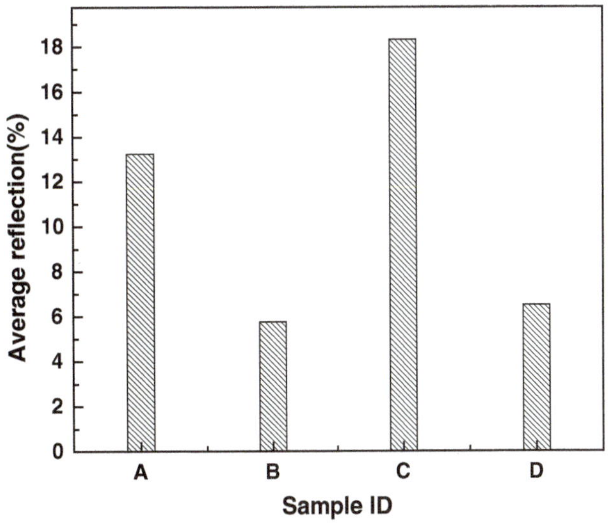

Fig. 11 Average reflectance of textured c-Si surface without silicon nitride (SiNx) ARC: NaOH-IPA textured [sample ID-A], NaOH–NaOCl textured [sample ID-C] and with SINx ARC: NaOH-IPA textured [sample ID-B] and NaOH–NaOCl textured [sample ID-D]

approach. One important point is that the reflectivity of the c-Si surface does not have such a serious effect on the conversion efficiency of the fabricated solar cell, because SiNx/TiO_2 film as AR (antireflective) coating is usually used for all commercial silicon solar cells after texturization. Moreover, after SiNx coating on NaOH–NaOCl-textured c-Si wafer, the average reflectivity drastically reduces from 18.31 % to around 6.49 % whereas for NaOH-textured *c-Si* wafer it is reduced from 13.23 to ~5.75 % after SiNx AR coating. as shown in Fig. 11.

Illuminated current–voltage characteristics of the fabricated c-Si cells with the NaOH–NaOCl as well as conventional NaOH-IPA approaches were observed under global spectrum of AM 1.5, which are shown in Fig. 12. It

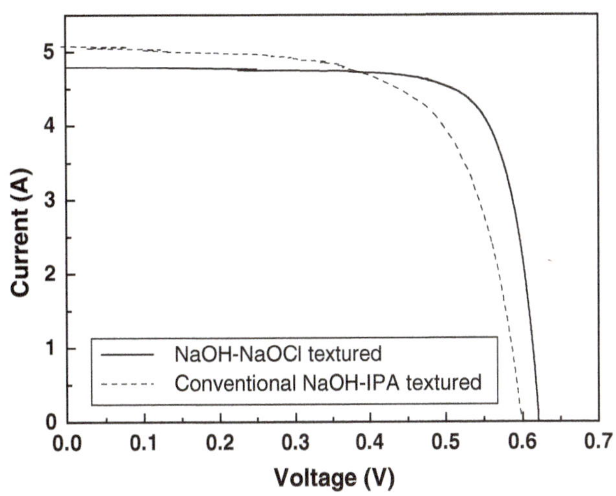

Fig. 12 LIV characteristics of NaOH-IPA-textured and NaOH–NaOCl-textured c-Si solar cells

Table 2 Performance parameters of NaOH–NaOCl-textured and NaOH-IPA-textured large-area (125 mm × 125 mm) c-Si solar cells

Texturization solution	V_{oc} (V)	I_{sc} (A)	V_m (V)	I_m (A)	P_m (W)	FF	R_{sh} (Ω)	R_s (mΩ)	Eff. (%)
NaOH–NaOCl	621.4	4.80	527.10	4.40	2.32	0.78	14.50	3.00	15.78
NaOH	598.7	5.08	472.54	4.29	2.03	0.67	6.43	5.03	13.78

V_{oc} open circuit voltage, I_{sc} short circuit current, V_m voltage at max. power point, I_m current at max. power point, P_m max. power output, *FF* fill factor, *Eff* efficiency, R_{sh} shunt resistance, R_s series resistance

was observed from these graphs that the overall efficiency achieved in the current approach of surface texturing by using NaOH–NaOCl was much higher than that of the conventional approach with NaOH-IPA. This was reflected in the electrical parameters of the solar cell for two different texturing approaches as shown in Table 2. The low open circuit voltage (V_{oc}) and fill factor (FF) of the c-Si solar cell fabricated using NaOH-IPA texturing might be due to its higher value of leakage current compared to the NaOH–NaOCl-textured solar cell. Reverse characteristics of dark voltage–current characteristic (DIV) as shown in Fig. 13 give a clear indication of significantly large leakage of current for NaOH-textured compared to NaOH–NaOCl-textured c-Si solar cells. For comparatively higher leakage current, the observed shunt resistance was lower (6.43 Ω) in the case of NaOH-textured than that of NaOH–NaOCl-textured c-Si solar cell (14.5 Ω) as shown in Table 2. Lower value of leakage current in the case of NaOH–NaOCl-textured cell might be due to uniform surface structure throughout the area (125 mm × 125 mm) of c-Si with relatively smaller steps of <5 μm. As a result, no surface contaminants were present after texturing of the contaminated c-Si wafer as shown in Fig. 8. This effectively improved the dark current and also other electrical parameters of solar cells as mentioned in Table 2. The decrease in series resistance (Rs = 3 mΩ) of NaOH–NaOCl-textured silicon solar cell also indicated the

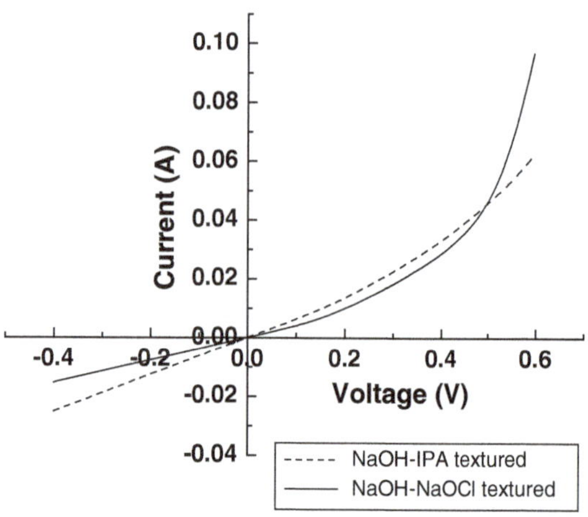

Fig. 13 DIV characteristics of NaOH-IPA-textured and NaOH–NaOCl-textured c-Si solar cells

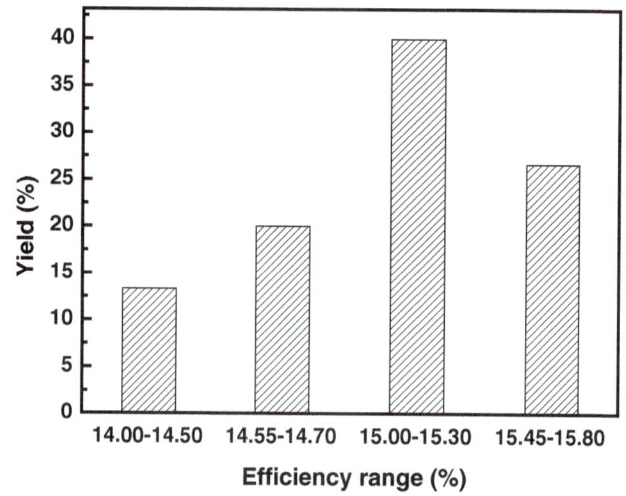

Fig. 14 Percentage of industrial yield of NaOH–NaOCl-textured large-area (125 mm × 125 mm) c-Si solar cell

reduction of surface contaminant present on the solar cell surface. Actually, series resistance and other electrical parameters of the solar cells were strongly dependent on front and back contacts. The presence of surface contaminants degrades the conventional NaOH-IPA-textured c-Si solar cell performance parameters, except for short circuit current as mentioned in Table 2 in spite of using a huge amount of costly IPA in the industrial process sequence. A slightly higher value of short circuit current in conventionally textured c-Si cells compared to that in NaOH–NaOCl-textured cells is due to its lower value of reflectance after texturing.

We successfully achieved around 85 % yield in the efficiency range 14.5–15.8 % using NaOH–NaOCl-texturing approach as shown in Fig. 14, whereas in conventional NaOH-IPA-textured solar cell the yield percentage in the efficiency range 13.5–14.0 % was around 80 % only, as shown in Fig. 15. Higher open circuit voltage (V_{oc}) of NaOH=NaOCl-textured c-Si solar cell compared to NaOH-IPA-textured solar cell had two reasons. The first one was due to less etch depth (<5 μm) generated on very large-area c-Si substrate with no significant surface contaminations after NaOH–NaOCl texturing and the second reason was the effect of chlorine attachment with silicon (Si–Cl bond) during treatment of c-Si with NaOH–NaOCl solution at 80 °C, as shown in the FTIR characteristics of Fig. 16.

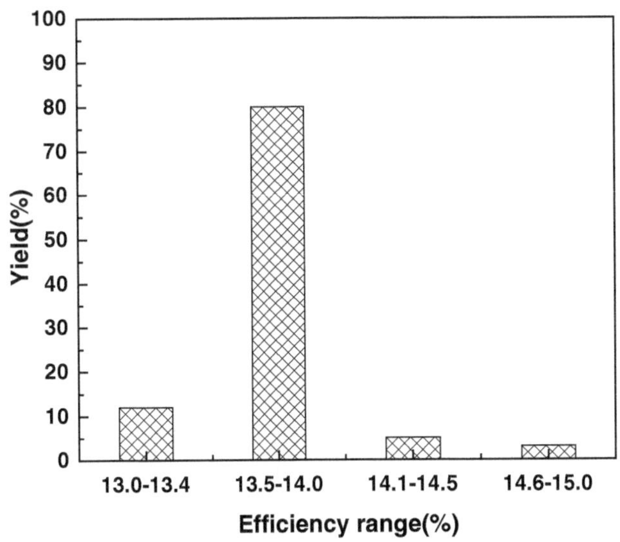

Fig. 15 Percentage of industrial yield of NaOH-IPA-textured large-area (125 mm × 125 mm) c-Si solar cell

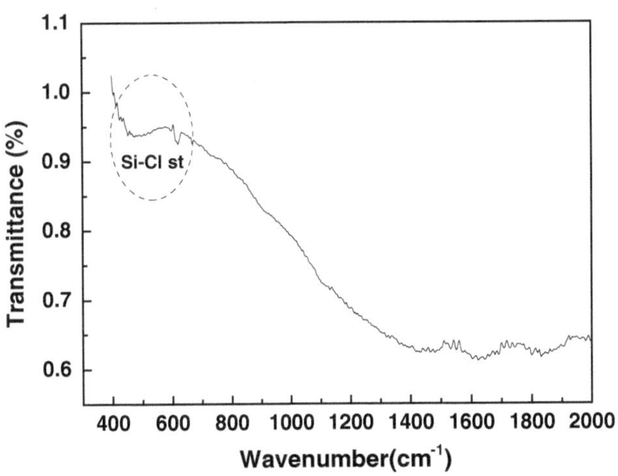

Fig. 16 FTIR analysis of NaOH–NaOCl-textured c-Si solar cell

This Si–Cl bond might have helped the gettering during diffusion of NaOH–NaOCl-treated wafers [10, 12]. It clearly indicates that this newly developed NaOH–NaOCl approach of c-Si texturing discussed in this article can easily be applied to c-Si wafer surface with additional effect of surface contamination removal for commercial production of very large-area c-Si solar cell of reasonably acceptable efficiency and contamination of c-Si wafers even without costly IPA addition.

Conclusion

The proposed texturization technique generated inverted pyramidal structure of etch depth 4 μm without any additional oxide masking step, which was normally needed for forming inverted pyramid structure during etching with any anisotropic etching solution. NaOCl in our proposed technique is a key material for growing in situ oxide masking layer on monocrystalline silicon (c-Si) surface during etching in NaOH–NaOCl bath. Our findings reported in this paper are based on comparative studies of optical/SEM imaging as well as FTIR spectroscopy of the c-Si sample before and after conventional NaOH and proposed NaOH–NaOCl texturization process along with the observed electrical parameters of solar cells. The comparative study of electrical parameters of the cells fabricated with conventional NaOH-IPA and the proposed cells brings out a significant improvement of parameters for solar cells. Open circuit voltage of 621.4 mV and a fill factor as high as 0.78 were the parameters achieved as a key for the high efficiency of (~16 %) crystalline silicon solar cell of area (125 mm × 125 mm) compared to conventional NaOH-IPA approach due to its additional surface contaminant removal behavior. Moreover, FTIR analysis indicated that the Si–Cl bond created during texturing with NaOH–NaOCl solution finally helped in gettering during phosphorus diffusion with $POCl_3$ dopant. This Cl gettering helps in the improvement of open circuit voltage of a c-Si solar cell. We are probably the first group to report on such a novel idea of using NaOH–NaOCl solution for monocrystalline silicon surface texturing with formation of maskless inverted structured for solar cell fabrication of area 125 mm × 125 mm and yield around 85 % in the conversion efficiency range 14.5–15.8 %, compared to 80 % yield in the efficiency range of 13.5–14.0 % of the conventional NaOH-IPA approach. This process can be applied as a key step to the industrial production of very large-area monocrystalline silicon solar cells.

Acknowledgments The author would like to thank the Meghnad Saha Institute of Technology, TIG for providing us the infrastructural support for carrying out research activity in this area. The authors also gratefully acknowledge the DST, Government of India for financial support for carrying out solar cell-related research activity.

References

1. Vazsonyi, E., De Clercq, K., Einhaus, R., Van Kerschaver, E., Said, K., Poortmans, J., Szlufcik, J., Nijs, J.: Sol. Energy Mater. Sol. Cells **57**, 179–188 (1999)
2. Chitre, S.R.: Proceedings of the 13th IEEE International Photovoltaic Specialists Conference, Washington, DC, pp. 152–154 (1978)

3. King, D.L., Buck, M.E.: Proceedings of the 22nd IEEE International Photovoltaic Specialists Conference, Las Vegas, pp. 303–308 (1991)
4. Sethi, C., Anand, V.K., Walia, K., Sood, S.C.: TECHNIA. Int. J. Comput. Sci. Commun. Tech. **5**, 785–788 (2012)
5. Nishimoto, Y., Namba, K.: Sol. Energy Mater. Sol. Cells **61**, 393–402 (2000)
6. Chaoui, R., Lachab, M., Chiheub, F., Seddiki, N.: Proceedings of the 14th European Photovoltaic Solar Energy Conference, Barcelona, pp. 812–814 (1997)
7. Sangeetha, V., Kanagathara, N., Sumathi, R., Sivakumar, N., Anbalagan, G.: J. Mat **2013**, Article ID 262094, 1–7
8. Gangopadhyay, U., Dhungel, S.K., Mondal, A.K., Saha, H., Yi, J.: Sol. Energy Mater. Sol. Cells **91**, 1147–1151 (2007)
9. Vazsonyi, E., De Clercq, K., Einhaus, R., Van Kerschaver, E., Said, K., Poortmans, J., Szlufcik, J., Nijs, J.: Sol. Energy Mater. Sol. Cells **57**, 179 (1999)
10. Gangopadhyay, U., Dhungel, S.K., Kim, K., Manna, U., Basu, P.K., Kim, H.J., Karunagaran, B., Lee, K.S., Yoo, J.S., Yi, J.: Semicond. Sci. Technol. **20**, 938 (2005)
11. Campbell, P., Green, M.A.: Sol. Energy Mater. Sol. Cells **65**, 369–375 (2001)
12. Rivillon, S., Chabal, Y.J., Webb, L.J., Michalak, D.J., Lewis, N.S., Halls, M.D., Raghavachari, K.: J. Vac. Sci. Technol. A, **23**(4) (2005)

12

An experimental investigation of mesoporous MgO as a potential pre-combustion CO_2 sorbent

Sushant Kumar[1] · Surendra K. Saxena[1] · Vadym Drozd[1] · Andriy Durygin[1]

Abstract We examined the CO_2 capture capacity of mesoporous MgO (325 mesh size, surface area = 95.08 ± 1.5 m^2/g) as a potential pre-combustion CO_2 sorbent. Our results show that 96.96 % of MgO was converted to $MgCO_3$ at 350 °C and 10 bars CO_2 pressure. The sorbent could be completely regenerated at 550 °C under argon flow. The sorption rate parameters such as surface area and pore size were investigated.

Keywords Capture capacity · Mesoporous · Pre-combustion · Sorbent · Surface area · Pore size

Introduction

Fossil fuel accounts for the world's major energy supply and its use is anticipated to be continued throughout the 21st century [1]. The use of fossil fuel is always accompanied with a vast emission of CO_2. The anthropogenic CO_2 emissions upset the natural carbon cycle leading to an increased atmospheric CO_2 concentration. No one can deny that there is an urgent need to develop methods for CO_2 mitigation.

Currently, a large focus is devoted on capturing CO_2 from coal-fired power plant flue streams, which continues to be the major CO_2 contributor [2].There are three main technological approaches for CO_2 capture- post-, oxy- and pre-combustion. Post-combustion systems capture CO_2 from N_2-rich flue gas stream, produced by burning fossil fuel in air. Oxy-combustion uses pure stream of O_2, instead of air, to combust coal and thus produces CO_2-rich gas stream. On the other hand, pre-combustion systems are designed mainly to remove CO_2 from the syngas ($CO + H_2$) prior to its combustion for power production [3]. Under pre-combustion conditions, after the water–gas shift reactor ($CO + H_2O \rightarrow CO_2 + H_2$), the gas stream mainly consists of CO_2, H_2O and H_2. The partial CO_2 pressure for pre-combustion capture conditions is around 20–30 bar and the temperature is between 250 and 450 °C [4].

Both the physical and chemical solvents can be used for pre-combustion CO_2 capture. Unlike chemical solvent, physical solvent (such as Selexol and Rectisol) selectively absorbs CO_2 without forming any chemical bonds. Thus, the physical solvent requires relatively less regeneration energy as compared to that of chemical solvent [5]. However, these physical solvent-based processes suffer severe disadvantages: (1) lose pressure during regeneration step, and (2) require a low operating temperature. Therefore, syngas needs to be cooled prior to CO_2 absorption step to attain a relatively low operating temperature. After CO_2 absorption, the remaining hydrogen gas stream requires to be reheated to the gas turbine inlet temperature. However, chemical solvents have the advantage of high mass transfer driving force into solution and better acid gas selectivity. Also, chemical solvents can be used in processes that utilize thermal swing regeneration and generate the CO_2 at elevated pressure [6]. But chemical solvents increase the energy and cost penalty and thus are downgraded as a future CO_2 sorbent [7].

✉ Sushant Kumar
skuma002@fiu.edu

[1] Center for the Study of Matter at Extreme Conditions, College of Engineering and Computing, Florida International University, Miami, FL 33199, USA

Currently, the focus is to develop advanced physical and chemical solvent systems that have the potential to provide significant improvements in both cost and performance as compared to the Selexol and Rectisol for pre-combustion CO_2 capture. The challenges are to modify regeneration conditions to recover the CO_2 at a higher pressure, improve selectivity to reduce H_2 losses, and develop a solvent that has a high CO_2 loading at a higher temperature, which would increase integrated gasification combined cycle (IGCC) efficiency. In the same line, another vital task is to develop a new sorbent which could be highly efficient for pre-combustion capture conditions. Consequently, the US Department of Energy (DOE) performed the thermodynamic modeling activities that included screening analyses for a number of different metal oxides, zirconates, silicates and titanates under various operating conditions to identify new solid sorbents for pre-combustion CO_2 capture. About 18 sorbents were modeled and finally seven candidates are chosen—magnesium oxide (MgO), calcium oxide (CaO), lithium zirconate (Li_2ZrO_3), calcium zirconate ($CaZrO_3$), barium zirconate ($BaZrO_3$), barium titanate ($BaTiO_3$) and barium silicate ($BaSiO_3$) for further investigations [8]. Among these solid sorbents, MgO and CaO are attractive because of their easy accessibility and favorable thermodynamic properties [9].

The alkaline earth metal oxides (such as CaO and MgO) combine with CO_2 to form thermodynamically stable carbonates. Metal carbonates, when heated, liberate pure stream of CO_2 gas and regenerate the oxides. Eventually, the generated pure CO_2 gas can either be sequestered underground or used for enhanced oil recovery [10]. CaO are abundant and thus relatively easily accessible than MgO. However, CaO as CO_2 sorbent suffers severe major drawbacks and a detailed discussion can be found elsewhere [11]. It is a fact that regeneration of oxides needs a lot of energy [12]. MgO and $Mg(OH)_2$ are known to be better candidates than CaO for CO_2 capture applications due to their low regeneration energy requirement and low operating temperature [13]. Thus, MgO (periclase), although occurs only rarely as an oxide, we choose it here for our study. One should note that MgO when recycled between naturally occurring magnesite or dolomite can cause relatively lesser energy or carbon emission penalty [14].

The CO_2 absorption capacity of MgO was studied as a function of particle size, surface area, temperature, pressure, support and concentration of water vapor. The CO_2 uptake capacity on different MgO sorbents at different conditions is listed in Table 1. Most of the experiments are restricted to low temperature and ambient or low pressure condition. Hence, the reported sorption capacities of CO_2 on MgO are not very high. However, the sorbents which can operate in the range of 300–350 °C would be ideal for the use in IGCC applications. Therefore, we study the CO_2 capture capacity of MgO in the relatively high temperature and pressure condition of 300–375 °C and 10–50 bars, respectively.

Experimental procedure

The reaction

$$MgO(s) + CO_2(g) \leftrightarrow MgCO_3(s)$$

was studied experimentally. The carbonation reaction was performed in a closed system which permits us to efficiently maintain high temperature and pressure for a long period of time. The decomposition reaction of carbonate was examined using Thermo gravimetric Analysis (TGA) technique.

About 0.2 g of 325-mesh-sized magnesium oxide (delivered by Alfa Aesar) was put inside a closed cylindrical vessel (1.25″ long and 0.35″ internal diameter). About 0.1 ml of water was also introduced on the vessel walls. Chemically pure CO_2 gas (Airgas) was passed into this system. Before experiments, CO_2 gas was flushed three times to ensure a pure CO_2 atmosphere inside the reactor. The reaction was performed for 30 min at a desired temperature and pressure condition. Once the reaction completed, the system was air-cooled. The product was then ground using mortar and pestle. The powder particle was again put back for another reaction at the same experimental condition. This cycle was repeated until no increment in weight of the product was observed.

Thermo gravimetric analysis (TGA) of the product was done using TGA 2950 Thermo gravimetric analyzer. The samples (10 mg) were heated under argon purge, at a heating rate of 10 °C/min to a final temperature of 800 °C. The TGA provides continuous measurements of the sample weight as a function of time and temperature. The amount of formation of $MgCO_3$ was analyzed by the percent loss in weight of the sample while heating up to 800 °C in an argon atmosphere.

The product characterization was performed using X-ray powder diffraction method. Bruker GADDS/D8 is equipped with Apex Smart CCD Detector and direct-drive rotating anode. The MacSci rotating anode (Molybdenum) operates with a 50 kV generator and 20 mA current. X-ray beam size can vary from 50 to 300 μm. The usual collection time is 1200 s.

An isothermal gas adsorption was employed to measure internal surface areas of the powder particles. Micromeritics Tristar II 3020 (surface area and porosimetry analyzer instrument) was used with N_2 as adsorptive gas at 77 K (liquid nitrogen bath). The samples were first degassed under 300 °C with a N_2 gas flow for 1 h to remove

Table 1 CO_2 uptake capacity of MgO obtained from the literature

	Sorbent	Gas stream	Carbonation temperature (°C)	Pressure (bar) or, flow rate	Particle Size	Regeneration temperature (°C)	CO_2 capture capacity (mmol/g) conversion, %	References
1	MgO	Pure CO_2	50–1000	100 mL/min (flow rate)	–	–	0.99	[15]
2	MgO/Al_2O_3 (10 wt. % MgO)	(13 v % H_2O, 13 v % CO_2)	30, 150	1	(20–40) mesh size	350	1.36	[16]
3	MgO	(11 v % H_2O, 1 v % CO_2)	50–100	0.01	–	150–400	1.05	[17]
4	K_2CO_3/MgO	(11 v % H_2O, 1 v % CO_2)	50–100	–	–	150–400	2.98	[17]
5	MgO	330/660 ppm in air	0, 100	0.2	–	–	0.64, 0.43	[18, 19]
6	MgO/MCM-41	Pure CO_2	25	1	–	–	1.06	[20]
7	Mesoporous MgO	Pure CO_2	25, 100	1	–	–	1.82, 2.27	[21]
8	Nonporous MgO	Pure CO_2	25	1	–	–	0.45	[21]
9	MgO–ZrO_2	Pure CO_2	30, 150	1	–	–	1.15,1.01	[22, 23]
10	MgO (31.7 wt%)/Al_2O_3 (22.4 wt%)	Pure CO_2	20, 200, 300	1	–	–	0.13, 0.24, 0.5	[24]
11	MgO (33.8 wt%)/ Al_2O_3 (20.8 wt%)	Pure CO_2	20, 200, 300	1	–	–	0.08, 0.12, 0.5	[24]
12	K_2CO_3/MgO/Al_2O_3	Flue gas	60	1	–	480	2.49	[25]
13	K_2CO_3/MgO	(9 v % H_2O, 1 v % CO_2)	60	40 mL/min (flow rate)	–	400	2.7	[26]
14	MgO nanocrystal	Flue gas	60	25sccm (flow rate)	5 nm	60–600	6.4	[27]
15	MgO	Pure CO_2	350	1.33, 3.33	–	–	0.089, 0.091	[28]
16	MgO	Pure CO_2	300–500	9–36	<44 μm	–	70–80 % (~200 min)	[29]
17	MgO	Pure CO_2	–	20–40	<44 μm	–	100 % (~120 min)	[30]

the moisture and other adsorbed gases before analysis. The internal surface area was calculated using the Brunauer–Emmett–Teller (BET) method. The pore volume was also calculated from the adsorbed nitrogen after complete pore condensation ($P/P_0 = 0.9925$) using the ratio of the densities of liquid and gaseous nitrogen. The pore size was calculated using the Barrett–Joyner–Halenda (BJH) method.

Results and discussion

Figure 1 confirms the formation of $MgCO_3$ at different temperatures and CO_2 pressures. MgO, $Mg(OH)_2$, $MgO{\cdot}2MgCO_3$ and $MgCO_3$ were identified conventionally by their corresponding Joint Committee Powder Diffraction Standard (JCPDS) card number 79-0612,82-2345,31-0804 and 86-2345, respectively.

One of the vital factors in gas–solid carbonation reaction is the presence of water and there have been numerous observations where water acts as a catalyst [31–34]. Therefore, we have also used water (0.1 ml) for the MgO–CO_2 reaction. In absence of water, no CO_2 was absorbed at these conditions due to the kinetic limitations. The CO_2 sorption capacity of MgO increases significantly in the presence of water vapor. Under humid condition, MgO rapidly locks CO_2 in the form of $MgCO_3$. Recently, Fagerlund et al. [29] proposed the reaction mechanism for MgO carbonation in the presence of steam:

$$MgO + H_2O \leftrightarrow MgO \cdot H_2O^*$$

$$MgO \cdot H_2O^* + CO_2 \leftrightarrow MgCO_3 + H_2O$$

$$MgO + CO_2 \leftrightarrow MgCO_3$$

Figure 2 depicts the adsorption/absorption model for MgO–CO_2–H_2O reaction. Based on the previous work, it could be reasonable to corroborate that water vapor surrounds MgO particles where CO_2 reacts to form CO_3^{2-} ions and H^+ ions [34–36]. Free Mg^{+2} ions could further react with the CO_3^{2-} ions to form $MgCO_3$. However, $MgCO_3$ forms an impervious layer around unreacted MgO particles and hinders the further diffusion of CO_2 molecules.

Here, we used mortar and pestle to grind the product. As mentioned earlier, we conducted each experiment for 30 min and ground the sample after that. Grinding helps in scrubbing off the outer nonporous layer of $MgCO_3$. And we performed grinding until we noticed no change in product weight after subsequent experiments. In general, after 3–4 cycles, we observed no change in the weight of product. It is certain that such intermittent grinding step is limited to the laboratory and cannot be seen as an industrial operation. Therefore, it is recommended to have an aggregative fluidization regime for a fluidized bed reactor while scaling up MgO–CO_2 reaction. Also, increasing the amount of water vapor cannot lead to the complete carbonate conversion of MgO. Thus, in addition to the amount of steam, surface properties of MgO (such as surface area, particle size, porosity) are also very crucial parameters for the carbonation process.

The thermal analysis curve does not show any significant differences in amounts of carbonate in the high-pressure (50 bars) experiments. We obtained almost similar TGA plots and XRD patterns for different temperatures 300–375 °C and 50 bars CO_2 pressure. However, the experiment performed at 300 °C and 10 bars did indicate that the product was not simply $MgCO_3$. The X-ray diffraction

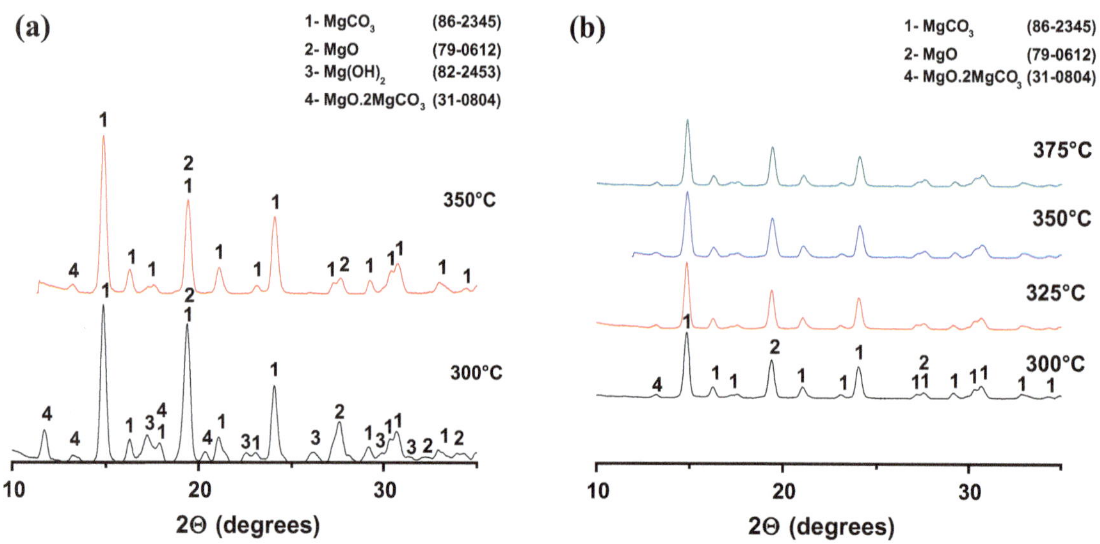

Fig. 1 XRD patterns for $MgCO_3$ formation after reaction at various temperatures and CO_2 pressure of **a** 10 bars and **b** 50 bars

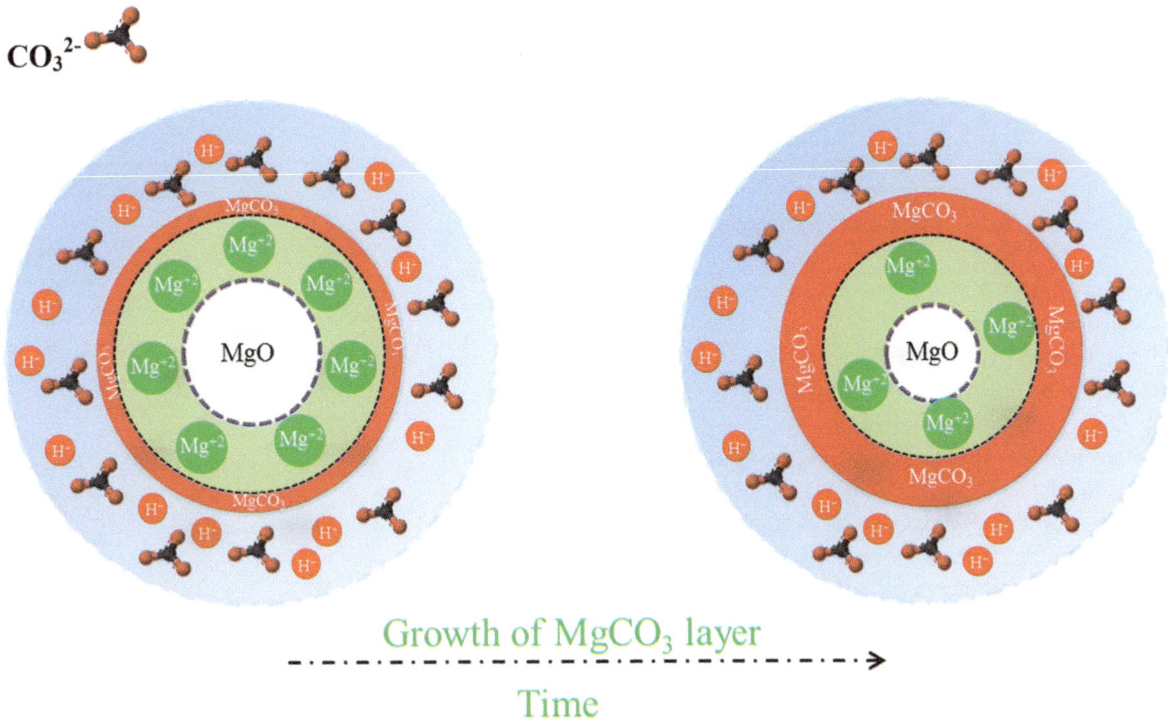

Fig. 2 The adsorption/absorption model for $MgO–CO_2–H_2O$ reaction

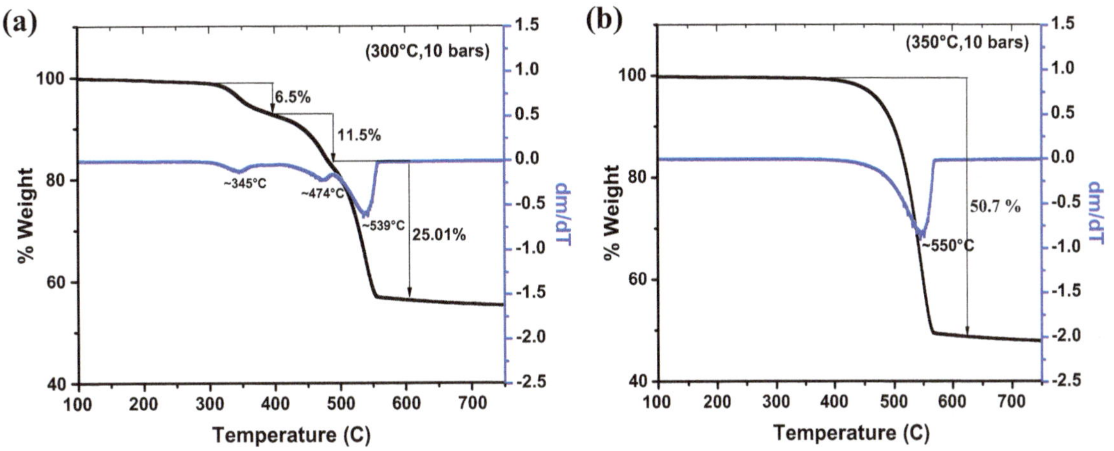

Fig. 3 DTG plots for product at **a** 300 °C and 10 bars and **b** 350 °C and 10 bars

pattern in Fig. 1a confirms the presence of Mg $(OH)_2$ and relatively high amount of $MgO{\cdot}2MgCO_3$ at 300 °C and 10 bars. DTG curve (Fig. 3a) also evidences the similar situation. The two peaks around 350 and 475 °C were attributed to the losses of water of crystallization and hydroxyl water, respectively. However, another peak at 550 °C corresponds to the complete decomposition of $MgCO_3$. But at a temperature and pressure of 350 °C and 10 bars, respectively, only $MgCO_3$ was formed. This is evidenced by both DTG (Fig. 3b) and X-ray diffraction analysis (Fig. 1a).

The steps to calculate % conversion of MgO to $MgCO_3$ is explained in supplementary section [S1]. A conversion of 30.54 and 96.96 % for MgO to $MgCO_3$ was observed at 300 and 350 °C, respectively. Here, we observed the formation of relatively high amount of oxymagnesite ($MgO{\cdot}2MgCO_3$) at 300 °C and 10 bars. It can be easily calculated that the CO_2 capture capacity of oxymagnesite is about two-third that of $MgCO_3$. Hence, formation of oxymagnesite can markedly reduce the overall uptake of CO_2 by MgO particles. However, it can be observed from X-ray patterns (Fig. 1) that at higher

Table 2 Surface properties of MgO particles before and after carbonation reaction 350 °C, 10 bars

Sample	Surface area (m^2/g)	Pore volume (cm^3/g)	Pore size (nm)	SA/PV(10^6 m^{-1})
As-received MgO	95.08 ± 1.5	0.22	9.09	439.59
After reaction at (350 °C, 10 bars)	4.15 ± 0.05	0.01	16.25	246.15

temperature or pressure, the formation of oxymagnesite is suppressed.

In the same line, recent study reveals that oxymagnesite forms as an intermediate during the thermal decomposition of hydrated magnesium carbonate [37]. Moreover, oxymagnesite can also be formed as a product of reaction between Mg $(OH)_2$ and CO_2 in anhydrous synthesis [29] and/or solid-state reaction of MgO and CO_2 using steam [38]. In this regard, Duan et al. [13] calculated the phase diagram of MgO–Mg$(OH)_2$–$MgCO_3$, which suggests that the transition temperature for direct conversion of $MgCO_3$ to $Mg(OH)_2$ increases with increase in P_{H2O}. Here, when temperature increases from 300 to 350 °C, partial pressure of water also increases and consequently transition temperature also increased. At 300 °C, a relatively less amount of MgO transforms to its carbonates. This can be attributed to a possible high conversion of $MgCO_3$ to Mg$(OH)_2$, which is in agreement with the phase diagram. Moreover, a significant formation of oxymagnesite at 300 °C and 10 bars also leads to a less direct conversion of MgO to $MgCO_3$. Thus, it can be deduced that a high amount of oxymagnesite forms at a low partial pressure of water. Therefore, the reaction mechanism is changed to:

$$MgO + CO_2 \leftrightarrow MgCO_3$$

$$MgO + 2MgCO_3 \leftrightarrow MgO \cdot 2MgCO_3$$

$$MgO \cdot 2MgCO_3 + CO_2 \leftrightarrow 3MgCO_3$$

Moreover, at 50 bars of CO_2 pressure and a temperature of 300 °C, MgO has the highest yield of 98.54 %. At high temperatures (300–375 °C) but constant pressure (50 bars), we observe a slight but continuous decrease in the absorption of CO_2, which is in congruence with previously reported results [30]. It is well known that at low temperature, the physisorption process dominates but at elevated temperature CO_2 chemisorbs on MgO and thus CO_2 uptake capacity gradually starts decreasing.

CO_2-uptake capacity of a metal oxide is primarily dominated by the factors such as surface area, pore volume, pore functionality and pore size [28]. BET surface area measurement, pore volume and average pore sizes presented in Table 2 indicates the significant role of CO_2 to influence the particle structure. As received, 325 mesh size mesoporous MgO has a pore size of 9.09 nm and a high surface area (95.08 ± 1.5 m^2/g). Figure 4 illustrates that the N_2 adsorption–desorption isotherm curves for as-

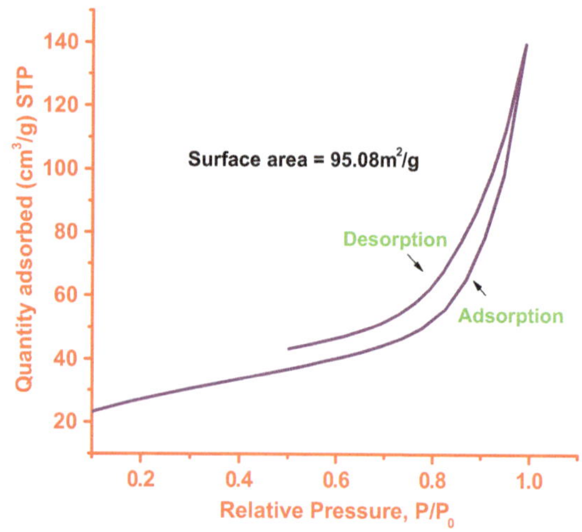

Fig. 4 N_2 adsorption–desorption isotherm of as-received MgO

received MgO follows Type IV isotherms (as per IUPAC classification); typical for mesoporous substances [39]. Also, the hysteresis pattern is H3 (following IUPAC classification) indicating the presence of slit-like pores. After capture of CO_2 at 350 °C and 10 bars, surface area dramatically decreased to (4.15 ± 0.05) m^2/g. The ratio of surface area to pore volume is also reduced by almost half and results in high diffusion paths. The significant decrease in surface area attributes to the basicity of mesoporous MgO. The mesoporous MgO is highly basic with well-ordered pores to hold high CO_2 at both lower and higher temperatures. It is well known that porous materials allow molecules to pass through their pore aperture for storage, separation or conversion [36]. MgO with a fine particle size (<44 μm) has a high content of mesopores, which leads to good mass transfer properties during the absorption process. The mechanism for metal oxide reaction with CO_2 gas has been discussed a lot [40–43]. CO_2 molecules diffuse through the pores of mesoporous MgO and the present large active sites hold these CO_2 molecules [19]. The trapped CO_2 molecules further react to form $MgCO_3$.

It can be observed from Table 2 that at 350 °C and 10 bars, 96.96 % of MgO was converted to $MgCO_3$ and almost all the pores have been utilized after reaction with CO_2 molecules. Henceforth, almost no CO_2 molecules could have diffused further in the pores. Noticeably, pore size increased to 16.25 nm. Thus, it is apparent that the porosity of MgO particles plays a very vital role for CO_2 uptake.

Previously, Beruto et al. has reported a very low CO_2 absorption capacity of MgO in absence of water vapor [27]. An uptake capacity of 0.089 and 0.091 mmol CO_2/g MgO was observed at 350 °C in 1.33 and 3.33 bars of CO_2, respectively. Thus, it was concluded that there is a high activation energy barrier to MgO recarbonation. On the contrary, Feng et al. [14] heated the sorbent (MgO) to 1000 °C in pure CO_2 and noted a low but almost constant (for 8 cycles) absorption capacity of 0.99 mmol CO_2/g MgO. Thus, unlike CaO-based sorbents, MgO does not show a fast decline in their CO_2 capture capacity over a large number of carbonation–calcination cycles. Bhagiyalakshmi et al. [20] synthesized basic mesoporous MgO (surface area of 250 m^2/g) using mesoporous carbon obtained from SBA-15 and obtained a maximum CO_2 adsorption of 2.27 mmol CO_2/g at 100 °C and nearly 1.81 mmol CO_2/g at 25 °C for a feed flow rate of 30 ml/min CO_2 gas (99.9 % purity). Therefore, it is evident that even for a relatively lower surface area (95 m^2/g compared to 250 m^2/g), MgO can have a better CO_2 uptake at a higher temperature (here, 300–375 °C) and higher CO_2 pressure (10–50 bars). Recently, Fagerlund et al. [28] reported a very high % conversion of MgO to $MgCO_3$ under high temperature and pressure conditions. However, MgO showed a slow conversion rate (50 % for ~7 h) for the condition [P_{CO2} (~18–19 bar), P_{H2O} (~1–2 bar) and 300–350 °C] which can be suitable for pre-combustion capture process. The reason could be the deposition of impervious carbonate layer on the surface of MgO which does not allow CO_2 molecules to further diffuse in the pores. Therefore, continuous removal of these nonporous layers of carbonates is needed. As mentioned earlier, we recommend to have an aggregative fluidization regime for a fluidized bed reactor while scaling up MgO–CO_2 reaction.

In summary, we demonstrate the increase in uptake of CO_2 by mesoporous MgO at high temperature and pressure in the humid environment. We observed the significant role of temperature, pressure, water vapor and porosity of the sorbent on the improvement of the CO_2 absorption capacity of mesoporous MgO.

Conclusion

MgO is a promising candidate for pre-combustion CO_2 capture due to their stable absorption capacity of CO_2, adequate mechanical strength, high attrition resistance and good regenerability. We observed a 96.96 % of MgO conversion to $MgCO_3$ at 350 °C and 10 bars CO_2 pressure. Moreover, CO_2 uptake capacity of MgO can be increased by performing the reaction in the presence of slight humid environment and scrubbing off the outer non-porous layer of $MgCO_3$. We also demonstrate that MgO has a low regeneration temperature (550 °C) and stable CO_2 capture capacity over wide range of temperature and pressure. Interestingly, the present study also illustrate that a high amount of oxymagnesite forms at a low partial pressure of water.

References

1. Kumar, S.: Clean hydrogen production methods, Springer Briefs in energy. Springer, New York (2015)
2. Merkel, T.C., Lin, H., Wei, X., Baker, R.: Power plant post-combustion carbon dioxide capture: an opportunity for membranes. J. Membr. Sci. **359**, 126–139 (2010)
3. Davison, J.: Performance and costs of power plants with capture and storage of CO_2. Energy **32**, 1163–1176 (2007)
4. Burton, B.P., Dupin, N., Fries, S.G., Grimvall, G., Fernández Guillermet, A., Miodownik, P., Oats, W. A., Vinograd, V., Z. Metallkd **92**, 514 (2001)
5. Miller, B., Tillman, D.: Combustion engineering issues for solid fuel systems. pp. 449–451. Elsevier (2008)
6. DOE/NETL: Carbon dioxide capture and storage RD&D roadmap, (2010)
7. Yu, C.H., Huang, C.H., Tan, C.S.: A review of CO_2 capture by absorption and adsorption. Aerosol Air Qual. Res. **12**, 745–769 (2012)
8. Evaluation of Dry Sorbent Technology; Presentation at the 2010 NETL CO_2 Capture Technology Meeting, Pittsburgh, PA, September 13–17, 2010
9. Kumar, S., Saxena, S.K.: A comparative study of CO_2 sorption properties for different oxides. Mater. Renew. Sustain. Energy **3**, 30 (2014)
10. Benson, S.M., Cole, D.R.: CO_2 sequestration in deep sedimentary formations. Elements **4**, 325–331 (2008)
11. Gruene, P., Belova, A.G., Yegulalp, T.M., Farrauto, R.J., Castaldi, M.J.: Dispersed calcium oxide as a Reversible and efficient CO_2-sorbent at intermediate temperatures. Ind. Eng. Chem. Res. **50**, 4042–4049 (2011)
12. Rackley, S.: Carbon capture and storage. Elsevier, Burlington (2010)
13. Duan, Y., Sorescu, D.C.: CO_2 capture properties of alkaline earth metal oxides and hydroxides: a combined density functional theory and lattice phonon dynamics study. J. Chem. Phys. **133**, 074508–074511 (2010)
14. Feng, B., An, H., Tan, E.: Screening of CO_2 adsorbing materials for zero emission power generation systems. Energy Fuels **21**, 426–434 (2007)
15. Li, L., Wen, X., Fu, X., Wang, F., Zhao, N., Xiao, F., Wei, W., Sun, Y.: MgO/Al_2O_3 sorbent for CO_2 capture. Energy Fuels **24**, 5773–5780 (2010)
16. Lee, S.C., Chae, H.J., Lee, S.J., Choi, B.Y., Yi, C.K., Lee, J.B., Ryu, C.K., Kim, J.C.: Development of regenerable MgO-based sorbent promoted with K_2CO_3 for CO_2 capture at low temperatures. Environ. Sci. Technol. **42**, 2436–2471 (2008)
17. Ward, S.M., Braslaw, J., Gealer, R.L.: Carbon dioxide sorption studies on magnesium oxide. Thermochim. Acta **64**, 107–114 (1983)

18. Choi, S., Drese, J.H., Jones, C.W.: Adsorbent materials for carbon dioxide capture from large anthropogenic point sources. Chemsuschem **2**, 796–854 (2009)
19. Fu, X., Zhao, N., Li, J., Xiao, F., Wei, W., Sun, Y.: Carbon dioxide capture by MgO-modified MCM-41 materials. Adsorpt. Sci. Technol. **27**, 593–601 (2009)
20. Bhagiyalakshmi, M., Lee, J.Y., Jang, H.T.: Synthesis of mesoporous magnesium oxide: its application to CO_2 chemisorption. Int. J. Greenhouse Gas Control **4**, 51–56 (2010)
21. Liu, S., Zhang, X., Li, J., Zhao, N., Wei, W., Sun, Y.: Preparation and application of stabilized mesoporous $MgO–ZrO_2$ solid base. Catal. Commun. **9**, 1527–1532 (2008)
22. Li, B., Wen, X., Zhao, N., Wang, X.Z., Wei, W., Sun, Y., Ren, Z.H., Wang, Z.J.: Preparation of high stability $MgO–ZrO_2$ solid base and its high temperature CO_2 capture properties. J. Fuel Chem. Technol. **38**, 473–477 (2010)
23. Yong, Z., Mata, V., Rodrigues, A.E.: Adsorption of carbon dioxide onto hydrotalcite-like compounds (HTlcs) at high temperatures. Ind. Eng. Chem. Res. **40**, 204–209 (2001)
24. Li, L., Li, Y., Wen, X., Wang, F., Zhao, N., Xiao, F., Wei, W., Sun, Y.: CO_2 capture over $K_2CO_3/MgO/Al_2O_3$ dry sorbent in a fluidized bed. Energy Fuels **25**, 3835–3842 (2011)
25. Lee, S.C., Choi, B.Y., Lee, T.J., Ryu, C.K., Soo, Y.S., Kim, J.C.: CO_2 absorption and regeneration of alkali metal-based solid sorbents. Catal. Today **111**, 385–390 (2006)
26. Ruminski, A.M., Jeon, K.J., Urban, J.J.: Size-dependent CO_2 capture in chemically synthesized magnesium oxide nanocrystals. J. Mater. Chem. **21**, 11486–11491 (2011)
27. Beruto, D., Botter, R., Searcy, A.W.: Thermodynamics of two, two–dimensional phases formed by CO_2 chemisorption on MgO. J. Phys. Chem. **91**, 3578–3581 (1987)
28. Lu, W., Yuan, D., Sculley, J., Zhao, D., Krishna, R., Zhou, H.C.: Sulfonate-grafted porous polymer networks for preferential CO_2 adsorption at low pressure. J. Am. Chem. Soc. **133**, 18126–18129 (2011)
29. Fagerlund, J., Highfield, J., Zevenhoven, R.: Kinetics studies on wet and dry gas-solid carbonation of MgO and $Mg(OH)_2$ for CO_2 sequestration. RSC Adv. **2**, 10380–10393 (2012)
30. Highfield, J., Bu, J., Fagerlund, J., Zevenhoven, R.: ICCDU-XI (2011)
31. Butt, D.P., Pile, S.P., Park, Y., Vaidya, R., Lackner, K.S., Wendt, C.H., Nomura, K.: Report LA-UR-98-7631 (1998)
32. Hassanzadeh, A., Abbasian, J.: Regenerable MgO-based sorbents for high-temperature CO_2 removal from syngas: 1. sorbent development, evaluation, and reaction modeling. Fuel **89**, 1287–1297 (2010)
33. Kato, Y., Yamashita, N., Kobayashi, K., Yoshizawa, Y.: Kinetic study of the hydration of magnesium oxide for a chemical heat pump. Appl. Therm. Eng. **16**, 853–862 (1996)
34. Liu, C., Shih, S.: Kinetics of the reaction of iron blast furnace slag/hydrated lime sorbents with SO2 at low temperatures: effects of the presence of CO_2, O_2, and NO_X. Ind. Eng. Chem. Res. **48**, 8335–8340 (2009)
35. Zevenhoven, R., Teir, S., Eloneva, S.: Heat optimisation of a staged gas-solid mineral carbonation process for long-term CO_2 storage. Energy **33**, 362–370 (2008)
36. Zevenhoven, R., Eloneva, S., Teir, S.: A study on MgO-based mineral carbonation kinetics using pressurized thermogravimetric analysis. GHGT-8, Norway, (2006)
37. Frykstrand, S., Stietzel, C., Forsgren, J., Angstrom, J., Potin, V., Stromme, M.: Synthesis, electron microscopy and X-ray characterization of oxymagnesite, $MgO{\cdot}2MgCO_3$, formed from amorphous magnesium carbonate. CrystEngComm **16**, 10837–10844 (2014)
38. Fagerlund, J., Nduagu, E., Romao, I., Zevenhoven, R.: CO_2 fixation using magnesium silicate minerals part 1: process description and performance. Energy **41**, 184–191 (2012)
39. Li, R., Chen, G., Dong, G., Sun, X.: Controllable synthesis of nanostructured TiO2 by CTAB-assisted hydrothermal route. New J. Chem. **38**, 4684–4689 (2014)
40. Deng, H., Grunder, S., Cordova, K.E., Valente, C., Furukawa, H., Hmadeh, M., Gándara, F., Whalley, A.C., Liu, Z., Asahina, S., Kazumori, H., O'Keeffe, M., Terasaki, O., Stoddart, F.J., Yaghi, O.M.: Large-pore apertures in a series of metal-organic frameworks. Science **336**, 1018–1023 (2012)
41. Bhatia, S.K., Perlmutter, D.D.: Effect of the product layer on the kinetics of the CO_2 –Lime Reaction. AIChE J. **29**, 79–86 (1983)
42. Butt, D.P., Lackner, K.S., Wendt, C.H., Conzone, S.D., Kung, H., Lu, Y.C., Bremser, J.K.: Kinetics of thermal dehydroxylation and carbonation of magnesium hydroxide. J. Am. Ceram. Soc. **79**, 1892–1898 (1996)
43. Kumar, S.: The effect of elevated pressure, temperature and particle morphology on the carbon dioxide capture using zinc oxide. J. CO_2 Util. **8**, 60–66 (2014)

13

Phase change materials (PCMs) integrated into transparent building elements: a review

Paris A. Fokaides[1] · Angeliki Kylili[1] · Soteris A. Kalogirou[2]

Abstract Phase change materials (PCMs) represent an innovative solution that can contribute to the improvement of the energy performance of buildings. Recently a trend towards integrating PCMs into transparent envelope components is observed. This study aims to present the main solutions proposed in the literature for applications in the past few years for PCMs integrated into transparent buildings elements. The temporal development of this application as well as the fundamental principles of its operation is described in detail. The concept of the existing transparent PCM systems is presented, and the rationale of selecting appropriate materials is discussed. This is followed by the current practices in testing the thermal performance of transparent PCMs. The future trends in terms of the current barriers and the potential improvements are discussed. To this end the future technologies of transparent PCMs are also considered.

Keywords Phase change materials (PCMs) · Transparent · Building elements

✉ Paris A. Fokaides
eng.fp@frederick.ac.cy; p.fokaides@frederick.ac.cy

[1] School of Engineering and Applied Sciences, Frederick University, Cyprus, 7., Y. Friderickou Str., 1036 Nicosia, Cyprus

[2] Department of Mechanical Engineering and Materials Science and Engineering, Cyprus University of Technology, Limassol, Cyprus

Introduction

The promotion of buildings' energy efficient design is one of the main priorities of the research community in the building sector. A novel approach, which in recent years gaining more and more ground, is the use of phase change materials (PCMs). PCMs have the ability to change their phase (typically from solid to liquid and vice versa) at room temperature. This results to the storage of the amount of latent energy which is required to change the phase of the material. PCMs represent a highly efficient solution, as their use improves building shell's energy management. In particular, the use of these materials enables the better management of the energy flow to and from the building, decelerating in this way the rate of thermal losses. PCMs thermal performance is a challenging scientific field as the phase change is accompanied by a change of the materials' key properties, such as the heat capacity and thermal conductivity.

PCMs are used in a range of applications in the technical and building sector. In the latter case, they are mainly used in opaque components, typically in walls and ceilings. Lately, the use of PCM-doped infrared reflective coatings was reported [1]. However, in the recent years, a trend towards integrating these materials into transparent envelope components or employing transparent PCM in building systems is being observed. This review work focuses on the transparent glazing systems that incorporate PCMs, and considering the fact that these building components, particularly in warm dominant climates, are more vulnerable to thermal losses due to their higher thermal transmittance and higher radiation losses, this incorporation is of special importance.

The synthesis and the characteristics of stearic acid (SA)/expanded graphite (EG) composites as thermal energy storage materials were reported in Fang et al. [2]. The authors denoted that the properties of these composites can be used in the exterior walls of buildings so that the cooling load of air-conditioning system is reduced. Similarly, the aim of the work of Chwieduk [3] was to point out ways of substituting thick and heavy thermal mass external walls, which are typical in buildings in high latitude countries, by thin and light thermal mass PCM-incorporated walls that meet the same comfort levels throughout the whole year.

Additionally, Kuznik and Virgone [4] investigated experimentally the thermal performances of a PCM copolymer composite wallboard in a full-scale test room. The results indicated that the PCM-incorporated wallboard achieves the lowering of the air temperature in the room by up to 4.2 °C and also enhances the natural convection in the room. The authors also stated that the specific PCM is particularly interesting for its use in renovation building projects.

The work of Medina et al. [5] concerned with the merging of two building technologies, structural insulated panel and PCM, into a new product. The results of the thermal performance evaluation of their new product indicated that the peak heat flux reductions achieved with 10 and 20 % PCM incorporation were 37 and 62 % respectively, while the average reductions in daily heat transfer were 33 and 38 % for concentrations of 10 and 20 % PCM, respectively. The thermophysical properties and the process of melting of a PCM, an epoxy resin/paraffin spheres composite in particular, were investigated numerically and experimentally, using a transient hot plate apparatus and differential scanning calorimetry respectively [6]. Based on their findings the authors concluded that this material can be used to improve comfort, save energy in buildings and reduce the weight of wallboards. Xu and Li [7] developed a paraffin/diatomite/multi-wall carbon nanotubes composite PCM for its incorporation into thermal energy storage cement-based composites. The use of this kind of composite was found to have beneficial effects for improving the thermal conductivity and heat storage, without affecting the thermal properties, chemical compatibility and thermal stability of the cement-based composites.

Several reviews have been published on the use of PCM in buildings. Zalba et al. [8] presented the history of thermal energy storage with solid–liquid phase change. Materials used by researchers as potential PCM and commercial PCM were described, together with their thermophysical properties. Farid [9] published a review about PCM building applications, which provides an insight to efforts to develop new classes of PCMs for use in energy storage in building applications. Tyagi and Buddhi [10] published a comprehensive review of various possible methods for heating and cooling in buildings. The thermal performance of various types of systems like PCM Trombe wall, PCM wallboards, PCM shutters, PCM building blocks, air-based heating systems, floor heating, ceiling boards, etc., was presented. The main focus of Kenisarin and Mahkamov [11] review was the assessment of the thermal properties of various PCM, methods for heat transfer enhancement and design configurations of heat storage facilities to be used as a part of solar passive and active space heating systems, greenhouses and solar cooking.

Sharma et al. [12] review summarizes the investigation and analysis of the available thermal energy storage systems incorporating PCM for use in different applications. Zhu et al. [13] presented an overview of the previous research work on dynamic characteristics and energy performance of buildings due to the integration of PCM. Wang et al. [14] presented the concept of ideal energy-saving building envelope, which is used to guide the building envelope material selection and thermal performance design. Finally, a review on the use of PCM for the free cooling of buildings is presented by Aroul et al. [15].

This review paper aims to present the latest developments in the application of PCMs in transparent components. This is not considered in any of the above reviews. A brief history of the PCMs and the fundamentals of operation of PCM-incorporated glazing systems will be primarily presented in sections "PCM-incorporated transparent glazing development and fundamentals of operation" and "Current practices of transparent PCM-incorporated systems", respectively. The main solutions that have been implemented in recent years and materials used in transparent phase change components as well as their main properties will be referred in section "Current practices in testing the thermal performance of transparent PCM-incorporated systems". A discussion regarding the current barriers, the considerations for their further development, as well as the near-future trends is presented in section "Future trends", while the findings of this review paper are given in the last section.

PCM-incorporated transparent glazing development and fundamentals of operation

Transparent PCM and PCM-incorporated transparent glazing development

The first documented use of a PCM as a form of latent heating was by Dr. Maria Telkes in 1948 [16]. The Hungarian research associate of Massachusetts Institute of Technology (MIT) became one of the pioneers in this field through the construction of the very first PCM heated house with architect Eleanor Raymond. The Solar One building housed approximately 4 m^2 of Glauber salts, which were ventilated with fans to deliver warm air inside the building in winter, and cool air in summer. In fact, this system was able to keep the house warm for approximately 11 sunless days. After 3 years, the house failed because the utilized PCM lost most of its energy storage capacity [17]. Since then, the development of PCMs for passive heating has been widely studied [18, 19]; however, their application developments had been slow. Harland et al. [16] attributed this slow development to the high costs of PCMs that was an obstacle to become a design option.

The earliest mention in literature regarding translucent PCM was found in Manz et al. [20]. The authors of this study proposed a two-layered passive wall system, combining a salt hydrate phase change material and a transparent honeycomb-type insulation material (Fig. 1). The PCM was filled into glass containers, which were commercially available glass blocks. A few years later, Ismail and Henriquez [21] were the first to report on the results of an experimental investigation on simple and composite glass windows filled with PCM.

PCM-incorporated transparent glazing fundamentals of operation

The passive solar mechanism of PCM glazing systems is offered by the outer and inner insulating glazing units. The outer insulated glazing unit with suspended prismatic filter between the panes of glass reflects the higher angle sunlight back out, transmitting the low-angle sunlight into the inner unit that is filled with sealed polycarbonates into which the PCM is encapsulated. As a result, the strong sunlight of the summer is kept out of the building, while the lower angle winter sunlight is allowed to be captured by the PCM as shown in Fig. 2 [22]. A PCM material with a (low) melting point close to the comfort or operational temperature in the building and high latent heat of fusion has the great latent heat storage potential. This potential is enabled by the phase transition between the liquid and the solid state and thus allows an increase of the thermal inertia of the glazing system [23].

During the melting process from solar radiation, the specific heat of PCM increases to more than one hundred times to absorb a large quantity of energy [24]. By nighttime the external temperature decreases below the melting point of the PCM, so that the PCM starts to solidify, releasing into the building its stored energy. By the next morning, the PCM typically solidifies completely to start the cycle again [22]. The net effect of this process is the reduction of heat flow from outdoor to indoor space during daytime [24].

Fig. 1 Layout of TIM–PCM external wall system (Manz et al. [20])

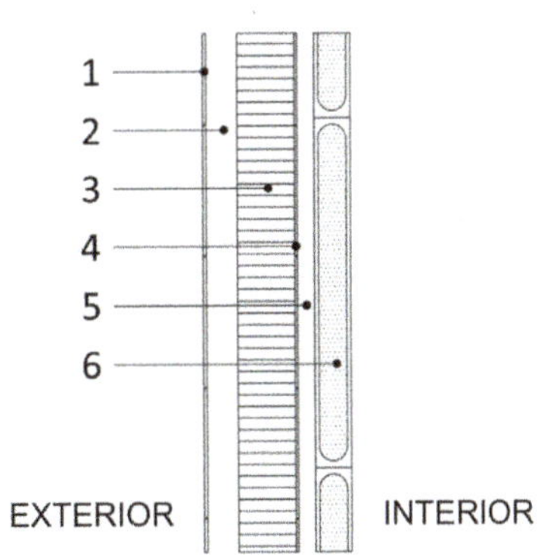

Layer Id	Description	Thickness [mm]
1	Glass pane	-
2	Air gap	-
3	Transparent insulation material	-
4	Glass pane	-
5	Air gap	-
6	PCM in a glass container	-

Fig. 2 Fundamentals of operation of transparent PCM (Figure reprinted with the written permission of the copyright holder)

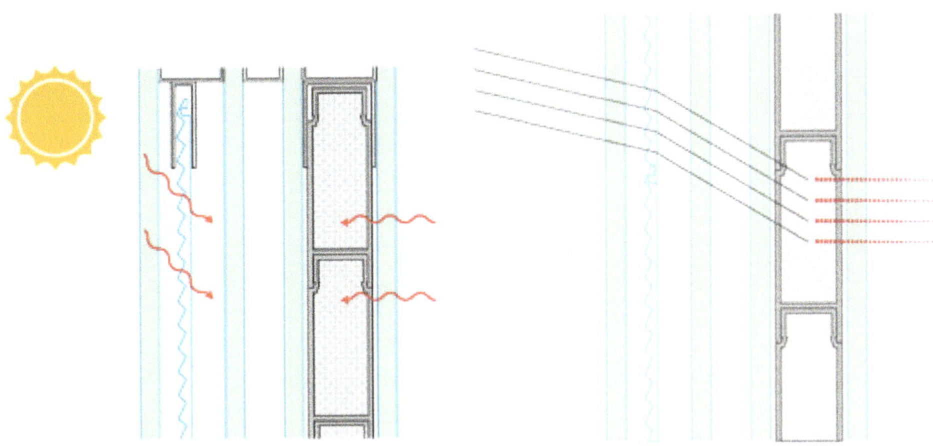

Current practices of transparent PCM-incorporated systems

Concept of existing transparent PCM-incorporated wall systems

Several transparent building elements incorporating translucent, or see-through, PCM have been proposed in the literature the recent years. Transparent PCM is a material that allows light to pass through but absorbs the infrared part of the spectrum and this increases the temperature of the PCM up to the point of melting, which as in all PCMs is done at constant temperature. The PCM can be transparent to the internal space, and act like a window, or it may be transparent to an opaque material (i.e., the wall element) in front of which it is installed. Ismail and Henriquez [21] presented a relatively simple and effective concept. The suggested layout consisted of a double glass window, separated by a gap of certain width, filled with a PCM of certain fusion temperature, as shown in Fig. 3.

Weinlaeder et al. [25] encapsulated PCMs in transparent plastic containers. The containers were placed behind a conventional double glazing with an air gap of about 10 mm as shown in Fig. 4. Numerical analysis of a double glazing containing PCM in the spacing between the two panes, shown in Fig. 5, was performed by Goia et al. [26]. A numerical procedure based on a 1D nodal model was

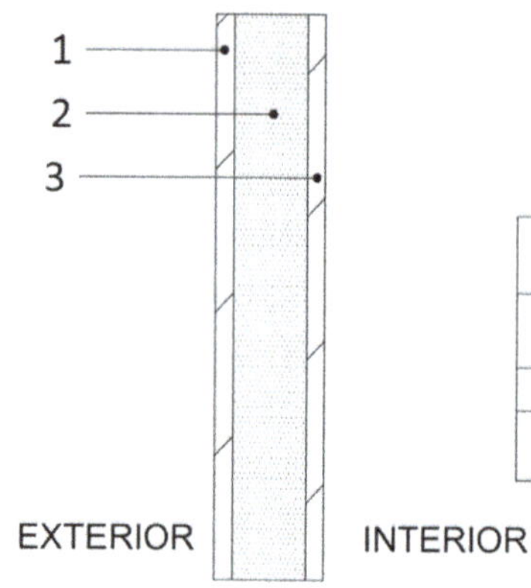

Layer Id	Description	Thickness [mm]
1	Glass	-
2	PCM Layer	-
3	Glass	-

Fig. 3 Layout of double glass window with PCM (Ismail and Henriquez [21])

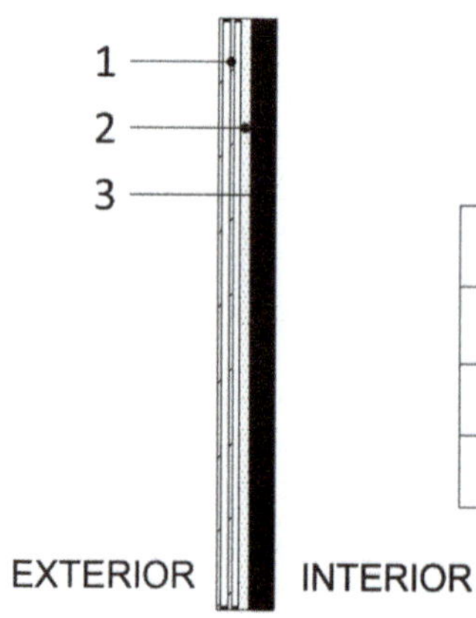

Layer Id	Description	Thickness [mm]
1	Double Glazing	-
2	Plastic container with PCM	16
3	Wall heating/cooling	-

Fig. 4 Layout of PCM-facade-panel (Weinlaeder et al. [25])

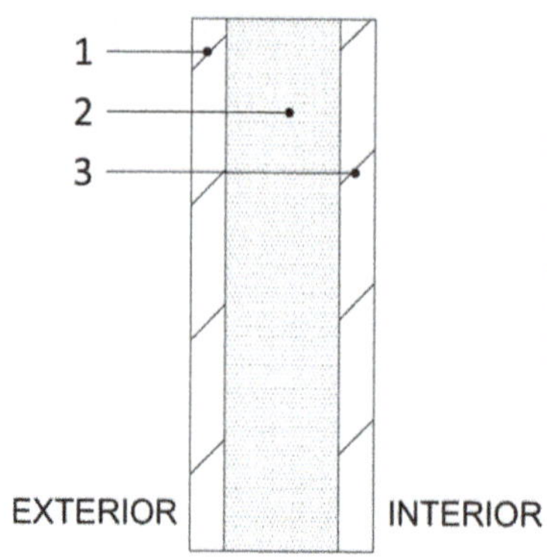

Layer Id	Description	Thickness [mm]
1	External glass pane	3/16
2	PCM Layer	10/16
3	Internal glass pane	3/16

Fig. 5 Layout of PCM glazing (Goia et al. [29])

developed, neglecting heat and shortwave radiation transmission along the x and y axes. In their standard model, three nodes were associated to each glass pane, while the PCM layer was represented by means of five nodes.

In Goia et al. [27], a prototype of a simple PCM glazing system was proposed, while its behavior was compared with that of a conventional reference double glazed unit. The analysis indicated that the PCM glazing system provided better indoor thermal conditions than the conventional system for most of the time during the various seasons. In addition, it was found that the higher the outdoor solar irradiance, the greater the benefits offered by PCM glazing unit.

The aim of the study in Gowreesunker et al. [28] was the evaluation of the radiation heat transfer properties and performance of PCM glazing systems. The optical properties were evaluated using spectrophotometry and the thermal properties were assessed using the T-history method.

The aim of the work in Goia [29] was to investigate the location of the PCM layer and the nominal melting temperature of the PCM, which are two variables that play significant role in PCM glazing systems. Different PCM glazing configurations had been tested by means of a numerical analysis to assess their behavior with respect to the energy efficiency and thermal performance. The results indicated that the position of the PCM layer has a relevant influence on the thermophysical behavior of the PCM glazing system. PCM glazing systems, especially those with the PCM layer inside the outermost cavity can be beneficial in terms of thermal comfort, while a more complex evaluation is needed when the energy performance is addressed. The results of the work undertaken in Goia et al. [30, 31] have highlighted a good ability of the PCM glazing to store solar energy and to smooth and delay peak values of the total heat flux.

Buddhi et al. [32] investigated the thermal performance of a wooden box having a PCM-integrated window in south direction. The PCM employed was commercial grade lauric acid, which enabled to keep the temperature of the experimental test cell higher during evening and night. An experimental and numerical simulation study of the application of PCMs in translucent building elements was also presented in Bontemps et al. [33]. In fact, the experimental study was conducted in a test cell split by a wall made of hollow glass bricks filled with PCM. The PCMs tested were fatty acid, paraffin, and salt hydrate.

Additionally, a theoretical thermal analysis of a window with a shutter containing PCM was presented in Alawadhi [24]. The results indicate that the melting temperature of PCM must be close to the upper temperature limit of windows during daytime, while the quantity of PCM should be adequate to absorb a lot of heat during the day, without the PCM melting. Also, the presented shutter of 0.03 m thickness was able to reduce the heat gain by 23.29 %. The potential of incorporating PCMs in structural cells of shading elements associated to southward façade window was also presented in Soares et al. [34]. It was concluded that numerical simulation can be used to define an optimal system configuration, with an optimal PCM melting temperature for any location and characteristic climatic data.

Heim [35] carried out simulations of a transparent insulation wall integrated with a PCM. His system comprised of an external glazing, a transparent insulation material (TIM) honeycomb structure, and a PCM-ceramic composite, which were separated by air gaps to form a triple zone building model. The results showed the properties of the PCM enabled 5 % more stable temperature for the internal space during winter. For the work of Weinlaeder et al. [36], an interior sun protection system consisting of vertical slats filled with PCM was monitored for more than 2 years. It was observed that the surface temperature on the interior side of the PCM-filled slats did not exceed the PCM melting temperature of 28 °C, while conventional systems often reached 40 °C. This proved the significant contribution of PCM technology in improving the thermal comfort of buildings.

Further to the incorporation of transparent PCM in transparent building elements, various PCMs have also been exploited as thermal storage materials for cooling, heating, thermal comfort, and thermal source in buildings [37]. Diarce et al. [38] investigated the thermal performance of a new type of ventilated active façade that includes a PCM in its outer layer, as shown in Fig. 6. Their prototype façade was designed to be suitable for use in building restoration. The external layer contains the PCM, inside an aluminum macroencapsulation system, and had a thickness of 20 mm. The ventilated air channel was situated between the external layer and the insulation material, and it was 60 mm thick. Insulation material (50-mm-thick plates of extruded polystyrene) was placed next to the air gap. The inner layer was a brickwork wall made of 70-mm-thick hollow common bricks, and a 2-mm-thick plasterboard lined the inner walls of the room. A new type of ventilated facade with macro-encapsulated PCM in its air cavity, as shown in Fig. 7, was presented by De Gracia et al. [39]. The ventilated facade presented an air channel of 15 cm thick which represented 0.36 m^2 of channel area. The inner layer was based on the alveolar brick constructive system while the outer envelope was made by a glass layer. An extra outer layer of expanded polyurethane panels was placed to cover the transparent glazing during the summer period since solar radiation inside the cavity was avoided for cooling purposes.

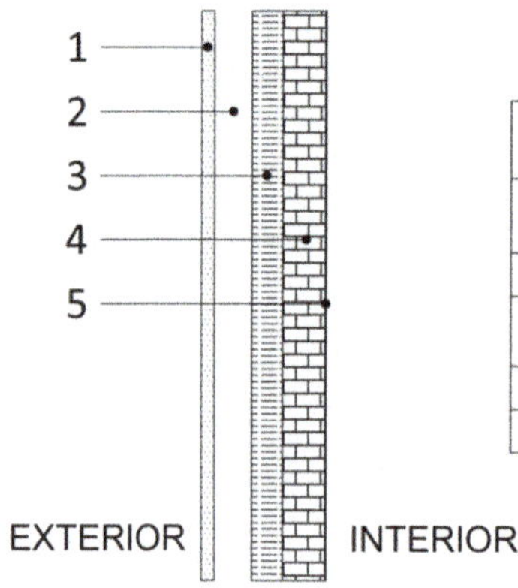

Layer Id	Description	Thickness [mm]
1	External sheet containing PCM	20
2	Ventilated air gap	60
3	Extruded polyestyrene insulation	50
4	Brickwork inner wall	70
5	Gypsum plasterboard	2

Fig. 6 Layout of ventilated active façade (Diarce et al. [38])

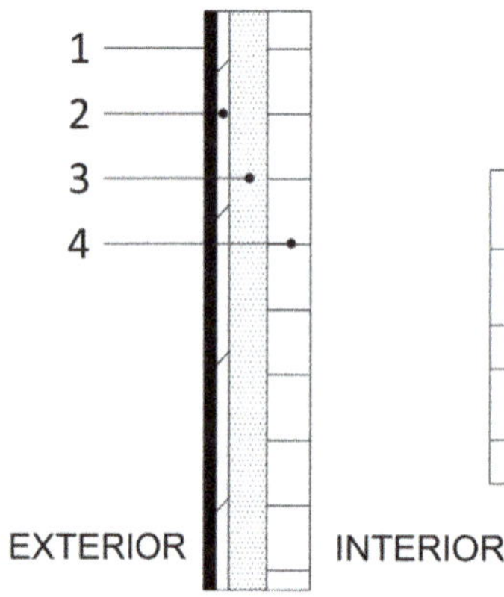

Layer Id	Description	Thickness [mm]
1	Opaque insulation panel	-
2	Glass layer	-
3	PCM panels and air channel	-
4	Alveolar brick	-

Fig. 7 Layout of ventilated facade with PCM (De Gracia et al. [39])

PCM materials used in transparent building elements

More than 50 % of incident solar radiation's energy is measured in the visible spectrum range. Based on this fact, the concept of a transparent element should allow visible radiation to be transmitted and invisible radiation in the infrared part of the spectrum to be absorbed and converted into heat. Therefore, the selected PCMs should not only have high heat capacity, but also favorable absorption and transmission under specific wave lengths. The light transmittance of water coincides perfectly with the sensitivity curve of the human eye. For this reason, high water content PCMs are mainly selected for transparent building elements. Subsequently, the same authors [20] used a salt hydrate PCM.

The categorization of PCMs as originally suggested by Abhat [40] is given in Fig. 8, and the main physical properties of the employed PCMs for transparent building elements are summarized in Table 1. The use of $CaCl_2$. $6H_2O$ as PCM was first proposed by Swanton [41]. Manz [42] gives an overview of the research work which has been carried out to improve the thermal properties of the pure substance with additives, especially in order to suppress supercooling and to reach long-term stability. Their experimental investigations were carried out using a commercially available PCM based on $CaCl_2{\cdot}6H_2O$ with additives of approximately 5 % of the mass. This material shows a melting enthalpy of 192 J g^{-1} and a melting temperature interval of 24–29 °C. Porisini [43] tested the long-term stability of this PCM in small quantities and found no change in thermal properties after 5650 melting/solidification cycles. Later, Buddhi and Sharma [44] presented the experimental measurements of the transmissivity of commercial grade stearic acid, which revealed that the low thermal conductivity and high transmissivity of the material enabled its employment as a transparent insulating material.

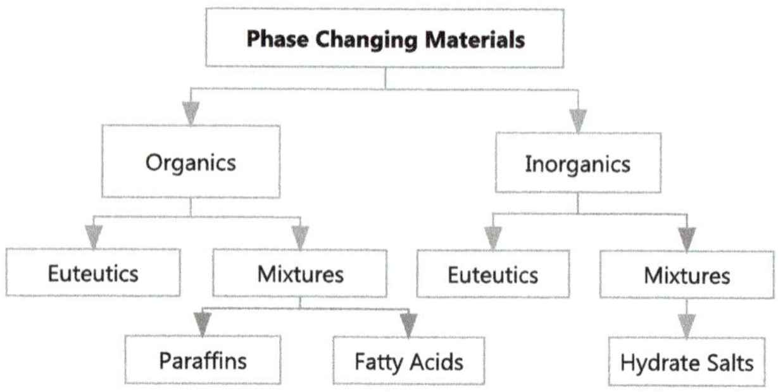

Fig. 8 Classification of PCMs (Zalba et al. [8])

Table 1 The main physical properties of the employed PCMs for transparent building elements

Work	PCM	Melting area (°C)	Congealing area (°C)	Heat storage capacity ($kJ\ kg^{-1}$)	Specific heat capacity ($kJ\ kg^{-1}\ K^{-1}$)	Density ($kg\ l^{-1}$)	Heat conductivity ($W\ m^{-1}\ K^{-1}$)
Manz et al. [20]	$CaCl_2{\cdot}6H_2O$	24–29		192			
Ismail and Henriquez [24]	Mixture of glycol						
Buddhi et al. [32]	Commercial grade lauric acid	42.2		181			
Heim [35]	Fatty acids	20/30/40		25			
Athienitis et al. [47]	Gypsum board (13 mm thick) soaked with butyl stearate	16.8–20.9		29.1			0.18
Weinlaeder et al. [25]	Paraffin wax RT25 [60]	22–26	26–22	148 ± 7.5 %	2	Sol. 0.88	0.2
						Liq. 0.76	
	$CaCl_2{\cdot}6H_2O$	27 ± 0.5		190 ± 19	Liq. 2.22 ± 0.11	Sol. 1.7 ± 0.17	Sol. 0.79 ± 0.03
					Sol. 1.50 ± 0.08	Liq. 1.53 ± 0.08	Liq. 0.48 ± 0.04
	$LiNO{\cdot}3H_2O$	30 ± 0.5		270 ± 27	Liq. 1.79 ± 0.09	Sol. 1.556 ± 0.156	Sol. 1.02 ± 0.05
					Sol. 1.23 ± 0.06	Liq. 1.4 ± 0.007	Liq. 0.56 ± 0.03
Bontemps et al. [33]	Fatty acid capric/myristic	21.4		162			
	Paraffin C 18–20 (Linpar)	25		170			
	Salt hydrate	27.5		192			
Goia [29]	Paraffin wax	Peak 33		140	2.5	0.8	0.2
Goia et al. [26]	Paraffin wax	Peak 25/30/35		140	2.5	0.8	0.2
Goia et al. [48]	Paraffin wax	35		160			
De Gracia et al. [39]	Salt hydrate SP-22	22	18	150			
Diarce et al. [38]	RT35 [60]	29–36	36–31	170	2	Sol. 0.86	0.2
						Liq. 0.77	
Goia et al. [27]	Paraffin wax	34		145			
Gowreesunker et al. [28]	RT27 organic paraffin [60]	25–28		184			
De Gracia et al. [49]	Salt hydrate SP-22	21.5	18				
Goia et al. [30]	Paraffin wax	35		170		Sol. 0.88	
Goia et al. [31]	RT35HC [60]	34–36	36–34	240 + 7.5 %	2	Sol. 0.77	0.2
						Liq. 0.67	

PCM incorporated in glazing products

Currently, there are commercially available types of glazing such as Delta®-Cool 28 or GlassX® that incorporate PCM. The PCM incorporated in the GlassX products is comprised of $CaCl_2{\cdot}6H_2O$ salt hydrates that are completely sealed in clear polycarbonate. The products have a latent thermal storage of up to 4268 kJ/m^2, meaning that a period of 8 h passes before heat is transmitted [45]. This property shifts the peak energy demand of the summer later at night and thus reduces the interior room temperatures by at least 5 °C. While in winter, the solar gains are maximized by charging up the PCM during the day in order to release it back into the building throughout the night as it solidifies. The manufacturers of GlassX® claim that their products reduce a building's heating and cooling loads by 30–50 % with only 1/3 coverage on a south-facing façade compared to opaque, insulating walls such as brick, wood, or metal panels [45]. Regarding Delta®-Cool 28, the manufacturers demonstrated its application in translucent Polymethyl Methacrylate (PMMA) panels in a glass facade system of a zero energy office building in Kempen, Switzerland. Every other window panel was equipped with PCM for reducing the solar heating of the interior [46].

Current practices in testing the thermal performance of transparent PCM-incorporated systems

Both laboratory and outdoor testing facilities as well as specialized equipment were employed for the definition of the thermal performance of the proposed transparent building elements incorporating translucent PCM. Manz et al. [20] tested the PCM optical properties indoors in a laboratory using a spectrophotometer and Abbe refractometer, while the experimental investigation of the TIM and transparent PCM prototype facade was conducted in outdoor test facilities consisted of a cooling/heating apparatus and heat flow sensors. Their laboratory equipment allowed them to define also the refraction index of the PCM, while their outdoor test facility employed enabled them to measure also liquid mass flow temperatures, the heat flux, and the solar irradiance.

Ismail and Henriquez [21] characterized optically the different glass sheets of different thicknesses and panels of double sheets of different spacings and fillings employing only one laboratory equipment—a spectrophotometer. Consequently, their measurements were limited to the determination of transmittance, reflectance, and absorbance of the PCM. Athienitis et al. [47] had developed an outdoor test facility with two configurations of the double facade with photovoltaic panels that incorporated a differential scanning calorimeter (DSC), which recorded the latent heat and the transition temperatures of the glazing system. Weinlaeder et al. [25] and Goia et al. [26] set up their experiments for the PCM-integrated glazing systems in outdoor test cells. Temperature, heat flux, and incoming and transmitted solar radiation measurements were taken in the work of Weinlaeder et al. [25], while the outside test cell of Goia et al. [26], equipped with thermocouples, heat flux meters, and pyranometers collected data regarding the indoor surface temperatures of the glazing, the outdoor solar irradiance, the indoor surface heat flux, and also the transmitted solar irradiance of the PCM glazing system. For a subsequent work, Goia et al. [48] employed the Characterization of Advanced Transparent Materials (CATRAM) facility for the investigation of the proposed PCM-incorporated transparent system consisting of a halogen lamp and three array spectrometers.

Bontemps et al. [33] also configured outdoor test cells for the testing of the wall made of glass bricks filled with PCMs. Their equipment included fluxmeters and thermocouple for the measurements of heat flux and indoor temperature. Additionally meteorological data were taken from a nearby meteorological station. De Gracia et al. [39] used house-like cubicles with thermocouples, a moisture/temperature transducer, an electrical network analyzer, a heat pump, an air flow velocity–temperature transmitter, a pressure transmitter, and heat flux sensors for the investigation of their ventilated facade with macro-encapsulated PCM. Additional equipment, including solar pyranometers and an anemometer, was used for the data collection of the outdoor weather conditions. Their testing facility and equipment allowed them to take a range of measurements to evaluate the performance of the ventilated facade with macro-encapsulated PCM, which are given in Table 2. The active ventilated facade with a PCM, suggested by Diarce et al. [38], was studied in a PASLINK test cell with temperature probes and also employed air velocity sensors and pyranometers. Thus, in their work Diarce et al. [38] presented data of the global vertical radiation, the outdoor temperatures, and the temperatures inside the PCM at the bottom and the top of the façade.

Furthermore Goia et al. [27, 30] employed the Testing Window INnovative System (TWINS) outdoor test cell facility equipped with heat flux sensors, thermocouples and pyranometers for measuring solar irradiance, while Goia et al. [27] also employed an infrared thermal camera for their measurements. In his research, Gowreesunker et al. [28] employed an environmental chamber with controlled air temperature along with thermistors, a spectrophotometer, a pyranometer, and a 150 W metal halide (iodide discharge) lamp to enable the comparison between a PCM-glazed unit and a traditional

Table 2 The employed testing facilities and measurements for the investigation of the thermal performance of PCMs for transparent building elements

Work	Investigated object	Test conditions	Equipment	Performed measurements
Manz et al. [20]	External wall system composed of transparent insulation material (TIM) and translucent phase change material (PCM)	Laboratory	Spectrophotometer	Spectral transmittance and reflectance
			Abbe refractometer	Refraction index of the PCM
		Outdoor test facility	Cooling/heating apparatus	Liquid mass flow temperatures
			Heat flow sensors	Heat flux
				Irradiance
Ismail and Henriquez [24]	PCM glass systems of thicknesses 3, 4, 5, 6 and 8 mm	Laboratory	Spectrophotometer	Transmittance
				Reflectance
				Absorptance
Athienitis et al. [47]	Double facades with phase change storage and photovoltaics	Outdoor test facility	Differential scanning calorimeter (DSC)	Latent heat
				Transition temperatures
Weinlaeder et al. [25]	Double glazings combined with PCM	Outdoor test facility		Temperature
				Heat flux
				Incoming solar radiation
				Transmitted solar radiation
Bontemps et al. [33]	Wall made of glass bricks filled with PCM	Outside test cells	Fluxmeters	Heat flux
			Thermocouple	Indoor temperature
			Meteorological station	Meteorological parameters
Goia et al. [26]	PCM glazing systems	Outside test cell	Thermocouples	Indoor surface temperatures of the glazing PCA
			Heat flux meters	Outdoor solar irradiance
			Pyranometers	Indoor surface heat flux
				Transmitted solar irradiance of PCM glazing
Goia et al. [48]	PCA incorporated transparent system	Characterization of Advanced TRAnsparent Materials (CATRAM) facility	Halogen lamp	Transmittance
			Three array spectrometers	Reflectance
				Absorptance
De Gracia et al. [39]	Ventilated facade with macro-encapsulated phase PCM in its air cavity	House-like cubicles	Thermocouples	Internal wall temperatures
			Moisture/temperature transducer	Indoor air temperature and humidity
			Electrical network analyzer	Electrical consumption of the HVAC systems
			Heat pumps	
			Thermocouples	Air temperature of the cavity
			Air flow velocity Temperature Transmitter	Air velocity of the cavity
			Pressure transmitter	Pressure drop across the air cavity
			Heat flux sensors	Outer and inner skin surface temperature
				External surface temperature
				Heat flux transferred to indoor
				Heat flux transferred to the front and back surface of a PCM panel
				Temperature of the PCM
				Front and back surface temperature of the PCM panels
			Solar pyranometers	Horizontal and vertical global solar radiation
			Moisture/temperature measuring transducer	Outer air temperature and humidity
			Anemometer	Wind speed and direction

Table 2 continued

Work	Investigated object	Test conditions	Equipment	Performed measurements
Diarce et al. [38]	Active ventilated facade with a PCM	PASLINK test cell	Temperature probes Air velocity sensors Pyranometers	Global vertical radiation Outdoor temperatures Temperatures inside the PCM at the bottom and the top of the facade
Goia et al. [27]	Prototype of a simple PCM glazing system	Test cell facility (TWINS—Testing Window INnovative System)	Thermocouples Heat flux meters Pyranometers Thermal infrared camera	Indoor surface temperature Surface heat flux Incident solar irradiance Spot measurements of the surface temperature field
Gowreesunker et al. [28]	PCM-glazed unit and traditional double glass unit	Environmental chamber with controlled air temperature	Thermistors Spectrophotometer Pyranometer 150 W metal halide (iodide discharge) lamp	Temperature measurements of the samples and the chamber temperature transmittance radiation heat flux
De Gracia et al. [49]	Ventilated facade with PCM			Thermal profiles of the PCM
Goia et al. [30]	PCM glazing prototype	Outdoor test cell facility (TWINS—Testing Window INnovative System)	Heat flux sensors thermocouples pyranometers	Surface heat fluxes Temperatures Incident and transmitted solar irradiance
Goia et al. [31]	Advanced dynamic glazing prototype integrating PCM and thermotropic layers	Outdoor test cell facility	Thermocouples heat flux meters pyranometers	Surface and air temperature measurements; Heat flux Outdoor and transmitted solar irradiance

double glass unit. Finally, the advanced dynamic glazing prototype integrating PCM and thermotropic layers proposed by Goia et al. [31] was also tested in an outdoor test cell facility, employing thermocouples, heat flux meters and pyranometers.

It should be noted that in Goia et al. [27], the use of the infrared thermal camera also allowed spot measurements of the surface temperature field. De Gracia et al. [49] also presented the thermal profiles of the PCM. The employed testing facilities, equipment and measurements for the investigation of the thermal performance of PCMs for transparent building elements in literature are summarized in Table 2.

Future trends

Current barriers

Although PCMs integrated into transparent building elements are very promising, their application in buildings is still considered rather new. There are still several barriers that these systems have to overcome, in order to make them more interesting and penetrate the construction industry. The main design issue associated with these glazing systems regards the change in transparency of the PCM as it changes phase. Currently, only organic PCMs have been found to be transparent in the solid state and fully transparent to visible light in the liquid phase for their employment in glazing systems [28]. However, they allow high amounts of visible light to pass through but do not offer the same visibility as regular windows [50].

An additional problem with passive PCM systems is their reliability. Weinlaeder et al. [36] reported that passive PCM systems might not work in cases such as during long hot days when the PCM could not be completely discharged during the night. Additionally, Soares et al. [51] stated that most of the studies have been focused on the performance of these systems for extreme summer or winter conditions, eliminating their functionality during the rest of the seasons. Another barrier reported in literature is the low thermal conductivity of many promising PCMs, which vary in the values between 0.15 and 0.2 W/mK for organic PCMs and around 0.5 W/mK for inorganic salts. The low thermal conductivity decreases the rate of heat absorption and releases throughout the PCM, so that the product may not fully utilize its full latent heat storage [50]. The results of Evola et al. [52] experiments indicated that the PCM wallboards utilized only the 45 % of the possible latent heat storage, due to several factors including thermal conductivity.

Further work for improvements

After the identification of the weak points of PCMs integrated into building elements, some areas of improvement are as follows:

- More accurate assessment of the performance and potential of PCM systems in real dynamic conditions is required [50, 53]
- Development of validated simulation tools for the modeling of PCM performance is required [53]
- Integration of the PCM solutions within the established buildings' thermal regulation codes through the development of methodologies that take into consideration the latent heat loads from the PCMs [50].
- Economic feasibility of PCM applications in the built environment needs to be carried out [53].
- Further research on the flammability of PCM; definition of the ratio of PCM fulfilling the building fire standards [53].
- Increasing the thermal storage capacity for a given volume of a PCM. Although microencapsulation reduces the risk of liquid state PCMs from leaking, it produces a lower latent heat storage capacity. Subsequently, new methods or materials need to be discovered for the encapsulation of the PCM, or thinner shell for the capsules need to be developed [50].
- Enhancing the heat transfer of the PCM [50]. Fan and Khodadadi [54] reviewed methods regarding this aspect, with porous materials with high thermal conductivity such as graphite-based PCM systems and metal foams receiving particular interest [55, 56]. Also, materials with high thermal conductivity to the PCM, such as various carbon-based nano-fillers or porous structures have been studied for the enhancement of the overall thermal conductivity [57–59].

Future technologies

Building systems incorporating PCMs that harness solar thermal energy for heating during winter and optimized to reduce the overheating problem during summer are the most promising. Soares et al. [51] reported that the development of rotary, portable, movable and reconfigurable systems, mainly those associated with glazed façades, will be the next milestone in PCMs. Nanoencapsulation of PCMs, which has a similar concept to microencapsulation, and the combination of PCMs with nanofluids are also considered promising to improve PCM systems. Smaller size capsules would provide higher heat transfer speeds [50]. Another future trend of PCMs regards the adjustability of the phase change temperature for the improvement of the efficiency. A controllable or adaptive system would enable a dynamic change of the phase change temperature in response to user preferences, different climates and different seasons.

Conclusions

PCMs represent an innovative solution that can contribute to the improvement of the energy performance of buildings. A promising current trend concerns the integration of PCMs into transparent envelope components. This work introduces the evolution of PCMs technology for building applications, and thoroughly explains the concept of PCMs integrated into transparent buildings elements. The paper also reviewed the main solutions proposed by the literature in the past few years for PCMs integrated into transparent buildings elements. It can be concluded that PCM glazing systems provide better indoor thermal conditions than the conventional system, as they alleviated the peak heat temperatures of summer days, and release their stored energy during nighttime, when the outdoor temperatures typically drop. Finally, this work analyzes the barriers of the current PCMs technologies, and what can be done for their further development, as well as the near-future trends. In the next few years, rotary, portable, movable and reconfigurable glazing systems, nanoencapsulation of the PCMs, and systems with adjustable phase change temperature are anticipated to emerge.

References

1. Karlesi, T.M., Santamouris, M.: Research on thermochromic and PCM doped infrared reflective coatings. In: Kolokotsa, D., Santamouris, M., Akbari, H. (eds.) Advances in the Development of Cool Materials for the Built Environment, pp. 83–103. Bentham Publisher, Sharjah (2013)
2. Fang, G.H., Li, Z., Chen, X., Liu, X.: Preparation and characterization of stearic acid/expanded graphite composites as thermal energy storage materials'. Energy **35**, 4622–4626 (2010)
3. Chwieduk, D.A.: Dynamics of external wall structures with a PCM (phase change materials) in high latitude countries. Energy **59**, 301–313 (2013)
4. Kuznik, F.J., Virgone, J.: Experimental assessment of a phase change material for wall building use. Appl. Energy **86**, 2038–2046 (2009)
5. Medina, M.A., King, J.B., Zhang, M.: On the heat transfer rate reduction of structural insulated panels (SIPs) outfitted with phase change materials (PCMs). Energy **33**, 667–678 (2008)
6. Aadmi, M.M., Karkri, M., Hammouti, E.I.: Heat transfer characteristics of thermal energy storage of a composite phase change

materials: numerical and experimental investigations. Energy, available online 10 June 2014
7. Xu, B., Li, Z.: Paraffin/diatomite/multi-wall carbon nanotubes composite phase change material tailor-made for thermal energy storage cement-based composites. Energy, available online 13 June 2014
8. Zalba, B., Marín, J.M., Cabeza, L.F., Mehling, H.: Review on thermal energy storage with phase change: materials, heat transfer analysis and applications. Appl. Therm. Eng. **23**, 251–283 (2003)
9. Farid, M.M., Khudhair, A.M., Razack, S.A.K., Al-Hallaj, S.: A review on phase change energy storage: materials and applications. Energy Convers. Manag. **45**, 1597–1615 (2004)
10. Tyagi, V.V., Buddhi, D.: PCM thermal storage in buildings: a state of art. Renew. Sustain. Energy Rev. **11**, 1146–1166 (2007)
11. Kenisarin, M., Mahkamov, K.: Solar energy storage using phase change materials. Renew. Sustain. Energy Rev. **11**, 1913–1965 (2007)
12. Sharma, A., Tyagi, V.V., Chen, C.R., Buddhi, D.: Review on thermal energy storage with phase change materials and applications. Renew. Sustain. Energy Rev. **13**, 318–345 (2009)
13. Zhu, N., Ma, Z., Wang, S.: Dynamic characteristics and energy performance of buildings using phase change materials: a review. Energy Convers. Manag. **50**, 3169–3181 (2009)
14. Wang, X., Zhang, Y., Xiao, W., Zeng, R., Zhang, Q., Di, H.: Review on thermal performance of phase change energy storage building envelope. Chin. Sci. Bull. **54**, 920–928 (2009)
15. Aroul, A., Ra, V., Velraj, R.: Review on free cooling of buildings using phase change materials. Renew. Sustain. Energy Rev. **14**, 2819–2829 (2010)
16. Harland, A., Mackay, C., Vale, B.: Phase change materials in architecture. In: Proceedings SB10 Wellington—innovation and transformation, held in May 2010, Wellington, New Zealand
17. Koekenbier, S.F.: PCM energy storage during defective thermal cycling design of the "Capacity Cube" and modelling of PCM pouches to trace the impact of incomplete thermal cycling. 2011. Delft University of Technology, Faculty of Mechanical, Maritime and Materials Engineering
18. Peippo, K., Kauranen, P., Lund, P.D.: A multicomponent PCM wall optimized for passive solar heating. Energy Build. **17**, 259–270 (1991)
19. Peippo, K., Kauranen, P., Lund, P.D.: An organic PCM storage system with adjustable melting temperature. Sol. Energy **46**, 275–278 (1991)
20. Manz, H., Egolf, P.W., Suter, P., Goetzberger, A.: TIM-PCM external wall system for solar space heating and daylighting. Sol. Energy **61**, 369–379 (1997)
21. Ismail, K.A.R., Henriquez, J.R.: Parametric study on composite and PCM glass systems. Energy Convers. Manag. **43**, 973–993 (2012)
22. GreenSpec. GreenSpec's website, high-tech glazing with phase-change material. (2014)
23. Grynning, S., Goia, F., Rognvik, E., Time, B.: Possibilities for characterization of a PCM window system using large scale measurements. Int. J. Sustain. Built Environ. **2**, 56–64 (2013)
24. Alawadhi, E.M.: Using phase change materials in window shutter to reduce the solar heat gain'. Energy Build. **47**, 421–429 (2012)
25. Weinlaeder, H., Beck, A., Fricke, J.: PCM-facade-panel for daylighting and room heating. Sol. Energy **78**, 177–186 (2005)
26. Goia, F., Perino, M., Haase, M.: A numerical model to evaluate the thermal behaviour of PCM glazing system configurations. Energy Build. **54**, 141–153 (2012)
27. Goia, F., Perino, M., Serra, V.: Improving thermal comfort conditions by means of PCM glazing systems. Energy Build. **60**, 442–452 (2013)
28. Gowreesunker, B.L., Stankovic, S.B., Tassou, S.A., Kyriacou, P.A.: Experimental and numerical investigations of the optical and thermal aspects of a PCM-glazed unit. Energy Build. **61**, 239–249 (2013)
29. Goia, F.: Thermo-physical behaviour and energy performance assessment of PCM glazing system configurations: a numerical analysis'. Front. Archit. Res. **1**, 341–347 (2012)
30. Goia, F., Perin, M., Serra, V.: Experimental analysis of the energy performance of a full-scale PCM glazing prototype. Sol. Energy **100**, 217–233 (2014)
31. Goia, F., Bianco, L., Cascone, Y., Perin, M., Serra, V.: Experimental analysis of an advanced dynamic glazing prototype integrating PCM and thermotropic layers. Energy Proc. **48**, 1272–1281 (2014)
32. Buddhi, D., Mishra, H.S., Sharma, A.: Thermal performance studies of a test cell having a PCM-window in south direction. In: IEA, ECESIA Annex 17, Advanced thermal energy storage through phase change materials and chemical reactions—feasibility studies and demonstration projects, 4th workshop, 21–24 March 2003, Indore, India, pp. 54–58
33. Bontemps, A., Ahmad, M., Johannès, K., Sallée, H.: Experimental and modelling study of twin cells with latent heat storage walls. Energy Build. **43**, 2456–2461 (2011)
34. Soares, N., Samagaio, A., Vicente, R., Costa, J.: Numerical simulation of a PCM shutter for buildings space heating during the winter. World Renewable Energy Congress 2011, Sweden, 8–13 May
35. Heim, D.: Whole year analysis of TIM–PCM solar thermal storage wall. In: Proceeding of SimBuild 2004, IBPSA-USA National Conference Boulder, CO
36. Weinlaeder, H., Koerner, W., Heidenfelde, M.: Monitoring results of an interior sun protection system with integrated latent heat storage. Energy Build. **43**, 2468–2475 (2011)
37. Deng, S., Wang, R.Z., Dai, Y.J.: How to evaluate performance of net zero energy building—a literature research. Energy, available online 2 June 2014
38. Diarce, G., Urresti, A., García-Romero, A., Delgado, A., Erkoreka, A., Escudero, C., Campos-Celador, A.: Ventilated active façades with PCM. Appl. Energy **109**, 530–537 (2013)
39. De Gracia, A., Navarro, L., Castell, A., Ruiz-Pardo, A., Álvarez, S., Cabeza, L.F.: Thermal analysis of a ventilated facade with PCM for cooling applications'. Energy Build. **65**, 508–515 (2013)
40. Abhat, A.: Low temperature latent heat thermal energy storage: heat storage materials. Sol. Energy **30**, 313–332 (1983)
41. Swanton, J.R.: U.S. Patent 2'544'474 (1951)
42. Manz, H.: Sonnenstrahlungsbeladene Latent Wärmespeicher in Gebäudefassaden 1996. PhD thesis, Swiss Federal Institute of Technology, Zurich, Switzerland
43. Porisini, F.C.: Salt hydrates used for latent heat storage: corrosion of metals and reliability of thermal performance. Sol. Energy **41**, 193–197 (1988)
44. Buddhi, D., Sharma, S.D.: Measurements of transmittance of solar radiation through stearic acid: a latent heat storage material. Energy Convers. Manag. **40**, 1979–1984 (1999)
45. GlassX North America. GlassX's website, Home. (2014)
46. Delta®. Delta®'s website, Project Profiles. (2014)
47. Athienitis, A.K., Zhang, J., Feldman, D.: A study of double facades with phase-change storage and photovoltaics'. International Conference "Passive and Low Energy Cooling for the Built Environment", May 2005, Santorini, Greece, pp. 855–859
48. Goia, F., Zinzi, M., Carnielo, E., Serra, V.: Characterization of the optical properties of a PCM glazing system. Energy Proc. **30**, 428–437 (2012)

49. De Gracia, A., Navarro, L., Castell, A., Cabeza, L.F.: Numerical study on the thermal performance of a ventilated facade with PCM'. Appl. Therm. Eng. **61**, 372–380 (2013)
50. Kalnæs, S.E.: State-of-the-art and future building envelopes: application of low emissivity materials, phase change materials and vacuum insulation panels 2014. Master thesis, Norwegian University of Science and Technology
51. Soares, N., Costa, J.J., Gaspar, A.R., Santos, P.: Review of passive PCM latent heat thermal energy storage systems towards buildings' energy efficiency. Energy Build. **59**, 82–103 (2013)
52. Evola, G., Marletta, L., Sicurella, F.: A methodology for investigating the effectiveness of PCM wallboards for summer thermal comfort in buildings'. Build. Environ. **59**, 517–527 (2013)
53. Pomianowski, M., Heiselberg, P., Zhang, Y.: Review of thermal energy storage technologies based on PCM application in buildings. Energy Build. **67**, 56–69 (2013)
54. Fan, L., Khodadadi, J.M.: Thermal conductivity enhancement of phase change materials for thermal energy storage: a review. Renew. Sustain. Energy Rev. **15**, 24–46 (2011)
55. Sedeh, M.M., Khodadadi, J.M.: Thermal conductivity improvement of phase change materials/graphite foam composites. Carbon **60**, 117–128 (2013)
56. Jiang, J., Zhu, Y., Ma, Y., Yang, D., Lu, F., Chen, J., Shi, J., Song, D.: Preparation and performance of bulk porous Al foams impregnated with phase-change-materials for thermal storage. Prog. Nat. Sci.: Mater. Int. **22**, 440–444 (2012)
57. Babaei, H., Keblinski, P., Khodadadi, J.M.: Improvement in thermal conductivity of paraffin by adding high aspect-ratio carbon-based nano-fillers. Phys. Lett. A **377**, 1358–1361 (2013)
58. Ji, P., Sun, H., Zhong, Y., Feng, W.: Improvement of the thermal conductivity of a phase change material by the functionalized carbon nanotubes. Chem. Eng. Sci. **81**, 140–145 (2012)
59. Yu, S., Jeong, S.G., Chung, O., Kim, S.: Bio-based PCM/carbon nanomaterials composites with enhanced thermal conductivity. Sol. Energy Mater. Sol. Cells **120**, 549–554 (2014)
60. Rubitherm. Rubiherm's website, RUBITHERM® PCM's. (2014)

14

Electrochemical and in situ ATR-FTIR studies of ethanol electro-oxidation in alkaline medium using PtRh/C electrocatalysts

E. H. Fontes · R. M. Piasentin · J. M. S. Ayoub ·
J. C. M. da Silva · M. H. M. T. Assumpção ·
E. V. Spinacé · A. O. Neto · R. F. B. de Souza

Abstract Pt/C, Rh/C and PtRh/C electrocatalysts with different Pt:Rh atomic ratios supported on Vulcan XC 72 carbon were prepared using borohydride as reducing agent and tested for ethanol electro-oxidation in alkaline medium. X-ray diffraction patterns showed the formation of PtRh alloy. Transmission electron micrographs showed the nanoparticles with particle sizes between 3 and 6 nm for all materials. Electrochemical experiments showed PtRh/C (50:50) the most promising material for ethanol electro-oxidation; however, in situ ATR-FTIR experiments was observed that the ethanol oxidation is incomplete due to the formation of acetate and carbonate.

Keywords PtRh/C electrocatalysts · Ethanol electro-oxidation in alkaline medium · In situ ATR-FTIR

Introduction

The use of ethanol directly as a fuel for Proton Exchange Membrane Fuel Cells (PEMFC) has been studied extensively over the last years due to its large production from renewable sources, low toxicity and high-energy density of 8 kWh/kg; however, the breaking of the C–C bond of the ethanol molecule at low temperatures continues to be a challenge, and acetaldehyde and acetic acid are the principal products formed [1–7].

E. H. Fontes · R. M. Piasentin · J. M. S. Ayoub ·
J. C. M. da Silva · M. H. M. T. Assumpção ·
E. V. Spinacé · A. O. Neto · R. F. B. de Souza (✉)
Instituto de Pesquisas Energéticas e Nucleares, IPEN/CNEN-SP,
Av. Prof. Lineu Prestes, 2242 Cidade Universitária, CEP
05508-900 São Paulo, SP, Brazil
e-mail: souza.rfb@gmail.com

Recently, the development of anion exchange membranes for alkaline fuel cells [8–12] has attracted the interest in studying the ethanol electro-oxidation in alkaline medium [13–15]. Alkaline medium presents several advantages when compared to acidic medium, as an oxygen reduction reaction takes place with faster kinetic than in acid media, increasing the electric potential of the fuel cell [15] and many non-noble metals are stable for electrochemical applications, either as anode or a cathode [14]. Despite this, the break of C–C bond of ethanol molecule in alkaline medium is also a difficult task [16–18].

In acid medium, the addition of Rh to Pt catalysts increases the CO_2 yield by facilitating C–C bond cleavage; however, the obtained currents do not increase [19–21]. Kowal et al. [4, 22] prepared $PtRhSnO_2/C$ electrocatalyst depositing Pt and Rh atoms on carbon-supported SnO_2 nanoparticles and tested for ethanol electro-oxidation. They found that this electrocatalyst breaks the C–C bond in acid medium facilitating the oxidation of ethanol molecule to CO_2. In a further work, Li et al. [4] studied Pt–Rh–SnO_2 catalysts and concluded that the role of Rh is to break the C–C bond of ethanol molecule whereas the roles of Pt and SnO_2 are to facilitate the oxidation of the reaction intermediates and avoid their bonding to Rh sites.

Comprehensive fundamental studies using in situ spectroscopic studies to provide molecular information about ethanol electro-oxidation on Pt-based catalysts are widely reported in the literature for the acid medium [3, 4, 23, 24], while only few studies were done involving the ethanol electro-oxidation in alkaline solutions [14, 15, 25]. On this basis, this work investigates the ethanol electro-oxidation in alkaline medium using PtRh/C electrocatalysts. Eletrochemical and in situ ATR-FTIR studies were performed to obtain information about the activity of the electrocatalysts and the intermediates and products formed.

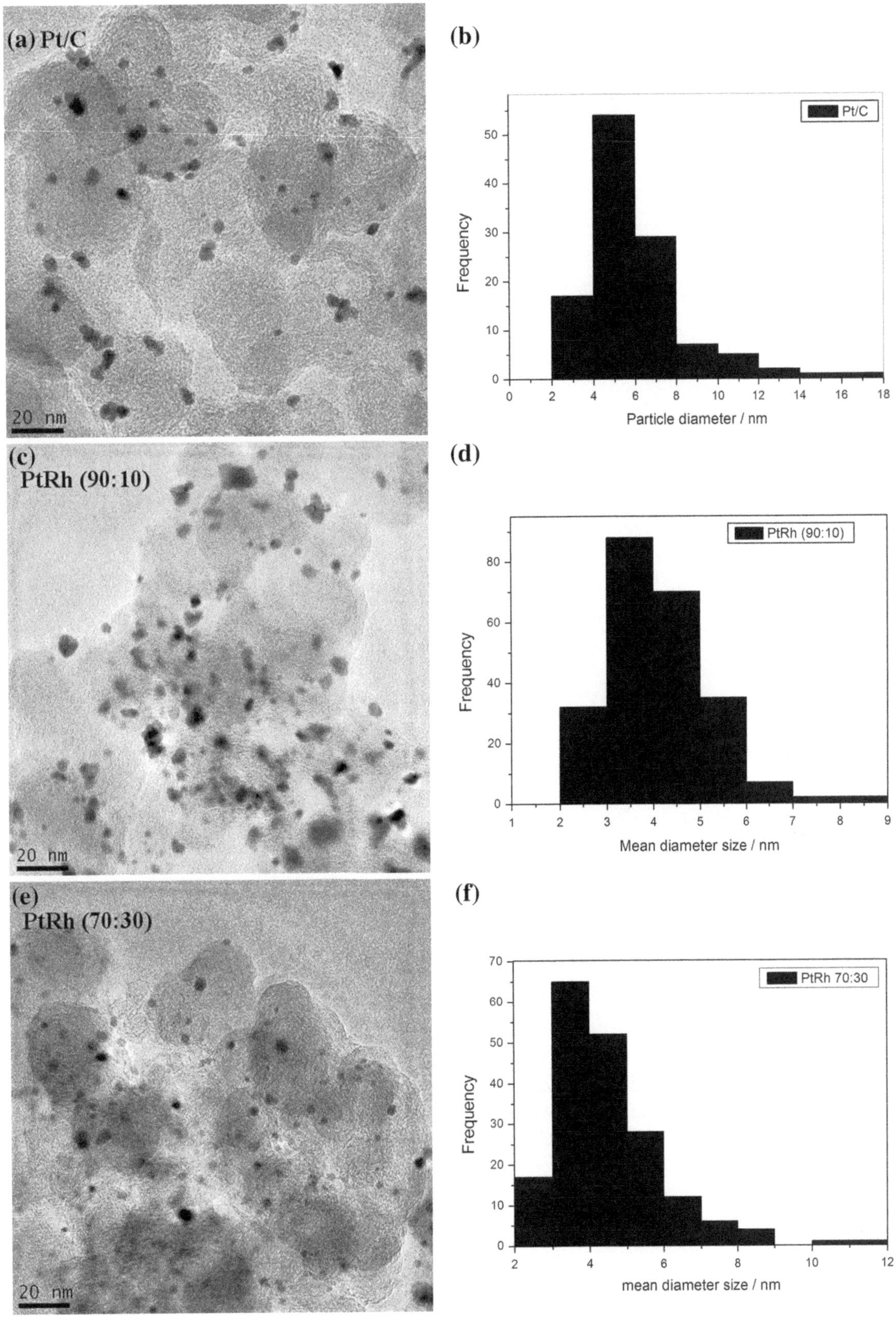

Fig. 1 TEM micrographs and histograms of the particle size distribution for Pt/C, Rh/C and PtRh/C

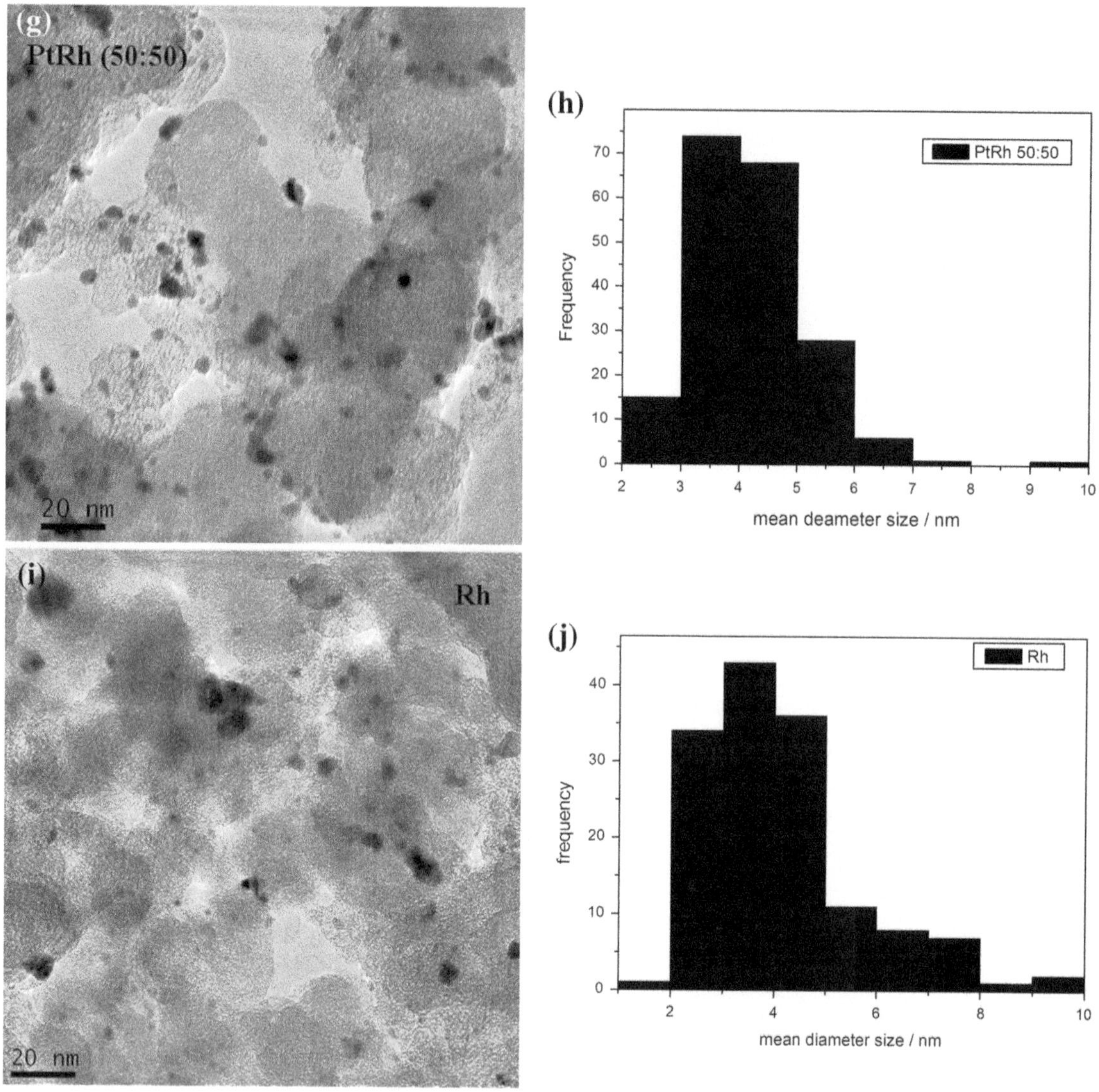

Fig. 1 continued

Experimental

PtRh/C electrocatalysts (20 % of metal loading, with Pt:Rh ratio atomics of 100:0, 90:10, 70:30, 50:50, and 0:100) were prepared using $H_2PtCl_6{\cdot}6H_2O$ (Chloroplatinic acid-Aldrich), and $RhCl_3{\cdot}xH_2O$ (rhodium chloride Aldrich) as metal sources, sodium borohydride (Aldrich) as reducing agent and Vulcan Carbon XC72 as support. In the borohydride reduction process, the metal sources were dissolved in a mixture of water/2-propanol (50/50, v/v) and the Vulcan XC72 support was dispersed in the solution. The resulting mixture was submitted to an ultrasonic bath for 10 min, where a solution of sodium borohydride was added under stirring in one portion at room temperature. After, the mixture was filtered and the solid was washed with water and dried at 70 °C for 2 h.

All electrocatalysts were characterized by transmission electron microscopy (TEM) analysis. These experiments were carried out using a JEOL JEM-2100 electron microscope operated at 200 kV, where it was determined the morphology, distribution, and size of the nanoparticles in the support. The mean particle sizes were determined by counting more than 200 particles from different regions of each sample. X-ray diffraction (XRD) analyses were performed on Rigaku diffractometer model Miniflex II using Cu Kα radiation source ($\lambda = 0.15406$ nm). The diffractograms were recorded from $2\theta = 20°$ to 90° with a step size of 0.05° and a scan time of 2 s per step.

The cyclic voltammetry (CV) and chronoamperometry (CA) measurements were carried at room temperature using a Microquimica potenciostat, where the working electrodes (geometric area of 0.3 cm^2 with a depth of 0.3 mm) were prepared using the thin porous coating technique [26], where an ink is prepared with 10 mg of catalyst and 100 μL of PTFE solution in 50 mL of water, then dispersed in ultrasonic bath for 5 min. After that, the ink is filtered and all the solids are deposited on the work electrode. The reference electrode was Ag/AgCl (3.0 mol L^{-1} KCl) and the counter electrode was a Pt plate. The electrochemical measurements were realized in the presence of 1.0 mol L^{-1} of ethanol + 1 mol L^{-1} KOH solutions saturated with N_2.

Direct alkaline ethanol fuel cell test was realized using a single cell with 5 cm^2 of area. The temperature was set to 60 °C for the fuel cell and 85 °C for the oxygen humidifier. All electrodes were constructed with 1 mg of metal per cm^2 in the anode and the cathode. In all experiments, a commercial Pt/C (BASF) was used as cathode. All the electrocatalysts were painted over a carbon cloth in the form of a homogeneous dispersion prepared using Nafion solution (5 wt%, Aldrich). After the preparation, the electrodes were hot pressed on both sides of a Nafion 117 membrane at 125 °C for 3 min under a pressure of 247 kgf cm^{-2}. Prior to use, the membranes were exposed to KOH 6 mol L^{-1} for 24 h as it has already been proposed by Hou et al. [27]. The fuel, ethanol 2.0 mol L^{-1} and 2.0 mol L^{-1} KOH were delivered at 1.0 mL min^{-1} and the oxygen flow was regulated at 150 mL min^{-1}. Polarization curves were obtained using a potentiostat/galvanostat PGSTAT 302NAutolab.

The spectro-electrochemical ATR-FTIR in situ measurements were performed with a Nicolet 6700 FT-IR spectrometer equipped with an MCT detector cooled with liquid N_2, ATR accessory (MIRacle with a Diamond/ZnSe Crystal Plate Pike®) and an electrochemical cell as explained better in the literature [23, 28, 29]. The working electrodes are the same that electrochemical experiments in presence of 1.0 mol L^{-1} ethanol + 1 mol L^{-1} KOH. The absorbance spectra were collected as the ratio R:R_0, where R represents a spectrum at a given potential and R_0 is the spectrum collected at −0.85 V. Positive and negative directional bands represent gain and loss of species at the sampling potential, respectively. The spectra were computed from 128 interferograms averaged from 3,000 to 850 cm^{-1} with the spectral resolution set to 8 cm^{-1}. Initially, a reference spectrum (R_0) was measured at −0.85 V,

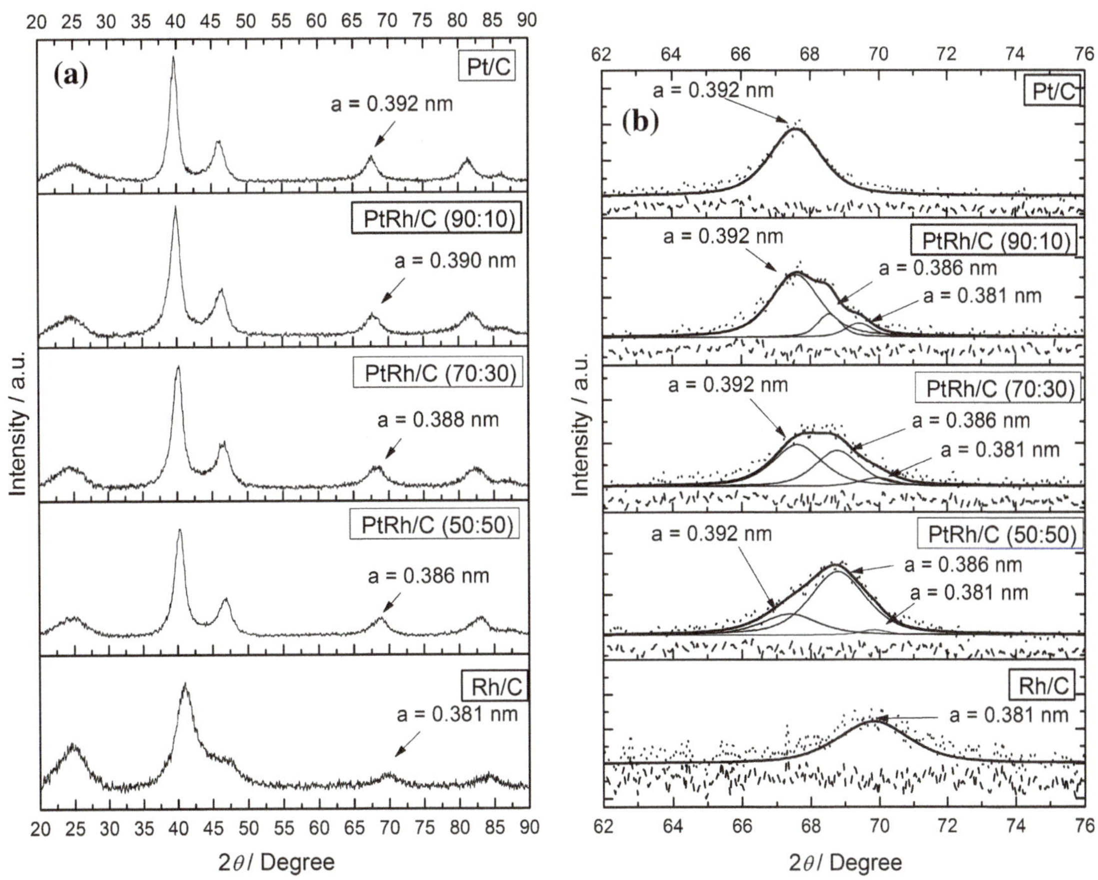

Fig. 2 **a** X-ray diffractograms of the Pt/C, Rh/C, and PtRh/C electrocatalysts. **b** Pawley refinement of peak (2 2 0) in all diffratograms with *dot* for experimental measure, *line* for modeled, and *dash* for residual line

and the sample spectra were collected after applying successive potential steps from 0.1 to −0.05 V.

Results and discussion

TEM micrographs and histograms of the particle size distribution for Pt/C, Rh/C and PtRh/C materials are shown in Fig. 1. All prepared materials showed the particles well dispersed on the support, even though some particle agglomerations can be observed. The nanoparticles' mean diameter was determined by counting about 200 randomly chosen particles from the relevant TEM images [30, 31] and the particles are with an average diameter of 4 nm for all Rh content catalyst, and 6 nm for Pt/C. The particles sizes measured are in accordance with similar materials described in literature [20, 24, 32].

Figure 2a shows XRD patterns of Pt/C, Rh/C and PtRh/C electrocatalysts. In all diffractograms, it can be seen four peaks at $2\theta = 40°$, 47°, 67° and 82°, which are associated with the (111), (200), (220) and (311) planes, characteristic of face-centered cubic (fcc) structure of Pt and Pt alloys. The diffraction peaks of Pt(fcc) and Rh(fcc) phases are very close; however, it was observed that the diffraction peaks of the fcc phase of PtRh/C electrocatalyts were slightly shifted to higher 2θ values compared to Pt/C electrocatalyst and this shift increases with the increase of Rh content [32]. According to Gupta et al. [20], such evidence accounts for the presence of PtRh alloy formation. To obtain more information about the material structure, the diffractograms were refined using the Pawley method [33] carried out with Fityk 0.98 software, as reported by Wojdyr [34]. The peaks in this region are related to Pt (220) (JCPDF #04 802), Rh (220) (JCPDF # 88 2334) and PtRh (220) (JCPDF # 65 7938). In Fig. 2b, it is possible to observe the changes in the peak shape as a result of the contribution of Pt(fcc) ($a = 0.392$ nm), Rh(fcc) ($a = 0.381$ nm) and $PtRh_{alloy}$(fcc) ($a = 0.386$ nm) phases. The presence of these three contributions was observed for all PtRh/C electrocatalysts and the peak related to PtRh alloy formation increases with the increase of Rh content.

Figure 3 shows the CV in alkaline media for Pt/C, Rh/C and PtRh/C electrocatalysts in 1 mol L^{-1} KOH. In the CVs, it was observed that with the increase of the Rh content the hydrogen adsorption–desorption region (−0.85 to −0.5 V) becomes less defined than the one observed for Pt/C electrocatalyst [35]. Also, the reverse scan of Pt/C electrocatalyst showed a peak around −0.3 V which was attributed to reduced PtOx species and with the increase in Rh content on PtRh/C materials, decrease the intensity of this peak. PtRh/C (50:50) and Rh/C electrocatalysts showed a redox pair in the current values of −0.3/−05 V and −0.3/−0.7 V, respectively, which may be attributed to the increase in the rhodium oxides species [35].

Figure 4 shows the liner sweep of Pt/C, Rh/C and PtRh/C electrocatalysts in the presence of 1.0 mol L^{-1} of ethanol and 1 mol L^{-1} of KOH. It was observed that the onset potential for ethanol electro-oxidation is very close (≈−0.65 V) for all eletrocatalysts as well as the current density values. It is interesting to note that Rh/C electrocatalyst showed a good activity for ethanol electro-oxidation in alkaline medium while in acid medium it was practically inactive [4, 22].

Figure 5 shows the chronoamperometry curves for Pt/C, Rh/C and PtRh/C electrocatalysts in the potential of −0.3 V for 30 min. For all electrocatalysts, it was observed a faster decay of current values during the first 2 min followed by a slower decay. The final current values at

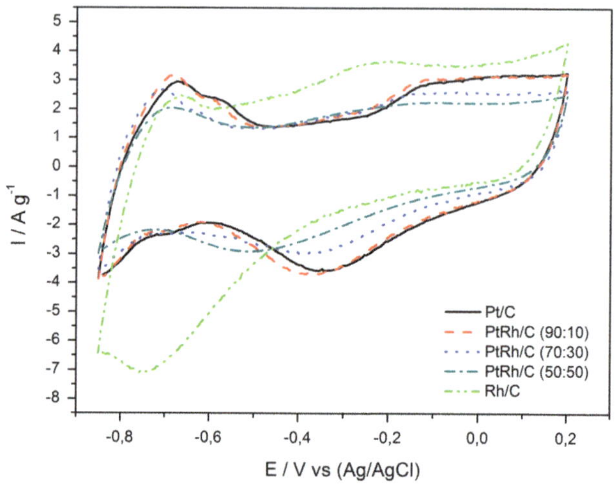

Fig. 3 Cyclic voltammograms of Pt/C, Rh/C and PtRh/C electrocatalysts in 1 mol L^{-1} KOH solution with a scan rate of 10 mV s^{-1} at 25 °C

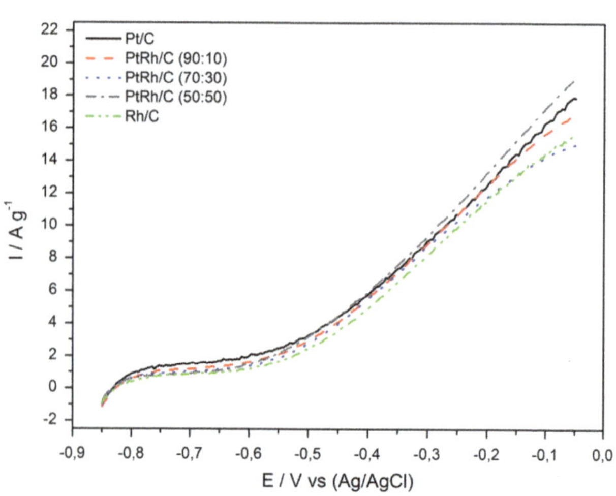

Fig. 4 Linear sweep voltammograms of Pt/C, Rh/C and PtRh/C electrocatalysts in 1 mol L^{-1} ethanol solution in 1 mol L^{-1} KOH with a scan rate of 10 mV s^{-1} at 25 °C

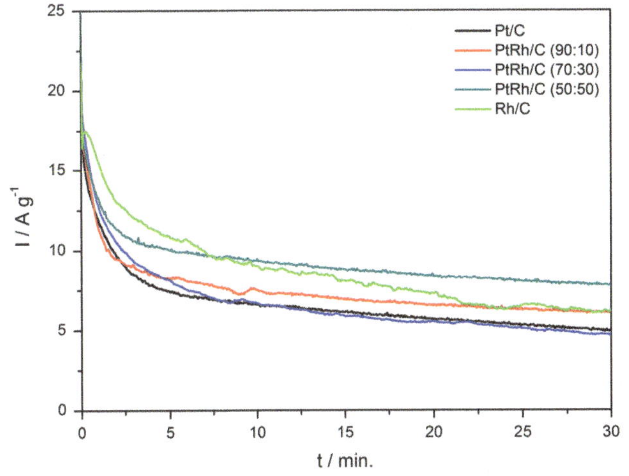

Fig. 5 Current–time curves at −0.3 V in 1 mol L^{-1} ethanol solution in 1 mol L^{-1} KOH for Pt/C, Rh/C and PtRh/C electrocatalysts at 25 °C

−0.3 V were: PtRh/C (50:50) (7.9 A g^{-1}) > Rh/C (6.2 A g^{-1}) ≈ PtRh/C (90:10) (6.1 A g^{-1}) > Pt/C (4.9 A g^{-1}) > PtRh/C (70:30) (4.6 A g^{-1}). It is worth noting that initially the Rh/C electrocatalyst presents higher current values than the other electrocatalysts; however, after 7 min its deactivation is faster than Pt/C and PtRh/C electrocatalysts,

The decay for PtRh is slower than that for Rh/C indicating superior stability. As the consequence of the Rh incorporation, some modifications of structural, electronic and kinetic features on the PtRh/C catalysts were observed, such as changes in lattice parameter and charge transfers from Rh to Pt (Pt d-band vacancies electronically [32]). However, this effect does not occur in Pt or Rh alone.

Figure 6 shows the performances of single alkaline DEFC with Pt/C, Rh/C and PtRh/C as anode electrocatalysts and the following order of maximum power density was observed: PtRh/C (50:50) (5.9 mW cm^{-2}) > Pt/C (5.5 mW cm^{-2}) > PtRh/C 70:30 (4.9 mW cm^{-2}) > PtRh (90:10) (4.7 mW cm^{-2}) > Rh/C (4.4 mW cm^{-2}).

A factor that can influence the activity of these materials may be caused by the manner in that ethanol adsorbs on the catalyst, as explained in the oxametallacyclic conformation model [36], which directly influences the kinetics of the reaction and the products formed. PtRh 50:50 alloy can be a conformation that favors the reactions kinetic [37], due to way of the ethanol adsorbs on catalyst surface, and influences too on OCV.

In situ ATR-FTIR experiments were performed to correlate the activity of ethanol electro-oxidation with the preferential mechanism (Fig. 7). The FTIR spectra in the region between 3,000 and 850 cm^{-1} were recorded during ethanol electro-oxidation for Pt/C, Rh/C and PtRh/C

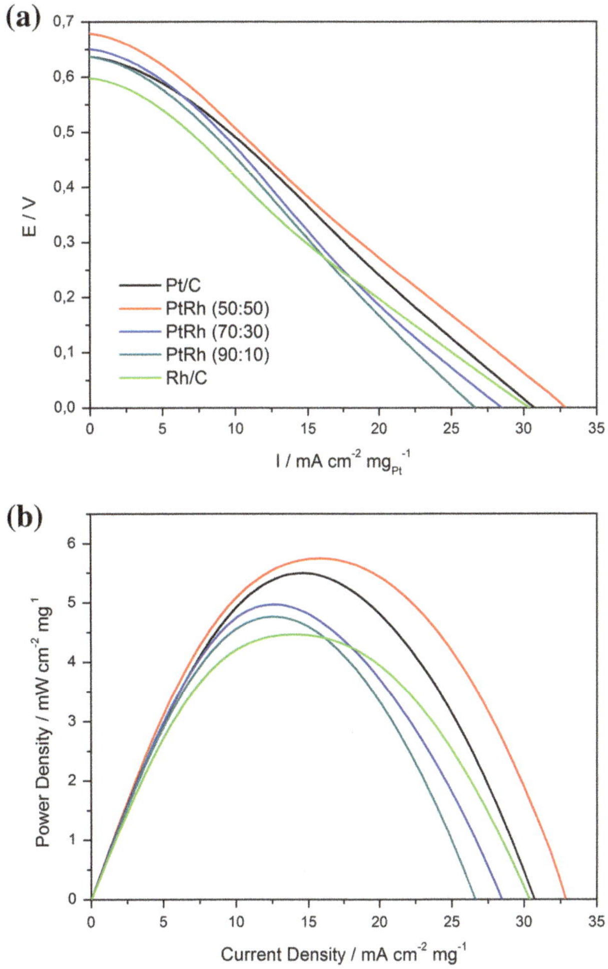

Fig. 6 *I–V* Curves and the power density at 60 °C of a 5 cm^2 DAEFC using Pt/C, Rh/C and PtRh/C electrocatalysts anodes (1 mg_{metal} cm^{-2} catalyst loading) and Pt/C E-TEK electrocatalyst cathode (1 mg_{Pt} cm^{-2} catalyst loading, 20 wt% Pt loading on carbon), Nafion® 117 membrane KOH treated, ethanol (2.0 mol L^{-1}) and oxygen pressure (2 bar)

electrocatalysts as a function of the potential in ethanol 1 mol L^{-1} and KOH 1 mol L^{-1}. Considering ATR-FTIR results, it was possible to observe the appearance of bands related to CO_2 at 2,343 cm^{-1} formation [25].

A carbonate ion band at 1,370 cm^{-1} [14] was not clearly visible in the spectra, due to overlapping with the acetate band at 1,410 cm^{-1} [14], although its presence can be inferred by the deconvolution of the spectra, as reported by Neto et al. [38]. The acetate was displayed as two intense peaks at ≈1,553 and ≈1,410 cm^{-1}, originating from asymmetric and symmetric C–O bond vibrations. The small peak at 926 cm^{-1} is associated with acetaldehyde C–C–O Stretching vibrations [15]. In the spectra, it is also possible to observe the presence of bands of ethanol that increase the intensity towards consumption of this species (1,080, 1,036, and 874 cm^{-1}) [25].

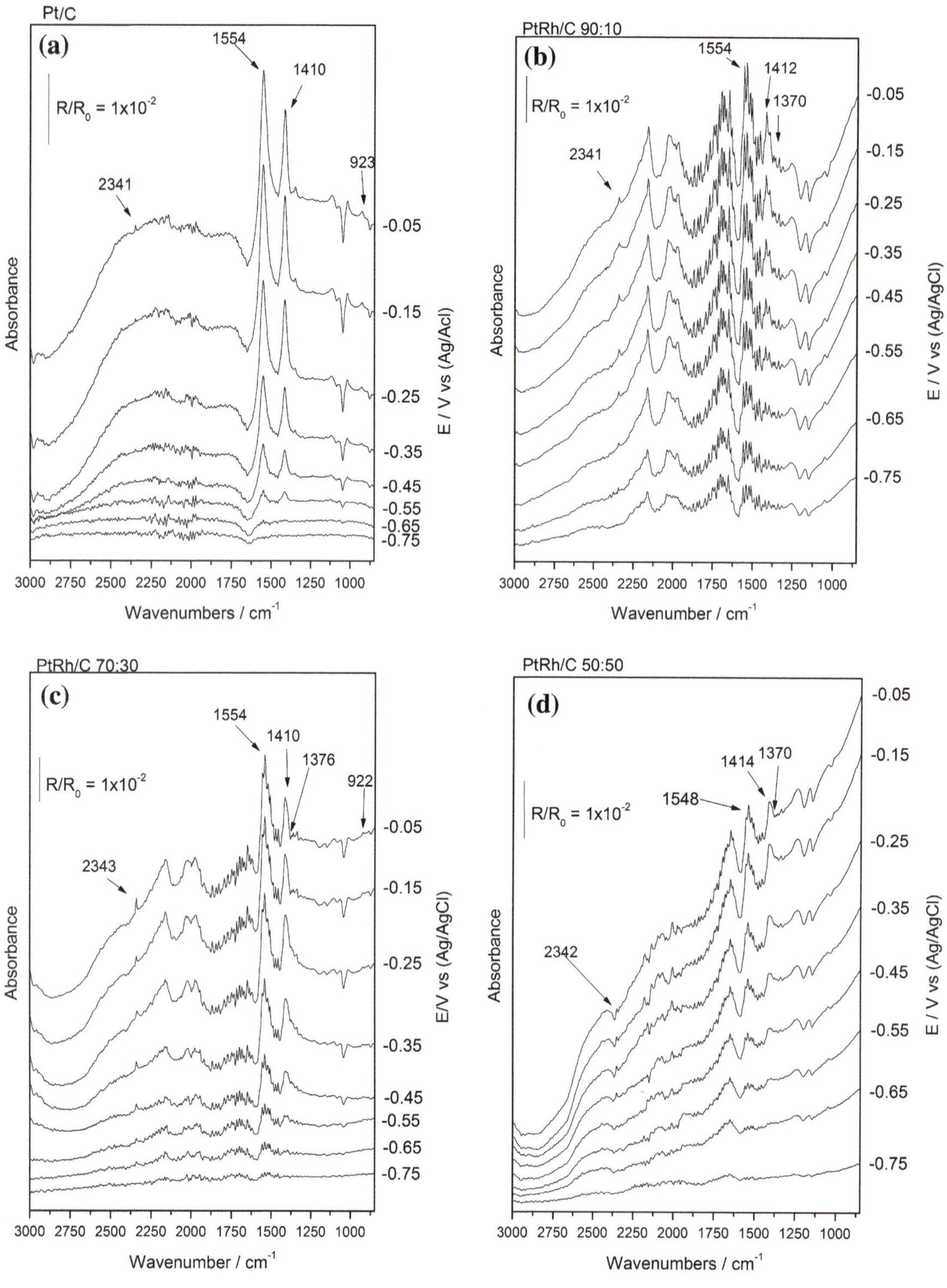

Fig. 7 In situ FTIR spectra taken at several potentials (indicated) in 1 mol L^{-1} KOH + 1.0 mol L^{-1} ethanol for Pt/C, Rh/C, and PtRh/C electrocatalysts. The backgrounds were collected at −0.85 V (Ag/AgCl scale)

The difference observed in the spectra can be seen more clearly after all the bands were deconvoluted to lorentzian line forms [38] of the CO_2, acetate, carbonate ions, methyl groups, and an acetaldehyde as shown in Fig. 8.

As shown in Fig. 8, the PtRh/C (70:30) catalyst shows the most intense band of CO_2 at all potentials, but it is important to note that PtRh/C 50:50 catalyst, the most active in EOR experiments, has an increase in signal

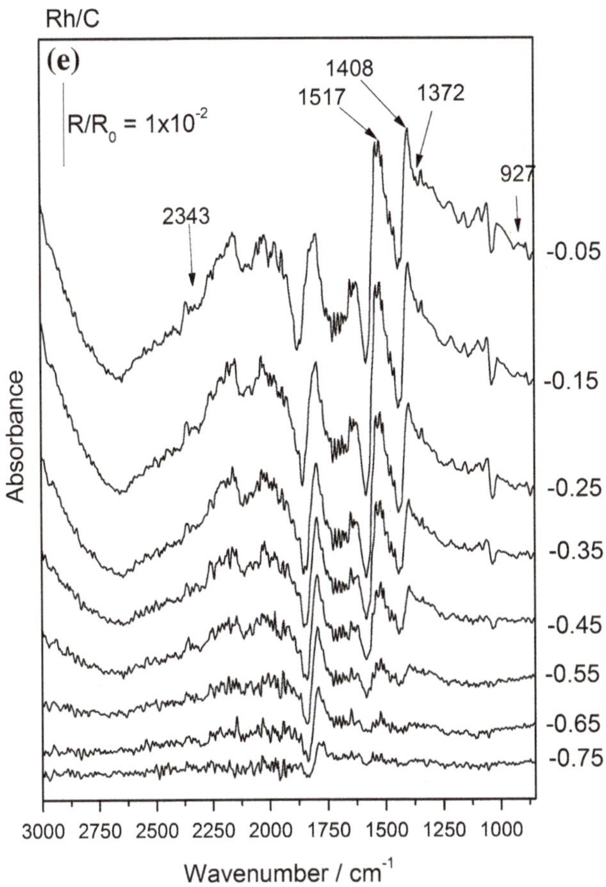

Fig. 7 continued

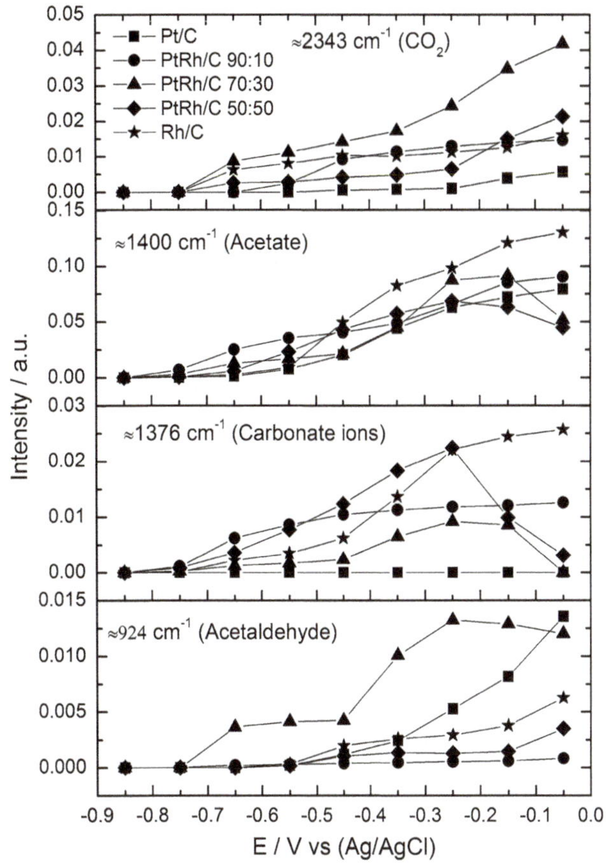

Fig. 8 Integrated CO_2, acetate, carbonate, methyl group, and acetaldehyde band intensity as a function of the electrode potential for Pt/C, Rh/C and PtRh/C electrocatalysts

intensity CO_2 production from −0.35 V potential that coincides with the decrease in the signal of acetate and carbonate ions.

When we analyzed the production of acetate, it is possible to see that from −0.45 V the Rh/C catalyst presents higher intensity for this band than other materials; can indicate that in alkaline medium, this metal was not completely selective to C–C broken.

The production of carbonate ions, present in Fig. 8c, brings clues that may explain the loss of activity of the catalyst Rh/C, since this material continuously produces these ions even at high potentials as −0.05 V. Carbonate ions can create a physical barrier on the catalytic sites, thus reducing the ability to get more alcohol to oxidize. The production of acetaldehyde, it is passed over on all materials tested, except for PtRh/C 70:30 and is a another indication that the products formed depends on the structure of the catalyst and the way of the fuel is adsorbed.

In summary, Rh is an active metal for EOR, however, suffers very rapidly with the catalyst poisoning in alkaline medium, as seen in ATR-FTIR measurements, ethanol is oxidized preferentially over it by way of formation of carbonate ions, that may physically block the catalytic sites, other way is the acetate formation, indicating that reaction over Rh is not selective for C–C cleavage, as reported for acid medium [4, 21, 22]. Though when associated with Pt, 50:50, the resulting alloy has greater activity than other catalysts studied, which lower potentials oxidize the ethanol preferably by same way of Rh/C, but in potentials higher than −0.35 V decrease carbonate ions productions and increase the CO_2 production, probably due to different types of ethanol adsorption on catalyst, this can be causes a increase in reaction kinetics as reported Shen et al. [37]. A factor that also can influence the activity of these materials may be caused by the manner that ethanol adsorbs on the active sites of the catalysts, as explained in the oxametallacyclic conformation model [36], which directly influences the kinetics of the reaction and the products formed. Therefore, PtRh/C 50:50 alloy could be a structure conformation that favors the kinetics reaction [37].

Conclusion

X-ray diffractograms of PtRh/C electrocatalysts showed the presence of Pt(fcc), Rh(fcc) and $PtRh(fcc)_{alloy}$ phases. The increase in Rh content on the PtRh/C electrocatalysts leads to an increase in $PtRh(fcc)_{alloy}$ phase. TEM micrographs showed a good distribution of the nanoparticles on the carbon support with average particle sizes in the range of 4–6 nm. Electrochemical and alkaline DEFC experiments showed that PtRh/C 50:50 was the best activity; however, Rh/C electrocalalyst showed a remarkable good activity in the first minutes but it deactivates more quickly than Pt/C and PtRh/C electrocatalysts. In situ ATR-FTIR experiments showed that ethanol oxidation reaction is incomplete; however, PtRu/C (50:50) is very promising due to the CO_2 production (from −0.35 V potential) that coincides with the decrease in the signal of acetate and carbonate ions.

Acknowledgments The authors thank the Laboratório de Microscopia do Centro de Ciências e Tecnologia de Materiais (CCTM) by TEM measurements, FAPESP (2012/03516-5, 2012/22731-4, 2013/01577-0 and 2011/18246-0) and CNPq (150639/2013-9) for the financial support.

References

1. Zhou, Y., Xu, Q.J.: A review of Pt-based anode catalysts preparation for direct ethanol fuel cell. Adv. Mater. Res. **860–863**, 797–800 (2014)
2. Beyhan, S., Léger, J.M., Kadirgan, F.: Pronounced synergetic effect of the nano-sized PtSnNi/C catalyst for ethanol oxidation in direct ethanol fuel cell. Appl. Catal. B **130–131**, 305–313 (2013)
3. García-Rodríguez, S., Rojas, S., Peña, M.A., Fierro, J.L.G., Baranton, S., Léger, J.M.: An FTIR study of Rh–PtSn/C catalysts for ethanol electrooxidation: effect of surface composition. Appl. Catal. B **106**(3–4), 520–528 (2011).
4. Li, M., Zhou, W.P., Marinkovic, N.S., Sasaki, K., Adzic, R.R.: The role of rhodium and tin oxide in the platinum-based electrocatalysts for ethanol oxidation to CO_2. Electrochim. Acta **104**, 454–461 (2013).
5. Xu, C., Shen, P.K.: Electrochemical oxidation of ethanol on Pt–CeO_2/C catalysts. J. Power Sources **142**(1–2), 27–29 (2005)
6. Ribeiro, J., dos Anjos, D.M., Kokoh, K.B., Coutanceau, C., Léger, J.M., Olivi, P., de Andrade, A.R., Tremiliosi-Filho, G.: Carbon-supported ternary PtSnIr catalysts for direct ethanol fuel cell. Electrochim. Acta **52**(24), 6997–7006 (2007)
7. Colmati, F., Antolini, E., Gonzalez, E.R.: Ethanol oxidation on a carbon-supported Pt75Sn25 electrocatalyst prepared by reduction with formic acid: effect of thermal treatment. Appl. Catal. B **73**(1–2), 106–115 (2007)
8. Wang, G., Weng, Y., Chu, D., Xie, D., Chen, R.: Preparation of alkaline anion exchange membranes based on functional poly(ether-imide) polymers for potential fuel cell applications. J. Membr. Sci. **326**(1), 4–8 (2009).
9. Wang, G., Weng, Y., Chu, D., Chen, R., Xie, D.: Developing a polysulfone-based alkaline anion exchange membrane for improved ionic conductivity. J. Membr. Sci. **332**(1–2), 63–68 (2009).
10. Herman, H., Slade, R.C.T., Varcoe, J.R.: The radiation-grafting of vinylbenzyl chloride onto poly(hexafluoropropylene-co-tetrafluoroethylene) films with subsequent conversion to alkaline anion-exchange membranes: optimisation of the experimental conditions and characterisation. J. Membr. Sci. **218**(1–2), 147–163 (2003).
11. Danks, T.N., Slade, R.C.T., Varcoe, J.R.: Alkaline anion-exchange radiation-grafted membranes for possible electrochemical application in fuel cells. J. Mater. Chem. **13**(4), 712–721 (2003).
12. Varcoe, J.R.: Investigations of the ex situ ionic conductivities at 30° C of metal-cation-free quaternary ammonium alkaline anion-exchange membranes in static atmospheres of different relative humidities. Phys. Chem. Chem. Phys. **9**(12), 1479–1486 (2007).
13. Cai, J., Huang, Y., Guo, Y.: PdTex/C nanocatalysts with high catalytic activity for ethanol electro-oxidation in alkaline medium. Appl. Catal. B **150–151**, 230–237 (2014)
14. Fang, X., Wang, L., Shen, P.K., Cui, G., Bianchini, C.: An in situ Fourier transform infrared spectroelectrochemical study on ethanol electrooxidation on Pd in alkaline solution. J. Power Sources **195**(5), 1375–1378 (2010).
15. Geraldes, A.N., Da Silva, D.F., Pino, E.S., Da Silva, J.C.M., De Souza, R.F.B., Hammer, P., Spinacé, E.V., Neto, A.O., Linardi, M., Dos Santos, M.C.: Ethanol electro-oxidation in an alkaline medium using Pd/C, Au/C and PdAu/C electrocatalysts prepared by electron beam irradiation. Electrochim. Acta **111**, 455–465 (2013)
16. Song, S.Q., Zhou, W.J., Zhou, Z.H., Jiang, L.H., Sun, G.Q., Xin, Q., Leontidis, V., Kontou, S., Tsiakaras, P.: Direct ethanol PEM fuel cells: the case of platinum based anodes. Int. J. Hydrogen Energy **30**(9), 995–1001 (2005)
17. Coutanceau, C., Brimaud, S., Lamy, C., Léger, J.M., Dubau, L., Rousseau, S., Vigier, F.: Review of different methods for developing nanoelectrocatalysts for the oxidation of organic compounds. Electrochim. Acta **53**(23), 6865–6880 (2008)
18. Neto, A.O., Farias, L.A., Dias, R.R., Brandalise, M., Linardi, M., Spinacé, E.V.: Enhanced electro-oxidation of ethanol using PtSn/CeO_2-C electrocatalyst prepared by an alcohol-reduction process. Electrochem. Commun. **10**(9), 1315–1317 (2008)
19. Lima, F.H.B., Gonzalez, E.R.: Ethanol electro-oxidation on carbon-supported Pt–Ru, Pt–Rh and Pt–Ru–Rh nanoparticles. Electrochim. Acta **53**(6), 2963–2971 (2008).
20. Sen Gupta, S., Datta, J.: A comparative study on ethanol oxidation behavior at Pt and PtRh electrodeposits. J. Electroanal. Chem. **594**(1), 65–72 (2006).
21. de Souza, J.P.I., Queiroz, S.L., Bergamaski, K., Gonzalez, E.R., Nart, F.C.: Electro-oxidation of ethanol on Pt, Rh, and PtRh electrodes. A study using DEMS and in situ FTIR techniques. J. Phys. Chem. B **106**(38), 9825–9830 (2002).
22. Kowal, A., Li, M., Shao, M., Sasaki, K., Vukmirovic, M.B., Zhang, J., Marinkovic, N.S., Liu, P., Frenkel, A.I., Adzic, R.R.: Ternary Pt/Rh/SnO_2 electrocatalysts for oxidizing ethanol to CO_2. Nat. Mater. **8**(4), 325–330 (2009).
23. De Souza, R.F.B., Silva, J.C.M., Simões, F.C., Calegaro, M.L., Neto, A.O., Santos, M.C.: New approaches for the ethanol

oxidation reaction of Pt/C on carbon cloth using ATR-FTIR. Int. J. Electrochem. Sci. **7**(6), 5356–5366 (2012)

24. Silva-Junior, L.C., Maia, G., Passos, R.R., de Souza, E.A., Camara, G.A., Giz, M.J.: Analysis of the selectivity of PtRh/C and PtRhSn/C to the formation of CO_2 during ethanol electrooxidation. Electrochim. Acta **112**, 612–619 (2013).

25. Zhou, Z.-Y., Wang, Q., Lin, J.-L., Tian, N., Sun, S.-G.: In situ FTIR spectroscopic studies of electrooxidation of ethanol on Pd electrode in alkaline media. Electrochim. Acta **55**(27), 7995–7999 (2010).
26. Neto, A.O., Brandalise, M., Dias, R.R., Ayoub, J.M.S., Silva, A.C., Penteado, J.C., Linardi, M., Spinacé, E.V.: The performance of Pt nanoparticles supported on $Sb_2O_5{\cdot}SnO_2$, on carbon and on physical mixtures of $Sb_2O_5.SnO_2$ and carbon for ethanol electro-oxidation. Int. J. Hydrogen Energy **35**(17), 9177–9181 (2010).
27. Hou, H., Wang, S., Jin, W., Jiang, Q., Sun, L., Jiang, L., Sun, G.: KOH modified Nafion112 membrane for high performance alkaline direct ethanol fuel cell. Int. J. Hydrogen Energy **36**(8), 5104–5109 (2011).
28. Silva, J.C.M., Parreira, L.S., De Souza, R.F.B., Calegaro, M.L., Spinacé, E.V., Neto, A.O., Santos, M.C.: PtSn/C alloyed and non-alloyed materials: differences in the ethanol electro-oxidation reaction pathways. Appl. Catal. B **110**, 141–147 (2011)
29. Henrique, R.S., De Souza, R.F.B., Silva, J.C.M., Ayoub, J.M.S., Piasentin, R.M., Linardi, M., Spinacé, E.V., Santos, M.C., Neto, A.O.: Preparation of Pt/C-In $2O_3.SnO_2$ electrocatalysts by borohydride reduction process for ethanol electro-oxidation. Int. J. Electrochem. Sci. **7**(3), 2036–2046 (2012)
30. Da Silva, S.G., Silva, J.C.M., Buzzo, G.S., De Souza, R.F.B., Spinacé, E.V., Neto, A.O., Assumpção, M.H.M.T.: Electrochemical and fuel cell evaluation of PtAu/C electrocatalysts for ethanol electro-oxidation in alkaline media. Int. J. Hydrogen Energy **39**(19), 10121–10127 (2014)
31. Herranz, T., García, S., Martínez-Huerta, M.V., Peña, M.A., Fierro, J.L.G., Somodi, F., Borbáth, I., Majrik, K., Tompos, A., Rojas, S.: Electrooxidation of CO and methanol on well-characterized carbon supported PtxSn electrodes. Effect of crystal structure. Int. J. Hydrogen Energy **37**(8), 7109–7118 (2012).

32. Kim, H.J., Choi, S.M., Nam, S.H., Seo, M.H., Kim, W.B.: Effect of Rh content on carbon-supported PtRh catalysts for dehydrogenative electrooxidation of cyclohexane to benzene over polymer electrolyte membrane fuel cell. Appl. Catal. A **352**(1–2), 145–151 (2009).
33. Pawley, G.S.: Unit-cell refinement from powder diffraction scans. J. Appl. Crystallogr. **14**(DEC), 357–361 (1981).

34. Wojdyr, M.: Fityk: a general-purpose peak fitting program. J. Appl. Crystallogr. **43**, 1126–1128 (2010).

35. Zhang, Y., Gao, X.P., Weaver, M.J.: Nature of surface bonding on voltammetrically oxidized noble-metals in aqueous-media as probed by real-time surface-enhanced Raman-spectroscopy. J. Phys. Chem. **97**(33), 8656–8663 (1993).

36. Mavrikakis, M., Doren, D.J., Barteau, M.A.: Density functional theory calculations for simple oxametallacycles: trends across the periodic table. J. Phys. Chem. B **102**(2), 394–399 (1998).

37. Shen, S.Y., Zhao, T.S., Xu, J.B.: Carbon supported PtRh catalysts for ethanol oxidation in alkaline direct ethanol fuel cell. Int. J. Hydrogen Energy **35**(23), 12911–12917 (2010).

38. Neto, A.O., Nandenha, J., Assumpção, M.H.M.T., Linardi, M., Spinacé, E.V., de Souza, R.F.B.: In situ spectroscopy studies of ethanol oxidation reaction using a single fuel cell/ATR-FTIR setup. Int. J. Hydrogen Energy **38**(25), 10585–10591 (2013).

15

Effects of acid concentration, temperature, and time on recycling of post-vehicle-application lithium-ion batteries of varying chemistries

Haikun Li[1] · Lindsay M. Corneal[1] · Charles R. Standridge[1]

Abstract The purpose of this project was to develop and validate a common process for separating the active materials from the copper and aluminum foils of post-vehicle-application lithium-ion batteries. Batteries from two different manufacturers were disassembled, and anode and cathode samples were used for recycling testing. The materials tested from a third manufacturer were scraps of coated foils from the manufacturing process that had never been assembled into cells. The cathodes from each manufacturer were aluminum foils with coatings of differing chemistries. The anode foils from each manufacturer were copper with a carbon coating. The process developed used acid baths to separate the active materials from the foils. To allow for separation of the differing active materials (depending on supplier and battery chemistry) from the foils, the acids were selected to react with the aluminum and copper foils. The processes were stopped once sufficient reaction had occurred so that the coatings no longer adhered to the foils. The optimum recycling condition was identified as the lowest acid concentration, the lowest temperature, and the shortest time required for full separation of the coatings from the foils for all cell chemistries (manufacturers). Full separation of the graphite coating from the copper foils of the anodes was achieved within 35 s by using 0.5 mol/L of H_2SO_4 (sulfuric acid) at 40 °C. Full separation of the active materials from the aluminum foils of the cathodes was achieved within 83 s by using 2.0 mol/L of HNO_3 (nitric acid) at 70 °C.

✉ Lindsay M. Corneal
corneall@gvsu.edu

[1] School of Engineering, Grand Valley State University, Allendale, MI 49401, USA

Keywords Lithium-ion batteries · Battery recycling · Material recovery · Electric vehicles

Introduction

Lithium-ion batteries (LIBs) have an efficient energy storage mechanism, whose use in vehicles will continue to expand with their electrification. The Corporate Average Fuel Economy (CAFE) standards were recently approved for 2017 through 2025, with a groundbreaking 54.5 miles per gallon fuel efficiency standard by 2025 [1]. Increased electrification of vehicles is one way that these aggressive fuel efficiency standards can be met.

A fundamental question is what to do with LIB post-vehicle application, which means that the battery has fallen below regulatory standards for use in on-road vehicles. Such a battery has additional economic value that can be reclaimed in one of three ways: remanufacturing for reuse in vehicles [4], repurposing the batteries into non-vehicle, stationary storage applications [5–7], and recycling when the cells are no longer able to hold a sufficient charge to support any application [8–12]. With the continued manufacturing and repurposing of LIBs, eventually each cell will no longer be useable and require recycling due to potential flammability and toxic cell components [13].

The lithium-ion batteries (LIBs) used in electric vehicles have a life span of 8–10 years [2]. Foster et al. [3] present a forecasting model which indicates that by 2035, there will be between 1.376 million (in a pessimistic forecast) to 6.759 million (in an optimistic forecast) post-vehicle-application LIBs. In addition, these authors suggest that the only economical way to process post-vehicle-application lithium-ion batteries is through an integrated process for repurposing and recycling, although this integrated approach has yet to appear in the marketplace.

When the batteries are no longer sufficient for vehicle applications, the vast majority of cells comprising the battery are still able to hold a charge and are thus suitable for remanufacturing for continued use in vehicles and being repurposed for non-vehicular applications. Repurposing of post-vehicle-application LIBs, however, is not currently available in the marketplace. The remaining cells that were not capable of further use must be recycled. Current industrial practice appears to be either to recycle all the cells in the battery, which is not keeping with the principles of sustainability, or to store the battery for an indefinite period as a method of disposition is perceived to be lacking.

Currently, battery recycling companies such as Toxco charge a fee to dispose of such batteries. Disposition appears to be accomplished by crushing and shredding the batteries, attempting to recover some materials that can be sold and burying the remainder. Again, this process is not in keeping with the principles of sustainability.

LIBs produced by different manufacturers contain different active materials, particularly for the cathodes. However, in all cases, the current collecting foils are copper and aluminum. Thus, the potential for a common recycling process, which would be highly desirable, exists.

Copper accounts for approximately 11–15 % of the battery by weight, and aluminum accounts for approximately 19–24 % of the battery by weight, depending on whether the battery is intended for an electric vehicle (EV), a hybrid electric vehicle (HEV), or a plug-in hybrid electric vehicle (PHEV) [14]. By breaking down the batteries and separating the coatings, copper and aluminum can be recycled. The other components such as steels, plastics, and the active materials can be either disposed of or recycled.

Most research being done in the recycling of lithium-ion batteries uses cells with $LiCoO_2$ as the cathode active material and focuses on the recovery of cobalt and lithium [15–20], with little attention to the copper and aluminum within the cells. The methods used incorporate various acid leaching and hydro- and pyrometallurgical processes [16–19] and bioleaching techniques [20].

This paper discusses the separation of the active materials from the copper and aluminum foils of post-vehicle-application LIBs. Batteries from three different manufacturers with differing chemistries/active materials were studied. A common process was developed and validated.

Materials and methods

The batteries utilized for this work were from three different manufacturers, identified as A, B, and C. Batteries from manufacturers A and B were disassembled, and the anode and cathode samples were used for the recycling testing. The materials tested from manufacturer C were scraps of coated foils from the manufacturing process that had never been assembled into cells. The cathodes were aluminum foils whose coatings are of differing chemistries for each manufacturer. Manufacturer A, B, and C use lithium iron phosphate ($LiFePO_4$), lithium nickel cobalt aluminum oxide (NCA), and lithium manganese oxide ($LiMn_2O_4$), respectively. The anode foils were the same for all manufacturers: copper with a carbon coating.

Acid baths were used to separate the active materials from the foils. The acids were selected to react with the aluminum and copper foils. Thus, the same acids were used for separation of foil coatings of varying chemistries which depended on the supplier. By focusing on the reaction with the aluminum and copper rather than the coating materials, it was hypothesized that the same process would be effective for batteries from the different suppliers.

Sulfuric acid (H_2SO_4) was selected because of its reactivity with copper, as shown in Eq. 1. It was hypothesized that the H_2SO_4 would react with the copper foil, thereby weakening the adhesion of the anode coatings to the foil.

$$Cu(s) + 2H_2SO_4(aq) \rightarrow 2H_2O(l) + SO_2(g) + CuSO_4(aq) \quad (1)$$

Nitric acid (HNO_3) was selected for the reaction with aluminum as shown in Eq. 2. The intent was for the HNO_3 to react with the aluminum foil, thereby weakening the adhesion of the carbon coating to the foil.

$$2Al(s) + 6HNO_3(aq) \rightarrow 2Al(NO_3)_3(aq) + 3H_2(g) \quad (2)$$

Because SO_2 and H_2 gases are products of the above reactions, all tests were conducted in a fume hood. Experiments were designed and conducted to find the lowest acid concentration, the lowest temperature, and the shortest time required for full separation of the coatings from the foils for each of the battery chemistries (manufacturers). Such a combination was hypothesized to result in the lowest cost of recycling. Experiments for each battery chemistry were used to determine the shortest separation time for the varying combinations of acid concentration and temperature.

Sulfuric acid has been shown to react with copper in concentrations as low as 0.5 mol/L [21]. Therefore, for separation of the carbon coatings from the copper foils, 50 mL of 2.0, 1.0, and 0.5 mol/L solutions of sulfuric acid (95.0–98.0 %, Sigma-Aldrich Co, St. Louis, MO) were prepared. While stirring, 5 g of the coated copper foils were placed in the sulfuric acid solution. The time taken for the coatings to separate from the foils was recorded. Tests were conducted at 25, 30, 40, and 50 °C. The tests were repeated with samples from each manufacturer.

For separation of the cathode active coatings from the aluminum foils, 50 mL of 2.0, 1.0, and 0.5 mol/L solutions of nitric acid (70 %, Sigma-Aldrich Co, St. Louis, MO) were prepared. While stirring, 3 g of the coated aluminum foils were placed in the nitric acid solution. The time taken for the coatings to separate from the foils was recorded. Tests were conducted at 25, 30, 40, 50, 60, and 70 °C. The testing combinations were repeated for each manufacturer.

The purpose of the testing was to identify a common process and process parameters for LIB recycling for all manufacturers. Thus, if it was found that the samples from a particular manufacturer were separating at faster times, the number of test combinations was reduced so that higher concentrations and temperatures were not tested if the lower concentrations and temperatures for a particular manufacturer were not limiting the separation time when compared to the time required for the samples from the other manufacturers.

Results and discussion

Separation of carbon coatings from copper foils

Upon immersion of the coated copper foils into the sulfuric acid solutions, bubbles of SO_2 gas formed on the surface. There was also minor heat dissipation (approximately 4 °C temperature rise) during the tests, and the solution turned slightly blue, indicating the formation of copper sulfate ($CuSO_4$) due to the reaction of the copper with the sulfuric acid. This reaction did cause a degradation of the adhesion of the carbon coating to the copper foils, as was intended. The H_2SO_4 concentrations, temperatures, and times required for full separation of the coating from the copper foils for the samples from manufacturers A, B and C are given in Tables 1, 2, and 3, respectively.

Increasing the H_2SO_4 concentration had very little effect on decreasing the separation times. At temperatures of 25 and 30 °C, increasing the sulfuric acid concentration from 0.5 to 1 mol/L and 2 mol/L resulted in separation times of 7–14 s faster for samples from manufacturer A and separation times of 5–23 s faster for samples from manufacturer C. However, when the temperature was raised to 50 °C, increasing the concentration had no effect on decreasing the separation times.

The separation of the carbon coatings from the copper foils occurred within 65 s for all concentration and temperature conditions tested. The fastest separation time of 5 s occurred with the samples from manufacturer B, with an H_2SO_4 concentration of 0.5 mol/L and temperature of 25 °C. For the samples from the other manufacturers, higher temperatures were required to achieve the fastest separation times. For the samples from manufacturer A, the shortest separation time was 35 s using an H_2SO_4 concentration of 0.5 mol/L at a temperature of 40 °C. For the samples from manufacturer C, the shortest separation time was 16 s at both the 1 and 0.5 mol/L concentrations, at a temperature of 50 °C.

It was noted that there were separation times for the samples from manufacturer C at lower temperatures that were still below the shortest separation time for the samples from manufacturer A. Using the same H_2SO_4 concentration of 0.5 mol/L and temperature of 40 °C, which was identified as the optimum condition for the samples from manufacturer A, the separation time for the samples from manufacturer C was 26 s. Using common conditions of an H_2SO_4 concentration of 0.5 mol/L at a temperature of 40 °C results in the separation of samples from all manufacturers within 35 s.

Table 1 Separation times for carbon coatings from copper foils from manufacturer A for varying H_2SO_4 concentrations and temperatures

H_2SO_4 concentration (mol/L)	Temperature (°C)	Separation time (s)
2	25	57
2	30	49
2	40	37
1	25	50
1	30	40
1	40	37
0.5	25	64
0.5	30	47
0.5	40	35

Table 2 Separation times for carbon coatings from copper foils from manufacturer B for varying H_2SO_4 concentrations and temperatures

H_2SO_4 concentration (mol/L)	Temperature (°C)	Separation time (s)
0.5	25	5

Table 3 Separation times for carbon coatings from copper foils from manufacturer C for varying H_2SO_4 concentrations and temperatures

H_2SO_4 concentration (mol/L)	Temperature (°C)	Separation time (s)
2	30	35
2	40	29
2	50	20
1	30	53
1	40	25
1	50	16
0.5	30	58
0.5	40	26
0.5	50	16

Table 4 Separation times for cathode active material coatings from aluminum foils from manufacturer A for varying HNO_3 concentrations and temperatures

HNO_3 concentration (mol/L)	Temperature (°C)	Separation time (s)
2	25	55
2	30	46
1	25	54
1	30	43
1	40	35
0.5	25	67
0.5	30	62
0.5	40	40

Table 5 Separation times for cathode active material coatings from aluminum foils from manufacturer B for varying HNO_3 concentrations and temperatures

HNO_3 concentration (mol/L)	Temperature (°C)	Separation time (s)
2	25	35
2	30	32
2	40	26
1	25	41
1	30	39
1	40	29
0.5	25	44
0.5	30	40
0.5	40	33

Separation of cathode active material coatings from aluminum foils

Upon immersion of the coated aluminum foils into the nitric acid solutions, bubbles of H_2 gas formed on the surface of the aluminum foils. There was also minor heat dissipation (approximately 2 °C temperature rise) during the tests. The reaction between the HNO_3 and aluminum resulted in a degradation of the adhesion of the cathode coatings to the foils, as was intended. The HNO_3 concentrations, temperatures, and times required for full separation of the coating from the aluminum foils for the samples from manufacturers A, B and C are given in Tables 4, 5 and 6, respectively.

For the conditions tested, the separation of the cathode active material coatings from the aluminum foils tended to take longer time than the time required for the separation of the carbon coatings from the copper foils for each manufacturer. The fastest separation time of 26 s was achieved with an HNO_3 concentration of 2 mol/L at 40 °C for the samples from manufacturer B. For the samples from manufacturer A, an HNO_3 concentration of 1 mol/L at 40 °C resulted in a separation time of 35 s. The samples from manufacturer C had much longer separation times than the samples from the other manufacturers. The shortest separation time for the samples from manufacturer C was 83 s, using an HNO_3 concentration of 2 mol/L at 70 °C.

Table 6 Separation times for cathode active material coatings from aluminum foils from manufacturer C for varying HNO_3 concentrations and temperatures

HNO_3 concentration (mol/L)	Temperature (°C)	Separation time (s)
2	50	180
2	60	161
2	70	83
1	50	273
1	60	210
1	70	120
0.5	50	>300.00
0.5	60	219
0.5	70	125

The common condition for the separation of the active materials from the aluminum foils would then be to use an HNO_3 concentration of 2 mol/L at 70 °C, and the separation would occur within 83 s for samples from all manufacturers. It would also be possible, however, to use lower temperatures which would work for the samples from manufacturers A and B but would approximately double the separation time for the samples from manufacturer C. An HNO_3 concentration of 2 mol/L at 50 °C would be able to achieve separation for the samples from manufacturer C in 180 s (3 min).

Future work

These results have shown that it is possible to identify a common acid concentration and temperature that will separate the differing active materials from the current collecting foils in a reasonable length of time (within 3 min). Once separated, the copper and aluminum foils can then be recycled. The next step will be to consider the other components of the cells such as steels, plastics, and the active materials. The recycling or disposal of each of these must be considered.

Summary and conclusions

The use of acid baths to separate the active material coatings from the collecting foils of post-vehicle-application LIBs of varying chemistries from three different manufacturers was evaluated. A process was developed and verified to separate the carbon coatings from the copper foils of the anode using sulfuric acid. The reaction between the H_2SO_4 and the copper resulted in the degradation of the adhesion of the carbon coatings to the foils. The combination of H_2SO_4 concentration of 0.5 mol/L and temperature of 40 °C resulted in the shortest time for full separation of the coating from the foil, within 35 s. A method was identified to separate the coatings from the aluminum foils of the cathodes using nitric acid.

The reaction between the HNO_3 and aluminum weakened the adhesion of the cathode coatings to the foils, resulting in their separation. The differences between the LIB chemistries from the three manufacturers resulted in greater variations in the conditions required for full separation of the cathode active material coatings than of the anode coatings. The results of testing various HNO_3 concentrations and temperatures identified that full separation of the coatings from the aluminum foils was possible within 83 s for samples from all manufacturers by using an HNO_3 concentration of 2 mol/L at 70 °C.

Given these results, the development of a common process with common parameter values for recycling the valuable materials in post-vehicle-application LIBs of varying chemistries that cannot be remanufactured or repurposed is shown to be possible.

Acknowledgments The authors acknowledge the Michigan Department of Transportation and the US Department of Transportation via the Mineta National Transit Research Consortium (MNTRC project#1137) for their financial support of this work. We would also like to thank Sybesma's Electronics for their support of this work.

References

1. Center for Climate and Energy Solutions, Federal Vehicle Standards.
(2014). Accessed 27 September 2014
2. Richa, K., Babbitt, C.W., Gaustad, G., Wang, X.: A future perspective on lithium-ion battery waste flows from electric vehicles. Resour. Conserv. Recycl. **83**, 63–76 (2014)
3. Foster, M., Isely, P., Standridge, C.R., Hasan, MdM: Feasibility assessment of remanufacturing, repurposing, and recycling of end of vehicle application lithium-ion batteries. J. Ind. Eng. Manag. **7**, 698–715 (2014)
4. Schneider, E.L., Kindlein, W., Souza, S., Malfatti, C.F.: Assessment and reuse of secondary batteries cells. J. Power Sour. **189**, 1264–1269 (2009)
5. Andrijanovits, A., Hoimoja, H., Vinnikov, D.: Comparative review of long-term energy storage technologies for renewable energy systems. Electron. Electr. Eng. **118**, 21–26 (2012)
6. Yang, Z., Liu, J., Baskaran, S., Imhoff, C.H., Holladay, J.D.: Enabling renewable energy and the future grid with advanced electricity storage. JOM **62**, 14–23 (2012)
7. Diaz-Gonzalez, F., Sumper, A., Gomis-Bellmunt, O., Villafafila-Robles, R.: A review of energy storage technologies for wind power applications. Renew. Sustain Energy Rev. **16**, 2154–2171 (2012)
8. Gratz, E., Sa, Q., Apelian, D., Wang, Y.: A closed loop process for recycling spent lithium ion batteries. J. Power Sour. **262**, 255–262 (2014)
9. Georgi-Maschler, T., Friedrich, B., Weyhe, R., Heegn, H., Rutz, M.: Development of a recycling process for Li-ion batteries. J. Power Sour. **207**, 173–182 (2012)
10. Paulino, J.F., Busnardo, N.G., Afonso, J.C.: Recovery of valuable elements from spent Li-Batteries. J. Hazard. Mater. **150**, 843–849 (2008)
11. Lain, M.J.: Recycling of lithium ion cells and batteries. J. Power Sour. **97–98**, 736–738 (2001)
12. Contestabile, M., Panero, S., Scrosati, B.: A laboratory-scale lithium-ion battery recycling process. J. Power Sour. **92**, 65–69 (2001)
13. Standridge, C.R., Corneal, L.: Remanufacturing, repurposing, and recycling of post-vehicle-application lithium-ion batteries. Mineta National Transit research consortium, San Jose, CA. transweb.sjsu.edu/project/1137.html (2014). Accessed 25 November 2014
14. Dunn, J.B., Gaines, L., Barnes, M., Sullivan, J., Wang, M.: Material and energy flows in the materials production, assembly, and end-of-life stages of the automotive lithium-ion battery life cycle. Energy. Sys. Division Argonne. Natl. Lab. Rep. (2012)
15. Lain, M.J.: Recycling of lithium ion cells and batteries. J. Power Sour. **97–98**, 736–738 (2001)
16. Zhang, P., Yokoyama, T., Itabashi, O., Suzuki, T.M., Inoue, K.: Hydrometallurgical process for recovery of metal values from spent lithium-ion secondary batteries. Hydrometallurgy **47**, 259–271 (1998)
17. Dorella, G., Mansur, M.: A study of the separation of cobalt from spent Li-ion battery residues. J. Power Sour. **170**, 210–215 (2007)
18. Li, L., Ge, J., Chen, R., Wu, F., Chen, S., Zhang, X.: Environmental friendly leaching reagent for cobalt and lithium recovery from spent lithium-ion batteries. Waste Manag. **30**, 2615–2621 (2010)
19. Georgi-Maschler, T., Friedrich, B., Weyhe, R., Heegn, H., Rutz, M.: Development of a recycling process for Li-ion batteries. J. Power Sour. **207**, 173–182 (2012)
20. Xin, B.P., Zhang, D., Zhang, X., Xia, Y., Wu, F., Chen, S., Li, L.: Bioleaching mechanism of Co and Li from spent lithium-ion battery by the mixed culture of acidophilic sulfur-oxidizing and iron-oxidizing bacteria. Bioresour. Technol. **100**, 6163–6169 (2009)
21. Benali, O., Larabi, L., Harek, Y.: Inhibiting effects of 2-mercapto-1-methylimidazole on copper corrosion in 0.5 M sulfuric acid. J. Saudi Chem. Soc. **14**, 231–235 (2010)

16

Investigations on thermal storage systems containing micron-sized conducting particles dispersed in a phase change material

S. Jegadheeswaran · S. D. Pohekar · T. Kousksou

Abstract The purpose of the present study is to investigate the performance enhancement of a shell and tube, latent heat thermal storage (LHTS) unit due to the addition of microsize copper particles in the phase change material (PCM). The thermo-physical properties of composites were experimentally obtained using Temperature-History technique. Numerical model was developed to explore the heat transfer characteristics of composites during both charging and discharging modes. The numerical model also features a transport equation for particle flux. The commercial computational fluid dynamics (CFD) code, FLUENT, was employed to solve the enthalpy-based two dimensional transient equations. In addition, exergy models were developed to demonstrate the effect of particle dispersion on exergy performance of LHTS systems. The thermal behavior and exergy performance in terms of exergy efficiency and total exergy stored of particle-dispersed PCM units are compared with those of pure PCM unit and nanoparticles composites. The heat transfer performance of PCM is significantly high throughout the discharging when microparticles are added. However, the enhancement could be observed only during the earlier stages of charging as particles are found to be hampering natural convection in the liquid PCM. When it comes to exergy performance, particles help in increasing the exergy efficiency of pure PCM system only when the natural convection is weak during charging. Also, the exergy recovered from composites is higher as compared to that of pure PCM and it is found that particles addition results in reduced storage capacity in terms of exergy.

Keywords Thermal storage · Phase change material · Conductivity particles · Exergy · Particle migration · Computational fluid dynamics

Nomenclature

a	Particle diameter (m)
A	Area (m^2), Coefficient in Eq. (19)
b	Tube wall thickness (m)
Bi	Biot number
C	Mushy zone constant
c_p	Specific heat (J/kg K)
e	Particle volume fraction
Ex	Total exergy stored/recovered (J)
$\dot{E}x$	Exergy rate (W)
f_l	Liquid fraction
g	Gravitational acceleration (m/s^2)
h	Enthalpy (J/kg) or heat transfer coefficient (W/m^2 K)
k	Thermal conductivity (W/m K)
K_c, K_μ	Constants in Eq. (13)
L	Length (m)
m	mass (kg)
$\dot{m}$	Mass flow rate of HTF (kg/s)

S. Jegadheeswaran (✉)
Department of Mechanical Engineering, Bannari Amman Institute of Technology, Sathyamangalam, Erode (DT) 638401, India
e-mail: jdees.2002@gmail.com

S. D. Pohekar
Mechanical Engineering Area, Tolani Maritime Institute, Induri, Pune 410507, India

T. Kousksou
Laboratoire des Sciences de l'Ingénieur Appliquées à la Mécanique et au Génie Electrique (SIAME), Université de Pau et des Pays de l'Adour, IFR, A. Jules Ferry, 64000 Pau, France

P	Pressure (Pa)
r	Radial co-ordinate (m)
r_i, R	Radius of test tube (m)
r_{int}	Radius of solid/liquid interface (m)
R_i	Tube radius (m)
R_o	Shell radius (m)
S	Source term
Ste	Stefan number
t	Time (s)
T	Temperature (°C or K)
T_{atm}	Atmospheric temperature (°C or K)
T_m	Melting temperature (°C or K)
u	Velocity component in x-direction (m/s)
V	Velocity vector (m/s)
x	Axial co-ordinate (m)

Greek letters

α	Thermal diffusivity (m^2/s)
β	Coefficient of volume expansion (1/K)
ε	Computational constant in Eq. (20)
ρ	Density (kg/m^3)
λ	Latent heat of fusion (J/kg)
μ	Dynamic viscosity (kg/m s)
$\dot{\gamma}$	Strain rate (1/s)
ψ	Exergy efficiency
Γ	Diffusive coefficient

Subscripts

f, final	Final state
HTF	Heat transfer fluid
i	Inflection point
in	Inlet
init	Initial state
l	Liquid PCM
out	Outlet
p, PCM	Phase change material
ref	Reference state
s	Solid PCM
t	Tube
w	Water

Introduction

Despite the abundant nature, solar thermal energy is not being utilized to its full potential, mainly due to mismatch between availability and need. Phase change material (PCM) loaded latent heat thermal storage (LHTS) systems seem to be an attractive solution to this problem. However, all the available PCMs possess very low thermal conductivity, which results in slower rates of charging and discharging. Therefore, thermal conductivity augmentation of PCM is emphasized [1], which is possible through introduction of high conductivity metal particles into the PCMs [2]. When the particles are dispersed into the base fluid (PCM), there would be conductivity enhancement as the particles cluster/aggregate. There is enough evidence in the literature for the significant improvement in the thermal conductivity of PCM even with low particle fraction [3–7].

Khodadadi and Hosseinizadeh [3] have shown that the solidification rate of pure PCM could be increased by adding copper nanoparticles. Ranjbar et al. [4] have also used copper nanoparticles to investigate the solidification of composites in rectangular container and have reported the heat transfer enhancement. However, the experimental results reported by Fan and Khodadadi [5] reveal that the increase in solidification rate is considerable only when the volume fraction of copper oxide particles is 0.5 and further increase in fraction could not accelerate the solidification linearly. The recent work by Kalaiselvam et al. [6] shows that both melting and solidification rates of PCM can be augmented through addition of aluminum/alumina particles. The enhanced heat transfer performance due to the presence of alumina particles in the PCM incorporated in the ceiling of buildings is also reported by Guo [7]. It is clear from the literature review that there has been a growing interest in particle-added PCMs. Nevertheless, only few works are reported. Moreover, only nanosize metal particles, which are very expensive, are considered.

Exergy analysis, based on second law of thermodynamics is an attractive way of evaluating the performance of thermal systems, especially when it comes to optimization [8]. Exergy analysis has been successfully carried out by many researchers for LHTS systems of different applications [9–11]. These studies have proved that the exergy analysis only can provide useful information on valuable energy during charging and discharging processes. When a heat transfer enhancement technique is employed for a LHTS system, the comparison between the systems in terms of exergy performance is more useful. This is because the design of LHTS systems musters a compromise between heat transfer rate, quantity and quality of stored/recovered energy [12]. To the best of authors' knowledge, the comparison between pure PCM unit and particle-dispersed unit is reported only in terms of thermal performance, except by the previous works of present authors [13, 14].

Our earlier works [13, 14] too have considered only nanosize particles for the thermal performance enhancement of PCMs. At present, the nanoparticles are not economically viable as the cost of nanoparticles is roughly 25 times that of microsize (1 μm = 10^{-6} m) particles of same material. This is due to mainly the processing cost. However, no work has investigated microsize particle-dispersed PCMs until now. Hence, the primary objective of this work is to examine the role of high conductivity microsize particles in augmenting the heat transfer performance of pure PCM as well as the exergy performance of the storage system. To the best of our knowledge, this is a maiden attempt investigating the cheaper form (microsize) of conductivity particles. Since microparticles are used in the present work as against the nanoparticles considered in our earlier works [13, 14], the methodology of investigation differs completely. Firstly, the thermo-physical properties of microparticles–PCM composites are evaluated experimentally as the analytical procedure employed in earlier works [13, 14] is valid for only nanoparticles–PCM composites. Secondly, the numerical modeling technique adopted here is quite different from that one used for nanoparticles system. This is because the formulation needs to take into account the migration of such microsize particles due to non-uniform shear fields in the liquid which is not the case when it comes to nanoparticles. Because of the expected particle migration, the present formulation also includes the momentum equation which of course is not considered for nanoparticle composites. As the particles are expected to continue migrating all the time, the present work aims at investigating the consequences of particle migration from performance enhancement point of view. This work is also intended to compare the relative benefits/shortcomings of microparticles dispersion in comparison with nanoparticles dispersion. The comparison is also extended to positive and adverse effects of particles dispersion on the exergy performance of conventional PCM.

Preparation and characterization of PCM–microparticles composites

Among the high conductivity particles like silver, nickel, gold, alumina, copper, aluminum etc., copper and aluminum are less expensive and easily available. As far as cost is concerned, there is not much difference between copper and aluminum. However, the thermal conductivity of copper is significantly higher than that of aluminum. The microcopper particles were prepared out of copper rods (99.9 % purity) using simple filing technique. Although it was difficult to control the size of the particles and size uniformity by manual filing, a smoothest file was used to obtain particles of smallest possible size. The size of the prepared particles was measured with electron microscope and the average size was found to be 250 μm. Prior to dispersing the copper particles (99.9 % purity), the same were coated with a surfactant, to enhance the stability of suspended particles. The PCM considered in this work is a hydrated salt and if the base fluid is such a polar solvent, ionic surfactants which are water soluble are generally recommended [15]. In view of this and availability, sodium acetate was chosen as a surfactant. The ratio of copper particles to the surfactant was 3:1 by mass. The surfactant-coated copper particles were then, dispersed into the liquid PCM (Product of Pluss Polymers, India-http://www.thermalstorage.in/plussproducts.html). The PCM (HS58) is a mixture of hydrated salts, additives and nucleating agents. Table 1 presents the properties of the PCM. Throughout the preparation, the PCM was kept at a temperature sufficiently higher than its melting point. A thermostat controlled electric heater was used to heat the mixing container.

While dispersing the particles, the liquid PCM in the mixing container was thoroughly stirred for about 15 min using a magnetic stirrer. After the stirring process, the mixture was treated with ultrasonic disruptor for more than 1 h. The amplitude and frequency of ultrasonic waves were set as 50 % and 50 Hz, respectively, during the sonication.

Measurement of latent heat and specific heats

For the measurement of thermo-physical properties of PCM, the most commonly used methods are drop calorimetry (DC), differential thermal analysis (DTA) and differential scanning calorimetry (DSC) methods. However, these methods are known for various shortcomings. They are,

- The sample PCM to be tested should be of very small size (1–10 mg). However, the properties of bulk PCM would be different from those of sample. Especially, when it comes to PCM composites, the heterogeneous of additives cannot be accounted for, if the sample size is very small. Moreover, the small sample size cannot reflect the degree of supercooling of the PCM as the small container would show higher degree of supercooling.
- Due to the small sample size, the phase change process cannot be visualized. Especially, the volume expansion of liquid PCM, which is critical for the computation of effective thermal conductivity during melting, cannot be observed.
- The related experimental units are complicated and expensive.

Table 1 Thermo-physical properties of HS-58

Properties	Values
Melting temperature (°C)	57–58
Latent heat (kJ/kg)	250
Thermal conductivity (W/m K)	0.65 (solid), 0.40 (liquid)
Density (kg/m^3)	1,400 (solid), 1,290 (liquid)
Specific heat (kJ/kg K)	2.500
Dynamic viscosity (kg/m s)	0.031
Coefficient of volume expansion (1/K)	0.00095

- These methods cannot measure the properties of several samples simultaneously.
- Thermal conductivity values cannot be obtained.

In view of the above-mentioned facts, the present work employs, a simple and cost effective method, namely *Temperature–History* method developed by Yinping et al. [16] for the measurement of all the thermo-physical properties. The liquid PCM, which is at a higher temperature than its melting point ($T_{init} < T_m$), in a test tube is suddenly exposed to atmosphere which allows the freezing of liquid PCM. The measurement of temperature of PCM at every small time interval would enable plotting the T-History curve. Typical T-History plots with three divided regions are shown in Fig. 1a, b.

The cooling of PCM can be termed as a conduction problem that involves surface convection effects. Hence, the Biot number (Bi = $hR/2\,k$, where h is the heat transfer coefficient of air outside the tube, R is the radius of tube and k is the thermal conductivity of PCM) plays an important role. In such problems, if the condition of Bi < 0.1 is satisfied, then the temperature distribution can be assumed as uniform and it allows employing the lumped capacitance method. In view of this, the tube of very smaller diameter as compared to the length has been chosen in the work (i.e. inner diameter = 13 mm, length = 155 mm). These dimensions along with surrounding atmosphere would keep Bi lower and hence, error associated with the employment of lumped capacitance method would be negligible.

When lumped capacitance method is applicable, the energy balance equations for liquid sensible cooling ($t_0 \leq t \leq t_1$), phase change ($t_0 \leq t \leq t_1$) and solid sensible cooling ($t_2 \leq t \leq t_3$) periods are given, respectively,

$$\left(m_{t,p}c_{p,t} + m_{PCM}c_{p,l}\right)\left(T_{init} - T_s\right) = hA_tA_1 \quad (1)$$

$$\left(m_{t,p}c_{p,t} + m_{PCM}\left(\frac{c_{p,l} + c_{p,s}}{2}\right)\right)\left(T_m - T_i\right) + m_{PCM}\lambda = hA_tA_2 \quad (2)$$

$$\left(m_{t,p}c_{p,t} + m_{PCM}c_{p,s}\right)\left(T_i - T_f\right) = hA_tA_3 \quad (3)$$

Along with PCM, another tube containing distilled water (reference fluid) at the same temperature of PCM is also exposed to the same atmosphere and T-History curve is shown in Fig. 1c. This curve can also be divided into three portions using the same temperature ranges used for PCM and the energy equations are,

$$\left(m_{t,w}c_{p,t} + m_wc_{p,w}\right)\left(T_{init} - T_s\right) = hA'_tA'_1, t'_0 \leq t \leq t'_2 \quad (4)$$

$$\left(m_{t,w}c_{p,t} + m_wc_{p,w}\right)\left(T_m - T_i\right) = hA'_tA'_2, t'_1 \leq t \leq t'_3 \quad (5)$$

$$\left(m_{t,w}c_{p,t} + m_wc_{p,w}\right)\left(T_i - T_f\right) = hA'_tA'_3, t'_3 \leq t \leq t'_4 \quad (6)$$

Finally, simple mathematical manipulation of Eqs. (1) through (6) results in expressions for liquid and solid specific heat and latent heat of PCM. They are,

$$c_{p,l} = \frac{m_{t,w}c_{p,t} + m_wc_{p,w}}{m_{PCM}}\frac{A_tA_1}{A'_tA'_1} - \frac{m_{t,p}}{m_{PCM}}c_{p,t} \quad (7)$$

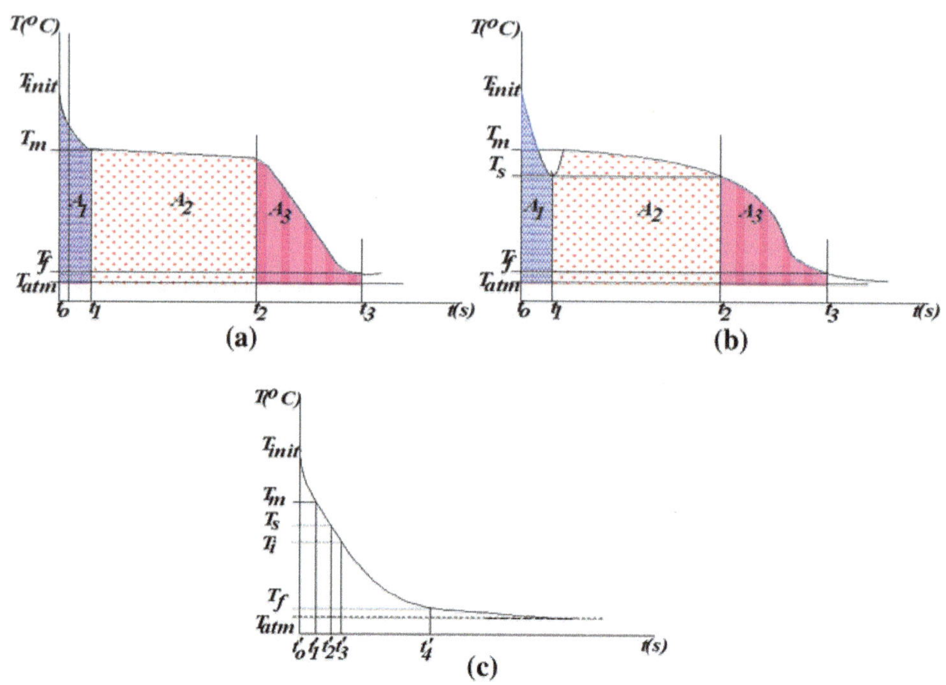

Fig. 1 T-History curves of
a PCM—without supercooling
b PCM—with supercooling
c reference fluid

$$c_{p,s} = \frac{m_{t,w}c_{p,t} + m_w c_{p,w}}{m_{PCM}} \frac{A_t A_3}{A_t' A_3'} - \frac{m_{t,p}}{m_{PCM}} c_{p,t} \tag{8}$$

$$\lambda = \frac{m_{t,w}c_{p,t} + m_w c_{p,w}}{m_{PCM}} \frac{A_t A_2}{A_t' A_2'} (T_m - T_i) - \left(\frac{m_{t,p}}{m_{PCM}} c_{p,t} + \left(\frac{c_{p,l} + c_{p,s}}{2} \right) \right) (T_m - T_i) \tag{9}$$

It should be noted that the selection of t_1 is straight forward as the release temperature (T_s) is starting point of solidification. However, it would be difficult to locate the end point of solidification as solidification may not take place isothermally. This problem is not addressed by Yinping et al. [16]. Hence, the present work adopts a new technique i.e. use of inflection point to indicate the end of solidification process. The inflection point is the one at which the first derivative of T-History curve becomes minimum [17]. This can be justified by the fact that the temperature of PCM remains constant or decreases gradually during phase change, but decreases exponentially once the phase change is completed.

Measurement of thermal conductivities

If the test tube containing molten PCM is suddenly put into the water, which is at a temperature lower than the melting temperature of PCM, solidification can be initiated. Because of smaller diameter to height ratio, the lumped capacitance method is applicable here also. Hence, the heat transfer from PCM to water during solidification is one dimensional and the one dimensional transient heat diffusion equation for cylindrical geometry can be written as,

$$\frac{k_s}{\rho c_{p,s}} \frac{1}{r} \frac{\partial}{\partial r} \left(r \frac{\partial T}{\partial r} \right) = \frac{\partial T}{\partial t} \quad r_{int} \leq r \leq r_i, \quad t > 0 \tag{10}$$

The initial and boundary conditions are, respectively,

$$k_s \left. \frac{\partial T}{\partial r} \right|_{r=r_i} = h_w (T_w - T(r_i, t)) \quad t > 0$$

$$T(r_{int} = r_i) = T_m \quad t = 0$$

At all instants, the conditions for solid/liquid interface are,

$$r_{int} = r_i \quad t = 0$$

$$\frac{k_s}{\rho_s \lambda} \left. \frac{\partial T}{\partial r} \right|_{r=r_{int}} = \frac{\partial r_{int}}{\partial t}$$

Now Eq. (10) along with initial and boundary conditions can be solved using perturbation method. The solution procedure of perturbation method can be found elsewhere [18]. Accordingly, the resulting equation for thermal conductivity of solid PCM is given as,

$$k_s = \frac{[1 + \mathrm{Ste}]}{4 \left[\frac{t_f (T_m - T_w)}{\rho_s r_i^2 \lambda} - \frac{1}{h_w r_i} \right]} \tag{11a}$$

where t_f is the time taken for complete solidification.

The above equation can further be simplified by neglecting the term, $\frac{1}{h_w r_i}$ as Yinping et al. [16] have stated that,

$$\frac{t_f (T_m - T_w)}{\rho_s r_i^2 \lambda} \gg \frac{1}{h_w r_i}$$

Thus, we have,

$$k_s = \frac{[1 + \mathrm{Ste}]}{4 \left[\frac{t_f (T_m - T_w)}{\rho_s r_i^2 \lambda} \right]} \tag{11b}$$

Similarly, the tube containing solid PCM is dipped into a hot water bath, which is at a temperature higher than the melting point of PCM. Once the time taken for the complete melting (t_m) is known, the expression for thermal conductivity of liquid PCM can be arrived following a similar procedure described above.

$$k_l = \frac{[1 + \mathrm{Ste}]}{4 \left[\frac{t_m (T_w - T_m)}{\rho_l r_i^2 \lambda} \right]} \tag{12}$$

In fact, Eqs. (11a, 11b) and (12) are obtained by neglecting second order term of the perturbation expansion. However, the error in the calculated values would be less significant i.e. <5 %, if the following conditions are satisfied.

- Bi <0.1—This is to ensure the one dimensional heat transfer
- Ste <0.5—Ste is taken as the smaller perturbation parameter in the perturbation method and the variables (temperature, interface location and time) are expanded as powers in terms of Ste

Mathematical formulation

In order to investigate the melting and solidification heat transfer characteristics of microparticles–PCM composites, a two dimensional shell and tube module is employed. The PCM is stored in the insulated shell and the tube carries heat transfer fluid (HTF). The vertical orientation of unit is considered to ensure the axisymmetric phase change around the inner tube and thus, only one half of the physical model with axis as symmetry line is sufficient for computational analysis (Fig. 2). All the physical boundaries along with the axis are adiabatic boundaries. In all numerical trails, the dimensions of the unit were kept constant and are given in Table 2.

As mentioned earlier, migration of microsize particles is generally observed due to non-uniform shear fields in the liquid. This means that the particles migrate from higher shear rate regions to lower shear rate regions. Hence, it becomes necessary to track the spatial variation of particle concentration at all times during the phase change processes. To evaluate particle concentration, this work incorporates a constitutive equation which is proposed by Phillips et al. [19]. The constitutive equation is obviously a transport equation for particle flux and is given as,

$$\frac{\partial e}{\partial t} + u.\nabla e = \nabla.\left[a^2 e\dot{\gamma}\left(K_c + K_\mu e \frac{1}{\mu}\frac{\partial \mu}{\partial e}\right)\nabla e\right] + K_c a^2 e^2 \nabla\dot{\gamma} \tag{13}$$

The first term on the right hand side of Eq. (13) is the diffusive term with diffusive coefficient $\Gamma = a^2 e\dot{\gamma}\left(K_c + K_\mu e \frac{1}{\mu}\frac{\partial \mu}{\partial e}\right)$ and the second term on the right hand side of Eq. (13) is nothing but source term (S_e).

Following the continuum approach, the conservative equations for mass and momentum are written as,

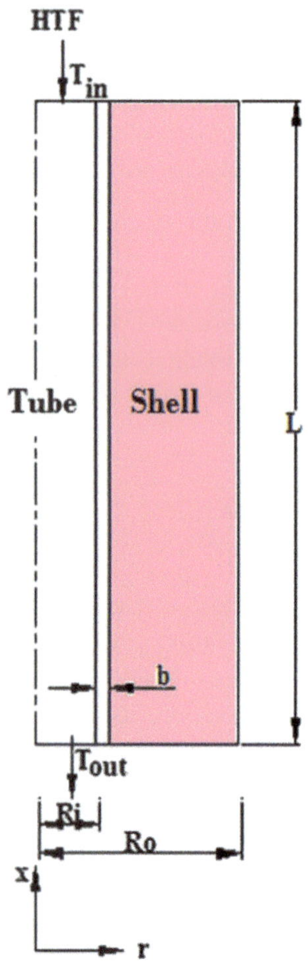

Fig. 2 Computational model of LHTS unit

Table 2 Dimensions of system

Parameters	Values (m)
R_i	0.015
R_o	0.07
L	1
B	0.001

$$\frac{\partial \rho}{\partial t} + \nabla.(\rho V) = 0 \tag{14}$$

$$\frac{\partial}{\partial t}(\rho u_i) + \frac{\partial}{\partial x_j}(\rho u_i u_j) = -\frac{\partial p}{\partial x_i} + \mu\frac{\partial^2 u_i}{\partial x_j^2} + \Delta\rho.g + S_i \tag{15}$$

The source term of Eq. (15) is buoyancy term which describes natural convection through density as a function of temperature. However, the Boussinesq model of natural convection defines density as a constant value irrespective of temperature in all equations except for the buoyancy term in the momentum equation as given below.

$$\Delta\rho.g = \rho\beta(T - T_0)g \tag{16}$$

Hence, the momentum equation becomes,

$$\frac{\partial}{\partial t}(\rho u_i) + \frac{\partial}{\partial x_j}(\rho u_i u_j) = -\frac{\partial p}{\partial x_i} + \mu\frac{\partial^2 u_i}{\partial x_j^2} + \rho\beta(T - T_0)g + S_i \tag{17}$$

The momentum equation of PCM is largely affected by the dispersion of particles as the particles addition changes the rheological behavior of the pure PCM. Hence, the viscosity needs to be defined as a function of particle fraction and Krieger equation [20] is applicable for the same.

$$\mu_{\text{eff}} = \mu\left(1 - \frac{e}{e_{\text{Max}}}\right)^{-1.82} \tag{18}$$

where $e_{\max}$ is the particle fraction when the effective viscosity tends to infinity and is 0.68 for hard spherical particles.

The source term in momentum equation allows the formulation to treat the solid and liquid phases of PCM as porous media and is given as,

$$S_i = -Au_i \tag{19}$$

In a computational cell, the velocity of liquid has a finite value and becomes zero when cell is full of solid. This can be formulated by defining the coefficient A as a function of liquid fraction of the computational cell. i.e.,

$$A(f_l) = \frac{C(1 - f_l)^2}{f_l^3 + \varepsilon} \tag{20}$$

The constant C is called '*mushy zone constant*' and is set as 10,000. The other constant ϵ is a small computational constant taken as 0.001 to avoid division by zero arising from zero liquid fraction.

Now the energy equation is,

$$\frac{\partial}{\partial t}(\rho h) + \frac{\partial}{\partial x_i}(\rho u_i h) = \frac{\partial}{\partial x_i}\left(k\frac{\partial T}{\partial x_i}\right) \tag{21}$$

For HTF, the following mass, momentum and energy conservative equations.

$$\frac{\partial \rho}{\partial t} + \nabla.(\rho V) = 0 \tag{22}$$

$$\frac{\partial(\rho V)}{\partial t} + \nabla.(\rho VV) = -\nabla P + \mu\nabla^2 V + \rho g \tag{23}$$

$$\frac{\partial T}{\partial t} + V\nabla T = \alpha(\nabla^2 T) \tag{24}$$

The numerical model considers only laminar flow of HTF as turbulent does not have any role in enhancing the phase change rate [13, 14].

For numerical solution of governing equations, the commercial computational fluid dynamics (CFD) code called FLUENT 6.2.36 is adopted. The code uses an iterative procedure in which the phase change rate is linearized as truncated Taylor series and then the old iteration values are used to compute the linear term. The mesh is created using quadrilateral elements and numerical trials involving pure PCM were performed with coarse and finer meshes, i.e. grid spacing ranging from 0.1 to 0.001. From the results of instantaneous heat fluxes and liquid fraction (Table 3), the grid spacing of 0.0015 seems to be the optimum one as further decrease does not result in any notable differences in the results.

Time step size of 0.001 s was selected for transient analysis as increase in time step led to convergence difficulties and decrease could not make any difference in the results. The difference between liquid fraction values obtained with time step size of 0.001 and that of obtained with 0.0005 was found to be in the range of 0.1–0.2 %.

For the convergence of the solution, convergence criterion of 10^{-5} for continuity and momentum equations and 10^{-6} for energy equations were fixed. To improve the rate of convergence, under relaxation factor was introduced and the optimum under relaxation factor for liquid fraction update was found to be 0.7. The solution controls used are:

- *SIMPLE* algorithm for pressure–velocity coupling
- *Standard scheme* (pressure equation) and *Second order upwind scheme* (momentum and energy equations) as discretization schemes

In Fluent solver, unique modeling requirements can be dynamically loaded in the form of separately developed codes written in "C" language. This sort of code is called user defined function (UDF). Hence, a UDF was developed

Table 3 Instantaneous heat flux and liquid fraction values with various grid sizes (Pure PCM, T_{in} = 340 K)

Time (h)	Heat flux (W)				Liquid fraction			
Grid size	0.1	0.01	0.0015	0.001	0.1	0.01	0.0015	0.001
1	767	778	791	791.2	0.15	0.18	0.20	0.201
2	699	713	725	725.3	0.302	0.37	0.43	0.432
3	606	628	645	645.2	0.54	0.59	0.66	0.662
4	491	502	517	517.2	0.71	0.784	0.85	0.853
5	269	286	304	304.3	0.84	0.878	0.98	0.982

to take into account the variation of thermal conductivity with respect to two solution variables, i.e. temperature and particle fraction, and the same was compiled in the solver. Similarly, another UDF which defines the particle fraction-dependent viscosity was compiled.

The major difference between modeling of nano and microparticles systems lies in the form of particle flux equation which is a customized equation. In the solver, only standard conservative equations are preloaded and thus, the solver considers the particle flux equation as an additional transport equation for an arbitrary scalar. Hence, the particle flux has to be defined as a scalar which is called user defined scalar (UDS). Similar to UDF, UDS has to be executed in the form of a separate code written in "C" language. For the execution of UDS, two UDFs were also required as they define the diffusion coefficient and source term of particle flux equation.

Exergy analysis

Following the authors' previous works [13, 14], the exergy parameters are computed as follows:

Charging process

Rate of exergy transferred by HTF,

$$\dot{\mathrm{E}}\mathrm{x}_{\mathrm{HTF}} = \dot{m}c_{\mathrm{p,HTF}}\left[(T_{\mathrm{in}} - T_{\mathrm{out}}) - T_{\mathrm{atm}}\ln\left(\frac{T_{\mathrm{in}}}{T_{\mathrm{out}}}\right)\right] \tag{25}$$

Rate of exergy stored,

$$\dot{\mathrm{E}}\mathrm{x}_{\mathrm{PCM}} = \dot{Q}\left[1 - \frac{T_{\mathrm{atm}}}{T_{\mathrm{PCM}}}\right] \tag{26}$$

Exergy efficiency,

$$\psi = \frac{\dot{E}x_{\mathrm{PCM}}}{\dot{E}x_{\mathrm{HTF}}} \tag{27}$$

Discharging process

Rate of exergy transferred by PCM,

$$\dot{E}x_{\mathrm{PCM}} = -\dot{Q}\left[1 - \frac{T_{\mathrm{atm}}}{T_{\mathrm{PCM}}}\right] \tag{28}$$

Rate of exergy recovered,

$$\dot{E}x_{\mathrm{HTF}} = \dot{m}c_{\mathrm{p,HTF}}\left[(T_{\mathrm{out}} - T_{\mathrm{in}}) - T_{\mathrm{atm}}\ln\left(\frac{T_{\mathrm{out}}}{T_{\mathrm{in}}}\right)\right] \tag{29}$$

Exergy efficiency,

$$\psi = \frac{\dot{E}x_{\mathrm{HTF}}}{\dot{E}x_{\mathrm{PCM}}} \tag{30}$$

Experimental set up

An experimental test rig was built in-house to verify the accuracy of the numerical model (Fig. 3). The main test section comprises two co-axial cylindrical tubes. The outer acrylic tube (9 cm inner diameter and 10 cm outer diameter) is having sealed bottom cover and removal top cover. The removal cover facilitates the easy filling and removal of PCM. Moreover, the unsealed top cover allows the air to escape during melting of PCM. The inner tube made of copper (diameter 1.9 cm and thickness 0.1 cm) allows the hot/cold water to flow downwards. The annular space between the inner and outer tubes serves as storage compartment for PCM–particle composites. The effective height of test section is 10 cm. The entire test section including the top and bottom surfaces was covered with glass wool insulating material.

A water bath located above the test section was used to supply water at constant temperature with an accuracy of ±2 °C. For flow rate measurements, a rotameter (accuracy (±3 %) was used. During the experiments, temperatures of PCM were measured at different radial and axial locations. Nine k-type thermocouples calibrated with accuracy of ±1.5 % were inserted at radial positions 0.5, 3 and 4 cm from centre of the test section and at axial positions 3, 6 and 9 cm from the top. Because of axisymmetric phase change processes, all thermocouples are placed only on the right side of PCM. In addition, two thermocouples were used to measure the inlet and outlet temperatures of water to confirm the change in temperature of water during charging and discharging.

Results and discussions

Validation of the numerical model

Figure 4 presents the comparison between numerical and experimental results for representative locations. As seen, certain discrepancy between numerical and experimental results is observed and the deviation of numerical results from experimental measurements is relatively higher during melting.

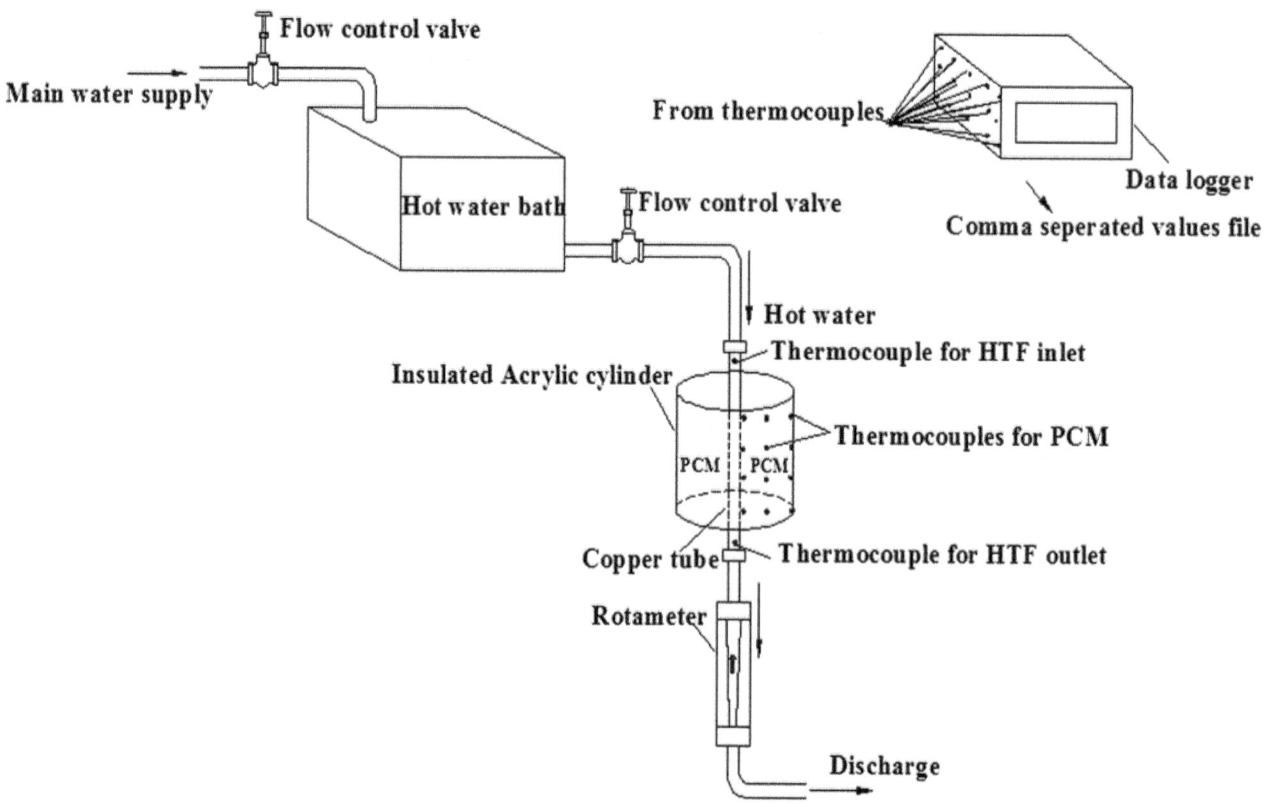

Fig. 3 Schematic of experimental setup

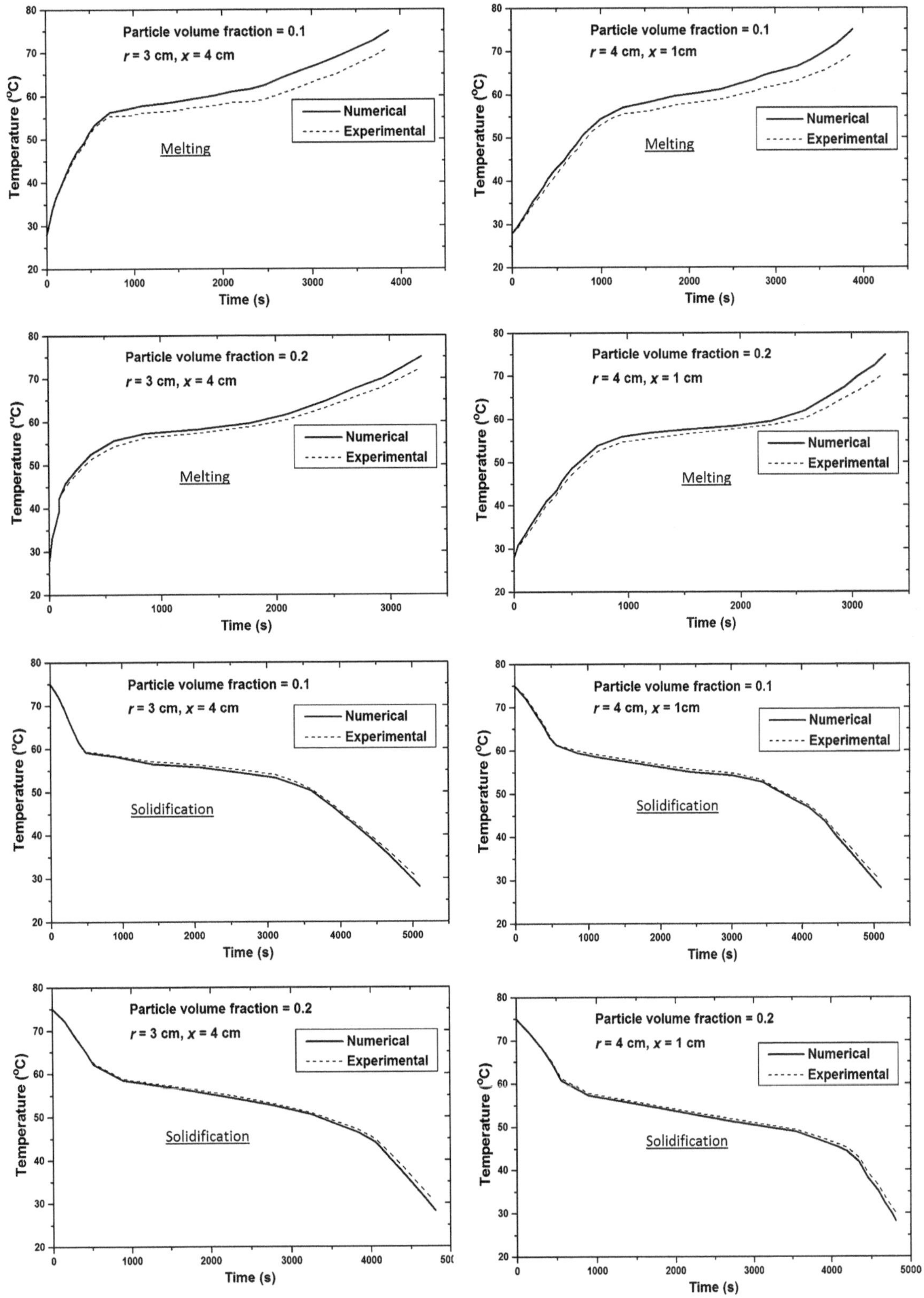

Fig. 4 Comparison of results of numerical model employed with those of experiments

As it is discussed in the later part of this section, the composite PCMs suffer from uneven distribution of particles due to particle migration from one region to other regions. The non-uniform distribution of particles calls for definition of thermal properties as functions of the variable, i.e. particle fraction. However, the CFD code used in this work permits only definition of thermal conductivity as a variable and the specific heat and latent heat values are to be defined as constants. This limitation results in some difference between numerical and experimental results. Since the particle migration is found to be prevailing only during melting process, the definition of constant specific heat and latent heat could not affect the solidification results. Similarly, the effect of particle migration is stronger in composites of lower particle fraction and thus, the deviation of numerical results from experimental measurements is more pronounced in case of composites with 0.1 particle fraction. In general, the deviation increases as the time progresses due to stronger particle migration at later stages of melting.

In case of solidification, discrepancy arises due to mainly the difficulty in maintaining the adiabatic wall conditions and variation of temperature of HTF at the inlet during experimentation. Experimental uncertainties have also contributed to a certain extent. Despite the deficiency, the predicted values are qualitatively correct and the results confirm the validity of the adopted numerical model.

Thermo-physical properties

For each composite, the T-History experiments were performed twice with different initial temperatures (around 10 and 30 °C above the melting point) to verify the accuracy of measurements. In general, the difference between two set of results was not more than 3 %. Similarly, to check the repeatability of measurements, the trial for any sample was conducted twice and it was found that each result differs from corresponding result of repeated trial only by <1 % (Table 4).

In order to establish a relationship between thermal conductivity and particle fraction, Fig. 5 is plotted. In the plot, symbols represent the experimental results of thermal conductivity and the curve is the generated linear regression corresponding to the experimental data. The curve fitting procedure subsequently led to the following function, which was used in the UDF of thermal conductivity.

$$k_l = 0.39686 + 1.57022e$$

Performance of LHTS unit employing PCM–microparticles

The heat transfer and exergy performances of microparticle-dispersed systems are presented in comparison with those of nanoparticle-dispersed systems. The nanoparticles composites were analyzed employing the numerical procedure reported in our earlier papers [13, 14]. Due to the enhanced thermal conductivity by microparticles, it is obvious that the heat transfer rate is much higher throughout the solidification process (Fig. 6a). However, it is also important to examine the relative strength of microparticles in enhancing the heat transfer rate.

According to Table 5, the microparticles composite of any particle fraction shows relatively higher increase in thermal conductivity as compared to nanoparticles composite of same particle fraction. Due to relatively larger size, microparticles can remain closer to each other than what is possible in nanoparticles composites. This means the particles network is stronger in microparticles composites than in their counter parts. However, it is known that the particles migration in microparticles composites leads to non-uniform distribution of particles. Thus, the heat transfer rate may be higher only at locations where more particles exist and there may not be much enhancement in the overall heat transfer rate. Moreover, as the particle fraction at any location may vary with time, the qualitative behavior of microparticles composites in terms of heat transfer rate may not be same as that of nanoparticles composites.

The comparison of nano and microparticles in terms of heat transfer rate enhancement is shown in Fig. 6b. Accordingly, the enhancement in the heat transfer rate due to microparticles dispersion is relatively higher than that of nanoparticles dispersion and this is true for any value of particle fraction. Moreover, microparticle composites maintain higher heat transfer rates at all times throughout solidification. This clears that there is no difference between microparticles and nanoparticle systems from qualitative perspective, although the enhancement by the former is quantitatively higher than that by the latter.

The higher heat transfer rate at all times highlights that the initial particle fraction, which is uniform everywhere remains same throughout the solidification. This can be verified from Fig. 7a–c, in which the time variation of particles fraction is shown. As it can be seen, the variation of particle fraction with time is negligible. Same is observed at all locations. Hence, the particle migration appears to be very weak during solidification. The reason for the insignificant particle migration is simple and straight forward. The onset of particle migration is mainly due to strain rate gradient and the strain rate gradient depends on the strength of fluid flow. During phase change, the only fluid flow is due to natural convection which is incomparable during solidification [1]. Obviously, the weaker fluid flow could not promote the particle migration and thus, microparticles exhibit higher solidification rates at all times as compared to their nano-counterparts.

In order to quantify the relative benefit of microparticles over nanoparticles, the decrease in solidification times due to microparticles is compared with that of due to nano particles. The results are presented in Table 6 and provide further evidence to the superiority of microparticles during solidification.

Table 4 Experimentally evaluated properties of microparticles composites

Particle fraction	Latent heat (kJ/kg)	Specific heat (kJ/kg K)	k_s (W/m K)	k_l (W/m K)
0.05	108.4	0.986	0.231	0.1288
0.1	73.2	0.6	0.276	0.1503
0.15	56.5	0.419	0.324	0.1749
0.2	40.5	0.31	0.359	0.1951

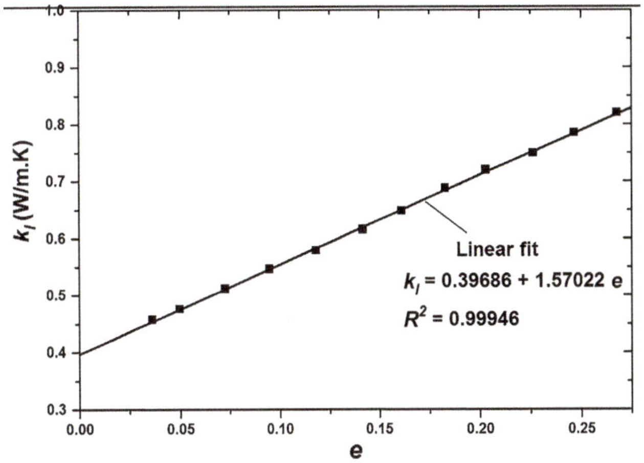

Fig. 5 Variation of thermal conductivity of liquid microparticles composites with particle fraction

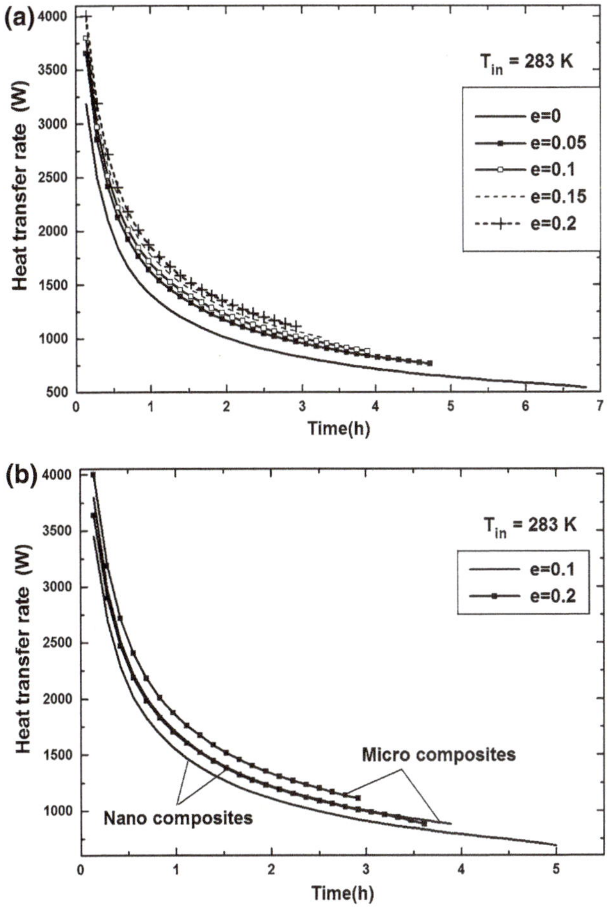

Fig. 6 Effect of microparticles on instantaneous heat transfer rate during solidification **a** for various particle fractions **b** in comparison with effects of nanoparticles

Unlike solidification, the flow of liquid PCM cannot be ignored during melting due to the existence of natural convection. Hence, to examine the particle distribution at various times during melting, the concentration profiles are plotted along the radial direction (Fig. 7d–f). Since composites of various particle fractions melt at different rates, in view of comparison, the presented profiles are the one those correspond to melt fractions 0.25, 0.5 and 0.75.

In the liquid region, the distribution of particles is such that the volume fraction is found to be the lowest, adjacent to the tube wall, and gradually increases in the radial direction. The increase in volume fraction continues up to the middle of the liquid region and thereafter the value starts decreasing to reach the minimum value at the liquid–solid interface. As it is known, the fluid flow due to natural convection is established as simultaneous upward and downward motions of liquid with the former along the tube wall and the latter along the interface. In such case, it can be assumed that the upward motion is bounded by the left half of the liquid region where as the downward motion exists within the second half. For the upward motion of hot fluid, the local shear rate at tube wall remains maximum and the maximum shear rate forces the particles moving to the centre region where the shear rate is zero. Similarly, maximum shear rate should be found at the interface as it acts as a wall. Hence, the particles are forced to migrate toward centre region due to the flow of cold liquid. This is the reason why the particle profiles appear as shown in Fig. 7d–f.

Although the above-mentioned trend is observed for all values of particle fractions, the variation along the radius becomes less as the particle fraction increases. This can be attributed to the fact that higher particle fraction weakens the natural convection, which in turn dampens the particle migration. The observed particle migration in this work is consistent with the results of Phillips et al. [18]. As the melting progresses, natural convection becomes stronger and hence, the particle migration would be on increasing mode. As a result, steeper particle concentration profiles are observed at later stages of melting.

The time variation of heat transfer rates during melting of microcomposites is compared with that of pure PCM in Fig. 8a. In general, heat transfer enhancement due to microparticles in the PCM is observed only for some initial period of melting. No enhancement but rather decreased heat transfer rate could be seen during the remaining period. In fact, enhanced heat transfer in the beginning and decreased heat transfer later are found also with nanoparticles composites. Hence, both the versions of particles enhance the heat transfer only during the conduction dominated period.

Table 5 Increase in thermal conductivity of pure PCM due to particles

Particle fraction	Nano		Micro	
	Solid (%)	Liquid (%)	Solid (%)	Liquid (%)
0.05	15.4	15	19.5	19
0.1	32.3	32.5	38.4	38.2
0.15	52.3	52.5	58.5	58
0.2	73.8	72.5	78	78.3

Since microparticles addition dampens the natural convection considerably, convection dominated melting becomes slow. However, there is significant difference between nanoparticles composites and microparticles composites as shown in Fig. 8b. First notable difference lies in the form of heat transfer enhancement period. Overall, microparticles of any fraction enhance the heat transfer rate for shorter period in comparison with its nano-counterpart. However, the gap narrows down with increase in particle fraction. Due to higher thermal conductivity, microparticles PCMs are supposed to exhibit higher heat

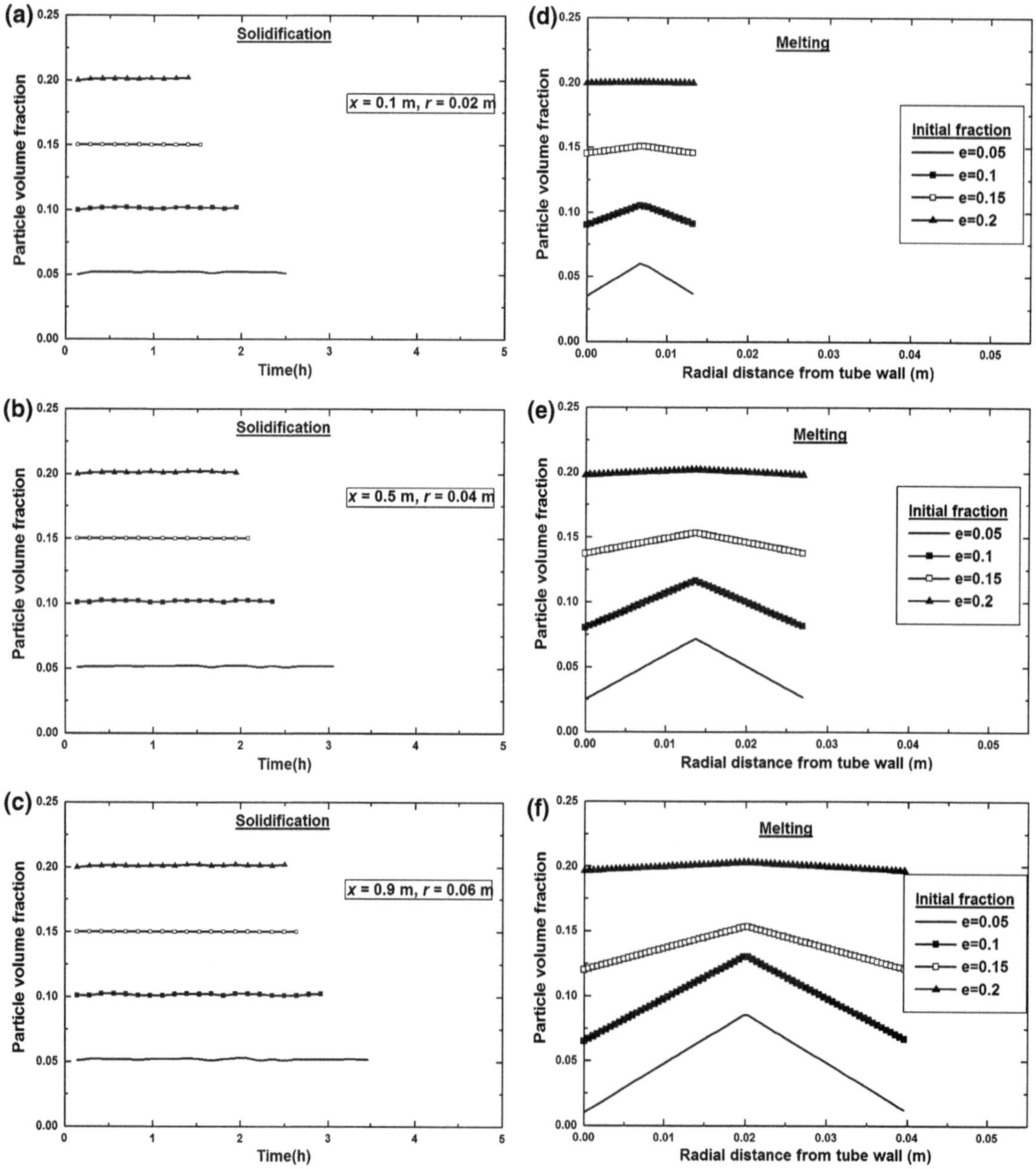

Fig. 7 Time variation of particle fraction (**a–c**) during solidification (**d–f**) during melting

Table 6 Decrease in complete solidification time due to nano and microparticles addition ($T_{in} = 283$ K)

Particle fraction	Decrease in time (%)	
	Nano	Micro
0.05	14	30.6
0.1	26.5	43
0.15	39	51
0.2	47	57.1

transfer rate for longer period as compared to nanoparticles PCMs. However, due to particle migration in the liquid PCM, particle fraction adjacent to heat transfer surface becomes less and keeps reducing as time progresses. Hence, effective thermal conductivity may not be considerably high so that the reduction in heat transfer owing to dampened natural convection could be compensated. As seen earlier, particle migration is not significant when particle fraction is higher and thus, higher particle fraction composites could exhibit higher heat transfer rate for

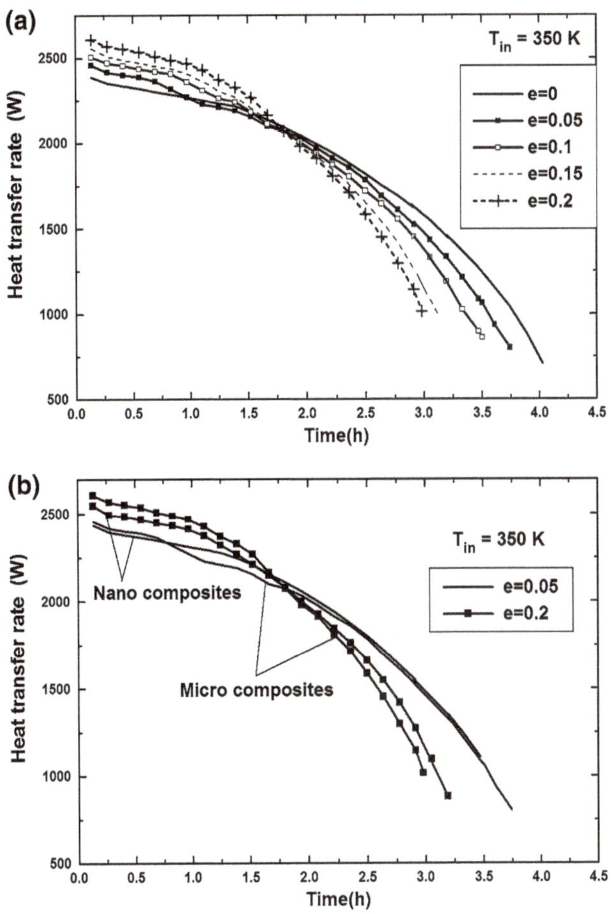

Fig. 8 Effect of microparticles on instantaneous heat transfer rate during melting **a** for various particle fraction **b** in comparison with effects of nanoparticles

longer period. Because of same reason, during enhanced heat transfer period, there is not much difference between the heat transfer rates of nano and microcomposites of lower particle fractions. The difference widens as the particle fraction increases. Nevertheless, relatively higher heat transfer rates are observed with microparticles PCMs of any fraction in comparison with nanoparticles composites of same volume fraction.

During convection-dominated period, both the types of composites show poor heat transfer rates as compared to pure PCM and microparticles composites seem to be suffering more irrespective of volume fraction. This is due to the combined effect of higher viscosity and particle migration. In case of nanoparticles PCMs, first of all there is no particle migration and thus, the effective thermal conductivity remains uniform, which in turn compensates the effect of dampened natural convection to a certain extent. On the other hand, significant particle migration is observed with microparticles of all particles fractions especially at higher times. The particle migration, as it is known, cannot help in enhancing the conduction heat transfer rate. Moreover, microparticles PCM of any particle fraction is more viscous than its nano-counter parts (Table 7). The higher viscosity obviously has a significant role in reducing the convection heat transfer.

Despite lower heat transfer rates during convection-dominated period, microparticles composite of any particle fraction needs less time for complete melting as compared to nanoparticle composite of same particle fraction/pure PCM. Higher melting achieved during conduction dominated period helps the microcomposites to dominate nanocomposites and pure PCM throughout the melting process and the effect of lower heat transfer rate is not pronounced in this regard. The comparison of micro and nanocomposites in terms of reduction in melting time is given in Table 8. It should be mentioned that the superiority of microparticles composites in reducing the melting time is more pronounced if the particle fraction is higher. Again this is related to particle migration. As seen earlier, higher heat transfer rates are observed for longer period when particle fraction is more because of weak particle migration. Lower heat transfer rate for a smaller period could not dampen the effect of high heat transfer rate, which prevails for longer period. As a result, higher particle fraction microcomposites are found to be reducing melting time significantly. On the other hand, only moderate reduction is observed when particle fraction is less as the effect of enhanced melting rate is dampened to a certain extent by lower rate of heat transfer during long period of melting.

As it is seen, both nano and microparticles enhance the solidification rate of pure PCM significantly. Similarly, the heat transfer between PCM and HTF takes place at higher

Table 7 Comparison of viscosities of nano and microparticles composites

Particle fraction	Dynamic viscosity (kg/m s)	
	Nano	Micro
0.05	0.035241	0.035623
0.1	0.040342	0.041408
0.15	0.046539	0.048792
0.2	0.054155	0.058434

Table 8 Decrease in complete melting time due to nano and microparticles addition (T_{in} = 350 K)

Particle fraction	Decrease in time (%)	
	Nano	Micro
0.05	6.8	7
0.1	10	13
0.15	17	22
0.2	20.5	26

temperatures when particles are dispersed. Therefore, the exergy efficiency of particles dispersed PCMs should be higher than that of pure PCM. Moreover, the solidification rates and temperatures at which heat transfer occurs are found to be better with microparticles and thus, it can be predicted that systems with microparticles–PCM composites could exhibit higher exergy efficiency than those of their nanoparticle counter parts. The exergy efficiency results displayed in Fig. 9a back the points discussed above.

During melting, microcomposites possess better exergy efficiency than nanocomposites or pure PCM only when the heat transfer rate is enhanced as shown in Fig. 9b. During this period, the difference between exergy efficiencies of micro and nanocomposites is more significant, if the particle fraction is higher. This is because of large gap between the heat transfer rates of those two composites. Similar behavior in terms of difference in exergy efficiencies can be seen even during the second part of melting, i.e. the period of decreased heat transfer rate by nano/microcomposites. However, now the exergy efficiency of the microcomposites is less in comparison with nanocomposites or pure PCM as the former is found to be inferior in terms of heat transfer rate during this period.

The reduction of latent heat of pure PCM is more pronounced when microparticles are added. Due to this fact, the ratio of total energy recovered from microparticle-added PCM and that from pure PCM is significantly less as compared to the ratio obtained from nanoparticles as seen from Fig. 10a. However, the microparticles appear to be superior over nanoparticles, if the comparison is made in

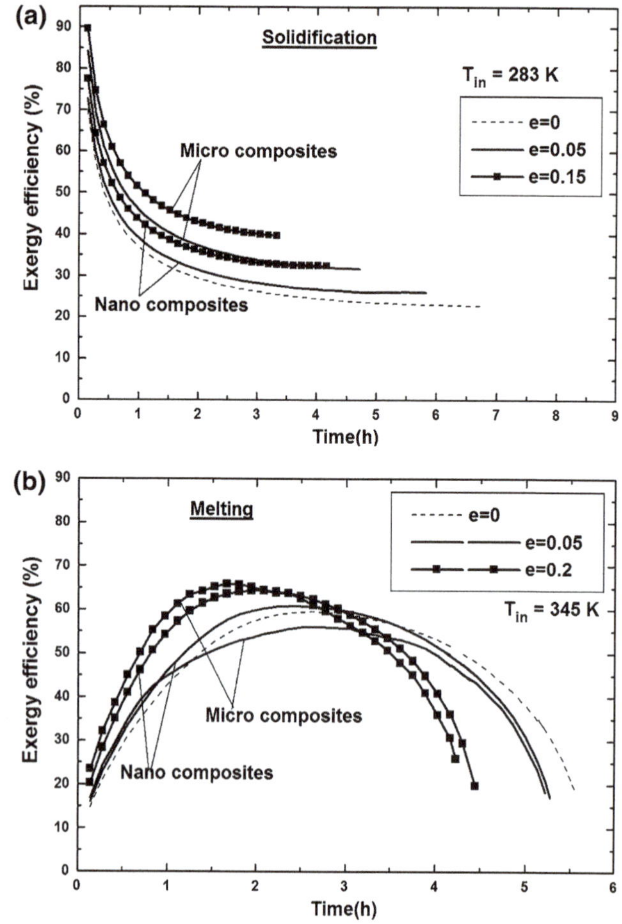

Fig. 9 Comparison of effects of nano and microparticles on exergy efficiency **a** during solidification **b** during melting

terms of ratio of total exergy recovered from composites and that from pure PCM (Fig. 10a). The superior exergy performance by microparticles composites is again attributed to the combined effect of higher heat transfer rates and lesser solidification time.

As observed during solidification, the effect of reduced latent heat due to the addition of microparticles is also manifested during melting. According to Fig. 10b, the ratio of total energy stored by microparticle-added PCM and that by pure PCM decreases with increase in particle fraction. As far as exergy ratio is concerned, the behavior of microparticles PCMs during melting is distinct from what is observed during solidifcation. Although the same trend is seen with nanoparticles, the decrease in energy/exergy ratio is relatively more in case of microparticle PCMs. As seen earlier, the latent heat of microparticles PCM of any particle fraction is less as compared to that of its nano-counterpart. Hence, it can be stated that the reduced latent heat exerts similar effect on both the energy and exergy storage capacities. Furthermore, this is true whether it is nano or microparticles.

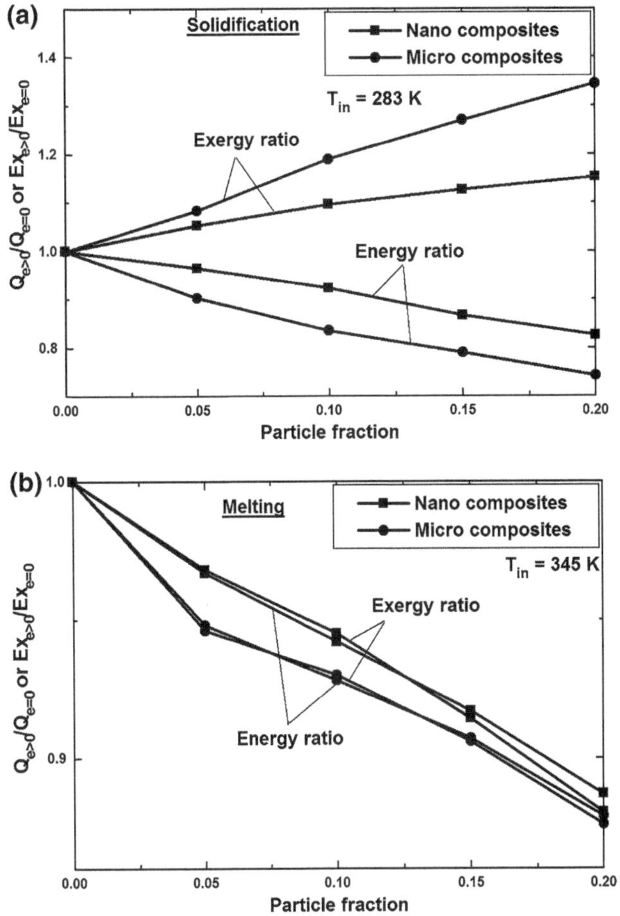

Fig. 10 Comparison of effects of nano and microparticles on **a** total energy recovered/total exergy recovered **b** total energy stored/total exergy stored

Conclusions

The present work investigates the influence of microsize conductivity particle dispersion on the heat transfer and exergy performances of PCM. The thermo-physical properties of composites are measured experimentally following Temperature-History technique. The phase change processes are mathematically modeled with an enthalpy-based method and the resulting system of conjugate governing equations is solved employing commercial CFD code, FLUENT. A constitutive equation for particle flux is included in the mathematical model to simulate particle migration in the PCM. The conclusions of the evaluation derived from investigations are:

- Due to the augmentation of thermal conductivity, the discharging rate of composite PCMs is significantly high and micro particles are superior to nanoparticles in enhancing solidification rate.
- The short coming of particles in the PCMs is exposed in the light of reduced latent heat and specific heat capacities of composites PCMs. Moreover, reduction in heat capacities is relatively high when microparticles are added.
- From exergy performance perspective, the microparticles are superior to nanoparticles. However, opposite trend is disclosed by microparticles during melting.
- Particles addition, however, dampens the natural convection during charging and the effect is more prominent when microparticles are dispersed as the viscosity of microparticles composite is higher than that of nanoparticles composite. However, in solar thermal applications, heat transfer enhancement is required mainly for discharging process rather than for charging. This is because heat is available at constant rate for longer period and the same has to be removed at faster rate. Hence, the microparticles addition can be stated as a promising enhancement technique for LHTS system of solar water heaters as producing hot water at faster rate in the evening and night is most critical. Moreover, using microparticles instead of nanoparticles would reduce the cost to larger extent.
- Due to relatively larger size as compared to nanosize particles, one may expect settlement of microsize particles. Since it is demonstrated that there is always migration of microparticles, the same may overcome the settlement problem. However, this was not verified in this work which requires testing of the system for number of thermal cycles. The thermal cycle testing would be carried out in future work.
- Hence, if non-settlement of particles is confirmed, then particle dispersion would be a better technique than other enhancement techniques i.e. improving heat exchanger design (using fins) and inserting non-moving high conductivity structures as these techniques would significantly reduce the volume of PCM in the system.

References

1. Jegadheeswaran, S., Pohekar, S.D.: Performance enhancement of latent heat thermal storage system: a review. Renew. Sustain. Energy Rev. **13**, 2225–2244 (2009)
2. Fan, L., Khodadadi, J.M.: Thermal conductivity enhancement of phase change materials for thermal energy storage: a review. Renew. Sustain. Energy Rev. **15**, 24–46 (2011)
3. Khodadadi, J.M., Hosseinizadeh, S.F.: Nanoparticle-enhanced phase change materials (NEPCM) with great potential for improved thermal energy storage. Int. Commun. Heat Mass Transf. **34**, 534–543 (2007)
4. Ranjbar, A.A., Kashani, S., Hosseinizadeh, S.F., Ghanbarpour, M.: Numerical heat transfer studies of a latent heat storage system

containing nano-enhanced phase change material. Therm. Sci. **15**, 169–181 (2011)
5. Fan, L., Khodadadi, J.M.: Experimental verification of expedited freezing of nanoparticle-enhanced phase change materials (NEPCM). In: Proceedings of ASME/JSME 8th Thermal Engineering Joint Conference, Hawaii (2011)
6. Kalaiselvam, S., Parameshwaran, R., Harikrishnan, S.: Analytical and experimental investigations of nanoparticles embedded phase change materials for cooling application in modern buildings. Renew. Energy **39**, 375–387 (2012)
7. Guo, C.: Application of nano-particle-enhanced phase change material in ceiling board. Adv. Mater. Res. **150–151**, 723–726 (2010)
8. Jegadheeswaran, S., Pohekar, S.D., Kousksou, T.: Exergy based performance evaluation of latent heat thermal storage system: a review. Renew. Sustain. Energy Rev. **14**, 2580–2595 (2010)
9. Ezan, M.A., Ozdogan, M., Gunerhan, H., Erek, A., Hepbasli, A.: Energetic and exergetic analysis and assessment of a thermal energy storage (TES) unit for building applications. Energy Buildings **42**, 1896–1901 (2010)
10. Pandiyarajan, V., Chinnappandian, M., Rghavan, V., Velraj, R.: Second law analysis of a diesel engine waste heat recovery with a combined sensible and latent heat storage system. Energy Policy **39**, 6011–6020 (2011)
11. Rosen, M.A., Pedinelli, N., Dincer, I.: Energy and exergy analyses of cold thermal storage systems. Int. J. Energy Res. **23**, 1029–1038 (1999)
12. Shabgard, H., Robak, C.W., Bergman, T.L., Faghri, A.: Heat transfer and exergy analysis of cascaded latent heat storage with gravity-assisted heat pipes for concentrating solar power applications. Sol. Energy **86**, 816–830 (2012)
13. Jegadheeswaran, S., Pohekar, S.D.: Exergy analysis of particle dispersed latent heat thermal storage system for solar water heaters. J. Renew. Sustain. Energy **2**, 023105 (2010)
14. Jegadheeswaran, S., Pohekar, S.D., Kousksou, T.: Performance enhancement of solar latent heat thermal storage system with particle dispersion—an exergy approach. Clean-Soil Air Water **39**, 964–971 (2011)
15. Schramm, L.L., Stasiuk, E.N., Marangoni, D.G.: Surfactants and their applications. Annu. Rep. Prog. Chem. Sect. C **99**, 3–48 (2003)
16. Yinping, Z., Yi, J., Yi, J.: A simple method, the T-history method, of determining the heat of fusion, specific heat and thermal conductivity of phase-change materials. Meas. Sci. Technol. **10**, 201–205 (1999)
17. Hong, H., Park, C.H., Choi, J.H., Peck, J.H.: Improvement of the T-history method to measure heat of fusion for phase change materials. Int. J. Air-Cond. Ref. **11**, 32–39 (2003)
18. Parang, M., Crocker, D.S., Haynes, B.D.: Perturbation solution for spherical and cylindrical solidification by combined convective and radiative cooling. Int. J. Heat Fluid Flow **11**, 142–148 (1990)
19. Phillips, R.J., Armstrong, R.C., Brown, R.A., Graham, A.L., Abbott, A.L.G.: A constitutive equation for concentrated suspensions that accounts for shear-induced particle migration. Phys. Fluids **4**, 30–40 (1992)
20. Krieger, I.M.: Rheology of monodisperse lattices. Adv. Colloid Interf. Sci. **3**, 111–136 (1972)

17

Effect of Ni concentration on the structural and hydrogen storage characteristics of Zr–Mn based laves phase system

Vinod Kumar · D. Pukazhselvan · A. K. Tyagi · S. K. Singh

Abstract The present communication reports the hydrogen storage characteristics of $ZrMn_{2-x}Ni_x$ laves phase hydrides. It is observed that promising hydrogen storage materials can be tailored when "x" is in the range of $1.5 \leq x \geq 1$. The materials formed within this range reversibly store ~2 to ~3 H/F.U. under manageable operating conditions. The composition $ZrMn_{0.5}Ni_{1.5}$ is found to absorb and desorb hydrogen under room temperature. The other two compositions $ZrMn_{0.75}Ni_{1.25}$ and ZrMnNi absorb ~3.6 H/F.U. and release ~3 and ~3.2 H/F.U., respectively. In each case, almost half of the effective storage capacity of total can be reversibly stored within 5 s under room temperature. Therefore, the materials falling within the range of $1.5 \leq x \geq 1$ are promising candidates for stationary device applications. The correlation between the compositions, structural–microstructural features and thermodynamic data are presented.

Keywords Laves phase · Hydrogen storage · Crystal structure · Thermodynamics

V. Kumar · D. Pukazhselvan · S. K. Singh (✉)
Hydrogen Energy Lab, Department of Physics, Deenbandhu Chhotu Ram University of Science and Technology, Murthal (Sonipat), Haryana, India
e-mail: sksingh2k6@gmail.com

D. Pukazhselvan
Department of Physics and Astronomy, Padova University, Padua, Italy

A. K. Tyagi
Solid State Chemistry Section, Babha Atomic Research Centre, Trombay, Mumbai, India

Introduction

$ZrMn_2$ laves phase alloy attracts much attention due to its higher storage capacity (~1.7 wt%) with rapid H absorption kinetics [1]. However, hydrogen desorption occurs only at elevated temperatures which restricts its application potential for variety of devices such as heat pumps, Ni/MH batteries, stationary hydrogen storage cylinders, etc. Researchers have shown that the desorption characteristics of $ZrMn_2$ system can be improved by doping/alloying with suitable metals [2–5]. The work of Matsumura et al. [6] revealed that the reason for the higher stability of the $ZrMn_2$ hydride is due to the higher bond strength between the stored H with Mn atoms. They have also shown that the bond strengths between Zr–Mn, Mn–Mn and Mn–H all get significantly modified by doping by other suitable metals. Both the Zr and Mn sites can be substituted with hyper/hypo-stoichiometry of the laves phase which often makes the presence of substitute atoms even in excess of 10 per formula unit [7–13]. It may be pointed out that incorporation of several metal atoms results in unwanted side reaction [14]. As for example, dissolution of constituent metal atoms of MH electrode in the electrolyte of an electrochemical device costs its durability.

In an earliest study, Suzuki and Nishimiya [15] have investigated hydrogen storage characteristics of $ZrMn_{1.6}Ni_{0.4}$, ZrMnNi and $ZrMn_{0.4}Ni_{1.6}$ systems and shown that the dissociation pressure of the $ZrMn_2$ system gets significantly enhanced with the incorporation of Ni content. They have also concluded that ZrMnNi is the most promising hydride with the hydrogen filling of ~3.5 hydrogen atoms per formula unit under near ambient conditions. Kim et al. [16] have studied several Zr–Mn–Ni–V alloys and observed that the equilibrium plateau pressure gets significantly enhanced with the concentration of Ni content.

However, there was significant reduction in the hydrogen capacity and the rate of reaction. The authors attributed the effect to the amount of reactive surfaces, since the presence of more Ni limits the amount of reactive surfaces while offering excellent catalytic activity. We have studied a series of Zr based systems, namely $ZrMn_{2-x}Ni_x$, where $x = 0$ to 2. XRD analysis revealed that the Mn rich compositions possess C14 hcp structure, whereas the Ni rich compositions exhibit C15 cubic structure. Hence, at a particular concentration of Zr–Mn–Ni the C14 structure transforms to C15 structure. We observed that except $ZrNi_2$ all the other composition studied in the present investigation absorbs above 3 H/F.U. under room temperature. The dehydrogenation from $ZrMn_2$ is found to be possible only above 250 °C, but Mn-Ni coexisting system exhibits very interesting desorption characteristics. Particularly, we observed that $ZrMn_{0.5}Ni_{1.5}$, $ZrMn_{0.75}Ni_{1.25}$ and $ZrMn_{1.0}Ni_{1.0}$ are the interesting compositions which can reversibly store ~2.2 H/F.U., ~3 H/F.U. and ~3.2 H/F.U., respectively. It is observed that $x = 1.0$–1.5 is a narrow range within which room temperature (and near STP) high capacity hydrides of this class can be tailored. The materials exhibit excellent thermodynamics, rapid absorption/desorption kinetics and good activation features.

Experimental techniques

All the required chemicals such as Zr, Mn and Ni were procured from Alfa Aesar Chemicals Ltd. The compositions of Zr/Mn/Ni were calculated as required for the formula $ZrMn_{2-x}Ni_x$, (where $x = 0$–2). The powder mixer taken in right stoichiometry was palletized by 100 KN load (pellet thickness: 5 mm and diameter: 15 mm) and the pellets were melted under inert ambient using a high frequency induction furnace (20 kW, 12 kHz) supplied by Inductotherm India pvt ltd, Ahmedabad. For hydrogen storage measurements, 2 g of samples were loaded in a metal-hydrogen high pressure reaction chamber. PCT absorption/desorption measurements have been performed by a digitalized Sievert's type apparatus. Heat of the reaction (ΔH) was calculated from the van't Hoff plots and the activation energy measurements were carried out using the standard Arrhenius relation. DSC measurements have been performed by PerkinElmer instrument and XRD analysis of all the samples were performed by Regaku mini Flex11 X-ray diffractometer employing CuK_α radiation ($\lambda = 1.541$ Å). XRD measurements have been performed within the 2θ range of 30°–80° with a scanning speed of 2 s for the every 0.05° angles (step size). The lattice parameters were calculated using the unit cell refinement technique and the refined lattice constant values were compared with the standard JCPDF data source files. The crystallite size was calculated using Scherrer method and the strain available on the crystals was calculated using the Williamson-Hall (W-H) method. Further details regarding the detailed material synthesis and processing protocols can be seen in our earlier report [17].

Results and discussion

Structural studies

Figure 1 shows the XRDs of the $ZrMn_{2-x}Ni_x$ system, x correspond to = 0, 0.5, 1.0, 1.25, 1.5 and 2, respectively. The starting phase $ZrMn_2$ corresponds to C14 structure and the end phase $ZrNi_2$ corresponds to C15 structure. The C14 structure retains up to the Mn/Ni concentration of 1.50/0.50 and any further substitution of Mn by Ni leads to its transformation to the C15 phase. Particularly, the Mn/Ni concentration range of 1.00/1.00–0.75/1.25 appears to be quite sensitive for this structural transformation. Extensive discussions are being made by researchers regarding the correlation between the phase abundance of C15 and C14 and the corresponding hydrogen storage/electrochemical performances [18–22]. In the present case, when the C15 phase abundance is dominant, the precipitation of the non-laves phase Zr_7Ni_{10} alloy does also occur. Since the Mn/Ni concentration region of 0.75/1.25–0.50/1.50 brings out the non-laves phase alloy in the system (see Fig. 1), we tried to understand the effect of minor variations (of concentration within the Mn/Ni concentration region of 0.75/1.25–0.50/1.50) in its phase abundance and hydrogen storage performances and the results were communicated elsewhere.

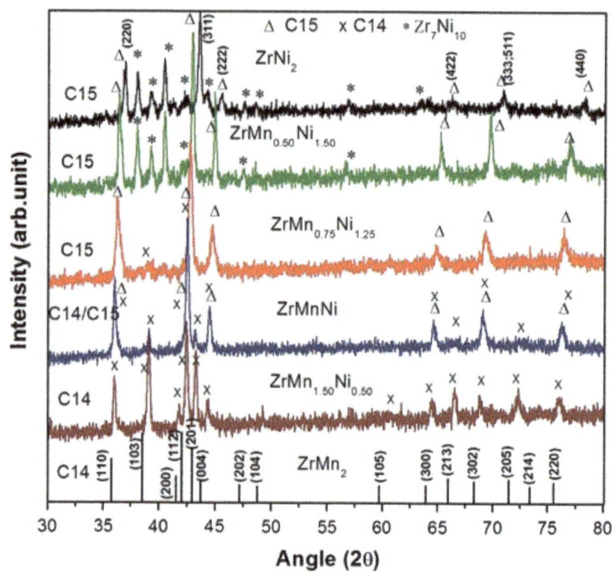

Fig. 1 XRDs of the $ZrMn_{2-x}Ni_x$ alloys ($x = 0$–2) prepared in the present study

As we decrease the amount of Mn in the composition, there is regular peak shifting toward the higher angle side. It shows that the lattice parameter of the system decreases with the increase of Ni content in the system. In a particular concentration (see Fig. 1), the contraction is followed by the precipitation of C15 phase also along with the C14 phase. Upon the further increase of Ni content the C14 phase gets eliminated and only C15 phase exists along with a non-laves phase of Zr_7Ni_{10}. The observed crystal structural details are summarized in the Table 1. The decrease in lattice parameters as we increase the Ni concentration can be attributed with the fact that both the covalent radius and atomic radius of Ni are smaller than that of Mn (Atomic radius of Mn and Ni are 1.79 and 1.62 Å, respectively; covalent radii of Mn and Ni are 1.17 and 1.15 Å, respectively).

The XRDs of hydrogenated ZrMnNi, $ZrMn_{0.75}Ni_{1.25}$ and $ZrMn_{0.5}Ni_{1.5}$ are shown in Fig. 2a, b and c. For effective comparison, the corresponding XRDs of the unhydrided samples are also given in the figure. As can be seen, the XRDs of hydrogenated samples of ZrMnNi and $ZrMn_{0.75}Ni_{1.25}$ exhibit a considerable shifting toward the lower angle side. The shifting arises due to the fact that the insertion of hydrogen in the interstitial sites does lead to expansion of lattice (the upper limit of expansion is known to be one-fourth of the actual lattice parameter) [23]. The peak shifting gives us a clue that hydrides of both the materials could be stable under room temperature. However, our subsequent hydrogen liberation experiments over a range of temperatures proved that hydrogen liberation with slow kinetics is still active at room temperature itself. Thus, the presence of a small peak at the 2θ value around 42.5 (highlighted by a rectangular box) can be attributed to the existence of some portion of the material in dehydrogenated form. The other two interesting information explored from this comparison is that (a) the phase abundance of C14 appears to increase during hydrogenation of both the ZrMnNi, $ZrMn_{0.75}Ni_{1.25}$ samples. It can be understood from the appearance of some dominant peaks of C14 after hydrogenation, (b) the non-laves phase Zr_7Ni_{10} which presents along with the $ZrMn_{0.75}Ni_{1.25}$ sample also gets hydrogenated and its usual orthorhombic phase itself undergoes phase transformation [24, 25]. Unlike the above two cases, in the case of $ZrMn_{0.5}Ni_{1.5}$ (Fig. 2c) the XRD of hydrogenated sample does not indicate any significant shifting from the usual XRD peak positions of its unhydrided sample. It clearly conveys us that there was no much hydrogen in the interstitial sites to cause any considerable lattice shifting. From this insight one may infer that the hydrogenated material is either not stable under room temperature or there is no hydrogenation at all. However, the subsequent experiments proved that in this case the former possibility exists (the low temperature experiments have proven that the equilibrium pressure of 1 atm occurs at temperatures lesser than room temperature).

One another noticeable behavior is that the crystallite size of these materials got reduced to a greater extent after its interaction with hydrogen (see Table 1). The reduction in the crystallite size in the case of $ZrMn_{0.5}Ni_{1.5}$ is even closer to 50 %. The crystallite size vs strain diagram for all the compositions of the present study is shown in Fig. 3. As seen in the figure, the system with higher crystallite size exhibits smaller strain and vice versa. It is also seen that the hydrided materials are incorporated with higher strain as compared to the corresponding intermetallic system.

Hydrogen storage characteristics

The hydrogen absorption kinetic curves of the as-prepared $ZrMn_{2-x}Ni_x$ alloys are shown in Fig. 4. Among all these

Table 1 The structural and microstructural details of the alloys prepared in the present study

Sr. no.	Composition	Lattice constant		Cell volume (Å)^3	Crystallite size (nm)
		a (Å)	c (Å)		
1	$ZrMn_2$ (Standard) (C14)	5.039	8.250	181.42	
2	$ZrMn_{1.5}Ni_{0.5}$ (C14)	5.001 ± 0.004	8.170 ± 0.012	176.960 ± 0.292	38.50
3	$ZrMn_{1.0}Ni_{1.0}$ (C14)	4.9921 ± 0.001	8.1458 ± 0.006	175.803 ± 0.152	41.69
	$ZrMn_{1.0}Ni_{1.0}$ (C15)	7.058 ± 0.002	7.058 ± 0.002	351.634 ± 0.225	
4	$ZrMn_{0.75}Ni_{1.25}$ (C15)	7.037 ± 0.005	7.037 ± 0.005	348.53 ± 0.711	27.82
5	$ZrMn_{0.5}Ni_{1.5}$ (C15)	7.021 ± 0.002	7.021 ± 0.002	345.92 ± 0.235	38.28
6	$ZrNi_2$ (C15)	6.902 ± 0.002	6.902 ± 0.002	328.75 ± 0.314	37.53
7	$ZrMn_{1.0}Ni_{1.0}$ (Hydride) (C15)	7.573 ± 0.004	7.573 ± 0.004	434.423 ± 0.717	28.88
	$ZrMn_{1.0}Ni_{1.0}$ (Hydride) (C14)	5.356 ± 0.003	8.728 ± 0.013	216.800 ± 0.381	
8	$ZrMn_{0.75}Ni_{1.25}$ (Hydride) (C15)	7.602 ± 0.002	7.602 ± 0.002	439.237 ± 0.315	21.73
	$ZrMn_{0.75}Ni_{1.25}$ (Hydride) (C14)	5.377 ± 0.002	8.795 ± 0.008	220.240 ± 0.238	
9	$ZrMn_{0.5}Ni_{1.5}$ (Hydride) (C15)	7.016 ± 0.002	7.016 ± 0.002	345.35 ± 0.248	30.70

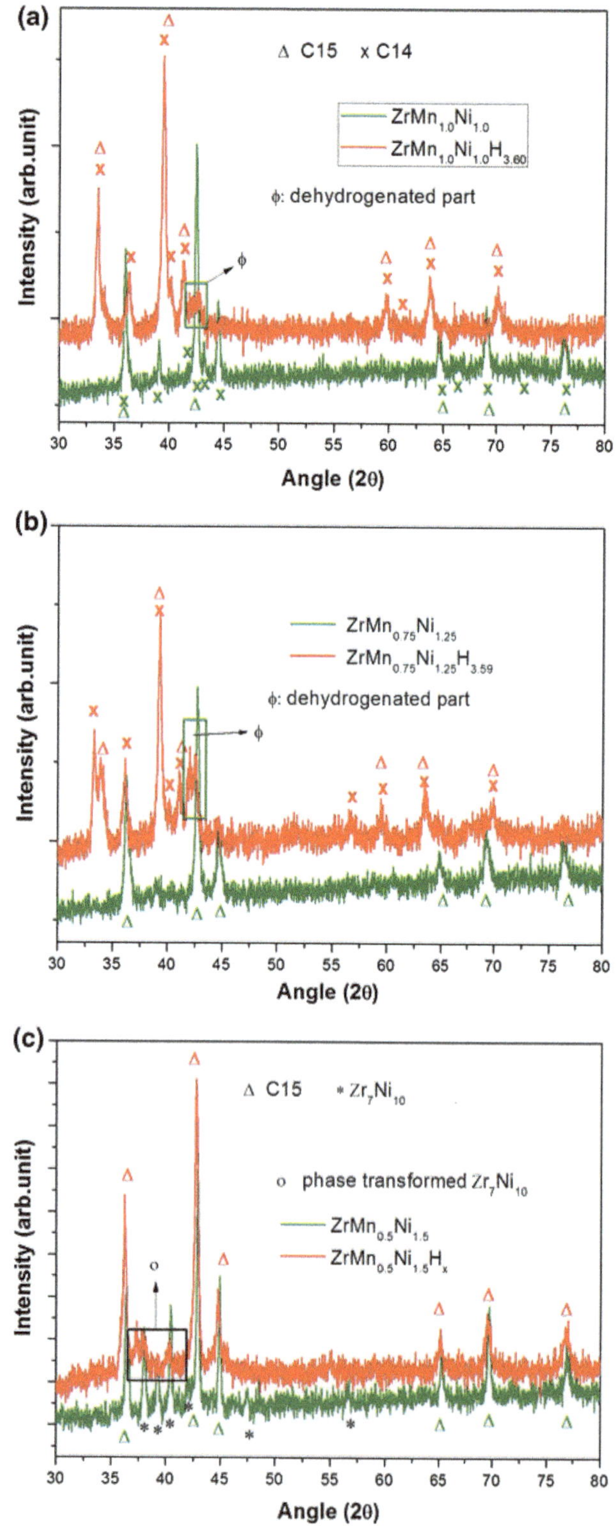

Fig. 2 Comparison of the XRDs of hydrogenated alloys, **a** ZrMnNi, **b** $ZrMn_{0.75}Ni_{1.25}$ and **c** $ZrMn_{0.5}Ni_{1.5}$. XRDs of the corresponding as-received materials are also given in the figure

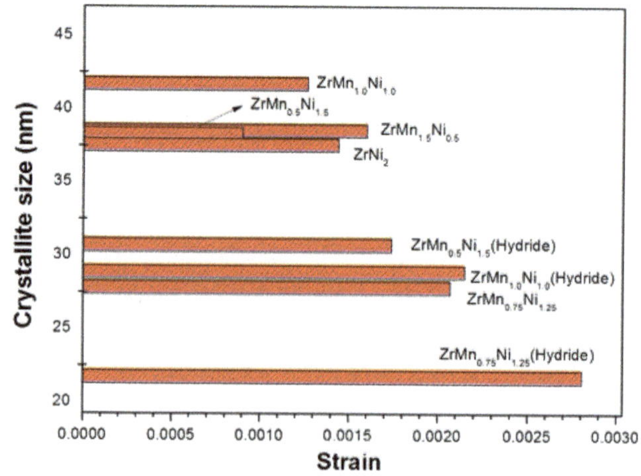

Fig. 3 The crystallite size vs strain analysis of all the alloys prepared in the present study. The hydrided materials are also given in the figure

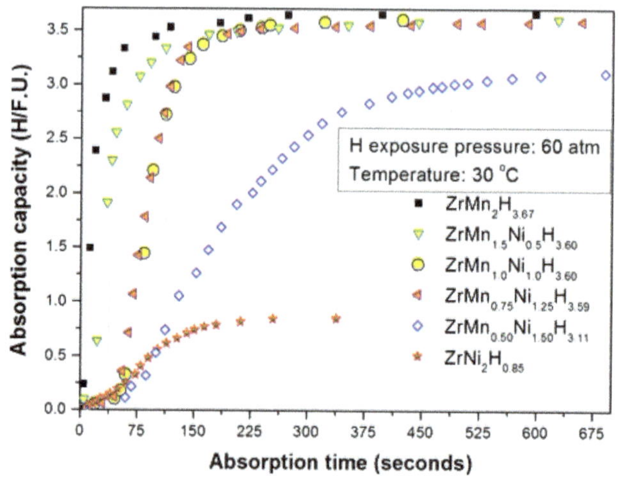

Fig. 4 The hydrogen absorption kinetics curves for all the materials prepared in the present study. The curves represent the kinetics obtained during the first time hydrogen exposure to the as-prepared samples. Hydrogenation temperature: 30 °C

alloys, the $ZrMn_2$ alloy absorbs hydrogen with rapid kinetics when exposed to hydrogen without any activation treatments. A capacity of ~3 H/F.U. could be obtained within few seconds and the maximum capacity reaches 3.67 H/F.U. As we incorporate Ni in place of Mn, the absorption kinetics decreases and only little decrease in storage capacity is witnessed up to the Mn/Ni concentration of 0.75/1.25. But, when we further reduce the Mn concentration, both the absorption kinetics and storage capacity get reduced and only 0.85 H/F.U. can be stored when all the Mn is replaced by Ni. However, during hydrogen desorption, the kinetics of the Ni rich compositions are better than that of Mn rich compositions. The end materials $ZrMn_2$, $ZrNi_2$ and the intermediate composition $ZrMn_{1.5}Ni_{0.5}$ liberate hydrogen only marginally with very poor desorption kinetics. In fact, just traceable amount of hydrogen can be observed under 100 °C and no significant improvement is observed even at ~200 °C (see Fig. 5). It

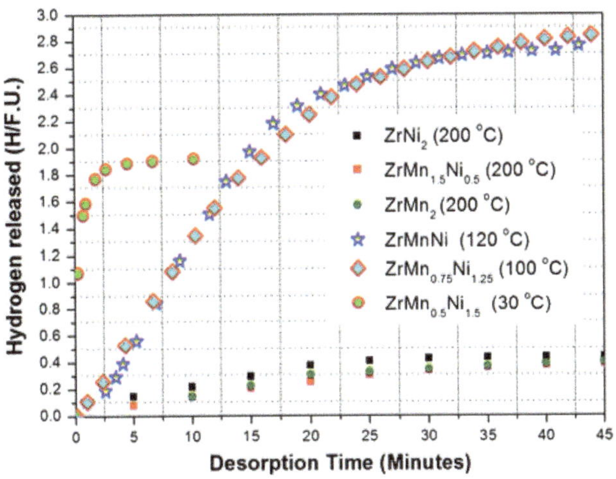

Fig. 5 The desorption kinetic curves of the hydrogenated samples. The calculated activation energy values are given in Table 2

may be due to the fact that the Mn rich compositions quickly make very stable hydrides and Ni rich composition makes only less stable or unstable hydrides. Further experiments suggested that temperature in excess of 200 °C is required for hydrogen liberation from the Mn rich phases. However, better hydrogen storage characteristics could be witnessed from some intermediate compositions such as ZrMnNi, $ZrMn_{0.75}Ni_{1.25}$ and $ZrMn_{0.5}Ni_{1.5}$. These hydrides absorb and liberate hydrogen under manageable temperatures. Particularly the alloy $ZrMn_{0.5}Ni_{1.5}$ liberates ~1.9 H/F.U. hydrogen at room temperature. The desorption activation energy values calculated for ZrMnNi, $ZrMn_{0.75}Ni_{1.25}$ and $ZrMn_{0.5}Ni_{1.5}$ hydrides are given in Table 2.

In the case of $ZrMn_{2-x}Ni_x$ we noticed that the x values in the range of 1–1.5 lead to promising Zr-based AB_2 hydrides suitable for commercial applications. Our thorough investigation revealed that x in the range of 0 to <1 and 1.5 to 2 does not lead to any interesting Zr–Mn–Ni system for reversible hydrogen storage applications. The former case materials interact too strongly with hydrogen and the latter case interact poorly with hydrogen. The PCI (desorption isotherm) curves of three interesting systems within the $x = 1–1.5$ range, such as ZrMnNi, $ZrMn_{0.75}Ni_{1.25}$ and $ZrMn_{0.5}Ni_{1.5}$ are shown in Fig. 6a, b and c, respectively. From the nature of the PCT desorption curves obtained in these compositions it is clearly understood that any little changes made in the Mn concentration within this small x value range lead to big changes in the thermodynamics of the system. Our estimation reveals that ZrMnNi restores ~3.2 H/F.U. and works within the temperature range of 120–180 °C, $ZrMn_{0.75}Ni_{1.25}$ restores ~3 H/F.U. and works well within the temperature range of 60–140 °C and $ZrMn_{0.5}Ni_{1.5}$ restores ~2.1 H/F.U. and works well within the temperature range of 10–60 °C. One can understand from the PCT curves that, when we decrease the concentration of Mn = 1 to Mn = 0.5, the stability of the hydride phase gets dramatically reduced and the plateau pressure varies significantly with respect to temperature. The van't Hoff plots for the said three compositions such as ZrMnNi, $ZrMn_{0.75}Ni_{1.25}$ and $ZrMn_{0.5}Ni_{1.5}$ are shown in Fig. 7 and the calculated thermodynamic data are given in Table 2. The equilibrium pressure of 1 atm in these three systems can be obtained at the temperatures of 120, 100 and −20 °C, respectively. The van't Hoff plots also prove that thermodynamics of the system varies considerably within a narrow range of Mn concentration, i.e. 0.75–1.5. Such behavior can strike considerable applications for heat storage and electrochemical devices. Further, the van't Hoff diagram clearly indicates that all these three system covers the working range of the commercial hydrogen fuel cells.

The PCT (re)absorption isotherm of the composition ZrMnNi, $ZrMn_{0.75}Ni_{1.25}$ and $ZrMn_{0.5}Ni_{1.5}$ are given in Fig. 8a. The composition $ZrMn_{0.5}Ni_{1.5}$ exhibits much higher plateau pressure as compared to $ZrMn_{0.75}Ni_{1.25}$ and $ZrMn_{0.5}Ni_{1.5}$. It clearly shows the large differences in their thermodynamic features between the x values of 1.25–1.5 in the $ZrMn_{2-x}Ni_x$ system. The reversible hydrogen absorption kinetic plots for the systems such as ZrMnNi, $ZrMn_{0.75}Ni_{1.25}$ and $ZrMn_{0.5}Ni_{1.5}$ are shown in Fig. 8b. As seen, ZrMnNi system restores hydrogen in excess of 3 H/F.U. and nearly half of this capacity is absorbed within 5 s. $ZrMn_{0.75}Ni_{1.25}$ also stores almost the similar capacity but $ZrMn_{0.5}Ni_{1.5}$ stores only 2.2 H/F.U. However, it should also be noted that unlike ZrMnNi and $ZrMn_{0.75}Ni_{1.25}$, $ZrMn_{0.5}Ni_{1.5}$ can absorb and release this capacity under room temperature. The DSC curves of all the $ZrMn_{2-x}Ni_x$ system are shown in Fig. 9. The DSC curves also confirm that the $ZrMn_{0.5}Ni_{1.5}$ sample behaves much differently as

Table 2 Summary of the observed heat of desorption and activation energy values

Sr. no.	Composition	$-\Delta H$ (kJ/mol H)	ΔS (J/(K mol H))	Desorption temp. at equilibrium pressure of 0.1 MPa	Activation energy (kJ/mol)	Effective H capacity (H/F.U.)
1	$ZrMn_{1.0}Ni_{1.0}$	32.14 ± 1.52	81.43 ± 3.45	119.10	21.03 ± 0.22	3.3
2	$ZrMn_{0.75}Ni_{1.25}$	27.97 ± 1.82	74.87 ± 4.32	100.01	17.91 ± 0.67	3.0
3	$ZrMn_{0.5}Ni_{1.5}$	18.61 ± 1.14	73.74 ± 3.60	−20.30	13.21 ± 0.16	2.2

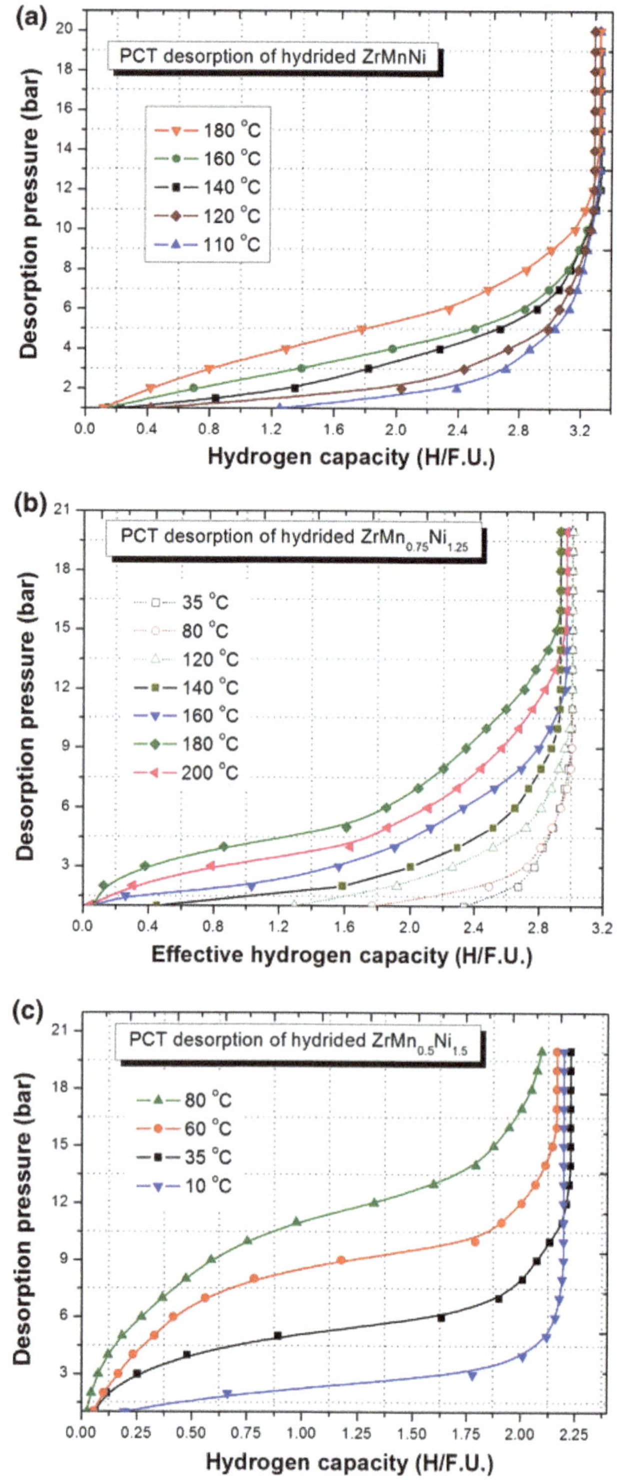

Fig. 6 **a**, **b**, **c** The PCT desorption isotherms of the alloys. **a** ZrMnNi, **b** $ZrMn_{0.75}Ni_{1.25}$ and **c** $ZrMn_{0.5}Ni_{1.5}$

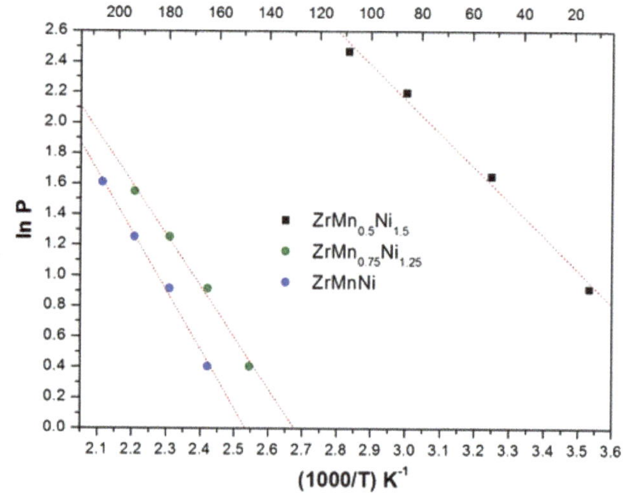

Fig. 7 The van't Hoff plots (dehydrogenation) for the compositions such as ZrMnNi, $ZrMn_{0.75}Ni_{1.25}$ and $ZrMn_{0.5}Ni_{1.5}$

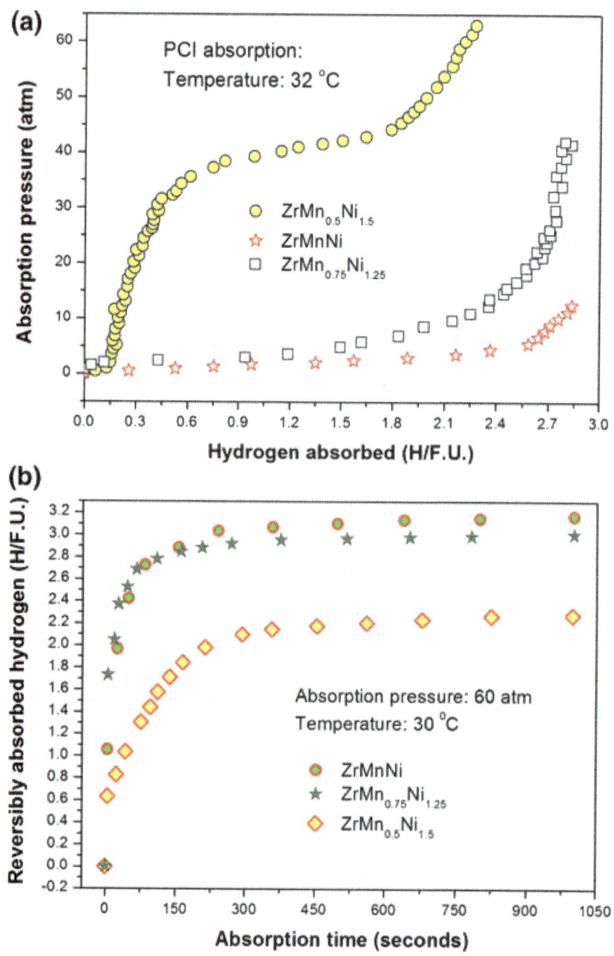

Fig. 8 **a** The PCT isotherms for reversible hydrogen absorption for ZrMnNi, $ZrMn_{0.75}Ni_{1.25}$ and $ZrMn_{0.5}Ni_{1.5}$ and **b** the reversible hydrogen absorption kinetics of ZrMnNi, $ZrMn_{0.75}Ni_{1.25}$ and $ZrMn_{0.5}Ni_{1.5}$ alloys

compared to all other compositions. This curious observation in the case of $ZrMn_{0.5}Ni_{1.5}$ may be related with the heat transfer associated with the evolution of residual hydrogen so as to transform as hydrogen free metal phase. Because, only in the case of $ZrMn_{0.5}Ni_{1.5}$ does the

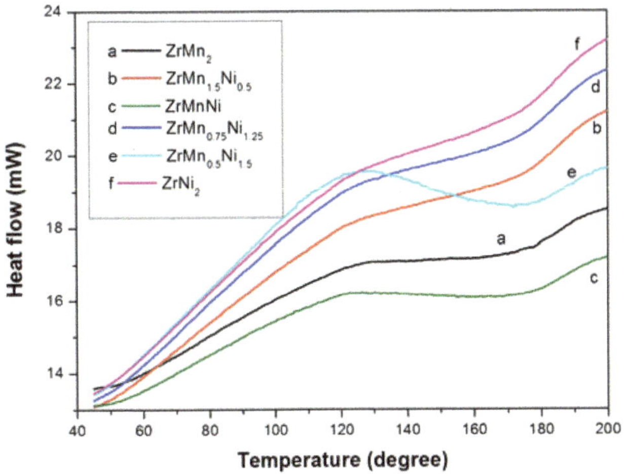

Fig. 9 DSC curves of hydrided $ZrMn_{2-x}Ni_x$ alloys ($x = 0$–2)

hydrogen liberation become effective at lower temperatures. Note that $ZrNi_2$ do not possess considerable hydrogen above room temperature and any composition above Ni/Mn concentration of 0.5/1.5 release only very small amount of hydrogen at the temperatures less than 200 °C (also refer desorption PCTs). Thus, the DSC profile of all other samples except $ZrMn_{0.5}Ni_{1.5}$ looks as similar as each other. The hydrogen absorption/desorption behavior of $ZrMn_{2-x}Ni_x$ for $x = 1$–1.5 makes it clear that these materials are suitable for varieity of stationary applications (as for example, large stationary hydrogen storage cylinders in hydrogen production/distribution spots, H isotope stores etc.) and even mobile applications. Particularly, in the case of $ZrMn_{0.75}Ni_{1.25}$, the reversible hydrogen storage of ~2 H/F.U. within few seconds shows that these are candidates par excellence for device applications.

In the present Zr–Mn–Ni alloy system, the hydrogen capacity of C14 phase is, in general, higher than the capacity of C15 phase, which is contrary to the observation of some earlier reports [21, 26, 27]. However, from performance point of view, particularly for low temperature desorption, C15 phase co-existing with the non-laves phase of Zr_7Ni_{10} is found to be better. Song et al. [28, 29] have earlier studied the $ZrMn_{0.9-x}V_xNi_{1.1}$ (x from 0.1 to 0.8) system and concluded that C15 phase with higher stacking faults performs better for electrochemical applications. In the present case, since Zr_7Ni_{10} concurrently presents with the C15 phase, the higher sorption performances cannot be solely related to the stacking faults itself. Moreover, there are reports suggesting that the Zr_7Ni_{10} offers catalytic activity for surface reactions and also acts as a medium for charge diffusion [30–33]. However, in spite of the presence of Zr_7Ni_{10} phase since few C15 compositions did not store hydrogen appreciably, it appears to be the distinct electronic structure of some composition of Zr–Mn–Ni makes the difference. Particularly, any compositions within the $ZrMn_{0.75}Ni_{1.25}$ and $ZrMn_{0.50}Ni_{1.50}$ could be promising.

Conclusions

In the case of $ZrMn_{2-x}Ni_x$ laves phase system, thermodynamically promising hydrogen storage alloys can be made within the range of $1.5 \leq x \geq 1$. $ZrMn_2$ and $ZrMn_{1.5}Ni_{0.5}$ absorb ~3.6 H/F.U., but the absorbed hydrogen cannot be liberated under manageable conditions. ZrMnNi, $ZrMn_{0.75}Ni_{1.25}$ and $ZrMn_{0.5}Ni_{1.5}$ hydrides reversibly store hydrogen capacity such as 3.2 H/F.U., 3 H/F.U. and 2.2 H/F.U., respectively. Each of these hydrides stores at least half of its effective capacity within 1 min of charging time. The working range of ZrMnNi, $ZrMn_{0.75}Ni_{1.25}$ and $ZrMn_{0.5}Ni_{1.5}$ are 120–180 °C, 60–120 °C and 10–60 °C, respectively. Interesting thermodynamic behavior could be witnessed when we proceed from $ZrMn_{0.75}Ni_{1.25}$ to $ZrMn_{0.5}Ni_{1.5}$ system which is also confirmed by DSC analysis.

Acknowledgments The authors thank Dr.V.C. Shahni of BARC for discussions. Dr. H.K. Singh and Dr. V.P.S. Awana of NPL are thanked for providing XRD facility. D.P.S thanks UGC, India for D.S. Kothari post doctoral fellowship and V.K thanks Dept. of Atomic Energy, India for Senior Research Fellowship. This work has been carried out under the financial assistance provided by BRNS, Department of Atomic Energy, India.

References

1. Fujii, H., Saga, M., Okamoto, T.: Magnetic, crystallographic and hydrogen absorption properties of YMn_2 and $ZrMn_2$ hydrides. J. Less Common Met. **130**, 25–31 (1987)
2. Fujitani, S., Yonezu, I., Saito, T., Furukawa, N., Akiba, E., Hayakawa, H., Ono, S.: Relation between equilibrium hydrogen pressure and lattice parameters in pseudobinary Zr–Mn alloy systems. J. Less Common Met. **172–174**, 220–230 (1991)
3. Young, K., Fetcenko, M.A., Koch, J., Morii, K., Shimizu, T.: Studies of Sn, Co., Al, and Fe additives in C14/C15 Laves alloys for NiMH battery application by orthogonal arrays. J. Alloys Compd. **486**, 559–569 (2009)
4. Matsumura, T., Yukawa, H., Morinaga, M.: Alloying effects on the electronic structure of $ZrMn_2$ intermetallic hydride. J. Alloys Compd. **279**, 192–200 (1998)
5. Gesari, S.B., Pronsato, M.E., Visintin, A., Juan, A.: Hydrogen Storage in AB2 Laves Phase (A = Zr, Ti; B = Ni, Mn, Cr, V): binding energy and electronic structure. J. Phys. Chem. C. **114**, 16832–16836 (2010)
6. Matsumura, T., Yukawa, H., Morinaga, M.: Alloying effects on the electronic structure of $ZrMn_2$ intermetallic hydride. J. Alloys Compd. **279**(2), 192–200 (1998)
7. Du, Y.L., Yang, X.G., Lei, Y.Q., Zhang, M.S.: Hydrogen storage properties of $Zr_{0.8}Ti_{0.2}$ $(Ni_{0.6}Mn_{0.3-x}V_{0.1+x}Cr_{0.05})_2$ (x = 0.0,

0.05, 0.15, 0.2) alloys. Int. J. Hydrogen Energy **27**, 695–697 (2002)

8. Pontonnier, L., Miraglia, S., Fruchart, D., Soubeyroux, J.L., Baudry, A., Boyer, P.: Structural study of hyperstoichiometric alloys $ZrMn^{2+x}$ and their hydrides. J. Alloys Compd. **186**, 241–248 (1992)
9. Lee, S.M., Lee, H., Kim, J.H., Lee, P.S., Lee, J.Y.: A study on the development of hypo-stoichiometric Zr-based hydrogen storage alloys with ultra-high capacity for anode material of Ni/MH secondary battery. J. Alloys Compd. **308**, 259–268 (2000)
10. Sun, J.C., Li, S., Ji, S.J.: The effects of the substitution of Ti and La for Zr in $ZrMn_{0.7}V_{0.2}Co_{0.1}Ni_{1.2}$ hydrogen storage alloys on the phase structure and electrochemical properties. J. Alloys Compd. **446–447**, 630–634 (2007)
11. Bendersky, L.A., Wang, K., Levin, I., Newbury, D., Young, K., Chao, B., Creuziger, A.: $Ti_{12.5}Zr_{21}V_{10}Cr_{8.5}Mn_xCo_{1.5}Ni_{46.5-x}$ AB_2-type metal hydride alloys for electrochemical storage application. J. Power Sources **218**, 474–486 (2012)
12. Jong-Man, P., Jai-Young, L.: Hydrogenation characteristics of the $Zr_{1-x}Ti_xCr_{1-y}Fe_{1+y}$ laves phase alloys. J. Less Common Metals. **160**(2), 259–271 (1990)
13. Yasuoka, S., Magari, Y., Murata, T., Tanaka, T., Ishida, J., Nakamura, H., Nohma, T., Kahara, M., Baba, Y., Teraoka, H.: Development of high-capacity nickel-metal hydride batteries using superlattice hydrogen-absorbing alloys. J. Power Sources **156**, 662–666 (2006)
14. Xiaoguang, Y., Yonquan, L., Jing, W., Qidong, W.: The electrochemical performances of Zr based hydrogen storage alloys. Trans. NFsoc. **5**, 61–65 (1995)
15. Suzuki, A., Nishimiya, N.: Thermodynamic properties of $Zr(Ni_xMn_{1-x})_2$-H_2 systems. Mat. Res. Bull. **9**, 1559–1571 (1984)
16. Kim, D.M., Lee, S.M., Jang, K.J., Lee, J.Y.: The electrode characteristics of over-stoichiometric $ZrMn_{0.5}V_{0.5}Ni_{1.4+y}$ ($y = 0.0$, 0.2, 0.4 and 0.6) alloys with C15 Laves phase structure. J. Alloys Compd. **268**(1–2), 241–247 (1998)
17. Kumar, V., Pukazhselvan, D., Singh, S.K.: Hydrogen Storage Characteristics of $MmNi_{5-x}M_x$ (M = Cu, Mn and Al). Adv. Sci. Lett. **18**, 164–169 (2012)
18. Du, Y.L., Yang, X.G., Zhang, Q.A., Lei, Y.Q., Zhang, M.S.: Phase structures and electrochemical properties of the Laves phase hydrogen storage alloys $Zr_{1-x}Ti_x(Ni_{0.6}Mn_{0.3}V_{0.1}Cr_{0.05})_2$. Int. J. Hydrogen Energy. **26**(4), 333–337 (2001)
19. Young, K., Ouchi, T., Huang, B., Chao, B., Fetcenko, M.A., Bendersky, L.A., Wang, K., Chiu, C.: The correlation of C14/C15 phase abundance and electrochemical properties in the AB_2 alloys. J. Alloys Compd. **506**(2), 841–848 (2010)
20. Young, K., Nei, J., Ouchi, T., Fetcenko, M.A.: Phase abundances in AB_2 metal hydride alloys and their correlations to various properties. J. Alloys Compd. **509**(5), 2277–2284 (2011)
21. Huot, J., Akiba, E., Ogura, T., Ishido, Y.: Crystal structure, phase abundance and electrode performance of Laves phase compounds (Zr, A)$V_{0.5}Ni_{1.1}Mn_{0.2}Fe_{0.2}$ (A = Ti, Nb or Hf). J. Alloys Compd. **218**(1), 101–109 (1995)
22. Song, X.Y., Zhang, X.B., Lei, Y.Q., Zhang, Z., Wang, Q.D.: Effect of microstructure on the properties of Zr–Mn–V–Ni AB2 type hydride electrode alloys. Int. J. Hydrogen Energy **24**(5), 455–459 (1999)
23. Figiel, H., Osuchowski, S., Paja, A.: Unit cell expansion in Laves phase-type compounds caused by interstitial hydrogen. Phys. Stat. Sol. (a) **204**, 3286–3290 (2007)
24. Takeshita, H.T., Fujiwara, N., Oishi, T., Noréus, D., Takeichi, N., Kuriyama, N.: Another unusual phenomenon for Zr_7Ni_{10}: structural change in hydrogen solid solution and its conditions. J. Alloys Compd. **360**, 250–255 (2003)
25. Takeshita, H.T., Kondo, S., Miyamura, H., Takeichi, N., Kuriyama, N., Oishi, T.: Re-examination of Zr_7Ni_{10} single-phase region. J. Alloys Compd. **376**, 268–274 (2004)
26. Zhang, Q.A., Lei, Y.Q., Yang, X.G., Ren, K., Wang, Q.D.: Annealing treatment of AB_2-type hydrogen storage alloys: II Electrochemical properties. J. Alloys Compd. **292**, 241–246 (1999)
27. Du, Y.L., Yang, X.G., Zhang, Q.A., Leo, Y.Q., Zhang, M.S.: Phase structures and electrochemical properties of the Laves phase hydrogen storage alloys $Zr_{1-x}Ti_x(Ni_{0.6}Mn_{0.3}V_{0.1}Cr_{0.05})_2$. Int. J. Hydrogen Energy **26**, 333–337 (2001)
28. Song, X., Zhang, X., Lei, Y., Zhang, Z., Wang, Q.: Effect of microstructure on the properties of Zr–Mn–V–Ni AB_2 type hydride electrode alloys. Int. J. Hydrogen Energy **24**, 455–459 (1999)
29. Yang, X.G., Zhang, W.K., Lei, Y.Q., Wang, Q.D.: Electrochemical properties of Zr–V–Ni system hydrogen storage alloys. J. Electrochem. Soc. **146**, 1245–1250 (1999)
30. Ruiz, F.C., Castro, E.B., Real, S.G., Peretti, H.A., Visintin, A., Triaca, W.E.: Electrochemical characterization of AB_2 alloys used for negative electrodes in Ni/MH batteries. Int. J. Hydrogen Energy **33**, 3576–3580 (2008)
31. Visintin, A., Peretti, H.A., Ruiz, F., Corso, H.L., Triaca, W.E.: Effect of additional catalytic phases imposed by sintering on the hydrogen absorption behavior of AB_2 type Zr-based alloys. J. Alloys Compd. **428**, 244–251 (2007)
32. Ruiz, F.C., Castro, E.B., Peretti, H.A., Visintin, A.: Study of the different Zr_xNi_y phases of Zr-based AB_2 materials. Int. J. Hydrogen Energy **35**, 9879–9887 (2010)
33. Bendersky, L.A., Wang, K., Boettinger, W.J., Newbury, D.E., Young, K., Chao, B.: Examination of multiphase (Zr, Ti)(V, Cr, Mn, $Ni)_2$ Ni-MH electrode alloys: part II solid-state transformation of the interdendritic B_2 phase. Metall. Mater. Trans. A **41**, 1891–1906 (2010)

18

Possibilities to decrease the absorber thickness reducing optical and recombination losses in CdS/CdTe solar cells

L. A. Kosyachenko

Abstract Based on the optical constants, calculations of optical losses in CdS/CdTe solar cells have been carried out taking into account reflections at the interfaces and absorption in the TCO and CdS layers. It has been shown that the losses caused by reflections at the interfaces result in lowering the short-circuit current by ~9 %, whereas absorption in the TCO and CdS layers with the typical thicknesses lead to losses of 15–24 %. Calculations of the *integrated* absorptive capacity of CdTe layer taking into account the spectral distributions of the standard AM1.5 solar radiation and the absorption coefficient of CdTe have been carried out. It is shown that in CdTe, the almost complete absorption of photons (≥99.9 %) in the $h\nu > E_g$ range is observed at a layer thickness of more than 20–30 μm, and the absorptive capacity of photons in a CdTe layer of thickness 1 μm is about 93 %. Based on the continuity equation and taking into account the recombination at the front and rear surfaces of the CdTe layer as well as in the space charge region, the restrictions imposed on the thickness of the absorber layer in CdS/CdTe heterojunction are studied. In all cases, along with the mobilities and lifetimes of charge carriers, the concentration of uncompensated impurities in CdTe plays a key role in the generation of photocurrent. When the CdTe absorber layer is thin (<1 μm), it is impossible to avoid a noticeable decrease of the short-circuit current density. The losses of the short-circuit current are equal to 19–20 % when the thickness of the CdTe layer is 0.5 μm whereas only 5 % for a typical thickness of 2–3 μm.

L. A. Kosyachenko (✉)
Chernivtsi National University, Chernivtsi 58012, Ukraine
e-mail: lakos@chv.ukrpack.net

Keywords CdTe solar cells · Optical losses · Recombination losses · Extra-thin absorber layer

Introduction

In recent years, photovoltaics (PV) demonstrates the highest growth rate in the entire energy sector despite the global financial crisis. For the next years the growth rate is expected to be also quite high. However, the PV contribution based mostly on silicon wafers by more than 90 % is less than 0.5 % in the global energy system. The contribution of PV solar power plants in generation capacity in the world will reach ~1 % only in the years 2018–2020 and may exceed 10–15 % in the years 2045–2050 [1–3].[1] The prospects for so desired rapid development of PV are not reassuring. Thus, solving the energy problems by developing Si PV seems too lengthy under the conditions of the approaching global warming.

The reason for the slow power growth of traditional Si solar modules lies in large consumption of materials and energy, high labor intensiveness and, as a consequence, in low productivity and high cost of modules with acceptable PV conversion efficiency for mass production [4, 5]. The problem is fundamental and lies in the fact that silicon is an *indirect* semiconductor and, therefore, the *total* absorption needs its significant thickness (up to 0.5 mm or more). As a result, to collect the charge photogenerated in a thick absorbing layer, considerable diffusion length of minority

[1] These rates are much higher in the European Union. Now the cumulative power of solar modules exceeds 1 % and will rise to 4–5 and 6–12 % of the EU's electricity demand by 2015 and 2020, respectively.

carriers and, therefore, *high quality material* is required for the solar module production.

A radical way to reduce the cost of solar modules and to increase drastically the volume of their production is the transition to thin-film technology based on direct-gap semiconductors deposited on a cheap large-area substrate (glass, metal foil, plastic). The direct-gap semiconductor can absorb solar radiation with a thickness, which is about 10^2 times less than the thickness of the mono- or polysilicon wafer. Thus, the transition from crystalline silicon to direct-gap semiconductors leads first of all to noncomparable less consumption of photoelectrically active material used in the solar cell.

High absorptivity of a direct-gap semiconductor has also an important consequence with respect to the quality of material used in solar cells. Since the direct-gap semiconductor can absorb solar radiation at its thickness much smaller than that of the Si wafer (or ribbon), the requirements for chemical purity and crystalline perfection of the absorber layer become much weaker. Indeed, to collect photogenerated charge carriers, it is necessary to have a diffusion length of minority carriers in excess of the thickness of the absorber layer. It follows from this that in the solar cell based on direct-gap semiconductor, the diffusion length L may be about 2 orders smaller, i.e. the carrier lifetime τ can be shorter by 4 (!) orders of magnitude ($L \sim \tau^{1/2}$) compared with silicon. Thus, the manufacture of thin-film solar modules based on the direct-gap semiconductors does not require high purification and crystallinity of the material as it is needed in the production of modules based on crystalline, polycrystalline, or ribbon silicon.

Thin-film technology has a number of other significant advantages. While Si devices are manufactured from wafers or ribbons and then processed and assembled to form modules, in thin-film technology many cells are simultaneously made and formed as a module. The layers of solar cells are deposited sequentially on moving substrates in a continuous highly automated production line (*conveyor system*) and, importantly, at temperatures not exceeding 200–650 °C compared with 800–1,450 °C for the main processes of c-Si or m-Si. This minimizes handling and facilitates automation leading to the so-called *monolithic integration*.

Thus, thin-film solar modules offer the lowest manufacturing costs and are becoming more prevalent in the industry because they allow improving manufacturability of the production at significantly larger scales than for wafer or ribbon Si modules. Therefore, it is generally recognised that the contribution of thin-film technology in solar energy will be to grow from year to year faster. Many analysts believe that it is only a matter of time before thin films would replace Si wafer-based solar cells as the dominant photovoltaic technology. It should be noted that experts and managers of some silicon PV companies long ago have come to conclusion that there will be serious limitations for growing their wafer silicon business by simply expanding in the future [6]. All of these companies are researching wafer-Si alternatives including the thin-film technologies and are already offering such commercial thin-film modules.

It is, however, necessary to agree that solar cells of all types on Si wafers (representatives of the so-called *first generation photovoltaics*) will maintain their market position in the future. In hundreds of companies around the world, one can always invest (with minimal risk) and implement the Si technology developed for microelectronics with some modifications. Mono- and polycrystalline Si wafers, which are used in the semiconductor industry, can be made into efficient solar cells with full confidence. It is also important that Si is very abundant, clean, nontoxic, and very stable. However, due to limitations in production in large volumes of Si for solar modules often-expressed projections for *significant* increase in their contribution to the world energy system in the coming years are highly questionable.

Unquestionable leaders in thin-film technologies are solar cells on amorphous silicon (a-Si), copper-indium-gallium diselenide ($CuIn_xGa_{1-x}Se_2$), and cadmium telluride (CdTe), whose market share is expanding every year [7]. It should be noted that compared with other thin-film materials, technology of CdTe solar modules is simpler and more suitable for large-scale production. Since the past decade, polycrystalline thin-film CdS/CdTe solar cells fabricated on glass substrates have been considered as one of the most promising candidates for large-scale applications in the field of PV energy conversion [8, 9]. The growth rate of solar PV industry based on thin-film CdTe solar cells is higher than any other PV technologies. Solar power plants based on CdTe with capacities of several tens of megawatts have already been built in Germany, Spain, Canada, the United States, and other countries. Agreements have been concluded on the construction of CdTe power plants with a capacity of an order of magnitude higher in the U.S. and even two orders of magnitude greater in China.

PV conversion efficiency of *small area* laboratory samples of CdS/CdTe solar cells exceeds 17 % [10]. The efficiency of *large area* CdTe solar modules is markedly inferior to that of silicon analogues until recently, but every year, this key indicator is continuously increasing and had already reached 14.4 % for some samples of modules in early 2012 [11]. Today the challenge facing the researchers and technologists is how to increase the efficiency of current large area CdS/CdTe modules decreasing the gap between the current efficiency and the theoretical limit of 28–30 % [12–14].

The present paper reports the theoretical study of optical and recombination losses, which are important causes of low efficiency of CdS/CdTe solar cells. The first part of the paper focuses on optical losses due to absorption and reflection at the interfaces in the device. It seems that the results of these calculations are interesting from a scientific and practical point of view, since they suggest possible ways to increase the efficiency of CdS/CdTe solar cells by reducing the optical losses, and vice versa they show in which cases a decrease in the optical losses gives only a small gain. The second part of the paper is devoted to a detailed study of recombination of charge carriers at the front and rear surfaces of the CdTe absorber layer, as well as in the space charge region of CdS/CdTe heterostructure. It is shown that the concentration of uncompensated impurities (defects), which determines the width of the space charge region is one of the key parameters of the material used as an absorber in the solar cell (of course, in addition to the lifetimes and mobilities of charge carriers). The most current practical issues with respect to minimizing the thickness of the absorber CdTe layer are also considered taking into account the influence of the above-mentioned factors. The importance of this aspect arises from the fact that one limiting factor for the terawatt production of CdTe modules in the future may be the scarcity of tellurium.

It should be noted that this study is a generalization of the results presented in some our previous publications [15–18]. The *quantitative* characteristics of optical and recombination losses are presented for the first time.

Optical losses in thin CdS/CdTe solar cells

In the literature there are a good amount of experimental data on the optical transmission of the glass/TCO(ITO or SnO_2:F) and glass/TCO/CdS structures that provide information about the losses caused by reflection from the interfaces and absorption in the glass, ITO, SnO_2:F, and CdS layers [19–21].[2] However, in the case of glass/TCO structure, the surface of TCO layer is in optical contact with air, rather than with CdS, as in the real case. In the case of glass/TCO/CdS structure, the surface of CdS is in contact with air rather than with the CdTe layer. As shown below, measuring the transmission of the glass/TCO and glass/TCO/CdS structures leads to the results, which differ significantly from those measured on the real structure of the solar cell. Further, in the literature, pronounced interference oscillations in the transmission curves are always observed, which are also relatively stronger than in the real device structure.

"Optical losses in thin CdS/CdTe solar cells" focuses on optical losses due to reflection at the interfaces and absorption in ITO, SnO_2:F, and CdS layers. Calculations have been carried out based on the optical constants of materials used (the refractive index and extinction coefficient).

Reflection losses

The sketch of a typical glass/TCO/CdS/CdTe solar cell is shown in Fig. 1. Before reaching the photoelectrically active CdTe absorber layer, solar radiation penetrates the glass plate, the TCO layer and the CdS window layer. Obviously, this is accompanied by optical losses upon reflection from the interfaces air–glass, glass–TCO, TCO–CdS, CdS–CdTe as well as absorption in glass plate, TCO and CdS. According to the Fresnel equations, when the light is at near-normal incidence, the reflection coefficient (reflectivity) from the interface between two contacting transparent materials is determined by their refractive indices n_1 and n_2:

$$R = \left(\frac{n_1 - n_2}{n_1 + n_2}\right)^2. \tag{1}$$

In the case of electrically conductive materials, the refractive index contains an imaginary part and can be written as $n^* = n - i\kappa$, where n is the refractive index, and κ is the extinction coefficient ($i = \sqrt{-1}$). The reflection coefficient from the interface is defined as the square of the modulus of $(n_1^* - n_2^*)/(n_1^* + n_2^*)$ and has the form [22]:

$$R = \frac{|n_1^* - n_2^*|^2}{|n_1^* + n_2^*|^2} = \frac{(n_1 - n_2)^2 + (\kappa_1 - \kappa_2)^2}{(n_1 + n_2)^2 + (\kappa_1 + \kappa_2)^2}. \tag{2}$$

At the air–glass interface we will find the reflection coefficient R_{12} (see Fig. 1) taking $n_1 = 1$, and $\kappa_1 = 0$ for

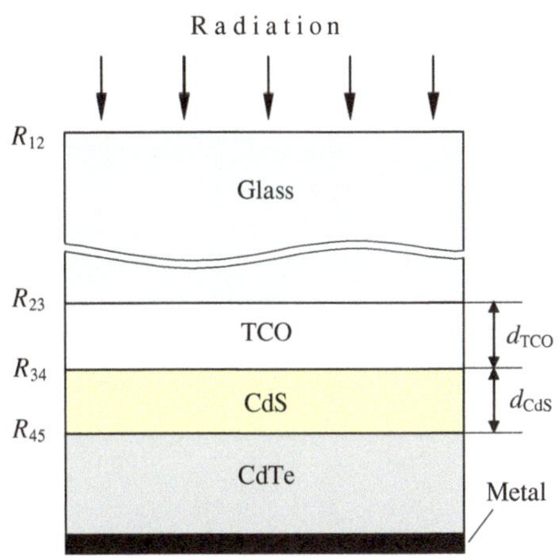

Fig. 1 Schematic cross-section of thin-film CdS/CdTe solar cell

[2] TCO is a transparent conducting oxide. ITO is a solid solution of indium oxide (In_2O_3) and tin oxide (SnO_2).

the air. For the convenience of presenting the main optical losses, we first assume that for glass $\kappa_2 = 0$. This is justified by the fact that in photovoltaic applications, specialized glass is often used with low iron oxide content, where the absorption is observed only in the ultraviolet region. As it will be shown later, absorption in the low-iron glass practically does not exhibit itself in short-circuit current (see "Effect of optical losses on short-circuit current").

For the refractive index of glass n_1 we will use the Sellmeier (Zelmeer) dispersion equation applied for quartz (SiO_2) [23]:

$$n^2 = 1 + \frac{a_1\lambda^2}{\lambda^2 - \lambda_1^2} + \frac{a_2\lambda^2}{\lambda^2 - \lambda_2^2} + \frac{a_3\lambda^2}{\lambda^2 - \lambda_3^2}, \tag{3}$$

where $a_1 = 0.6962$, $a_2 = 0.4079$, $a_3 = 0.8974$, $\lambda_1 = 68$ nm, $\lambda_2 = 116$ nm, and $\lambda_3 = 9{,}896$ nm.

To find the reflection coefficients at the interfaces glass–TCO, TCO–CdS, and CdS–CdTe it is necessary to know the values of the refractive index and extinction coefficient of TCO, CdS, and CdTe in the spectral range 300–850 nm. In the discussion below we consider the two most common structures of CdS/CdTe solar cell in which indium tin oxide (ITO) and F-doped tin oxide (SnO_2:F) are used as TCO. Figure 2 shows the spectral dependences of n and κ for ITO (typically 90 % In_2O_3, 10 % SnO_2) taken from Ref. [24, 25] and for SnO_2:F taken from Ref. [26]. Figure 2 also shows the data on CdS [27] and CdTe [28].

Note that the data on the optical constants of the materials cited in various sources differ somewhat, depending on the method adopted to grow the crystal or film. However, with a few exceptions, these differences manifest themselves weakly in the calculation results of the *integral* characteristics of a multilayer solar cell, which is short-circuit current J_{sc}. For example, calculations of J_{sc} using data for n and κ, obtained by ellipsometry for CdTe single crystal and films [28] lead to almost identical results.

Figure 3 shows the spectral dependence of the reflection coefficients $R(\lambda)$ at the interfaces calculated by substituting n and κ from Fig. 2 into Eq. (2). Emphasis is given on obtaining low reflection coefficients (less than 0.05–0.06 in the entire spectral range). The explanation for this is a relatively small difference between the optical constants of contacting materials. The calculation results shown in Fig. 3b for reflection at the interfaces of the same materials *with air* give the usually observed values of R, for example, $R = 0.037$–0.034 for the air/glass interface, $R = 0.25$–0.35 for the air/CdTe interface, etc.

As a consequence of the conservation of energy, the transmission coefficient (transmission) in each interface is given by $T = 1 - R$. Therefore, for reflection losses at all interfaces before the solar radiation reaches the CdTe absorber layer, the transmission $T(\lambda)$ can be calculated by the formula

$$T(\lambda) = (1 - R_{12})(1 - R_{23})(1 - R_{34})(1 - R_{45}), \tag{4}$$

where R_{12}, R_{23}, R_{34}, and R_{45} are the reflection coefficients of the interfaces: air–glass, glass–ITO (or glass–SnO_2), ITO–CdS (or SnO_2–CdS), and CdS–CdTe, respectively (see Fig. 1). As stated earlier, we consider normal incidence of the rays. As will be explained below, the multiple reflections of rays have an insignificant effect due to the relatively close values of the refractive indices of the contacting materials.

It follows from Fig. 4 that the overall reflection losses at the interfaces air–glass, glass–ITO, ITO–CdS, and CdS–CdTe amount to about 9 % in the entire spectral range

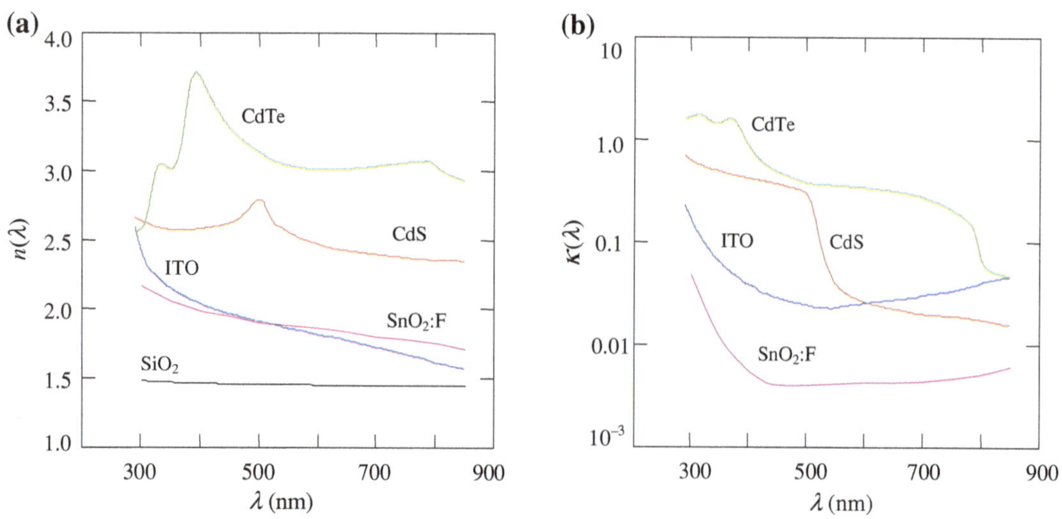

Fig. 2 Refractive index (**a**), and extinction coefficient (**b**) of SiO_2, ITO, SnO_2:F, CdS and CdTe as a function of wavelength (the curves $\kappa(\lambda)$ for glasses can be seen in Fig. 7)

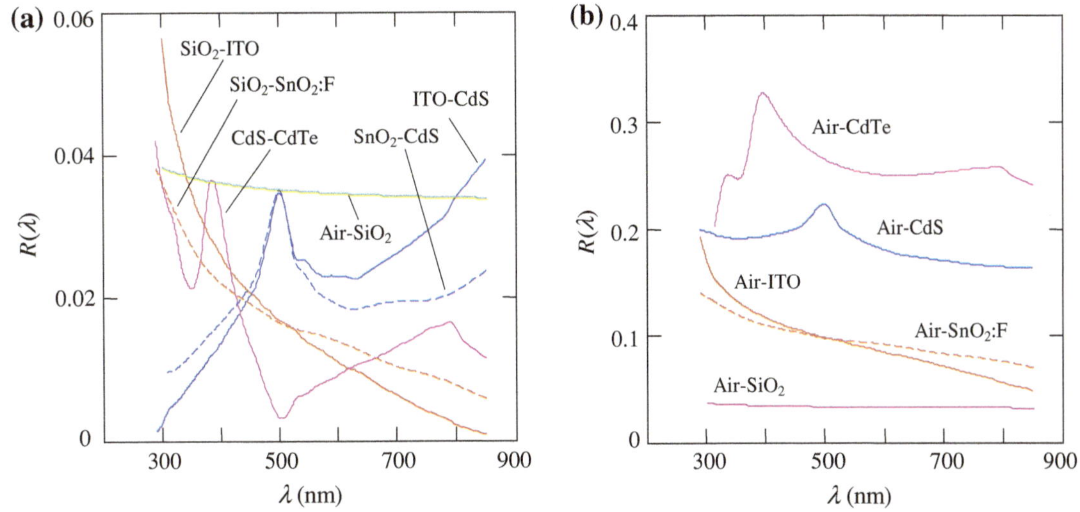

Fig. 3 Reflection coefficients R for the interfaces: air–glass (SiO_2), glass–ITO, glass–SnO_2, ITO–CdS, SnO_2–CdS, CdS–CdTe (**a**); and air–SiO_2, air–ITO, air–SnO_2, air–CdS, air–CdTe (**b**)

300–850 nm. This implies that the reflection losses in the case of thin-film CdS/CdTe structure are not as critical as in the case of a silicon solar cell. In fact, the coefficient of reflection from the mirror surface of silicon is at least 30 %, while in the region $\lambda < 400$ nm they are even higher than 50 %.

Accounting for absorption losses

The data for the extinction coefficient $\kappa(\lambda)$ shown in Fig. 2b allows to find the absorption coefficient as

$$\alpha(\lambda) = \frac{2\omega}{c}\kappa = \frac{4\pi}{\lambda}\kappa, \quad (5)$$

where $\omega = 2\pi\nu$ is the angular frequency, and c is the speed of light in free space.

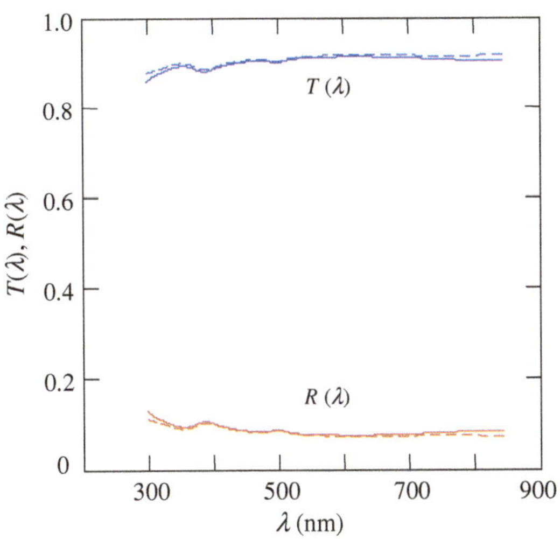

Fig. 4 Calculated transmittance $T(\lambda)$ and reflectance $R(\lambda)$ of the glass/ITO/CdS (*solid line*) and glass/SnO_2/CdS (*dashed line*) structures considering only reflections from all interfaces

Knowing $\alpha(\lambda)$ for the CdS and ITO (SnO_2), the absorption in these layers can be taken into account by incorporating in Eq. (4) the addition factors $\exp[-\alpha_{CdS}(\lambda)d_{CdS}]$, $\exp[-\alpha_{ITO}(\lambda)d_{ITO}]$, $\exp[-\alpha_{SnO}(\lambda)d_{SnO}]$:

$$T(\lambda) = (1 - R_{12})(1 - R_{23})(1 - R_{34})(1 - R_{45}) \times \exp(-\alpha_{SnO}d_{SnO})\exp(-\alpha_{CdS}d_{CdS}), \quad (6)$$

or

$$T(\lambda) = (1 - R_{12})(1 - R_{23})(1 - R_{34})(1 - R_{45}) \times \exp(-\alpha_{ITO}d_{ITO})\exp(-\alpha_{CdS}d_{CdS}), \quad (7)$$

where d_{ITO}, d_{SnO}, and d_{CdS} are the thicknesses of the ITO, SnO_2:F, and CdS layers, respectively.

Figure 5 shows the transmission of the glass/ITO/CdS structure (solid lines) and the glass/SnO_2/CdS structure (dashed lines) calculated with various thicknesses of the ITO, SnO_2:F, and CdS layers.

As can be seen in Fig. 5, in addition to reflection, appreciable losses (10–15 %) are observed in the wavelength range $\lambda > 500$–550 nm caused by absorption in SnO_2:F and greater losses (15–35 %) in the case of ITO. Much more losses are observed in the region $\lambda < 500$–550 nm caused by absorption in CdS together with ITO or SnO_2:F. In the wavelength region $\lambda > 500$–550 nm, by thinning the ITO it is possible to reduce the optical losses by 12–15 %; however, the effect of SnO_2:F thickness is only nominal. Greater effect can be achieved in the $\lambda < 500$–550 nm range by decreasing the thickness of the CdS layer. However, to obtain uniform CdS layer without discontinuities (known as 'pinholes') is difficult. The thickness of CdS about 50–60 nm is apparently close to the minimum possible value with the current technology used in the manufacture of CdS/CdTe solar cells [29].

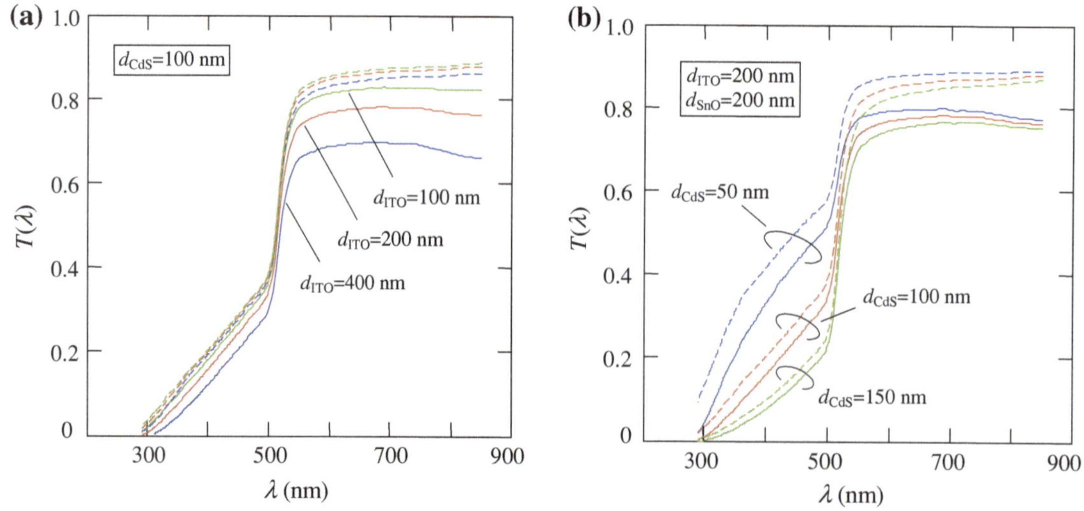

Fig. 5 Transmission of the glass/ITO/CdS (*solid lines*) and glass/SnO_2/CdS (*dashed lines*) structures calculated **a** at $d_{CdS} = 100$ nm and different thickness of ITO and SnO_2:F and **b** at $d_{ITO} = d_{SnO} = 200$ nm and different thickness of the CdS layer

Let us consider the errors caused by neglecting the multiple reflections and the interference effects in the TCO and CdS layers leading to the oscillations observed typically in the transmission curves of the glass plate covered with thin TCO or CdS layer. As is known, the transmission of a transparent film of thickness d with refractive index n at normal incidence taking into account the interference is given by the equation widely known in optics as the Airy formula for multiple beam interference in a system of two parallel plane reflecting surfaces for the case of no absorption [30]:

$$T(\lambda) = \frac{(1-R)^2}{(1-R)^2 + 4R\sin^2(4\pi nd/\lambda)}, \quad (8)$$

where R is the reflection coefficient of the film surface.

In Eq. (8), it is assumed that the film is in contact with identical materials at both sides. From our point of view, interference effects appear considerably weaker if the film is in contact with materials having refractive index higher than that of air. As an example, Fig. 6 shows the calculation results obtained with Eq. (8) for the transmission of ITO layer, when it is in contact with air (solid oscillating line) and in contact with SiO_2 (dashed oscillating line).

As can be seen, a pronounced oscillating interference behavior of $T(\lambda)$ is observed in both cases. However, as expected, in the case of reflection from ITO/SiO_2 interface, the amplitude of the oscillation is 5–7 times smaller, which is explained by a smaller difference in the refractive index and extinction coefficient of the materials at the interface.

In Fig. 6, transmission of the ITO layer (dotted line 3, without oscillation) is calculated as $T(\lambda) = (1 - R_{23})^2$ (R_{23} is the reflection coefficient at the ITO–SiO_2 interface), i.e. without taking into account multiple reflections. It is seen that

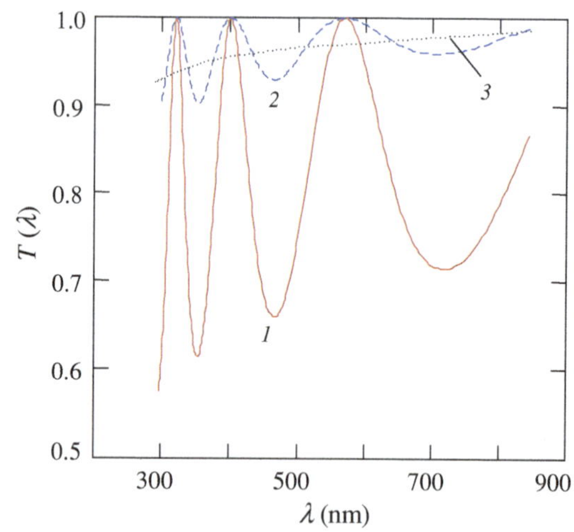

Fig. 6 Transmittance of the ITO film ($d = 200$ nm) (*1*) *in air* calculated by Eq. (8) (oscillating *solid line 1*), (*2*) contacted from both sides with SiO_2 and calculated by Eq. (8) at $R = R_{23}$ (oscillating *dashed line 2*), (*3*) the same but calculated as $[1 - R_{23}(\lambda)]^2$ (monotonically varying *dotted line 3*)

the average transmission of the ITO film in contact with SiO_2 (dotted line 3) calculated using Eq. (8) and the curve $T(\lambda) = (1 - R_{23})^2$ is comparable in magnitude. It can be assumed that the calculation results of short-circuit current density (presented below) as an *integrated* characteristic calculated over a broad spectral range, may vary slightly in these two cases. Thus, the interference effects in the solar cells under study appear to be rather weak, i.e. they contribute only a minor error in the results of the short-circuit current calculation.

Effect of optical losses on short-circuit current

Calculation of short-circuit current J_{sc} can quantitatively account for the effect of optical losses in the whole spectral

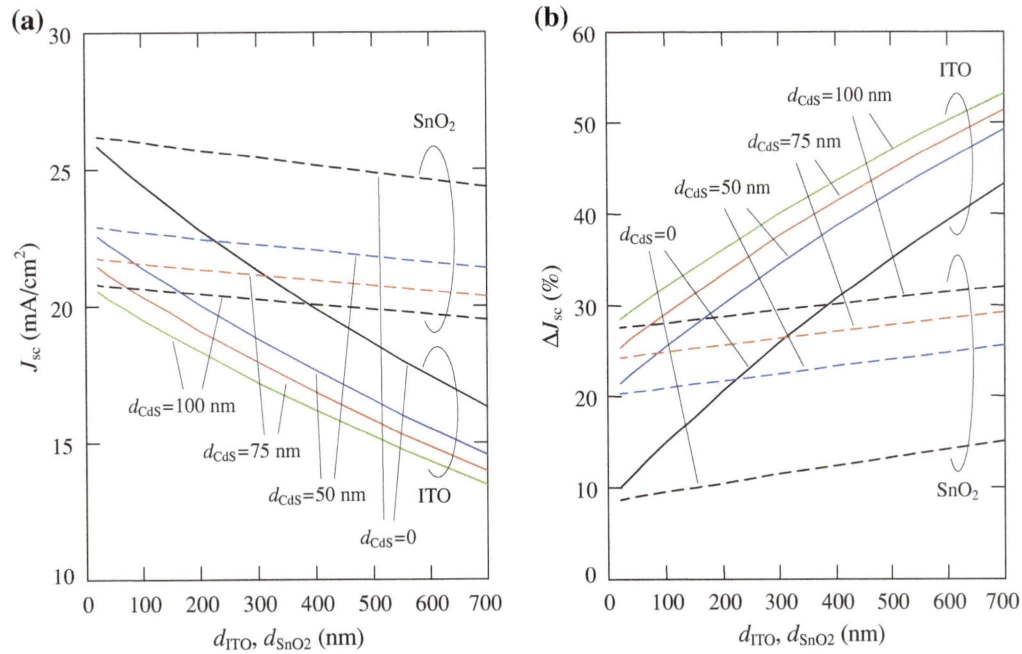

Fig. 7 **a** Effect of optical losses on short-circuit current depending on the thickness of ITO and SnO_2:F layer calculated for different thicknesses of CdS layer (100 % photoelectric conversion efficiency in the CdTe absorber layer is assumed); **b** the same but expressed in percents

range. We will calculate J_{sc} for AM1.5 total solar radiation using the Table ISO 9845-1:1992 [31].

If Φ_i is the spectral radiation power density and hv is the photon energy, the spectral density of the incident photon flux is Φ_i/hv_i can be written J_{sc} as

$$J_{sc} = q\sum_i T(\lambda_i)\frac{\Phi_i(\lambda_i)}{hv_i}\Delta\lambda_i, \tag{9}$$

where $\Delta\lambda_i$ is the wavelength range between the neighboring values of λ_i in the table and the summation is over the spectral range $\lambda = 300$ nm to $\lambda = \lambda_g = hc/E_g \approx 840$ nm.

It should be noted that solar radiation reported in the Table ISO 9845-1:1992 is with 0.5 nm interval in the range 300–400 nm and 1 nm interval in the range 400–1,500 nm. Therefore, replacing Φ_i and α_i in Eq. (9) with their average values in the interval $\Delta\lambda_i$, i.e. $(\Phi_i + \Phi_{i+1})/2$ and $(\alpha_i + \alpha_{i+1})/2$, leads to almost imperceptible changes in the results of calculations of J_{sc}.

Figure 7a shows the dependence of short-circuit current J_{sc} on the thickness of ITO and SnO_2:F calculated for different thickness of CdS with the assumption that the efficiency of photoelectric conversion in the CdTe layer is equal to 100 %. Figure 7b shows the percentage decreases of J_{sc}. In this case, the value of the optical losses is found using the expression

$$\Delta J_{sc}(\%) = \frac{J^{o}_{sc} - J_{sc}}{J^{o}_{sc}} \times 100\,\%, \tag{10}$$

where J^{o}_{sc} is the short-circuit current density found by Eq. (9) at $T(\lambda) = 1$ ($J^{o}_{sc} = 29.7\,\mathrm{mA/cm^2}$ according to the Table ISO 9845-1:1992).

As can be seen in Fig. 7, optical losses caused by TCO layer with thickness of 200 nm are about 20 and 10 % for ITO and SnO_2:F, respectively. When the thickness of the TCO layer is close to 0, the optical losses are caused only by reflection and become equal to 8.5 and 8.8 % for the SnO_2:F and ITO layers, respectively. Thicker ITO layers of 500–700 nm leads to unacceptably high optical losses (~40 % at $d_{CdS} = 0$), while using SnO_2:F the losses are only ~15 % (at $d_{CdS} = 0$). In both cases, the presence of CdS layer, even when its thickness is 50 nm (close to the practically minimum possible value [29]) leads to an additional increase in the optical losses by approximately 10 %. When the CdS thickness is 100 nm, the additional losses for ITO and SnO_2 are ~12 and 17 %, respectively. In the case of SnO_2:F, the presence of very thin CdS layer limits J_{sc} at ~23 mA/cm^2. The achievement of this value in ITO/CdS/CdTe solar cell is possible only with an ultrathin ITO layer.

It is evident from the data obtained that a short-circuit current density above ~25 mA/cm^2 is practically impossible with the use of CdS as the window layer since the band gap is only 2.42 eV (300 K). However, slightly higher current ($J_{sc} \approx 26$ mA/cm^2) can be obtained with the use of modified CdS-based films with wider band gap. In TCO/Zn_2SnO_4/CdS/CdTe devices, for example, interdiffusion between the CdS and Zn_2SnO_4 (ZTO) films effectively "consumes" the relatively thick CdS film during device fabrication [32, 33].

Another option for improving the charge collection in the short-wavelength region is the use of a wider band gap

$Cd_{1-x}Zn_xS$ ternary alloy to replace CdS as the window layer [34]. Results of the calculations show that integrating a ZTO buffer layer into a CdTe cell with the intention of modifying the TCO layer (rather than CdS) results in a very insignificant increase in J_{sc}, since the optical losses in the TCO are mainly due to reflection rather than absorption.

Until now, losses due to absorption in the glass were not considered because it was assumed that they are insignificant. Let us estimate the error, which we allowed.

It is well known that the most common glasses contain iron impurities in the form of iron salts within the silicon oxide that impede the transmission of light through the material. Because of this, apart from a fairly strong absorption in the UV, there is also a noticeable absorption in the visible and infrared spectral range. However, there are glasses with low iron oxide content, the so-called *low-iron solar glass* with high transmittance above 91.6 % and the virtual absence of color. Figure 8 shows the optical transmission of low-iron glass produced by Stanley Glass Co. Ltd. for PV modules [35].

Knowing the transmission of the glass $T(\lambda)$, one can find the spectral dependence of its absorption coefficient α_{glass}. Since the coefficient of reflection from the glass surface R_{12} is small ($\sim$4 %), to find α_{glass} the formula which does not take into account multiple reflections from the surfaces of the plate can be used [36]:

$$\alpha_{glass} = \frac{1}{d_{glass}} \ln\left[\frac{(1-R_{12})^2}{T(\lambda)}\right], \tag{11}$$

where d_{glass} is the glass thickness.

The spectral dependence of the absorption coefficient α_{glass}, found from the transmission curve using the formula (11), is shown in Fig. 8. As can be seen, the absorption

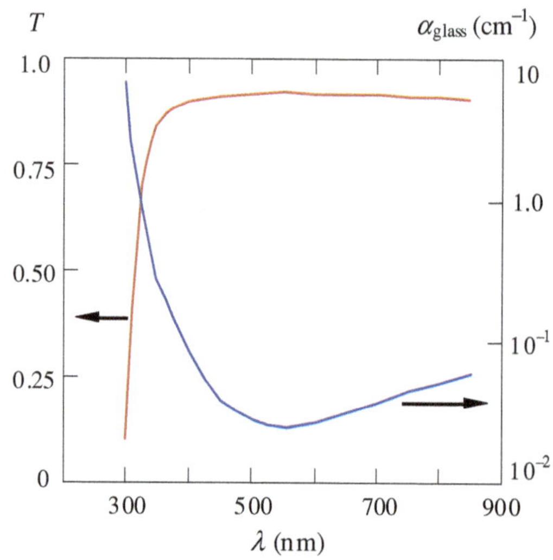

Fig. 8 Spectral distribution of the transmissions T and absorption coefficient α_{glass} of low-iron glass

coefficient of low-iron solar glass is lower by several orders than those for ITO, SnO_2, CdS, and CdTe. But one should keep in mind that the thickness of the glass in the CdS/CdTe solar cell (typically 3.2 mm) is at least three orders of magnitude greater than that of the thin-film device components.

To account for the absorption of glass in the calculation of short-circuit current density J_{sc} it is necessary to introduce in the expressions for the transmissions Eq. (6) and (7) a factor $\exp[-\alpha_{glass}(\lambda)d_{glass}]$. The calculations of J_{sc} show that taking into account the optical absorption in the 3.2-mm thick low-iron glass reduces J_{sc} by no more than $\sim$0.5 %. This corresponds to a decrease in the value of J_{sc} by $\sim$0.1 mA/cm^2 at a thickness of 300 μm for ITO or SnO_2. Thus, optical losses due to absorption in the low-iron glass can be ignored without introducing appreciable error that was made in the above calculations.

As seen in Fig. 3, the reflection coefficient R_{12} is very weakly dependent on λ averaging about 3.5 % and the contribution of the extinction coefficient ($\kappa = \alpha_{glass}\lambda/4\pi = 10^{-7}$–$10^{-5}$) to R_{12} is negligible according to Eq. (2) (see Fig. 7b). Therefore, we can say that the application of anti-reflective coating on the front surface of the glass can result in the ideal case in an increase in the short-circuit density of $\sim$3.5 %, which corresponds to $\sim$0.7 mA/cm^2.

Limitation of efficiency due to insufficient absorptive capacity of the CdTe layer

Calculating the short-circuit current in the previous sections, we assumed 100 % efficiency of photoelectric conversion in the CdTe layer. In fact, the efficiency of the solar cell is limited by the processes in CdTe. In addition to recombination at the surfaces and in the space charge region (SCR), the loss of an optical nature related to the insufficient absorptive capacity of the CdTe layer also takes place.

In this section, calculations are made for the exact values of *integrated* absorptive capacity of the CdTe layer when the spectral distribution of solar radiation and the absorption coefficient of the material are taken into account. It should be noted that the literature does not pay adequate attention to this factor. Meanwhile, consideration of the absorbance becomes particularly relevant in connection with the desire to reduce the thickness of the absorber layer in CdS/CdTe solar modules, i.e. to develop the devices with ultra-thin CdTe layer [37–41]. In this case, the optical losses in the absorber layer can be significant. As shown below, the losses due to the lack of absorptive capacity of the CdTe layer with a thickness of, for example, 0.5 μm are about 13 %.

As earlier, the losses of pre-penetrating radiation in CdTe layer can be taken into account by the introduction of

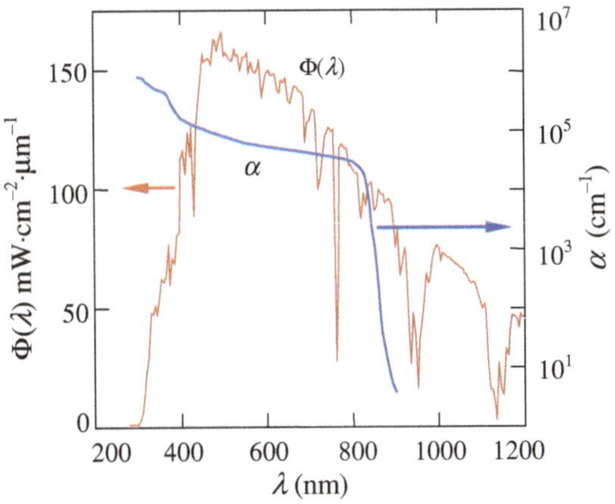

Fig. 9 Power spectral density of solar radiation $\Phi(\lambda)$ under standard conditions of AM1.5 and the absorption curve $\alpha(\lambda)$ for CdTe

the factor $T(\lambda)$, which represents the optical transmission of the glass substrate with layers of TCO and CdS. Thus, the expression for the integrated (rather than at a certain wavelength) absorptivity of solar radiation, which has reached the CdTe layer with a thickness d, can be represented as (at normal incidence of the rays):

$$A_\Phi(d) = \frac{\sum_i T(\lambda)\Phi_i[1 - \exp(-\alpha_i d)]\Delta\lambda_i}{\sum_i T(\lambda)\Phi_i\Delta\lambda_i}, \qquad (12)$$

where α_i is, as earlier, the absorption coefficient of CdTe at wavelength λ_i; Φ_i is the spectral power density of radiation at wavelength λ_i [31].

The AM1.5 spectral distribution of solar radiation $\Phi(l)$ and the absorption curve $a(l)$ of CdTe are shown in Fig. 9, where the wide range of variation of $\Phi(l)$ and $a(l)$ in the working ranges of l are illustrated. For $\alpha(\lambda)$ we used the data given in [41] and supplemented by the results of our measurements at the absorption edge [42]. For the optical transmission $T(\lambda)$ of SnO_2:F ($d = 200$ nm) and CdS ($d = 50$ nm) the date obtained in "Accounting for absorption losses" can be used, i.e. the $T(\lambda)$ transmission is given by Eq. (6). The summation in Eq. (12) should be taken from $\lambda = 300$ nm to $\lambda = \lambda_g = hc/E_g$, where E_g is the band gap of CdTe equal to 1.47 eV at 300 K [42].

Figure 10a shows the absorptivity of solar radiation A_Φ as a function of the thickness of CdTe calculated by Eq. (12) using the data on $\alpha(\lambda)$, $\Phi(\lambda)$ and the transmission of the glass/TCO/CdS structure $T(\lambda)$. Let us emphasize that the solid line shows how the absorptivity of solar radiation *power* increases with increasing the thickness of CdTe. In this case, one does not take into account that an electron–hole pair arises independently of the photon energy absorbed in the range from 1.47 to 4.1 eV (the latter corresponds to a wavelength of 300 nm). For this reason, the number of electron–hole pairs, and hence the short-circuit current generated in the solar cell is not proportional to the power of solar radiation. Obviously, in the context of the discussed issue, it is more correct to handle with the *flow of photons* rather than the *power* of solar radiation.

As mentioned earlier, the photon number in the spectral interval $\Delta\lambda_i$ can be obtained by dividing the power of radiation in this spectral range $\Phi_i\Delta\lambda_i$ by the photon energy $h\nu_i$. As a result, the expression for the *absorptivity of the photon flux* in the solar radiation $A_{h\nu}$ takes the following form:

$$A_{h\nu}(d) = \frac{\sum_i T(\lambda)\frac{\Phi_i}{h\nu_i}[1 - \exp(-\alpha_i d)]\Delta\lambda_i}{\sum_i T(\lambda)\frac{\Phi_i}{h\nu_i}\Delta\lambda_i}. \qquad (13)$$

The calculated dependence of the absorptivity on the thickness of CdTe for the photon flux $A_{h\nu}$ in the solar radiation AM1.5 is shown in Fig. 10a by the dashed line. As expected, for small d the fraction of absorbed photons in the solar radiation $A_{h\nu}$ is less than the fraction of absorbed radiation power A_Φ. This is understandable, because in a thin layer the probability of high-energy photon absorption is greater than that of the low-energy photons and there are less high-energy photons per unit of radiation power. With increasing d the curves $A_\Phi(d)$ and $A_{h\nu}(d)$ converge and coincide at a thicknesses d, where the total absorption of radiation energy and all of the photons occurs ($h\nu \geq E_g$).

It follows from the results shown in Fig. 10a that the *total* photon absorption in CdTe is observed at a layer thickness of 60–70 μm (99.9 % at a thickness of 20–30 μm), although very often in the papers devoted to CdTe solar cells it is assumed that this occurs at a thickness of the CdTe layer of several microns. Such an assumption usually rests on the argument that in CdTe, as a direct-gap semiconductor, the value of α in the fundamental absorption region exceeds 10^4–10^5 cm^{-1}, i.e. effective penetration depth of radiation α^{-1} is less than 1 μm (see, e.g. [42–44]). Our calculations, taking into account the spectral distribution of the absorption coefficient of CdTe and the standard solar radiation AM1.5, show that in fact the absorptivity of photons in the CdTe layer with a thickness of 2–3 μm is about 97 % and only ~93 % for a layer thickness of 1 μm.

It should also be borne in mind that most often a metal electrode is applied to the back surface of the CdTe absorber layer with the result that radiation with low absorption coefficient is reflected from such an interface. In this case with 100 % reflectance from the back surface the absorptivity is the same as for the double thickness of the absorber layer, i.e. the $A_{h\nu}(d)$ value in such a solar cell can be found by doubling the thickness d in Eq. (13). The calculation results in Fig. 10b show that in this case to ensure complete absorption of photons in the solar radiation, the CdTe layer should also be thick enough, i.e. greater than 40–50 μm. If, for example, 95 % absorptivity

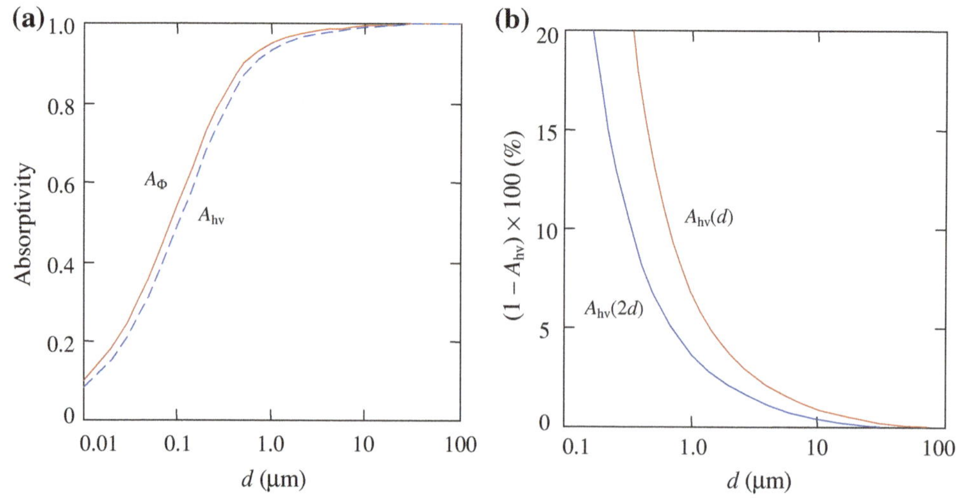

Fig. 10 **a** Dependence of absorptivity of solar radiation power A_Φ and absorptivity of photon flux A_{hv} in the solar radiation AM1.5 ($hv \geq E_g$) on the thickness of the CdTe layer. **b** Optical losses due to incomplete absorption depending on the thickness of the CdTe layer. Curves $A_{hv}(d)$ and $A_{hv}(2d)$ are without and with taking into account 100 % reflections from the back surface, respectively

in solar cell is to be accepted as an acceptable value, the thickness of the CdTe layer with a mirror back surface should be equal to ~0.7 μm rather than ~1.4 μm as in the case when the reflection from the back surface is not used.

It is pertinent to note that the absorptivity of CdTe even at 100 % reflection from the rear surface of the CdTe layer is significantly reduced with a further thinning the CdTe layer. For example, at $d = 0.5$ μm the optical loss is equal to 7 and 12 % at $d = 0.3$ μm.

Recombination losses and requirements imposed on thickness of the CdTe layer

In papers devoted to the CdS/CdTe solar cells with extremely thin CdTe absorber layer, the limitations imposed on short-circuit current density, open circuit voltage, fill factor, and, ultimately, the efficiency of the solar cell are discussed. One of the omissions in the analysis of CdS/CdTe solar cells is disregard such an important parameter as the width of the space charge region (SCR). One cannot argue against the fact that the width of the SCR (rather than just the thickness of CdTe) has a significant effect on the carrier collection and hence the photocurrent in the CdS/CdTe heterostructure as in any solar cell or photodiode. In fact, the electric field in the SCR ensures efficient collection of carriers photo-generated in the SCR as well as the charge reaching the SCR by diffusion from the neutral part of the CdTe. In addition, when the SCR is narrow, the strong electric field acting near the front surface (close to CdS) of the CdTe layer prevents recombination of carriers generated near the surface; however, when the SCR is wide, there can be significant reduction in photocurrent. Finally, the losses due to recombination of photogenerated electron–hole pairs in very SCR depend on its width. The results presented in our previous publications show that for a given thickness of the CdTe absorber layer, the changes of the SCR width truly can significantly affect the processes in the CdS/CdTe solar cell [45, 46]. In this section, the most practical issues with respect to minimizing the thickness of the absorber CdTe layer are considered taking into account the influence of the above-mentioned factors.

Like the optical losses, a quantitative description of losses caused by recombination at the front and back surfaces of the CdTe layer, as well as in the space charge region (SCR), can be obtained by calculating the short-circuit current density J_{sc}. In general, the J_{sc} is described by

$$J_{sc} = q \sum_i T(\lambda_i) \frac{\Phi_i(\lambda_i)}{hv_i} \eta(\lambda_i) \Delta\lambda_i, \tag{14}$$

where $\eta(\lambda_i)$ is the monochromatic photoelectric quantum efficiency at wavelength λ_i, and $T(\lambda)$ is, as earlier, the optical transmission of the glass substrate with layers of TCO and CdS.

It is known that the thin CdS film in CdS/CdTe heterostructure does not contribute to the photoelectric

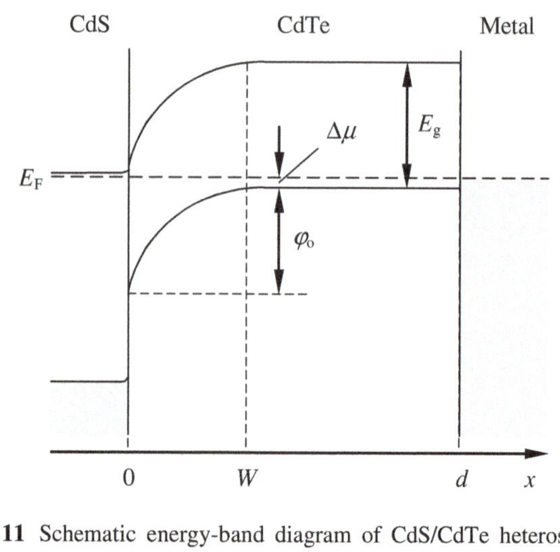

Fig. 11 Schematic energy-band diagram of CdS/CdTe heterostructure without electron reflector near the back surface of CdTe

conversion, but only serves as a "window", through which the radiation is introduced into the CdTe absorber layer. Therefore, the photocurrent in the CdS/CdTe solar cell is generated only in the CdTe layer. It should also be noted that CdS is a rather thin low-resistivity low-compensated *n*-type semiconductor, so that the SCR is located practically only in the p-CdTe layer (see, e.g. [8, 47]). So, for the CdS/CdTe heterostructure, a schematic energy-band diagram of an abrupt asymmetric p-n junction or a Schottky diode can be applied (Fig. 11).

Quantum efficiency of a solar cell always includes the *drift* and *diffusion* components, which are result of photogeneration of electron–hole pairs in the SCR and the neutral part(s) of the diode structure, respectively.

Considering the *drift* component, it is necessary to take into account losses due to recombination at the interface of CdS with CdTe, i.e. at the *front* surface of the CdTe layer. The solution of the continuity equation in this case is cumbersome, but with respect to the considered heterostructure with sufficient accuracy it can be reduced to a simpler expression [45]:

$$\eta_{\text{drift}} = \frac{1 + \frac{S_f}{D_p}\left(\alpha + \frac{2}{W}\frac{\varphi_o - qV}{kT}\right)^{-1}}{1 + \frac{S_f}{D_p}\left(\frac{2}{W}\frac{\varphi_o - qV}{kT}\right)^{-1}} - \exp(-\alpha W), \tag{15}$$

where S_f is the velocity of recombination at the front surface of the CdTe layer, φ_o is the height of the barrier at the contact with CdS (Fig. 11) related to the diffusion potential V_{bi} by the reration $\varphi_o = qV_{bi}$, D_n is the diffusion coefficient of electrons related to their mobility μ_n by the Einstein relation $qD_n/kT = \mu_n$, and W is the width of the SCR, which can be represented as [46]

$$W = \sqrt{\frac{2\varepsilon\varepsilon_o(\varphi_o - qV)}{q^2(N_a - N_d)}}, \tag{16}$$

where ε is the relative permittivity of the semiconductor, ε_o is the electric constant, V is the applied voltage, and $N_a - N_d$ is the concentration of uncompensated acceptors in the CdTe layer, which can differ by several orders of magnitude from the hole concentration in the valence band [45].

The *diffusion* component of the quantum efficiency η_{dif} is also found from the continuity equation. Taking into account recombination at the back surface of the CdTe layer and the boundary condition $\Delta n = 0$ at $x = W$, it can be written [44–48]

$$\eta_{\text{dif}} = \frac{\alpha L_n}{\alpha^2 L_n^2 - 1}\exp(-\alpha W) \times \left\{ \alpha L_n - \frac{\frac{S_b L_n}{D_n}\left[\cosh A - \exp(-\alpha(d - W))\right] + \sinh A + \alpha L_n \exp(-\alpha(d - W))}{\frac{S_b L_n}{D_n}\sinh A + \cosh A} \right\}, \tag{17}$$

where $A = (d - W)/L_n$, D_n and L_n is the electron diffusion coefficient and diffusion length, S_b is the velocity of recombination at the rear surface of the CdTe layer, and d is the thickness of the CdTe layer.

The quantum efficiency η, appearing in Eq. (14) is the sum of drift and diffusion components:

$$\eta = \eta_{\text{drift}} + \eta_{\text{dif}}. \tag{18}$$

The presence of W in Eqs. (15) and (17) reflects the significant impact of the SCR width (rather than only the thickness of the CdTe layer d) on the generation of the photocurrent in the CdS/CdTe heterostructure. As noted in the introduction to this section, the dependence of the efficiency of CdS/CdTe solar cell on W is usually not discussed in the papers dealing with the problem of minimizing the thickness of the CdTe absorber layer. The issue of uncompensated acceptor concentration $N_a - N_d$ in a layer of CdTe, which determines according to Eq. (16) the width of the SCR, is not discussed as well.

As is known, the Poisson equation implies that the width of space-charge region in the diode structure is determined by the spatial distribution of space charge due to band bending. In the case of a semiconductor doped with *one type of shallow impurity*, the space charge density is independent of the x coordinate and determined by the concentration of majority carriers (n or p) in the bulk of the diode structure. However, CdTe always contains a lot of uncontrolled impurities and defects of the donor and acceptor types and is a semiconductor compensated to a greater or lesser degree. The conductivity of CdTe in CdS/CdTe solar cell is determined by the position of the impurity or defect energy levels in the band gap, near which the Fermi level is located. In contrast, the width of the SCR is determined by the concentration of uncompensated impurities (defects) $N_a - N_d$, which cannot be identified with the concentration of holes in the valence band p. The $N_a - N_d$ and p values, as already noted, can differ by several orders of magnitude.

By varying the lifetimes of charge carriers and the concentration of uncompensated impurities, it is possible using Eqs. (15)–(17) to describe in detail all the variety of observed photoresponse spectra of CdS/CdTe solar cells. These equations also reflect the impact of the recombination of charge carriers at the CdTe surface layers on the photoelectric conversion efficiency. Below we show how recombination losses affect the value of short-circuit current density depending on the parameters of CdS/CdTe structure. In "Recombination losses at front surface of CdTe layer" and "Recombination losses in the SCR", the dependence of the recombination losses on the width of the SCR and the material of CdTe absorber are studied. In order to eliminate the influence of the thickness of CdTe and recombination at its rear surface, the layer is considered to be sufficiently thick in this case. After determining the optimal width for the SCR, the effect of thickness of the CdTe layer on the short-circuit current density is investigated in "Losses due to decrease in the thickness of the CdTe layer and generalization of the results".

Recombination losses at front surface of CdTe layer

Figure 12 shows the dependence of short-circuit current density of CdS/CdTe heterostructure on the width of the SCR calculated using Eqs. (15)–(17) under the conditions of AM1.5 (100 W/cm^2) solar irradiation. Calculations were made for the surface recombination velocities $S_f = 10^7$ cm/s and $S_f = 0$ at the front surface of the CdTe layer. The lifetimes of the charge carriers are assumed to be 10^{-10}, 5×10^{-10} and 2×10^{-9} s based on the fact that, according to published data, the lifetimes of the charge carriers in thin-film CdTe are usually in the range from 10^{-10} s to 2×10^{-9} s [48–50]. In calculations we use the mobilities of electrons and holes, which in thin-film CdTe are often taken to be 320 and 40 cm^2/(V s), respectively [51, 52]. The thickness of CdTe layer d is accepted to be 10 μm, when, as will be shown below, the thickness and the rear surface recombination have practically no effect on results of the calculations.

For comparison, the curves obtained by taking the lifetimes of electrons and holes as 2×10^{-6} s, and their mobilities 1,100 cm^2/(V s) and 100 cm^2/(V s), respectively, which are typical for the most pure and perfect CdTe single crystals, are also shown in Fig. 12 [53]. The curves illustrate the major impact of the lifetime of charge carriers on the short-circuit current in CdS/CdTe solar cell.

The results in Fig. 12 show that the short-circuit current density J_{sc} increases with W in the region, where W is well below 1 μm for the carrier lifetimes of 10^{-10}, 5×10^{-10} and 2×10^{-9} s. This is because with an increase of W more and more of the radiation is absorbed in the SCR, where the photogenerated electrons and holes are effectively separated by the electric field. For $W \rightarrow 10$ μm and without recombination at the front surface (solid lines in Fig. 12a), the short-circuit current density J_{sc} continues to increase approaching the maximum possible value of 23.2 mA/cm^2, which corresponds to absorption in the SCR of almost all radiation passed through a glass substrate,

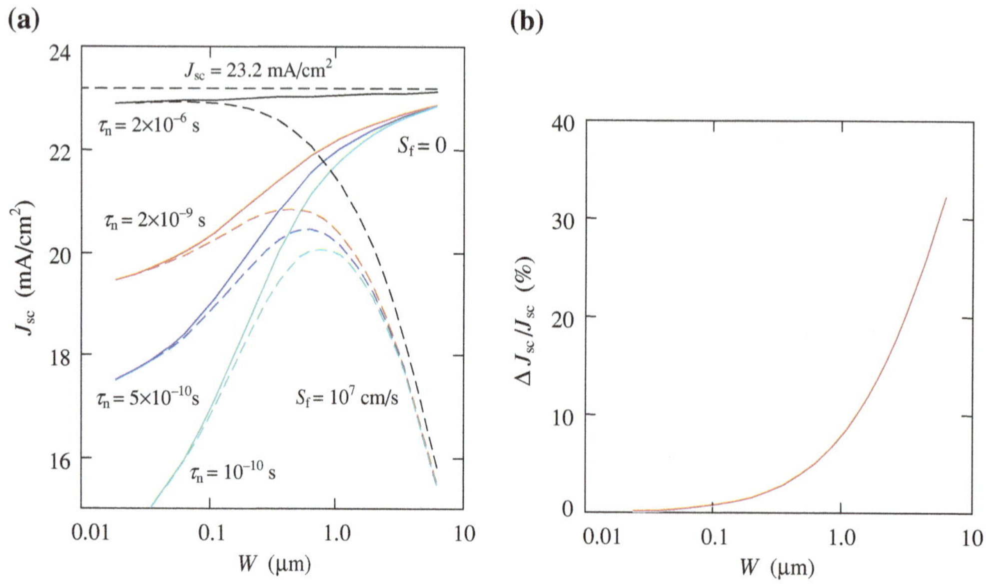

Fig. 12 **a** The dependence of short-circuit current density J_{sc} on the width of the space-charge region W calculated for different lifetimes of the electrons τ_n and the velocity of recombination at the front surface $S_f = 0$ (*solid lines*) and $S_f = 10^7$ cm/s (*dashed lines*). **b** Recombination losses at the front surface in percentage ($S_f = 10^7$ cm/s)

SnO_2:F and CdS layers (Fig. 1b). Of course, the maximum possible value of the short-circuit current density can be higher than 23.2 mA/cm^2, if it is provided a more high transparency $T(\lambda)$ of TCO and CdS layers.

When W is less than 1 µm (it is relevant in the light of the discussed problem), the short-circuit current density is much lower than the maximum possible value 23.2 mA/cm^2 even in the absence of surface recombination. The reason for this is the low mobility of minority carriers (electrons) and, especially, their short lifetime. Indeed, even if the lifetime of electrons is $\tau_n = 2 \times 10^{-9}$ s, their diffusion length $L_n = (\tau_n D_n)^{1/2}$ is approximately equal to 1 µm, i.e. photoelectrically beneficial region constitutes a small portion of the thickness of CdTe. If $\tau_n = 2 \times 10^{-6}$ s and $D_n = 27.5$ cm^2/s (as in single crystal CdTe), the value of L_n exceeds 70 µm and the value of J_{sc} becomes close to the maximum value 23.2 mA/cm^2 (at $S_f = 0$).

Taking into account the surface recombination (dashed lines in Fig. 12a), the increase in J_{sc} at $W > 0.2$–0.3 µm with widening the SCR begins to saturate and decrease with a further increase in W, since the electric field in the SCR is reduced and recombination at the front surface of CdTe is intensified. With the presence of surface recombination ($S_f = 10^7$ cm/s) and the real lifetimes of charge carriers (0.1–2) × 10^{-9} s, maxima on the curves are observed when the width of the SCR is equal to 0.4–0.8 µm. Evidently, with increasing W the diffusion component of the photocurrent is playing a lesser role, because the electron diffusion length L_n, even at $\tau_n = 2 \times 10^{-9}$ s is ~1 µm. Therefore, when $W > 3$–5 µm the calculated values of J_{sc} for different lifetimes of the charge carriers are close to each other. It should be noted that such an important fact as the presence of maximum on the dependence of the short-circuit current J_{sc} on the SCR width (or on the concentration of uncompensated impurities $N_a - N_d$) was substantiated in our previous works [44, 45].

To compare the effect of surface recombination with other kinds of losses, reduction of J_{sc} is conveniently being represented in percentages. This is shown in Fig. 12b. Notice that the losses in percentage do not depend on the lifetime of charge carriers. Of course, when the SCR expands ($W > 0.1$ µm), the recombination losses increase rapidly. At the maxima on the $J_{sc}(W)$ curves ($W = 0.4$–0.8 µm) calculated for $\tau_n = (0.1$–$2) \times 10^{-9}$ s, the losses account for 4–6 %.

Recombination losses in the SCR

Recombination of photogenerated charge carriers in the SCR of CdS/CdTe heterostructure can be taken into account using the Hecht equation [54]:

$$\eta_H(x) = \frac{\lambda_n}{W}\left[1 - \exp\left(-\frac{x}{\lambda_n}\right)\right] + \frac{\lambda_p}{W}\left[1 - \exp\left(-\frac{W-x}{\lambda_p}\right)\right], \quad (19)$$

where x is the coordinate, where an electron–hole pair arises (x is measured from the front surface of the CdTe layer), and λ_n and λ_p are the drift length of electrons and holes in the SCR, respectively:

$$\lambda_n = \mu_n F \tau_{no}, \quad (20)$$

$$\lambda_p = \mu_p F \tau_{po}. \quad (21)$$

In Eqs. (20) and (21), F is the electric field strength, τ_{no} and τ_{po} are the lifetimes of electrons and holes in the SCR, respectively.

In the Schottky diode, an electric field is not uniform, but consideration of the nonuniformity is simplified, since the field strength decreases linearly with the x coordinate. In this case, the field strength F in the expressions (20) and (21) for λ_p and λ_n can be replaced by the mean values of F in the sections (0, x) and (x, W) for electrons and holes, respectively [15]:

$$F_{(0,x)} = \frac{(\varphi_o - eV)}{eW}\left(2 - \frac{x}{W}\right), \quad (22)$$

$$F_{(x,W)} = \frac{(\varphi_o - eV)}{eW}\left(1 - \frac{x}{W}\right). \quad (23)$$

Evidently, charge collection efficiency in the SCR is determined by

$$\eta_{coll} = \int_0^W \eta_H(x)\alpha_\gamma \exp(-\alpha_\gamma \cdot x)\, dx, \quad (24)$$

where it is understood that the generation rate of electron–hole pairs in terms of one incident photon is $\alpha \cdot \exp(-\alpha x)$ [46].

Figure 13a shows the short-circuit current density J_{sc} as a function of the width of the SCR, calculated taking into account the recombination losses in the SCR by the formula:

$$J_{sc} = q \sum_i \eta_{coll}(\lambda_i) \frac{T(\lambda_i)\Phi_i(\lambda_i)}{h\nu_i} \Delta\lambda_i. \quad (25)$$

When calculating, the lifetimes of charge carriers in the SCR is accepted as earlier to be $\tau = \tau_{no} = \tau_{po} = 10^{-10}$, 5×10^{-10} and 2×10^{-9} s.

As in the case of recombination at the front surface, the short-circuit current density at $W < 0.2$–0.3 µm increases with W, because a larger part of the solar radiation is absorbed in the SCR. The J_{sc} values for different lifetimes at $W < 0.2$–0.3 µm are the same, because recombination in such a narrow space-charge region does not occurs. When the SCR extends, the $J_{sc}(W)$ curves calculated for different

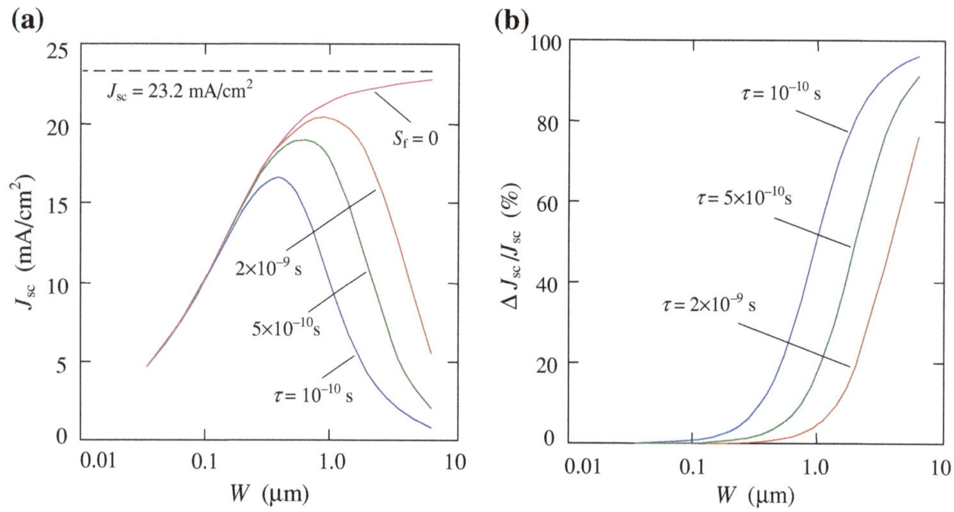

Fig. 13 **a** Short-circuit current density J_{sc} taking into account the recombination losses in the SCR as a function of its width W for different carrier lifetimesτ. **b** Recombination losses, expressed as percentage

lifetimes τ diverge. This is due to weakening the electric field leading to inefficient charge collection. When $W > 0.5$–1 μm, the recombination rate increases so that J_{sc} begins to decrease (at $W > 0.4$ μm even at $\tau = 2 \times 10^{-9}$ s). Of course, if one excludes the recombination in the SCR, J_{sc} for $W > 10$ μm approaches the maximum possible value 23.2 mA/cm^2 for the transmission $T(\lambda)$ of the glass/SnO_2:F/CdS structure.

Figure 13b shows the losses due to recombination in the SCR expressed as percentage. As can be seen, when $\tau = 10^{-10}$ s and the $W = 0.8$ μm (corresponding to the maximum J_{sc}) losses are too large (~40 %) and increase even more for greater width of the SCR. In contrast, the recombination losses decrease rapidly with narrowing the SCR and increasing the lifetime of charge carriers. If the SCR is not narrow enough ($W > 0.2$–0.4 μm) and the lifetime of charge carriers is small (e.g. 10^{-10} s), unacceptably high losses take place in the CdS/CdTe solar cell due to recombination in the SCR.

Losses due to decrease in the thickness of the CdTe layer and generalization of the results

When considering the recombination at the front surface of the CdTe layer in "Recombination losses at front surface of CdTe layer", it was established that in trying to create the conditions so that the major part of radiation is absorbed in the SCR, one cannot extend the SCR more than 0.6–0.8 μm giving preference to the largest carrier lifetime $\tau = 2 \times 10^{-9}$ s. As can be seen in Fig. 12a, the maximum in the dependence of $J_{sc}(W)$ in this case is observed at $W = 0.4$–0.6 μm, where recombination losses amount to 3–4 %. When SCR width is reduced to 0.3 μm the short-circuit current density J_{sc} decreases very slightly, but the losses are reduced to 2.4 %. Thus, the width of the SCR $W = 0.3$ μm can be considered as nearly optimal to minimize losses due to recombination at the front surface of the absorber.

Such a choice of the W value is not contrary to the requirements relating on the recombination losses in the SCR of CdS/CdTe heterojunction. As can be seen from Fig. 13, with a width of the SCR equal to 0.3 μm, the losses due to recombination in the SCR are very low at the carrier lifetimes 2×10^{-9} s. According to Eq. (5), $W = 0.3$ μm corresponds to the concentration of uncompensated acceptors close to 10^{16} cm^{-3}.

An obvious graphic comparison of the recombination losses on the front surface of CdTe and in the SCR is demonstrated in Fig. 14. As can be seen, at $W = 0.3$ μm the losses due to recombination in the SCR is only ~0.4 %, while at the front surface the loss is ~2.4 %. Thus, the summarized losses due to recombination at the

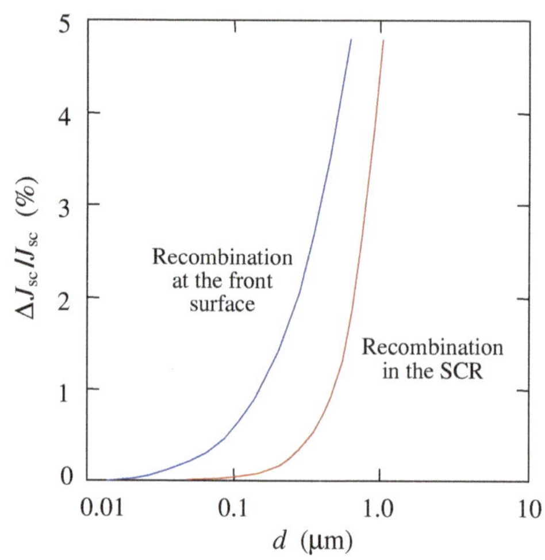

Fig. 14 Relative change of J_{sc} in percent due to recombination at the front surface of the CdTe layer and in the SCR of CdS/CdTe heterojunction

front surface layer and in the SCR with its width of 0.3 μm amount to 2.8 %. The short-circuit current density is reduced due to these losses from 21.3 to 20.7 mA/cm^2.

These results were obtained with CdTe layer thickness $d = 10$ μm. Having fixed the value of $W = 0.3$ μm and the recombination velocity at the rear surface of the CdTe layer $S_b = 10^7$ cm/s let us consider how the J_{sc} value will change with decreasing the film thickness d. In the following calculations, we will accept the charge carrier lifetime equal to 2×10^{-9} s (which is a realistic assumption [48–50]) because its shortening leads to a significant reduction of the short-circuit current.

Figure 15 shows the results of such calculations of relative decreasing J_{sc} (in percentage). As can be seen, the effect of the rear surface recombination on J_{sc} begins to appear when the thickness of the CdTe layer is less than ~5 μm.

Thinning the CdTe layer reduces the current density J_{sc} primarily, because of the removal of the photocurrent diffusion component which is created outside the space-charge region ($d > W$). In fact, the electron diffusion length $L_n = (\tau_n \ D_n)^{1/2}$ in CdTe with the lifetime $\tau_n = 2 \times 10^{-9}$ s is about 1 μm. So the decrease in CdTe thickness begins to reveal itself, when the thickness d is less than several L_n, i.e. several microns.

Of course, reduction in J_{sc} further intensifies due to the recombination of charge carriers at the rear surface of the CdTe layer. As can be seen in Fig. 15, reduction in J_{sc} due to removal of the photocurrent diffusion component and recombination at the rear surface of the absorber layer (labeled as "Photoelectric losses") amounts to ~5 % at the thickness of the CdTe layer $d = 1$ μm and increases up to ~12 % at $d = 0.5$ μm.

It follows from the results presented in "Limitation of efficiency due to insufficient absorptive capacity of the CdTe layer" that at the thicknesses of CdTe less than a few microns the losses associated with the lack of absorptivity A_{hv} of the CdTe layer reveal themselves perceptibly. Figure 15 shows the reduction of J_{sc} on the CdTe layer thickness due to a decrease in A_{hv} taking into account the reflection from the back surface (labeled as "Optical losses"). As it turns out, the reduction of J_{sc} due to both kinds of losses in the CdTe layer is comparable at $d \sim 1.5$ μm, but for smaller d the optical losses amount to ~7 % for $d = 0.5$ μm is lower than the photoelectric losses amounting ~12 % for the same d. We emphasize that "Optical losses" are due to the lack of absorptive capacity of the CdTe layer. As earlier noted, the optical losses due to *absorption* in the glass substrate, SnO_2:F and CdS layers as well as a *reflection* on the interfaces taken into account by the factor $T(\lambda_i)$ in Eq. (9) can be about 16 %.

It is known that to reduce the negative impact of recombination at the rear surface of the CdTe absorber, a heavily doped layer adjacent to the back contact is created. As a result, the potential energy barrier (*electron reflector*) tends to confine minority carriers in the more lightly doped region. This configuration is equivalent to a cell having a very small recombination velocity S_b at the back surface of the CdTe layer ($S_b < 100$ cm/s) [46, 48]. Figure 16 illustrates the quantitative effect of recombination on the rear surface of the CdTe layer with changing the film thickness d without electron reflector. The reduction of J_{sc} is found as the difference between its value at $S_b = 10^7$ cm/s and $S_b = 0$ in Eq. (14).

As one can see in Fig. 16, if $d > 1$ μm, the influence of recombination at the rear surface of the CdTe layer

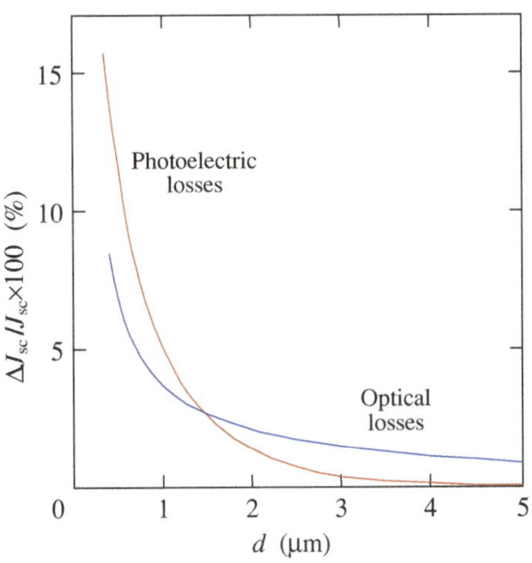

Fig. 15 Effect of the diminishing of photoelectrically active part of the CdTe layer and the optical losses on the short circuit current J_{sc} (because of insufficient absorptive capacity) as a result of reducing the thickness d of the CdTe layer

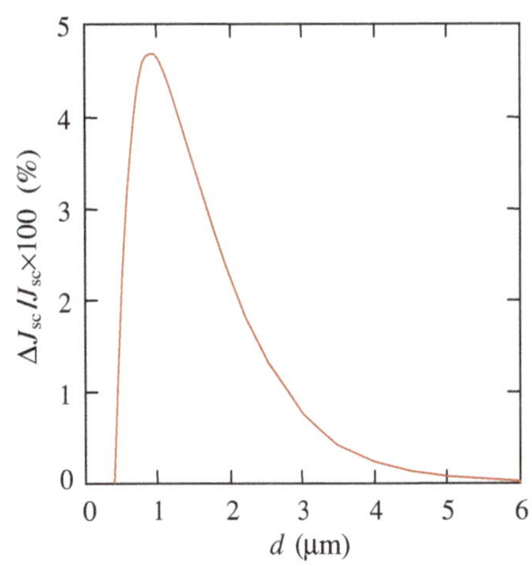

Fig. 16 The change of the short-circuit current density due to recombination at the rear surface of the CdTe layer as a function of its thickness

increases with decreasing d. This is because the concentration of photogenerated excess electrons Δn increases with decreasing d. In the range $d < 1$ μm, Δn decreases because $\Delta n = 0$ at $x = W$, which is one of the boundary conditions for solving the continuity equation for the neutral part of the CdTe layer [46]. Note that the maximum losses due to recombination at the rear surface of the CdTe layer do not exceed 4–5 %. At $d = 2$–3 μm, this kind of losses accounts for ~1 % and only 2–5 % in a narrow range $d = 0.5$–2 μm. In this regard, the question of reasonability to complicate the technology by creating an electron reflector arises. Obviously, when $d \geq 0.5$ μm, the creation of electron reflector makes no sense at all.

Completing the discussion on the dependence of short-circuit current density J_{sc} on the CdTe layer thickness d, we can summarize the obtained results taking into account all the losses assuming that the width of the space-charge region W is equal to 0.3 μm as its optimum value. Losses due to recombination at the front surface of the CdTe layer and in the SCR can be considered as independent of d and, in accordance with the above discussion, it should be assumed to be 2.4 and 0.4 %, respectively. For losses caused by thinning the absorber layer, one can use the calculation data presented in Fig. 15.

Figure 17 shows the relative change of J_{sc} as a function of the thickness of the CdTe layer d taking into account *all the types of losses* discussed above. It follows from the presented data that when $d = 2$–3 μm (can be considered as typical for the large-scale production of the CdTe modules) the total losses are about 5 % larger than those for a thick layer. When $d \geq 5$ μm, the total losses are 3–4 % but while thinning the CdTe layer up to 0.5 μm the losses increase to ~20 %. In absolute values of the short circuit current density J_{sc}, this corresponds to a decrease of J_{sc} from 19.7 mA/cm^2 at $d = 2$–3 μm to $J_{sc} = 16.6$ mA/cm^2 at $d = 0.5$ μm.

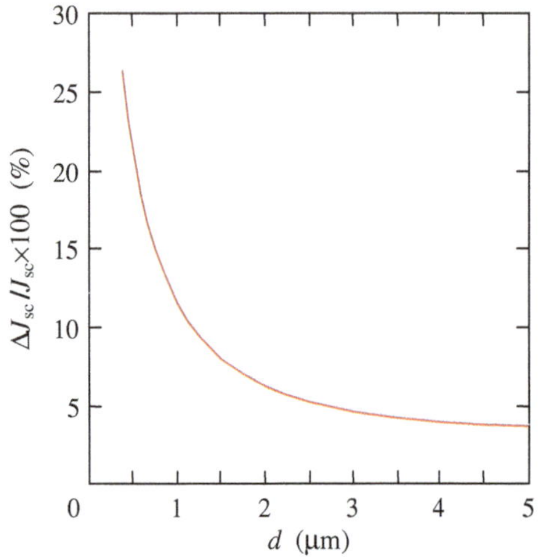

Fig. 17 Relative change of the short circuit current J_{sc} depending on the layer thickness d taking into account all types of losses

Evidently, the thinning of the absorber layer up to 0.5 μm reduces the consumption of CdTe, as a material for the manufacture of solar modules, up to 4–6 times compared with those for modules with the thickness of CdTe layer of 2–3 μm. Therefore, one must select "the golden means" to provide the most successful combination of material consumption and efficiency of CdS/CdTe modules.

Thus, the results presented above show in detail how the parameters of the CdTe absorber layer and the CdS window layer affect the characteristics of CdS/CdTe solar cell. Another aspect of the studies is that the comparison of the calculation results with the measured quantum efficiency spectra can quite accurately determine the parameters of solar cells. Among the most important of them are the concentration of uncompensated acceptors and the electron lifetime in the CdTe absorber, as well as the thicknesses of the CdTe and CdS layers. In addition, such a comparison allows formulating the recommendations for corrections of these parameters to increase the efficiency of photoelectric conversion of the device.

Let us demonstrate this by analyzing a fairly typical spectrum of CdS/CdTe solar cell presented in Fig. 18, where the results of measurements and calculations using Eqs. (15)–(18) for various parameters of the CdS/CdTe structure are shown by circles and solid lines, respectively.

As can be seen in Fig. 18, a good agreement of the calculation results with the measured spectral curve takes place for the concentration of uncompensated acceptors in the CdTe layer $N_a - N_d = 6 \times 10^{15}$ cm^{-3}, the lifetime of electrons in the neutral part of the layer $\tau_n = 4 \times 10^{-10}$ s, the thickness of the CdTe layer $d_{CdTe} = 2.5$ μm, and the thickness of the CdS film $d_{CdS} = 60$ nm. It is also shown in Fig. 18a, b, how the calculated spectrum $\eta(\lambda)$ is changed when $N_a - N_d$ and τ_n increase or decrease by one order of magnitude. As seen, such changes of $N_a - N_d$ and τ_n lead to a substantial modification of the spectrum. It should especially be noted that these parameters of the CdTe layer affect the $\eta(\lambda)$ spectrum *in different ways*.

When extending or shortening the electron lifetime τ_n, the quantum efficiency increases or decreases, respectively, only in the $\lambda > 500$ nm spectral range (high-energy photons with large absorption coefficient α do not reach the neutral part of the CdTe layer). With a decrease in $N_a - N_d$ (extension of the SCR), the efficiency is improved, but only in the $\lambda > 600$ nm range. At $\lambda < 600$ nm, the efficiency decreases because recombination on the front surface of the CdTe layer is intensified due to the weakening of the electric field strength. With an increase in $N_a - N_d$, the rise of $\eta(\lambda)$ with increasing λ in the $\lambda = 550$–800 nm range is replaced by its decline, i.e. qualitative changes in the

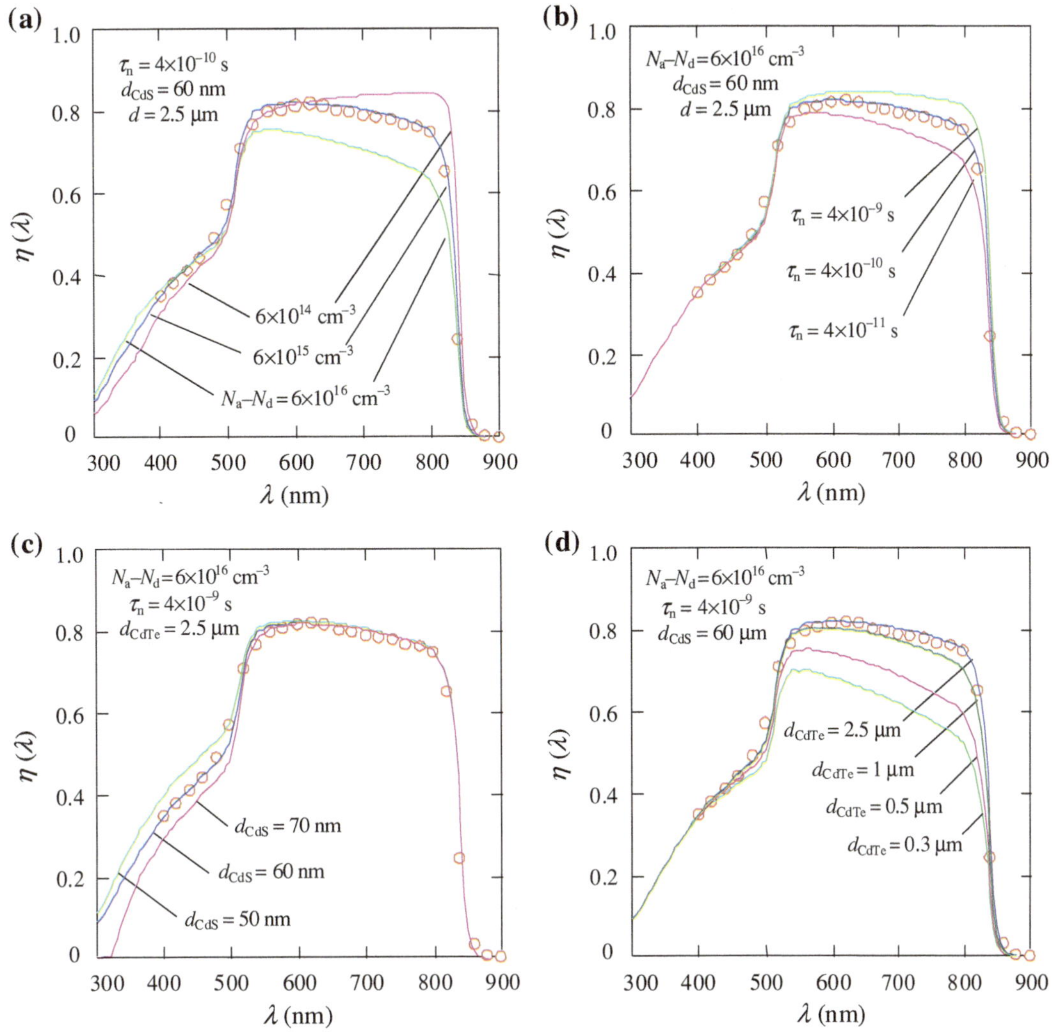

Fig. 18 Comparison of the measured spectrum (*circles*) with the calculations (*solid lines*) performed for different parameters of CdS/CdTe solar cell

spectrum occur. Although in the long-wave part of spectrum, the efficiency decreases with increasing $N_a - N_d$ (narrowing of the SCR), at $\lambda < 500$ nm, the intensity of surface recombination reduces and the efficiency increases slightly as a result of the enhanced electric field.

The features of the influence of $N_a - N_d$ and τ_n on the spectral distribution of $\eta(\lambda)$ lead to the fact that the best fit of the calculated results with the measured spectrum is achieved only for a single combination of $N_a - N_d$ and τ_n. Note that the deviation of 20–30 % from the selected value of $N_a - N_d$ or τ_n causes noticeable changes in the calculated spectra, i.e. the accuracy of determining these parameters by comparing the calculated and experimental results is quite high.

The data shown in Fig. 18c confirm the expected result that the thinning of the CdS layer leads to a decrease in the efficiency of solar cell only in the $\lambda < 500$ nm range, where the fundamental absorption ($h\nu \geq E_g$) takes place in CdS. Finally, as can be seen in Fig. 18d, when thinning the CdTe absorber layer the efficiency $\eta(\lambda)$ decreases, but there is almost no change in the $\lambda < 500$ nm range since high-energy photons ($\alpha > 5 \times 10^5$ cm^{-1}) even in a layer of thickness 0.3 μm are almost completely absorbed.

Analyzing only the changes in the quantum efficiency spectra shown in Fig. 18 and neglecting the spectral distribution of solar radiation, one can made too high or too low estimation of the influence of the discussed above parameters on the efficiency of solar cell. More reliable information on the impact of these parameters can be obtained by finding the short-circuit current density J_{sc} using Eq. (14), which takes into account the spectral distribution of solar radiation. In fact, an increase or decrease of the concentration of uncompensated acceptors $N_a - N_d$ in the CdTe layer by an order of magnitude results in *visually comparable changes* in the spectrum in Fig. 18a. At the same time, the short-circuit current density J_{sc}

calculated by Eq. (14) increases to 0.6 mA/cm^{-2} and decreases to 2.5 mA/cm^{-2} at increasing and decreasing $N_a - N_d$, respectively. This discrepancy is explained by the fact that in the $\lambda = 500$–800 nm range, the power of solar radiation significantly decreases with increasing λ and, therefore, increase of $\eta(\lambda)$ at $\lambda > 700$ nm markedly attenuates. In this sense, the effect of the electron lifetime τ_n is intermediate, since both increase and decrease of τ_n cause more or less uniform change in the $\eta(\lambda)$ spectrum and hence similar in magnitude changing of $J_{sc} \approx \pm 1$ mA/cm^{-2}.

Thus, the foregoing consideration of the physical processes in CdS/CdTe heterostructure can be used to develop recommendations for the proper optimization of the parameters (purposefully and with sufficient accuracy) for improving the photoelectric conversion efficiency in thin-film CdS/CdTe solar cells.

Conclusions

The results of the calculations performed in this study and their analysis allow us to formulate some important practical conclusions and recommendations concerning the optical and recombination losses in CdS/CdTe thin-film solar cell.

Reflection losses are about 8 % over a wide spectral range, and only in the region $\lambda < 400$ nm they increase to 10–11 %. Certainty, reflection losses can be reduced by ~4 % using antireflection coating on the front surface of the glass sheet. Reflection losses in solar cells with ITO and SnO_2:F transparent electrodes differ very little.

Due to *absorption* in the ITO layer of 200 nm thickness, the short-circuit current J_{sc} decreases by 12 %; however, the decrease can be 25–26 % when the thickness of ITO layer is 500 nm. In contrast, absorption in the SnO_2:F layer is much lower, causing only a decrease in J_{sc} of about 2 % at a thickness of 200 nm, and 4–5 % at a thickness of 500 nm. Decrease in J_{sc} caused by absorption in the CdS layer with a thickness of 50 nm is about 10 % and further increase of 5–6 % occurs at a thickness of 100 nm.

The calculations of *absorptivity* taking into account the power spectral distribution of the AM 1.5 solar spectrum and the absorption coefficient of CdTe show that in CdTe the total absorptivity (>99.9 %) of photons with energies $h\nu > E_g$ occurs at a layer thickness more than 20–30 μm. For the CdTe layer thickness of 2–3 μm the absorptivity of photons is about 97 % and reduces to ~93 % at a layer thickness of 1 μm (with no reflection from the rear surface).

Recombination at the front surface of the CdTe layer and *in the SCR* strongly depends on the width of the space charge region (SCR) W, which is determined by the concentration of uncompensated impurities (defects) in CdTe. The dependence of J_{sc} on W follows *a curve with a maximum* at $W = 0.4$–0.8 μm. The losses due to recombination at the front surface layer and in the SCR with its width of 0.3 μm amount to 2.8 %. When $W > 0.3$–0.4 μm the recombination losses in the SCR become *significant*, especially at short lifetimes of charge carriers.

The photoelectric losses outside the SCR caused by thinning the CdTe layer amount to about 1 % if the layer thickness d is 2–3 μm, but increases up to 12–15 %, when $d = 0.5$ μm. The losses due to lack of absorption capacity for these values of d are 2 and 7 %, respectively. The losses due to recombination at the rear surface of the CdTe layer *with no reflection* from the rear surface do not exceed 4–5 %.

From the viewpoint of reducing the *losses of all types*, the width of the SCR of about 0.3 μm is optimal, which corresponds to the concentration of uncompensated acceptors in CdTe close to 10^{16} cm^{-3}. When the thickness of CdTe layer is 2–3 μm, the total losses are about 5 % greater than in a layer of thickness $d > 5$ μm. Thinning of the CdTe layer to 0.5 μm decreases J_{sc} up to ~20 %.

Comparison of the calculation results with the measured quantum efficiency spectra can determine quite accurately the important parameters of solar cells such as the concentration of uncompensated acceptors, the electron lifetime in the CdTe absorber, and the thicknesses of CdTe and CdS layers. Such a comparison allows also formulating the recommendations for corrections of the parameters to increase the efficiency of photoelectric conversion in the device.

Acknowledgments The study is supported by the State Foundation for Fundamental Investigations of Ukraine within the Agreement Φ40.7/014. Obtaining the results presented in this paper would have been impossible without the participation of V.M. Sklyarchuk, V.Ya. Roshko, X. Mathew in the study.

References

1. EPIA Global market outlook for photovoltaics until 2015. European Photovoltaic Industry Association. (2011)
2. EUR 24344 PV Status Report 2010. Institute for Energy. Joint Research Centre. European Commission. (2010)
3. Jager-Waldau, A.: Research, solar cell production and market implementation of photovoltaics. Office for Official Publications of the European Union, Luxembourg (2010)
4. Szlufcik, J., Sivoththaman, S., Nijs, J.F., Mertens, R.P., Overstraeten, R.V.: Low cost industrial manufacture of crystalline

silicon solar cells. In: Markvart, T., Castaner, L. (eds.) Practical Handbook of Photovoltaics: Fundamentals and Applications, pp. 155–184. Elsevier, New York (2003)
5. Ferrazza, F.: Silicon: manufacture and properties. In: Markvart, T., Castaner, L. (eds.) Practical Handbook of Photovoltaics: Fundamentals and Applications, pp. 137–154. Elsevier, New York (2003)
6. Von Roedern, B.: Thin-film PV module review: changing contribution of PV module technologies for meeting volume and product needs. Renew Energy Focus **7**, 34–39 (2006)
7. Hegedus, S., Luque, A.: Achievements and challenges of solar electricity from photovoltaics. In: Luque, A., Hegedus, S. (eds.) Handbook of Photovoltaic Science and Engineering, 2nd edn, pp. 2–38. Wiley, New York (2011)
8. Goetzberger, A., Hebling, C., Schock, H.-W.: Photovoltaic materials, history, status and outlook. Mater Sci Eng **R40**, 1–46 (2003)
9. Romeo, A., Terheggen, M., Abou-Ras, D., Bőtzner, D.L., Haug, F.-J., Kőlin, M., Rudmann, D., Tiwari, A.N.: Development of thin-film Cu(In, Ga)Se2 and CdTe solar cells. Prog. Photovolt: Res. Appl. **12**, 93–111 (2004)
10. CdTe photovoltaic solar-cell efficiency record of 17.3% set by First Solar. Accessed 29 July 2012
11. First solar sets another world record for CdTe solar PV efficiency. Accessed 29 July 2012
12. Britt, J., Ferekides, C.: Thin-film CdS/CdTe solar cell with 15.8% efficiency. Appl. Phys. Lett. **62**, 2851–2853 (1993)
13. Meyers, P.V., Albright, S.P.: Photovoltaic materials, history, status and outlook. Prog. Photovolt.: Res. Appl. **8**, 161–168 (2000)
14. Bonnet, D.: CdTe thin-film PV modules. In: Makkvart, T., Castaner, L. (eds.) Practical Handbook of Photovoltaic: Fundamentals and Applications, pp. 333–366. Elsevier, Oxford (2003)
15. Kosyachenko, L.A., Grushko, E.V., Motushchuk, V.V.: Recombination losses in thin-film CdS/CdTe photovoltaic devices. Sol. Energy Mater. Sol. Cells **90**, 2201–2212 (2006)
16. Kosyachenko, L.A., Savchuk, A.I., Grushko, E.V.: Dependence of the efficiency of a CdS/CdTe solar cell on the absorbing layer's thickness. Semiconductors **43**(8), 1023–1027 (2009)
17. Kosyachenko, Grushko, E.V., Mathew, X.: Quantitative assessment of optical losses in thin-film CdS/CdTe solar cells. Sol. Energy Mater. Sol. Cells **96**(1), 231–237 (2012)
18. Kosyachenko, L.A., Grushko, E.V., Mikityuk, T.I.: Absorptivity of semiconductors used in the production of solar cell panels. Semiconductors **46**(4), 466–470 (2012)
19. Synowicki, R.A.: Spectroscopic ellipsometry characterization of indium tin oxide film microstructure and optical constants. Thin Solid Films **313–314**, 394–397 (1998)
20. Rose, D.H., Hasoon, F.S., Dhere, R.G., Albin, D.S., Ribelin, R.M., Li, X.S., Mahathongdy, Y., Gessert, T.A., Sheldon, P.: Fabrication procedures and process sensitivities for CdS/CdTe solar cells. Prog. Photovolt. Res. Appl. **7**, 331–340 (1999)
21. Archbold, M.D., Halliday, D.P.: Novel tubular geometry CdTe/CdS devices. In: Record of the 33rd IEEE Photovoltaic Specialists Conference 2008, pp. 1–4. ISSN:0160-8371 (2009)
22. Moss, T.S., Burrel, G.J., Ellis, D.: Semiconductor Opto-Electronics. Butterworth Publishers, London (1973)
23. Kasap, S.O.: Optoelectronics and Photonics: Principles and Practice. Prentice Hall, New Jersey (2000)
24. Fred Schubert, E.F.: Educational resources. Refractive index and extinction coefficient of materials. Rensselaer Polytechnic Institute. NY USA. (2007). Accessed 29 July 2012
25. RefractiveIndex.INFO. Accessed 29 July 2012
26. Compaan, A., Collins, R., Karpov, V., Giolando, D.: Fabrication and Physics of CdTe Devices by Sputtering. Annual Technical Report. 1 June 2006 to 31 May 2007. Contract No. RXL-5-44205-01. The University of Toledo, pp 1–87
27. Ninomiya, S., Adachi, S.: Optical properties of wurtzite CdS. J. Appl. Phys. **78**, 1183–1190 (1995)
28. Paulson, P.D., Mathew, X.: Spectroscopic ellipsometry investigation of optical and interface properties of CdTe films deposited on metal foils. Sol. Energy Mater. Sol. Cells **82**, 279–290 (2004)
29. Bonnet, D.: Cadmium telluride solar cells. In: Archer, M.D., Hill, R. (eds.) Clean Electricity from Photovoltaics, pp. 245–275. Imperial College Press, New York (2001)
30. Born, M., Wolf, E., Bhatia, A.B.: Principles of Optics, 7th edn. Cambridge University Press, Cambridge (1999)
31. Reference solar spectral irradiance at the ground at different receiving conditions. Standard of International Organization for Standardization ISO 9845-1:1992
32. Wu, X.: High-efficiency polycrystalline CdTe thin-film solar cells. Sol. Energy **77**, 803–814 (2004)
33. Ferekides, C.S., Balasubramanian, U., Mamazza, R., Viswanathan, V., Zhao, H., Morel, D.L.: CdTe thin film solar cells: device and technology issues. Sol. Energy **77**, 823–830 (2004)
34. Jones, E.W., Barrioz, V., Irvine, S.J.C., Lamb, D.: Towards ultra-thin CdTe solar cells using MOCVD. Thin Solid Films **517**, 2226–2230 (2009)
35. Stanley Glass Co., Ltd.
36. Pankove, J.I.: Optical Processes in Semiconductors. Prentice-Hall, New Jersey (1971)
37. Amin, N., Sopian, K., Konagai, M.: Numerical modeling of CdS/CdTe and CdS/CdTe/ZnTe solar cells as a function of CdTe thickness. Sol. Energy Mater. Sol. Cells **91**, 1202–1208 (2007)
38. Gupta, A., Parikh, V., Compaan, A.D.: High efficiency ultra-thin sputtered CdTe solar cells. Sol. Energy Mater. Sol. Cells **90**, 2263–2271 (2006)
39. Marsillac, S., Parikh, V.Y., Compaan, A.D.: Ultra-thin bifacial CdTe solar cell. Sol. Energy Mater. Sol. Cells **91**, 1398–1402 (2007)
40. Amin, N., Matin, M.A., Aliyu, M.M., Alghoul, M.A., Karim, M.R., Sopian, K. Prospects of back surface field effect in ultra-thin high-efficiency CdS/CdTe solar cells from numerical modeling. Int. J. Photoenergy. Article ID 578580 (2010) (8 pages)
41. Toshifumi, T., Adachi, S., Nakanishi, H., Ohtsuka, K.: Optical constants of $Zn_{1-x}Cd_xTe$ ternary alloys: experiment and modeling. J. Appl. Phys. **32**, 3496–3501 (1993)
42. Kosyachenko, L.A., Sklyarchuk, V.M., Sklyarchuk, O.V., Maslyanchuk, O.L.: Band gap of CdTe and $Cd_{0.9}Zn_{0.1}Te$ crystals. Semiconductors **45**, 1247–1250 (2011)
43. McCandless, B.E., Sites, J.R.: Cadmium telluride solar cells. In: Luque, A., Hegedus, S. (eds.) Handbook of Photovoltaic Science and Engineering, pp. 617–662. John Wiley, England (2003)
44. Kosyachenko, L.A.: Problems of efficiency of photoelectric conversion in thin-film CdS/CdTe solar cells. Semiconductors **40**, 710–727 (2006)
45. Kosyachenko, L.A., Savchuk, A.I., Grushko, E.V.: Dependence of efficiency of thin-film CdS/CdTe solar cell on parameters of absorber layer and barrier structure. Thin Solid Films **517**, 2386–2391 (2009)
46. Sze, S.M., Ng, K.K.: Physics of Semiconductor Devices, 3rd edn. Wiley-Interscience, New Jersey (2006)
47. Roussillon, Y., Giolando, D.M., Karpov, V.G., Shvydka, D., Compaan, A.D.: Reach-through band bending in semiconductor thin films. Appl. Phys. Lett. **85**, 3617 (2004)

48. Fahrenbruch, A.L.: The relationship of CdS/CdTe cell band profiles to J-V characteristics and bias-dependent quantum efficiency. In: Conferencse Record of the IEEE 4th World Conference on Photovoltaic Energy Conversion, Waikoloa, USA, pp. 376–379 (2006)
49. Sites, J., Pan, J.: Strategies to increase CdTe solar-cell voltage. Thin Solid Films **515**, 6099–6102 (2007)
50. Gessert, T.A., Metzger, W.K., Dippo, P., Asher, S.E., Dhere, R.G., Young, M.R.: Dependence of carrier lifetime on Cu-contacting temperature and ZnTe:Cu thickness in CdS/CdTe thin film solar cells. Thin Solid Films **517**, 2370–2373 (2009)
51. Gloeckler, M., Fahrenbruch, A.L., Sites, J.R.: Numerical modeling of CIGS and CdTe solar cells: setting the baseline. In: Proceedings of 3rd World Conference on Photovoltaic Energy Conversion, Osaka, Japan, 18 May 2003, vol. 1, pp. 491–494 (2003)
52. Hossain, M.S., Amin, N., Matin, M.A., Aliyu, M.M., Razykov, T., Sopian, K.: A numerical study on the prospects of high efficiency ultra thin $Zn_xCd_{1-x}S$/CdTe solar cell. Chalcogenide Lett. **8**, 263–272 (2011)
53. Acrorad Co. Ltd. Company. Comparison of semiconductor detectors. Accessed 29 July 2012
54. Hecht, K.: Zum Mechanismus des lichtelektrischen Primärstromes in isolierenden Kristallen. Zeits. Phys. **77**, 235–243 (1932)

19

Feasibility of an oxygen-getter with nickel electrodes in alkaline electrolysers

Daniel Symes · Connie Taylor-Cox · Leighton Holyfield · Bushra Al-Duri · Aman Dhir

Abstract Alkaline electrolysis is the long-established technology for water splitting to produce hydrogen and has been industrially used since the nineteenth century. The most common materials used for the electrodes are nickel and derivatives of nickel (e.g. Raney nickel). Nickel represents a cost-effective electrode material due to its low cost (compared to platinum group metals), good electrical conductivity and exhibits good resistance to corrosive solutions. The steady degradation of the nickel electrodes over time is known as a result of oxide layer formation on the electrode surface. Reducing oxide layer growth on the electrode surface will increase the efficiency and lifetime of the electrolyser. Titanium has a higher affinity to oxygen than nickel so has been introduced to the electrolyser as a sacrificial metal to reduce oxide layer formation on the nickel. Two identical electrolysers were tested with one difference: Cell B had titanium chips present in the electrolyte solution, whilst Cell A did not have titanium present. SEM results show a reduction of 16 % in the thickness of the Cell B oxide layer on nickel compared to the Cell A nickel, which is supported by the large increase in oxide layer build-up on the titanium in Cell B. EDX on the same samples showed on average a 59 % decrease in oxygen on the Cell B nickel compared to Cell A. XPS surface analysis of the same samples showed a 17 % decrease in the oxygen on Cell B nickel. These results support the hypothesis that adding titanium to an alkaline electrolyser system with nickel electrodes can reduce the oxide layer formation on the nickel.

Keywords Oxygen-getter · Hydrogen production · Water electrolysis · Alkaline electrolysis · Electrode degradation · Titanium · Electrolyser

D. Symes (✉) · C. Taylor-Cox · L. Holyfield · B. Al-Duri · A. Dhir
Centre for Hydrogen and Fuel Cell Research, School of Chemical Engineering, University of Birmingham, Edgbaston, Birmingham, West Midlands B15 2TT, UK
e-mail: dxs654@bham.ac.uk

Introduction

Hydrogen is most commonly produced from fossil fuel sources (approx. 95 %), which is neither sustainable (depletion of fossil fuel resources) and environmentally friendly (produces greenhouse gases as side products) [1–4]. Water electrolysis represents only 4 % of current hydrogen production, but is a growing area of research and the beginning of commercialisation of water electrolysers has started. The high cost can be attributed to electrical energy prices and the use of platinum group metals (PGMs) as catalysts to improve the efficiency for hydrogen generation [1, 3–6].

A high electrical potential is required to split pure water directly, so an electrolyte is added to reduce the resistance in the system. The most commonly used is a liquid electrolyte (NaOH, KOH), and there is increasing research on using solid electrolyte in the form of a polymer membrane. This only requires pure water, as well as an electrical energy input [1, 3, 7–11].

The use of caustic solution (typically 25–30 weight %) in alkaline electrolysers requires the electrodes to exhibit good electrochemical conductivity, high mechanical strength and a strong resistance to corrosion from the caustic environment. Nickel, or a derivative of nickel, is the most commonly used electrode material in commercial alkaline electrolysers, due to its high electrical conductivity and its robustness to prolonged exposure in caustic alkaline solution. Although nickel exhibits good electrode lifetime, this duration can be improved and a method for increasing the lifetime of a nickel electrode is investigated herein [3, 6, 9–12].

The mechanisms and characteristics of electrode degradation are a vastly researched area, especially in the field of alkaline electrolysis [8–16]. These mostly involve the reduction in electrolyte concentration and the use of electrocatalysts on the electrode surface to keep the high level electrochemical performance, whilst limiting electrode deactivation [15–19].

One potential method of degradation reduction which has not been utilised in the field of electrolysis is the use of an 'oxygen-getter'. An oxygen-getter is titanium or an alloy of titanium in the form of either sponge, chips or bar stock. The purpose of the oxygen-getter is to provide cathodic protection to the electrode material from the corrosive medium during the electrolysis process [20].

Titanium exhibits a strong affinity to oxygen compared to nickel as it has a much lower electronegativity χ; hence, it attracts electrons more freely which is why titanium has been chosen over other metals and it will protect the electrodes from cathodic attack by oxygen molecules in the solution. This enables the concentration and temperature of the aqueous solution to be increased, which in turn will increase the rate of electrolysis, without corroding the electrode itself [20].

In this case, titanium chips (10 mm $\times$ 10 mm $\times$ 1 mm) must be added (240 cm^2 total surface area) to ensure that the electrode material is protected from cathodic corrosion. The titanium chips will be held at the bottom of the reaction vessel under a nickel mesh to ensure they do not interact with the electrochemical reaction. The specified dimensions were chosen for this feasibility study to maximise the surface area in the small volume available at the underside of the vessel, and if proven, further research could be conducted on varying geometries and loadings of titanium in the system. A larger size of titanium chips would limit the surface area available in the vessel and the subsequent electronegativity of the titanium. It is vital that the titanium does not participate in the electrochemical reaction and is only present to attract O^{2-} ions away from the nickel and to the titanium.

Experimental materials and procedure

Two electrolysers (Cells A and B) were built in a Perspex casing consisting of six nickel (201) electrodes (100 mm $\times$ 100 mm $\times$ 1 mm) (total electrode surface area 1,206 cm^2) connected in parallel with a fixed separation distance of 6 mm. There is a clearance of 20 mm from the bottom of the Perspex vessel to the bottom of the nickel electrodes. Cell A is the above-mentioned electrolyser stack without titanium [Grade I (ASTM B265)] chips, whilst Cell B has the titanium chips in the bottom of the vessel. A schematic of the electrolyser is shown in Fig. 1.

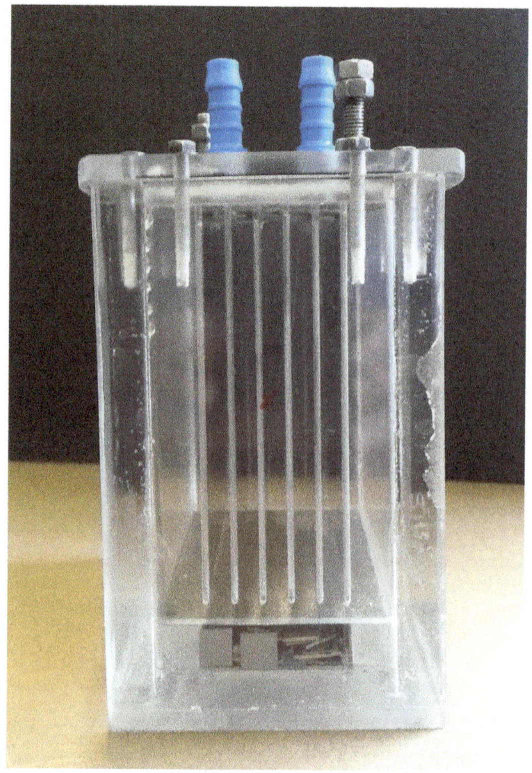

Fig. 1 Custom built alkaline electrolyser

The potassium hydroxide ($\geq$85 % KOH basis, pellets) was mixed with a specific volume of Type II Deionised Water [Millipore Distilled Water Purification System (Rios Doi3)], to produce a predetermined concentration of KOH solution. The potential was controlled, and the subsequent current measured using a direct current power supply (Thurlby Thandar Instruments QPX1200SP model).

Following the electrolysis reaction, the electrodes were prepared for SEM (JEOL JSM-6610LV), EDX (Oxford Instruments Inca 300 EDS system) and XPS (Thermo Scientific K-Alpha) analysis, were washed thoroughly with Type I (18 MΩ) water to remove any residual KOH solution and then subsequently dried.

Experimental results

This section is divided into two sections for determine the performance variation between Cell A and Cell B. The first section compares the differences in electrochemical performance, and the second section compares the surface characteristics between the two electrolysers.

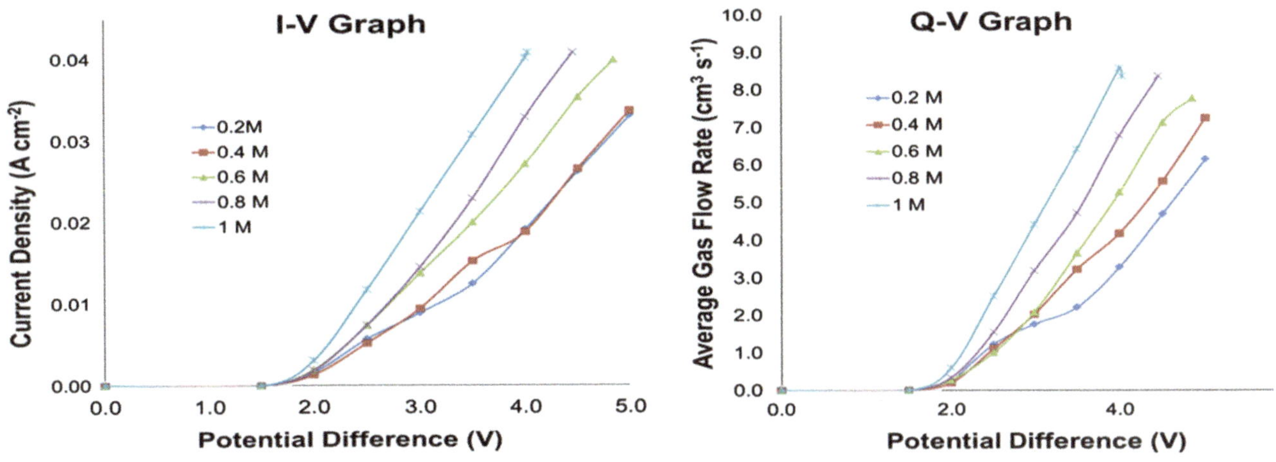

Fig. 2 IV characterisation and gas productivity curves

Electrochemical analysis

Cells A and B underwent electrochemical analysis to characterise current flow, gas productivity and long term degradation testing to quantify loss of electrochemical performance from being subject to a caustic alkaline environment on a 2-week basis.

IV analysis and gas productivity curves

The voltage was varied and the current flow recorded to characterise the electrochemical performance of the electrolyser. This analysis shows the resistances, known as overpotentials, in the electrolyser. The standard potential of electrolysis (1.23 V) and the thermo-neutral voltage (1.48 V) must be overcome for a current to flow between the electrodes. In reality overpotentials such as activation (the energy needed to overcome the activation energy needed for the reaction to occur), ohmic (energy needed to overcome the resistance to ion transfer in the electrolyte) and bubble (energy needed to overcome the reduction in nucleation sites on the electrode due to bubble evolution), the exist and addition energy is required in the system to overcome these.

This analysis was completed for various concentrations (0.2–1 M) of KOH solution. Figure 2 shows the results for both Cells A and B which are the same, thus proving that the two electrolysers are identical in electrochemical performance and suffer the same overpotential.

Using the same methodology for 0.2–1.0 M KOH solution, the gas production rates were measured to quantify the enhancement in the conductivity of the electrolyte solution and the corresponding production of hydrogen and oxygen (HHO) gas.

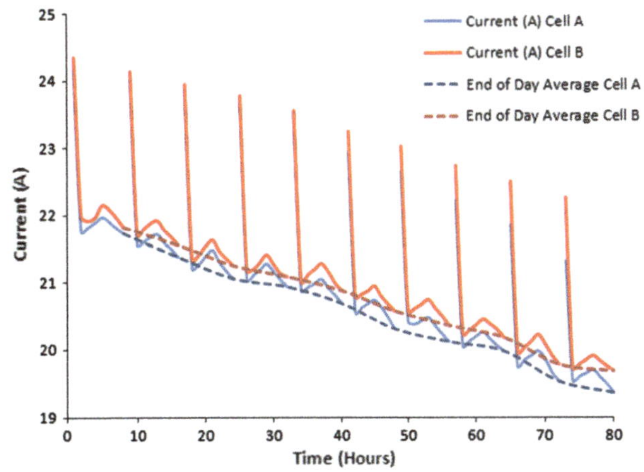

Fig. 3 Two-week electrochemical testing

Degradation testing

The aim was to test both cells for a total of 10 days (2 weeks), to mimic real-world utilisation of an electrolyser (8 h × 5 days a week). Figure 3 shows the constant voltage performance for both Cells for the 10-day duration. Note that the electrolyte solution was not changed throughout testing of each cell and were left in solution and in standby mode for 2 days between Days 5 and 6 (weekend).

Through 8 h of testing, the initial current spike quickly decreases after a few minutes before increasing again through the rise in temperature of the electrolyte as a result of the electrochemical reaction. The current reaches a constant when the temperature of the electrolyte solution reaches its maximum (dependant on the input potential), and then the current subsequently decreases, whilst the temperature remains constant, due to oxide layer formation on the electrode surface.

In Fig. 3, the general trend shows that the current reduces over time and is an expected result due to the degradation of the nickel electrodes. Throughout the testing, Cell B, the modified cell, shows degradation to a lesser extent than Cell A and the current is consistently higher. This helps to support our hypothesis that the titanium oxygen-getter concept could be used in this application to help protect the nickel electrodes due to possessing a higher affinity for oxygen from degradation.

The biggest difference that the titanium chips, in Cell B, have appeared to make is between Days 5 and 6 where the electrodes were left in the KOH solution over the weekend to further simulate industrial use. After the 2 days of no use, the Cell B current was 0.3 A higher than Cell A. This further supports the hypothesis showing that titanium has protected the electrodes even when there is no current flow.

Although in reality electrolysers operate for much longer than 2 weeks, the variation in electrochemical performance between Cell A and B is evident after 2 weeks operation. Whilst the experimental duration could be extended, the author feels that the feasibility of the principle which is being investigated here has been proven. Further study will analyse longer durations of electrolyser operation.

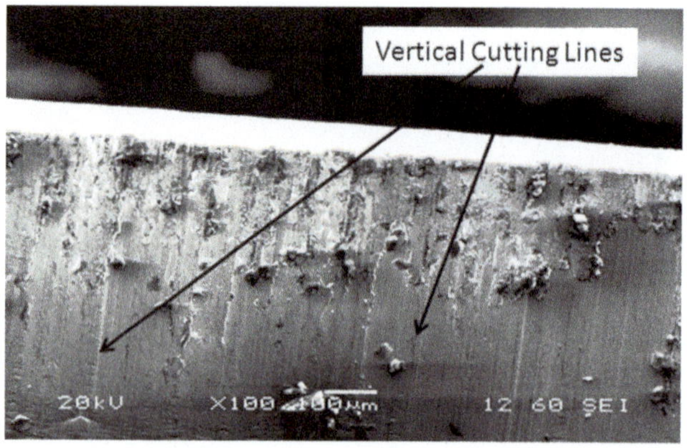

Fig. 4 New nickel cross-sectional view (×100 magnification)

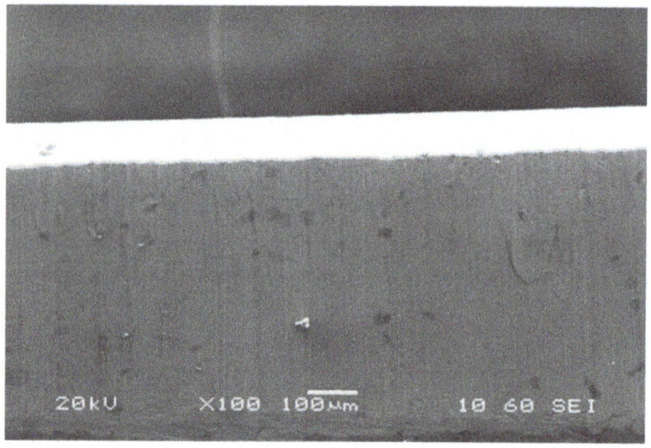

Fig. 5 Nickel (Cell A) cross-sectional view (×100 magnification)

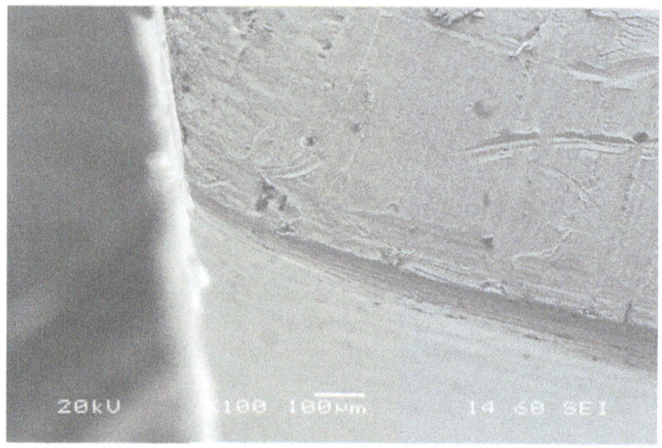

Fig. 6 Nickel (Cell B) cross-sectional view (×100 magnification)

Surface analysis

Analysis of the nickel electrode surface is important to quantify oxide layer formation on the surface. The oxide layer coverage limits the number of nucleation sites available for bubble evolution to occur, thus reducing the efficiency of the electrolyser. The nickel and titanium from the electrolysers will undergo scanning electron microscopy (SEM) and energy dispersive X-ray spectroscopy to analyse any potential reduction in oxide layer formation as a result of introducing titanium to the electrolysis environment.

The metallic samples were split into five categories: (1) new nickel, (2) nickel from Cell A, (3) nickel from Cell B, (4) new titanium and (5) titanium from Cell B. These were each analysed by the before mentioned techniques.

SEM—nickel (1, 2 and 3)

The SEM images of the new nickel were taken to provide a control sample against which the used electrode surfaces could be compared to see how the action of the electrolysis reaction affects the nickel surface.

The vertical line patterning is believed to be caused by the cutting of the sample. The thick bright strip on the edge of the sample in Fig. 4 is believed to be the side of the sample (caused by the cutting of the large plate into chips) reflecting light, rather than an indication of the presence of any foreign substance on the surface (to be confirmed by EDX).

On the cross-sectional images of the nickel used in Cell A give very clear evidence of oxide layer formation. From observing the cross-sectional view of the sample, the oxide layer build-up is visible. Figure 5 shows a striking visual contrast between the nickel in the body of the sample and the edge of the sample.

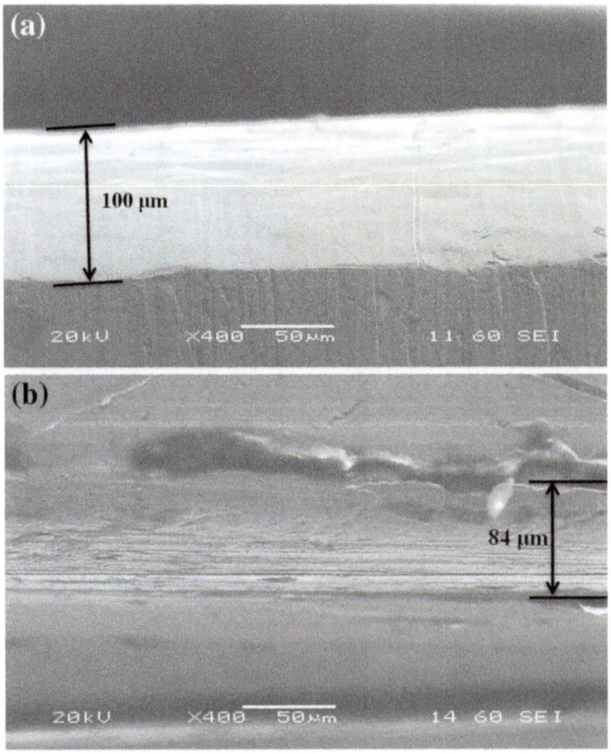

Fig. 7 **a** (Cell A) High magnification nickel cross-sectional view. **b** (Cell B) High magnification nickel cross-sectional view

The cross-sectional view of Cell B nickel (Fig. 6) shows a similar contrast between the nickel body and the edge of the sample to Cell A. Whilst in Cell A, the potential 'oxide layer' and the nickel body were different colours, in this image that is not the case, and the image is far more homogenous (Fig. 6). Despite this, it is believed that an oxide layer is evident in this image due to the strong horizontal marking on the lower side of the metal edge in both images consistent with Fig. 5 and different from the new nickel sample (Fig. 4).

There is also horizontal marking within the white region of Fig. 6, suggesting that it is laid on not all at once, but through a series of layers, very likely induced by the repeated exposure to the electrolysis reaction. The thickness of the oxide layers for Cell A and Cell B can be found in Fig. 7a, b.

The irregular build-ups seen in the middle of Fig. 7a indicate that material has been deposited, rather than the material being of homogeneous construction. In Fig. 7b, the oxide layer starts on the faint horizontal line between the anomalous deposits, which is highlighted by the thick black line in Fig. 7b. Using these markings, the oxide layer thickness measured from Fig. 7b is 84 µm. This indicates a 16 % reduction in oxide layer thickness from Cell A.

Titanium—(4 and 5) SEM

Similarly to the nickel, a new titanium sample was imaged to have a control sample against which the used titanium could be compared. The cross-sectional view of the titanium control sample (Fig. 8) shows very clearly where the sample has been cut, exhibiting a smooth edge on the outer side, but most of the edge is made up of small peaks where the titanium has deformed during the cutting process. In terms of any potential oxide layer, there is no marking or colour change to potentially indicate that the surface of the new titanium has been compromised.

Fig. 8 New titanium cross-sectional view (×100 magnification)

Fig. 9 Titanium (Cell B) cross-sectional view (×100 magnification)

Evidence of oxide deposition is shown in the cross-sectional images of Cell B titanium on the lower magnification image (Fig. 9); there is a clear colour change between the titanium and the oxide layer, as well as there being very strong horizontal marking in the white oxide layer; this is despite the fact that there is faint vertical marking on the titanium.

These effects are more noticeable in the higher magnification image (Fig. 10). The much stronger horizontal marking in the oxide layer seem to show titanium's greater affinity for oxygen, as does the oxide layer thickness,

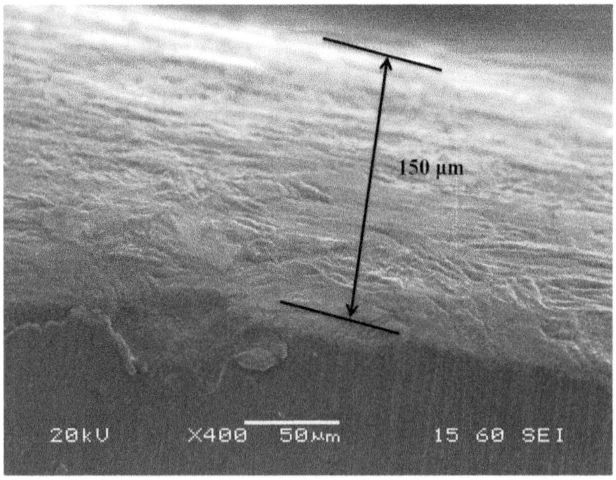

Fig. 10 Titanium (Cell B) cross-sectional view (×400 magnification)

Table 1 EDX surface composition on nickel samples

Sample	Atomic%		
	Nickel	Titanium	Oxygen
New nickel	0.40	0.00	18.2
Cell A nickel	0.21	0.00	31.1
Cell B nickel	0.27	0.01	25.9

which from Fig. 10 is shown to be 150 µm, nearly twice that of the corresponding nickel sample.

EDX—nickel (1, 2 and 3)

For all nickel samples, analysis carried out at 500× magnification focused on the planar surface of each sample. Damage to the surface from the electrolysis process may have allowed other metals to be present on the surface and any deposits found (predominantly after use) would be investigated to see if this was a large increased build-up of oxide layer. The atomic% for nickel, titanium and oxygen for the nickel samples is shown in Table 1.

When comparing the flat surface spectra data for the control sample, Cell A and B, the main differences are the levels of nickel and oxygen. A 12.9 atomic% increase in oxygen content appears on Cell A nickel compared to the unused sample, which coincides with a 0.19 atomic% decrease in nickel content between the two samples.

There is a 5.2 atomic% decrease in oxygen content on the nickel surface of Cell B compared to Cell A showing that the presence of the titanium has reduced the quantity of oxide layer formation, helping to reduce the rate of nickel electrode degradation. This is further evident with the increase of 0.06 atomic% nickel content on the Cell B nickel relative to Cell A nickel.

Table 2 EDX surface composition on titanium samples

Sample	Atomic%		
	Nickel	Titanium	Oxygen
New titanium	0.00	0.22	27.5
Cell B titanium	0.01	0.19	32 2

Table 3 XPS Surface analysis of nickel and titanium samples

Sample	Atomic%		
	Nickel	Titanium	Oxygen
New nickel	0.36	0.01	15.0
Cell A nickel	0.19	0.02	29.5
Cell B nickel	0.26	0.02	24.5
New titanium	0.01	0.12	25.0
Cell B titanium	0.02	0.09	29.5

EDX—titanium (4 and 5)

The EDX data for new and used titanium is shown in Table 2.

Cell B titanium shows an increase in oxygen content when compared to the control sample (4.7 atomic%). This shows that an oxygen layer is present, which is consistent with titanium having a higher affinity for oxygen than nickel. Analysis on the used titanium content of the samples shows a 0.3 atomic% decrease, consistent with the increased oxygen content on the titanium Cell B sample.

X-ray photoelectron spectroscopy (XPS)

To support the EDX evidence provided, the samples (1–5) also underwent XPS analysis for further verification of the surface composition. Both nickel (1–3) and titanium (4–5) samples were analysed for nickel, titanium and oxygen content on the samples. The peaks were then integrated to determine the surface elemental composition in atomic%. This is shown in Table 3.

Table 3 shows a 94 % increase in oxygen concentration on the nickel surface from new nickel to nickel in Cell A and then a 17 % decrease in when compared to nickel from Cell B. This is consistent with the analysis shown in the SEM and EDX analysis.

The metallic composition shows a 43 % decrease in nickel content on the Cell A nickel compared to the new nickel samples and a 25 % increase in nickel content of Cell B compared to Cell A nickel. The titanium content on the titanium samples decreases by 23 % on the Cell B titanium when compared to the new titanium sample, consistent with the hypothesis of this paper.

Conclusions

The research has analysed the effect of introducing titanium to the alkaline electrolysis system as a sacrificial metal. The electrochemical performance of Cell A decreased at a faster rate than that of Cell B, specifically after a 2-day stationary period. The only variable between the cells is the titanium; therefore, this is the determining factor in the improved electrochemical performance of Cell B when compared with that of Cell A.

The conclusions drawn from the electrochemical analysis is backed up with the findings from the surface analysis. This also points towards a decrease in oxide layer formation on the nickel electrode in Cell B when compared to that in Cell A. The elemental composition of the surface of nickel and titanium from Cells A and B has shown an increased presence of oxygen on the electrode surface. The thickness in oxide layer on the nickel from Cell B is 17 % less than in Cell A. This limits the number nucleation sites on the electrode surface and links to reduced electrochemical performance in Cell A than in Cell B.

It is concluded that the presence of titanium as a sacrificial metal in an alkaline electrolysis system with nickel electrodes increases the electrical efficiency and decreases the rate of electrode degradation over 2 weeks of operation.

Future work in this field will investigate the use of different titanium geometries, for example sponges, plates and spheres, and the effect that varying titanium loadings has on the thickness of oxide layer formation on nickel electrodes in an alkaline electrolyser.

Acknowledgments I would like to acknowledge supervisors and colleagues at the Centre for Hydrogen and Fuel Cell Research, University Birmingham, UK. I would also like to give thanks the Engineering and Physical Sciences Research Council for funding my Ph.D. as part of the Doctoral Training Centre in Hydrogen, Fuel Cells and their Applications. I also express my gratitude to the National EPSRC XPS Users Service (NEXUS) for their assistance with XPS analysis completed in this research.

References

1. Bičáková, O., Straka, P.: Production of hydrogen from renewable sources and its effectiveness. Int. J. Hydrogen Energy **37**(16), 11563–11578 (2012)
2. Holladay, J., Hu, J., King, D., Wang, Y.: An overview of hydrogen production technologies. Catal. Today **130**, 244–260 (2009)
3. Lu, P., Srinivasan, S.: Advances of water electrolysis technology with emphasis on use of the solid polymer electrolyte. J. Appl. Electrochem. **9**(3), 269–283 (1979)
4. Marini, S., Salvi, P., Nelli, P., Pesenti, R., Villa, M., Berrettoni, M., Zangari, G., Kiros, Y.: Advanced alkaline water electrolysis. Electrochim. Acta **82**, 384–391 (2012)
5. Zoulias, E., Varkaraki, E., Lymberopoulos, N., Christodoulou, C., Karagiorgis, G.: A review on water electrolysis. TCJST **4**(2), 41–71 (2004)
6. Harrison, K., Levene, J.: Electrolysis of water. In: Rajeshwar, K., McConnell, R., Licht, S. (eds.) Solar Hydrogen Generation: Toward a Renewable Future, pp. 41–63. New York: Springer (2008)
7. Symes, D., Al-Duri, B., Dhir, A., Bujalski, W., Green, B., Shields, A., Lees, M.: Design for on-site hydrogen production for hydrogen fuel cell vehicle refuelling at University of Birmingham, U.K. Energy Proc **29**, 606–615 (2012)
8. Zeng, K., Zhang, D.: Recent progress in alkaline water electrolysis for hydrogen production and applications. Prog. Energy Combust. Sci. **36**(3), 307–326 (2010)
9. Nikolic, V., Tasic, G., Maksic, A., Saponjic, D., Miulovic, S., Kaninski, M.: Raising efficiency of hydrogen generation from alkaline water electrolysis—energy saving. Int. J. Hydrogen Energy **35**(22), 12369–12373 (2010)
10. Zhang, H., Lin, G., Chen, J.: Evaluation and calculation on the efficiency of a water electrolysis system for hydrogen production. Int. J. Hydrogen Energy **35**(20), 10851–10858 (2010)
11. Mazloomi, S., Sulaiman, N.: Influencing factors of water electrolysis electrical efficiency. Renew. Sustain. Energy Rev. **16**(6), 4257–4263 (2012)
12. Bird, R., Steward, W., Lightfoot, E.: Transport Phenomena, 2nd edn. New York: Wiley (2002)
13. Santos, D.M.F., Sequeira, C.A.C., Macciò, D., Saccone, A., Figueiredo, J.L.: Platinum-rare earth electrodes for hydrogen evolution in alkaline water electrolysis. Int. J. Hydrogen Energy **38**(8), 3137–3145 (2013)
14. Matsushima, H., Fukunaka, Y., Kuribayashi, K.: Water electrolysis under microgravity part II: description of gas bubble evolution phenomena. Electrochim. Acta **51**(20), 4190–4198 (2006)
15. Mauer, D., Kirk, D., Thorpe, S.: The role of iron in the prevention of nickel electrode deactivation in alkaline electrolysis. Electrochem. Acta **52**(11), 3503–3509 (2007)
16. Bocca, C., Barbucci, A., Cerisola, G.: The influence of surface finishing on the electrocatalytic properties of nickel for the oxygen evolution reaction (OER) in alkaline solution. Int. J. Hydrogen Energy **23**(4), 247–252 (1998)
17. Abouatallah, R., Kirk, D., Thorpe, S., Graydon, J.: Reactivation of nickel cathodes by dissolved vanadium species during hydrogen evolution in alkaline mixture. Electrochim. Acta **47**(4), 613–621 (2001)
18. Marceta Kaninski, M., Maksic, A., Stojic, D., Milljanic, Š.: Ionic activators in the electrolytic production of hydrogen—cost reduction-analysis of the cathode. J. Power Sources **131**(1–2), 107–111 (2004)
19. Eigeldinger, J., Vogt, H.: The bubble coverage of gas-evolving electrodes in flowing electrolyte. Electrochim. Acta **45**(27), 4449–4456 (2000)
20. Faison, J.A., Paton, N.E., Thornton, T.J.: Method for removing ceramic material from castings using caustic medium with oxygen getter, U.S. Patent 5,679,270, filed July 1, 1996, and issued October 21, 1997

20

Self-standing electrochemical double layer capacitors for operation in severe temperature conditions

C. Lecoeur · B. Daffos · R. Lin · L. Divay · P. Le Barny · M. Pham Thi · P.-L. Taberna · P. Simon

Abstract Safe and ecological electrochemical double layer capacitors (EDLCs) are nowadays needed as energy sources for wearable electronics working in hard temperature conditions. Here, we report a method based on the use of an organic solvent without any surfactant to synthesize a self-supported paper-like electrode composed solely by multiwall carbon nanotubes, named buckypapers. Using an eutectic mixture of ionic liquids (1:1 in molar ratio), *N*-methyl-*N*-propylpiperidinium bis(fluorosulfonyl)imide and *N*-butyl-*N*-methylpyrrolidinium bis(fluorosulfonyl) imide, as electrolyte, EDLCs based on buckypapers were assembled and tested up to 3 V. Ionic liquids are well known as green solvents for the environment and are safe for humans in their use thanks to their low vapor pressure. In addition to these problems of security solved, this electrolyte allows working in a wide temperature range, i.e. from −50 to +100 °C, delivering capacitances similar to those obtained in usual organic solvents.

Keywords Energy storage · Electrochemical double layer capacitor · Buckypapers · Carbon nanotube · Ionic liquid

C. Lecoeur · B. Daffos · R. Lin · P.-L. Taberna (✉) · P. Simon
Centre Interuniversitaire de Recherche et d'Ingénierie des Matériaux CIRIMAT, UMR CNRS 5085, Université Paul Sabatier, 118 route de Narbonne, 31062 Toulouse Cedex 09, France
e-mail: taberna@chimie.ups-tlse.fr

C. Lecoeur · B. Daffos · R. Lin · P.-L. Taberna · P. Simon
Réseau sur le Stockage Electrochimique de l'Energie (RS2E), FR CNRS 3459, Toulouse, France

L. Divay · P. Le Barny · M. P. Thi
Thales Research and Technology, Campus Polytechnique, 1 avenue Augustin Fresnel, 91767 Palaiseau Cedex, France

Introduction

Electrochemical double layer capacitors (EDLCs) fill the gap between conventional capacitors and batteries. They show higher energy densities but lower power densities than capacitors, while exhibiting higher power densities but lower energy densities than batteries [1]. EDLCs are used in a wide range of applications, alone or in combination with batteries [2–5]. With the development of portable and wearable electronic devices, energy storage systems (ESS) like batteries and EDLCs must become smaller, lighter, thinner and indeed more flexible while maintaining the same level of performance. One of the limitation in the use of ESS like EDLCs lies in the operating temperature range that is today limited between −25 and +70 °C for carbonate-based electrolytes and from −40 and +70 °C for acetonitrile-based solutions [6, 7]. Widening this temperature range would open the use of EDLCs in power electronics in severe environments such as one can find in aeronautics, space, automotive or military applications for outdoor operations [8, 9].

Recently, we showed that thanks to the combination of exohedral carbons offering a fully accessible surface area with a low melting point ionic liquids (ILs) eutectic mixture, an operation temperature range between −50 and +100 °C could be achieved [10]. However, the specific capacitance of the carbon nanotubes (CNTs) grown from an Al foil was low, at about 3 mF cm^{-2} of electrode, mainly because of the small mass loading (0.2 mg cm^{-2}).

The flexibility of CNTs buckypapers (BPs), widely studied at room temperature [11–14], confers an interesting feature for using these EDLCs in smart-textile applications [15, 16]. In the present paper, we report results on the design of self-supported electrodes for small-size EDLCs for operation within a large temperature range, based on the use of buckypapers (BPs).

Experimental

Buckypaper synthesis

Nanotube networks also called buckypaper are generally obtained by special infiltration of aqueous nanotube suspension using a surfactant [17, 18]. We have developed a method based on the use of an organic solvent without any surfactant. A mixture of 150 mg of carbon nanotubes (MWCNT >95 %, Aldrich) in 500 mL of *N*-methylpyrrolidine (NMP, 98 %, Alfa-Aesar) is sonicated for 3 h. The obtained suspension is centrifuged at 3,000 rpm for 20 min so that the MWCNT bundles are discarded. The suspension is then filtered off under a pressure of 3.5 bars, through a 0.45 μm polytetrafluoroethylene filter (PTFE, Sartorius Stedim). As pores of the filter are bigger than single nanotubes, only the network of nanotubes formed during the filtration is retained. This confers a real self-standing behavior to our BPs as the excess of MWCNT are thus removed. The BP thus formed is washed with acetone, removed from the supporting filter membrane and finally dried under vacuum at 150 °C overnight.

Buckypaper structural characterizations

Gas sorption isotherms were recorded using Tristar II 3020 Micromeritics equipment under N_2 gas atmosphere at 77 K. Pore size distribution and mean pore size were calculated using the Barrett, Joyner and Halenda (BJH) method. Calculations of the specific surface area (SSA) were achieved using the Brunauer–Emmett–Teller (BET) equation. SEM observations were made using a FEG-SEM (Jeol JSM 6700F).

Cell assembly and electrochemical characterizations

Cells were assembled in a glove box (under inert Ar atmosphere, 5.0 purity grade) with both water and oxygen content lower than 0.1 ppm. Two-electrode cells were assembled by inserting a 520-μm thick glass fiber separator (GF/B, Whatman) between the two electrodes, whose diameter were 9 mm and mass 5.6 mg, in a symmetric two-electrode Swagelok® configuration (Fig. 1). As no current collectors were used, BP electrodes were directly in contact with cylindrical stainless steel plungers (diameter = 12 mm). Two electrolytes were tested: (1) tetraethylammonium tetrafluoroborate (NEt_4BF_4, 99 %, Acros Organics) 1 M dissoluted in acetonitrile (AN, 99.9 % extra dry Acroseal®, Acros Organics) and (2) an eutectic mixture of ionic liquids (1:1 in molar ratio), *N*-methyl-*N*-propylpiperidinium bis(fluorosulfonyl)imide ($PIP_{13}FSI$, 99.9 %, Solvionic) and *N*-butyl-*N*-methylpyrrolidinium bis(fluorosulfonyl)imide ($PYR_{14}FSI$, 99.9 %, Solvionic) [10].

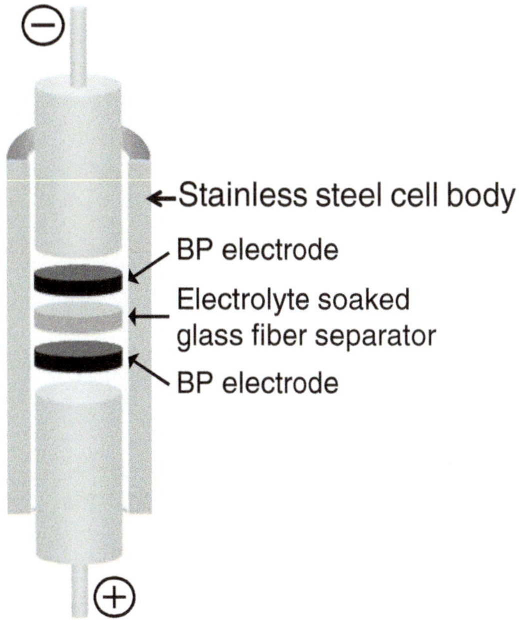

Fig. 1 Schematic of a symmetric two-electrode Swagelok® (stainless steel plungers) electrochemical double layer capacitor: the electrolyte (NEt_4BF_4 1 M in AN or ILs eutectic mixture) soaked glass fiber separator (GF/B, Whatman) is sandwiched between to BP electrodes, whose diameter is 9 mm and mass is 5.6 mg

Electrochemical measurements were done with a Biologic VMP3 potentiostat for room temperature experiments and a Biologic SP-200 potentiostat for low and high temperatures measurements. Swageloks® were directly connected to Biologic® instruments as channels' wires were inserted in plungers' holes. Electrochemical impedance spectroscopy (EIS) experiments were achieved at a bias voltage of 0 V. A sinusoidal voltage perturbation ($\Delta V = \pm 10$ mV) was applied in a frequency range from 200 kHz to 10 mHz for AN-based electrolyte, and to 200 μHz for ILs mixture. Cyclic voltammetries (CVs) were performed at various scan rates. To compare capacitances obtained at different scan rates, intensities were divided by the corresponding scan rate. Temperature control was achieved using a Vötsch climatic chamber with the temperature being additionally checked by a thermocouple. If not otherwise specified, reported capacitances concern the BP material by itself, and not of the full EDLC.

Results and discussion

SEM observations and gas sorption measurements

SEM micrograph shows randomly organized multi-walled carbon nanotubes (MWCNTs) (Fig. 2a) that confers flexibility to the material (inset Fig. 2a). The adsorption isotherm (Fig. 2b) is of type IV, with a progressive increase in

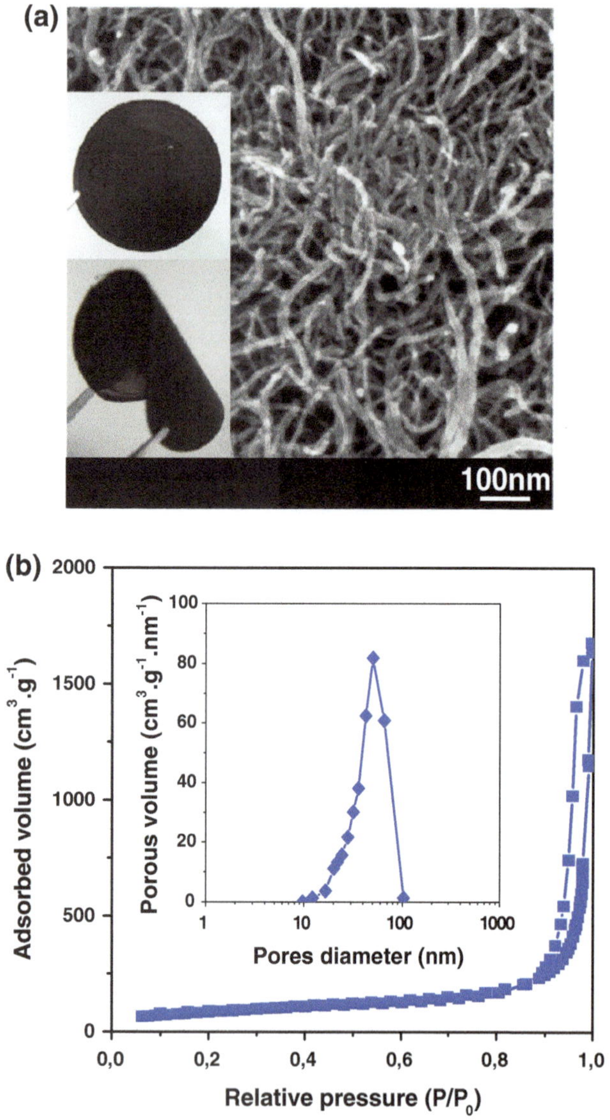

Fig. 2 **a** SEM micrograph of a BP electrode highlighting the entanglement of carbon nanotubes allowing its (*inset*) self-standing and flexible behaviors; **b** Type V isotherm characteristic of a mesoporous material with its derived pore size distribution

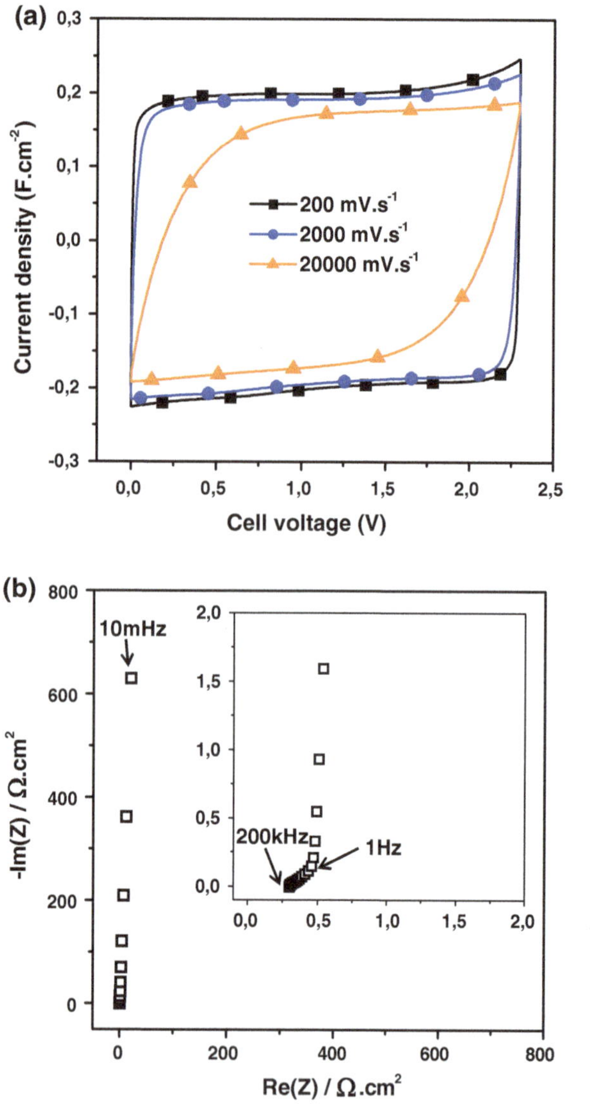

Fig. 3 **a** CVs of symmetric BPs/NET_4BF_4-AN cell obtained at room temperature (+25 °C) in the 0–2.3 V potential range at (*filled squares*) 200 mV s^{-1}, (*filled circles*) 2,000 mV s^{-1}, (*filled triangles*) 20,000 mV s^{-1} scan rates (*y* axis is plotted in F cm^{-2}); **b** the corresponding Nyquist plot

the gas adsorbed volume until 0.9 P/P_0. The isotherm shows an inflection and hysteresis loop H1 (adsorption and desorption curves are parallel and almost vertical) for $P/P_0 > 0.9$ which is typical from a mesoporous material with a narrow pore distribution. This is confirmed by the BJH calculation that gives a mean pore size of 40 nm (inset). The calculated SSA, calculated by the BET equation, is about 310 $m^2\ g^{-1}$.

Electrochemical measurement in NEt_4BF_4 1 M in AN electrolyte

In a first step, buckypaper electrodes have been tested in a conventional electrolyte: 1 M NEt_4BF_4 dissolved in acetonitrile (AN) that allows operation temperature range from −40 to +70 °C. Figure 3a shows the room temperature CV plots of a two-electrode cell assembled with our BP electrodes at various scan rates. All the CVs show the rectangular-shape electrochemical signature typical for a capacitive behavior within the 0–2.3 V potential range. Besides, high power and flexible applications are possible as the capacitive behavior is preserved at scan rates as high as 20 V s^{-1}. The Nyquist plot (Fig. 3b) confirms this good capacitive behavior, with a fast increase of the imaginary part at low frequencies. The behavior shown is a series—resistance/capacitor—circuit with clear evidence of porous electrode behavior. Ionic resistance in the porous network and the equivalent series resistance (ESR at 1 kHz) are comparable in values.

However, the operation temperature range of acetonitrile-based EDLC is limited by the boiling point of AN (81 °C), as well as the toxicity of potential degradation products and its low flash point (≈5 °C) [6, 7]. Replacing AN by propylene carbonate (PC) drastically limits the power performances because of the lower ionic conductivity of PC-based electrolytes. Additionally, operating temperature using PC-based electrolytes is restricted between −25 and +70 °C [6, 7] which is not sufficient for outdoor operation in severe environment, like the use of flexible EDLCs in smart-textile uses, for military applications. Working temperatures lower than −60 °C were reported by Brandon and co-workers [19–21], by adding different co-solvents with low boiling points such as methyl formate, methyl acetate or 1,3 dioxolane, in AN-based electrolytes. However, the maximum operation temperature was limited to +50 °C [19–21].

Electrochemical measurement in eutectic mixture of ionic liquids

Ionic liquids (ILs) are solvent free that can be used at high temperatures because they show negligible vapor pressure, are non-flammable and non-toxic. However, despite their large electrochemical stability window beyond 3 V leading to important energy density increase (E proportional to V^2), their low ionic conductivity and high viscosity still limit their use in electrochemical energy storage devices [22–24]. Room temperature IL electrolytes for EDLCs were developed in combination with activated carbon [25], carbon nanotubes [26] or graphene electrodes [27]. To extend the temperature operation range of our buckypapers-based EDLCs, we used an eutectic ILs mixture we recently developed [10].

Figure 4 shows CVs obtained at room temperature for two-electrode EDLC system in the 50 % $PIP_{13}FSI$/50 % $PYR_{14}FSI$ mixture. The rectangular shape of the CVs evidences a capacitive behavior. However, as expected, the lower conductivity of the mixture (Table 1) limits the rate capability of the electrode to 500 mV s^{-1}. Cyclic voltammetries were performed at +100, +25, −40 and −50 °C and are reported (Fig. 5a, b). As shown in Fig. 5a), the CV recorded at high temperature (+100 °C) for a scan rate of 100 mV s^{-1} shows a nice capacitive behavior too. More, the improvement in conductivities at high temperature allows a wider range of speed rates up to 2 V s^{-1} before being limited by the ohmic drop. The additional advantage of this eutectic mixture of ionic liquids is the wide range of temperatures, and more at low temperature, compared to pure ionic liquids used separately. The limitation is due to conductivities becoming too poor under −50 °C (Table 1)

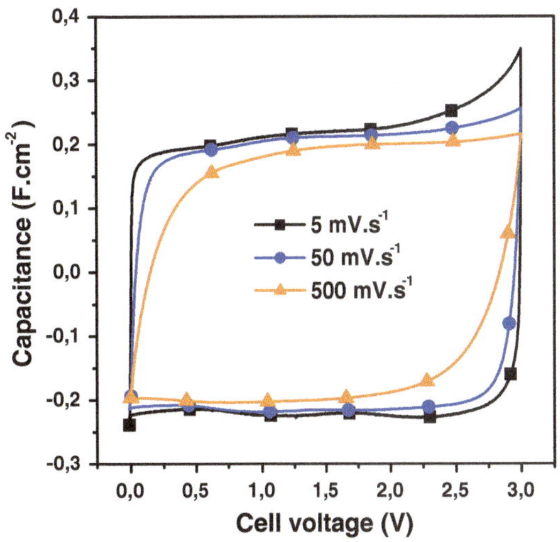

Fig. 4 CVs of symmetric BPs/ILs mixture cell obtained at room temperature (+25 °C) in the 0–3 V potential range at (*filled squares*) 5 mV s^{-1}, (*filled circles*) 50 mV s^{-1}, (*filled triangles*) 500 mV s^{-1} scan rates (y axis is plotted in F cm^{-2})

Table 1 Conductivity of the ILs eutectic mixture as a function of temperature

Temperature (°C)	Conductivity (S cm^{-1})
+100	2.9×10^{-2}
+25	5.9×10^{-3}
−40	5.7×10^{-5}
−50	1.5×10^{-5}
−70	6.8×10^{-7}

[28]. Figure 5b shows nice capacitive behaviors still preserved at −40 and −50 °C for a scan rate of 1 mV s^{-1}. The lower the working temperature, the lower the conductivity and, consequently, the potential scan rate is decreased down to 5 and 2 mV s^{-1} at −40 and −50 °C, respectively. A comparison of the different capacitances obtained in different electrolytes and temperatures will be discussed and compared with literature in "Capacitances vs. electrolytes and scan rates".

Figure 6 shows Nyquist plots for the symmetric BPs cell at different temperatures assembled using the ionic liquid mixture electrolyte. All plots confirm results obtained by CVs as they show a classical capacitive behavior, with a high increase of their imaginary part at low frequencies. Going from +100 °C down to −50 °C, the huge difference is seen at medium frequencies where the low-frequency capacitive behavior is shifted to higher resistances. A main difference with AN-based electrolyte is the presence of a semi-circle at high frequencies, showing an evolution too with temperature. Both differences, at high and medium frequencies, are due to the high difference of ILs mixture viscosity, which is temperature dependent (Table 1),

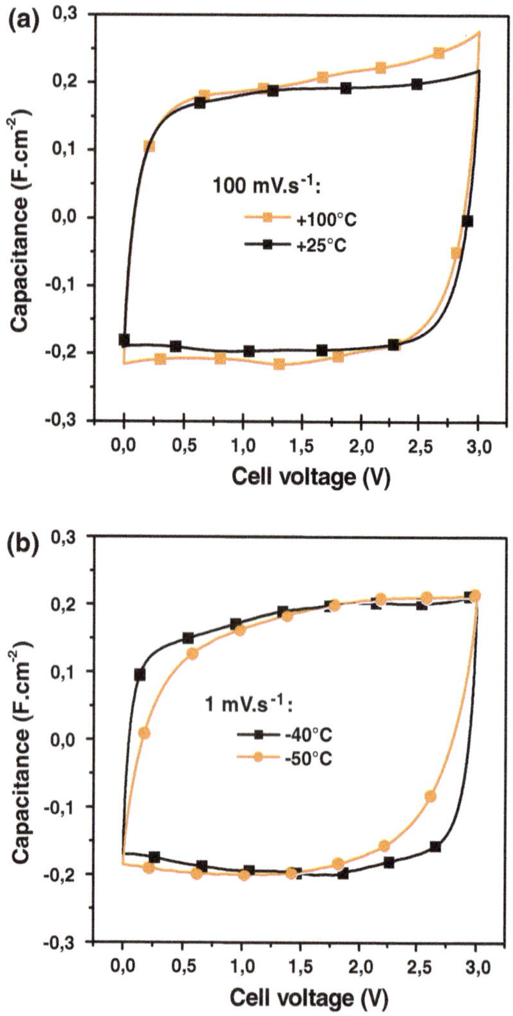

Fig. 5 CVs of symmetric BPs/ILs mixture cell cycled in the 0–3 V potential range **a** at 100 mV s^{-1} obtained at (*filled circles*) +100 °C and (*filled squares*) +25 °C and **b** at 1 mV s^{-1} obtained at (*filled circles*) −40 °C and (*filled squares*) −50 °C (*y* axis is plotted in F cm^{-2})

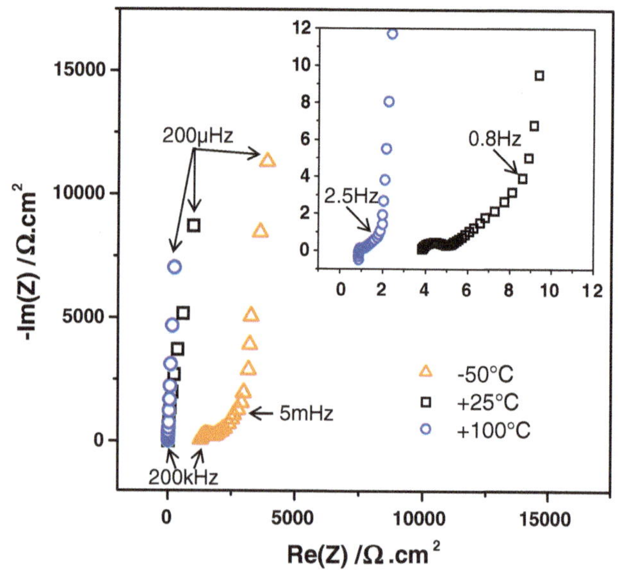

Fig. 6 Nyquist plots of symmetric EDLCs made with the ILs mixture as electrolyte at (*open circles*) +100 °C, (*open squares*) +25 °C and (*open triangles*) −50 °C

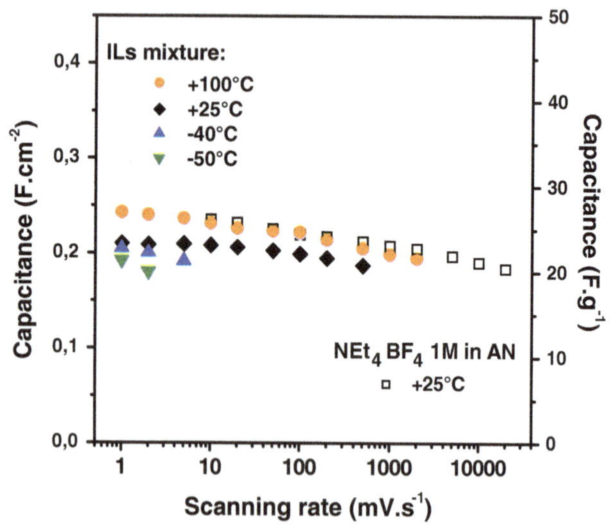

Fig. 7 Capacitances depending on the working temperature delivered by BPs electrodes in ILs mixture and AN-based electrolyte as a function of scan rates

leading to a difference in the ionic conductivity of the electrolyte inside the porous structure of the BPs, compared to AN-based electrolyte [29].

Capacitances vs. electrolytes and scan rates

Figure 7 summarizes results of cycling experiments, where delivered capacitances are reported as a function of scan rates. Buckypapers show excellent power performance and similar delivered capacitances, whatever electrolytes and temperatures used. Using the eutectic ILs mixture, the maximum capacitance was obtained at +100 °C and 1 mV s^{-1} with 0.24 F cm^{-2} (131.9 F cm^{-3}, 26.8 F g^{-1}). The capacitance values ranged from this maximum down to 0.18 F cm^{-2} (98.9 F cm^{-3}, 20.0 F g^{-1}), obtained at −50 °C and 2 mV s^{-1}. As scan rates range is lower than the one possible to use with AN-based electrolyte, our EDLC cannot be used for high power applications.

Previous results were obtained with onion-like carbons (OLC) and vertically aligned CNT (VA-CNT) [10]. VA-CNTs were grown directly on aluminum foils via plasma-enhanced chemical vapor deposition in a cold wall reactor. OLCs, mixed with 5 wt % PTFE, can be visualized as multi-shell fullerenes with concentric graphitic shells. Both types of materials were tested in Swagelok® cells using two layers of a 25-μm thick porous alumina separator.

Using our EDLC made of BP electrodes, ILs mixture electrolyte and glass fiber separator, surfaced capacitances were increased by a factor 2 compared to OLC (0.1 F cm^{-2}) and by a factor 50 compared to VA-CNT (0.04 F cm^{-2}) [10]. Those previous results together with the ones presented here definitely show that with this ILs electrolyte mixture, the use of mesoporous carbon electrodes is recommended, because of the high viscosity and the low conductivity of such kind of electrolytes. Furthermore, an excellent affinity is also obtained thanks to hydrophobic nature of those two compounds, namely the carbon and the IL mixture. Thus, these results definitely confirm that best performances can be achieved when electrolyte and electrode fit perfectly each other (chemically and physically speaking).

Conclusion

With this study, we could show the feasibility to build EDLCs able to work safely under severe conditions for smart-textile applications. We were able to cycle our EDLC between −50 and +100 °C without losing performance. The key point of using buckypapers is to get a standalone electrode without the use of any binders or current collectors. This reduces significantly the total mass of the EDLC, what is a benefit for energy densities but also for its incorporation in textiles. An effort has to be made to decrease the resistivity of whole system BP-electrolyte at higher scan rates. Meantime, at suitable temperatures, the AN-based electrolyte can be used for power applications.

Acknowledgments Authors would like to thank the ANR for financial support for this work (Progelec program). C.L., B.D., R.L., P.L.T. and P.S. thank the THALES company (Research & Technology department) for supporting this work.

References

1. Simon, P., Gogotsi, Y.: Materials for electrochemical capacitors. Nat Mater **7**(11), 845–854 (2008)
2. Faggioli, E., Rena, P., Danel, V., Andrieu, X., Mallant, R., Kahlen, H.: Supercapacitors for the energy management of electric vehicles. J. Power Sources **84**(2), 261–269 (1999)
3. Miller, J.R., Simon, P.: Electrochemical capacitors for energy management. Science **321**(5889), 651–652 (2008)
4. Kötz, R., Carlen, M.: Principles and applications of electrochemical capacitors. Electrochim. Acta **45**(15–16), 2483–2498 (2000)
5. Burke, A.: Ultracapacitors: why, how, and where is the technology. J. Power Sources **91**(1), 37–50 (2000)
6. Dash, R., Chmiola, J., Yushin, G., Gogotsi, Y., Laudisio, G., Singer, J., Fischer, J., Kucheyev, S.: Titanium carbide derived nanoporous carbon for energy-related applications. Carbon **44**(12), 2489–2497 (2006)
7. Dzubiella, J., Hansen, J.P.: Electric-field-controlled water and ion permeation of a hydrophobic nanopore. J. Chem. Phys. **122**(23), 234706–234714 (2005)
8. Miller, J.R., Burke, A.F.: Electrochemical capacitors: challenges and opportunities for real-world applications. Interface **17**(1), 53–57 (2008)
9. Cherenack, K., Zysset, C., Kinkeldei, T., Munzenrieder, N., Troster, G.: Woven electronic fibers with sensing and display functions for smart textiles. Adv. Mater. **22**(45), 5178–5182 (2010)
10. Lin, R., Taberna, P.L., Fantini, S., Presser, V., Perez, C.R., Malbosc, F., Rupesinghe, N.L., Teo, K.B.K., Gogotsi, Y., Simon, P.: Capacitive energy storage from −50 to 100 °C using an ionic liquid electrolyte. J. Phys. Chem. Lett. **2**(19), 2396–2401 (2011)
11. Izadi-Najafabadi, A., Yamada, T., Futaba, D.N., Yudasaka, M., Takagi, H., Hatori, H., Iijima, S., Hata, K.: High-power supercapacitor electrodes from single-walled carbon nanohorn/nanotube composite. ACS Nano **5**(2), 811–819 (2011)
12. Liu, J., Rinzler, A.G., Dai, H., Hafner, J.H., Bradley, R.K., Boul, P.J., Lu, A., Iverson, T., Shelimov, K., Huffman, C.B., Rodriguez-Macias, F., Shon, Y.S., Lee, T.R., Colbert, D.T., Smalley, R.E.: Fullerene pipes. Science **280**(5367), 1253–1256 (1998)
13. Endo, M., Muramatsu, H., Hayashi, T., Kim, Y.A., Terrones, M., Dresselhaus, M.S.: Nanotechnology: 'buckypaper' from coaxial nanotubes. Nature **433**(7025), 476 (2005)
14. Bahr, J.L., Yang, J.P., Kosynkin, D.V., Bronikowski, M.J., Smalley, R.E., Tour, J.M.: Functionalization of carbon nanotubes by electrochemical reduction of aryl diazonium salts: a bucky paper electrode. J. Am. Chem. Soc. **123**(27), 6536–6542 (2001)
15. Hu, L., Chen, W., Xie, X., Liu, N., Yang, Y., Wu, H., Yao, Y., Pasta, M., Alshareef, H.N., Cui, Y.: Symmetrical MnO_2—carbon nanotube—textile nanostructures for wearable pseudocapacitors with high mass loading. ACS Nano **5**(11), 8904–8913 (2011)
16. Jost, K., Perez, C.R., McDonough, J.K., Presser, V., Heon, M., Dion, G., Gogotsi, Y.: Carbon coated textiles for flexible energy storage. Energy Environ. Sci. **4**(12), 5060–5067 (2011)
17. O'Connel, M.J., Boul, P., Ericson, L.M., Huffman, C., Wang, Y., Haroz, E., Kuper, C., Tour, J., Ausman, K.D., Smaley, R.E.: Reversible water-solubilization of single-walled carbon nanotubes by polymer wrapping. Chem. Phys. Lett. **342**(3–4), 265–271 (2001)
18. Zhang, M., Yudaska, M., Koshio, A., Iijima, S.: Effect of polymer and solvent on purification and cutting of single-wall carbon nanotubes. Chem. Phys. Lett. **349**(1–2), 25–30 (2001)
19. Brandon, E.J., Smart, M.C., West, W.C.: Low-temperature supercapacitors. NASA Tech Briefs **32**, 32 (2008)
20. West, W.C., Smart, M.C., Brandon, E.J., Whitcanack, L.D., Plett, G.A.: Double-layer capacitor electrolytes using 1,3-D ioxolane for low temperature operation. J. Electrochem. Soc. **155**(10), A716–A720 (2008)
21. Brandon, E.J., West, W.C., Smart, M.C., Whitcanack, L.D., Plett, G.A.: J. Power Sources **170**, 225–232 (2007)
22. Armand, M., Endres, F., MacFarlane, D.R., Ohno, H., Scrosati, B.: Ionic-liquid materials for the electrochemical challenges of the future. Nat. Mater. **8**(8), 621–629 (2009)
23. Torimoto, T., Tsuda, T., Okazaki, K., Kuwabata, S.: New frontiers in materials science opened by ionic liquids. Adv. Mater. **22**(11), 1196–1221 (2010)
24. Balducci, A., Dugas, R., Taberna, P.L., Simon, P., Plée, D., Mastragostino, M., Passerini, S.: High temperature carbon–carbon supercapacitor using ionic liquid as electrolyte. J. Power Sources **165**(2), 922–927 (2007)

25. Wei, D., Ng, T.W.: Application of novel room temperature ionic liquids in flexible supercapacitors. Electrochem. Commun. **11**(10), 1996–1999 (2009)
26. Lu, W., Qu, L., Henry, K., Dai, L.: High performance electrochemical capacitors from aligned carbon nanotube electrodes and ionic liquid electrolytes. J. Power Sources **189**(2), 1270–1277 (2009)
27. Zhu, Y., Murali, S., Stoller, M.D., Ganesh, K.J., Cai, W., Ferreira, P.J., Pirkle, A., Wallace, R.M., Cychosz, K.A., Thommes, M., Su, D., Stach, E.A., Ruoff, R.S.: Carbon-based supercapacitors produced by activation of graphene. Science **332**(6037), 1537–1541 (2011)
28. Kunze, M., Jeong, S., Paillard, E., Winter, M., Passerini, S.: Melting behavior of pyrrolidinium-based ionic liquids and their binary mixtures. J. Phys. Chem. C **114**(28), 12364–12369 (2010)
29. Taberna, P.L., Simon, P., Fauvarque, J.F.: Electrochemical characteristics and impedance spectroscopy studies of carbon–carbon supercapacitors. J. Electrochem. Soc. **150**(3), A292–A300 (2003)

21

Betalain pigments as natural photosensitizers for dye-sensitized solar cells: the effect of dye pH on the photoelectric parameters

Kasim Uthman Isah · Umar Ahmadu · Adamu Idris · Mohammed Isah Kimpa · Uno Essang Uno · Muhammed Muhammed Ndamitso · Noble Alu

Abstract Dye-sensitized solar cells (DSSC) were fabricated using red *bougainvillea glabra* flower dye extracts as natural dye sensitizers at three dye pH values of 1.23, 3.0, and 5.7. Water was used as dye-extracting solvent. DSSCs from dye extract of pH 3.0 had the highest photocurrent density J_{sc} of 3.72 mA/cm^2 and fill factor FF of 0.59. While the DSSCs from dye sensitizer pHs of 1.23 and 5.7 had J_{sc} of 1.13 and 2.27 mA/cm^2, and fill factors of 0.43 and 0.61, respectively. The maximum powers P_{max} of the DSSCs were 0.50, 1.64, and 0.94 mW/cm^2 for dye sensitizer pH of 1.2, 3.0, and 5.7, respectively.

Keywords Dye-sensitized solar cell · Dye sensitizer · *Bougainvillea glabra* · PH

Introduction

Dye-sensitized solar cell (DSSC) developed by Gratzel et al. [1] has attracted considerable attention due to its environmental friendliness and relative low cost of production [2]. Sensitizer dye in DSSC plays a key role in harvesting sunlight and transforming solar energy into electric energy and thus of paramount importance to photovoltaic cell performance and efficiency [3]. It attaches to the surface of a wide band-gap mesoporous semiconductor serving as electron transporter [4]. Ruthenium-based dye sensitizers are very expensive and hard to prepare, which restricts their large-scale applications in solar cells, stimulating the search for alternatives such as natural dyes.

Anthocyanins from natural pigments and carotenoids have been used as natural dye sensitizers in DSSC and have shown low to modest solar energy conversion efficiencies [2, 5–7]. These natural dyes are easy to prepare, cheap, non-toxic, environmentally friendly, and easily biodegradable.

Betalain is another interesting class of pigments, whose purified extracts from commercial sources have been subjected to photoelectrochemical study [8]. Betalains consisting of the yellow betaxanthins and red–violet betacyanins are a group of water-soluble nitrogen-containing alkaloid pigments characteristic of certain members of plant sub-order *chenopodineae* within *caryophyllales* and some higher fungi [9]. They absorb visible radiation over the range 476–600 nm, they are immonium derivatives of betalamic acid (the chromophore of all betalains) and divided into two structural groups: the red–violet betacyanins that have a maximum absorptivity at $\lambda \approx 535$ nm and the yellow–orange betaxanthins with maximum absorptivity at $\lambda \approx 480$ nm [10–13]. Betalains are characterized by high molar extinction coefficients in the visible region and pH-dependent redox properties [14, 15]. They are relatively stable over pH range 3–7, however, at pH less than 3.5 the absorption maximum shifts toward lower wavelength [16]. Acidic conditions are known to induce recondensation of betalamic acid with amine group of the addition residue [17] and favors betalain-sensitized photo-electrodes with high optical densities capable of

K. U. Isah (✉) · U. Ahmadu · A. Idris · M. I. Kimpa · U. E. Uno
Department of Physics, Federal University of Technology, Minna, Nigeria
e-mail: kasim309@futminna.edu.ng

M. M. Ndamitso
Department of Chemistry, Federal University of Technology, Minna, Nigeria

N. Alu
Physics Advanced Laboratory, Sheda Science and Technology complex Sheda, Abuja, Nigeria

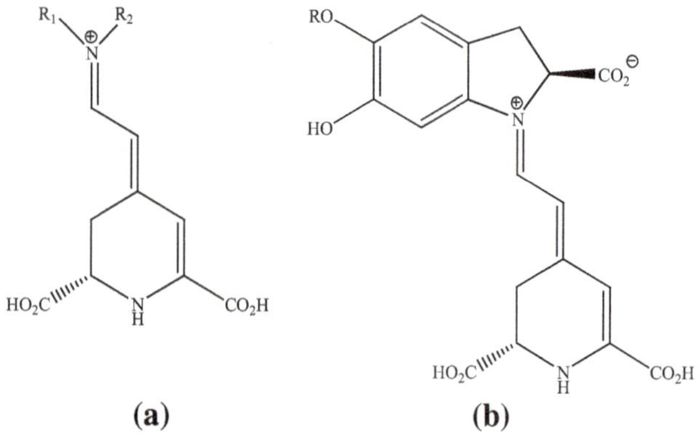

Fig. 1 General structures of the main betalain **a** betaxanthin ($R_1 = H$ and R_2 = amine or amino acid group), **b** betacyanin (R = β-D-glucose)

complete absorption in the visible range of 400–600 nm [6]. Solar energy conversion efficiency being a function of photocurrent density J_{sc}, open-circuit voltage V_{oc}, and fill factor FF [18], suggest their improvement is essential to raising conversion efficiency.

This work reports the use of water extracts of red *Bougainvillea glabra* flower a source of betalain as natural dye sensitizer and the influence of sensitizer dye pH on the photoelectric parameters of the DSSCs.

Bougainvillea is a member of the *Nyctaginaceae* family comprising of 18 shrubby species which though indigenous of South America are distributed widely in the world, with flowers ranging from white, yellow, orange, and various shades of red to purple and violet color [12–14]. The Betalain pigment in *Bougainvillea* is different from that in other sources due to the presence of saccharide type present in the betanidin (an aglycone of most betacyanins) which is produced 90 % from betacyanins, and the remaining 10 % comes from indicaxanthin a common betaxanthin [15]. *Bougainvillea glabra* has been found to have eleven violet–red pigments having maximum absorbance in wavelength range of 522–551 nm and is known to have two epimeric betacyanins; bougainvillein-r-1 and isobougainvillein-r-1 [15].

The general structure of Betalain is shown in Fig. 1 [10]; it contains carboxylic functional group which facilitates TiO_2 surface binding.

Experimental

The red *Bougainvillea glabra* flower dye pigments were extracted by crushing 20 g of the flower in a porcelain mortar with a pestle and adding 50 ml of water to the crushed sample. This was filtered to get the raw dye extracts of the flower which was divided into three and used as sensitizing dye at three different pHs. The first, the as-extracted dye had a pH of 5.70, the pH of the second was adjusted to 3.0 by adding 4 drops of 20 % HCl to it, and the third after adding 8 drop of HCl had a pH value 1.23.

The transparent conducting oxide was fluorine-doped tin oxide (FTO) with a sheet resistance of 15 Ω/cm^2 (SOLARONIX). The TiO_2 film was prepared by blending commercial TiO_2 powder (Degussa, P25) of Ca. 0.2 g, nitric solution (0.1 M) of 0.4 ml, polyethylene glycol (MW 10,0000) of ca. 0.08 g, and one drop of a non-ionic surfactant, Triton x-100. The mixture was well blended using an ultrasonic bath for 1 h. Squeegee was used to screen-print the resulting TiO_2 paste on the conducting layer/FTO glass substrate. It was left for 30 min so that the paste could settle to reduce the irregularity of the surface. The substrates were then left to dry. This screen-printing procedure was repeated to obtain the TiO_2 working electrode of appropriate thickness of 9 μm.

Prior to sensitisation, the TiO_2 photo-anodes were preheated at 150 °C for 30 min, allowed to cool. It was then subjected to $TiCl_4$ treatment by immersing in 40 mM of $TiCl_4$ solution for 30 min at a temperature of 70 °C and subsequently sintered at 450 °C for 45 min. After cooling to 80 °C, the TiO_2 electrodes were immersed into the three dye solutions of pH 1.23, 3.0, and 5.7 for 16 h. The photoelectrode were then rinsed with 99 % ethanol solution, to remove excess dye and thereafter left to dry. The counter electrode was made from platinum catalyst T/SP product by SOLARONIX which was screen-printed using a polyester mesh of 90 and dried. The substrates were sintered for 1 h at 45 °C. The DSSC was assembled from the various components following the procedure reported elsewhere [19].

Characterization

The absorption spectra of the dyes were determined using an AVASPEC 2048 UV–Visible spectrophotometer, while the morphology and composition characterization of the nanoporous TiO_2 film was carried out using Carl Zeiss EVO MA-10 scanning electron microscope (SEM), with EDX model ISIS 300 Oxford. Simulated Solar irradiation was provided by a Solar Simulator, Model 4200-SCS Semiconductor Characterization System under the irradiation of AM 1.5 (100 $mWcm^{-2}$), and the Current–Voltage curves were recorded by a digital Keithley multimeter Model 2400 coupled to a computer.

Results and discussion

Figure 2 shows the UV–Vis optical absorption spectra of the *bougainvillea* flower dye extracts at three different pHs.

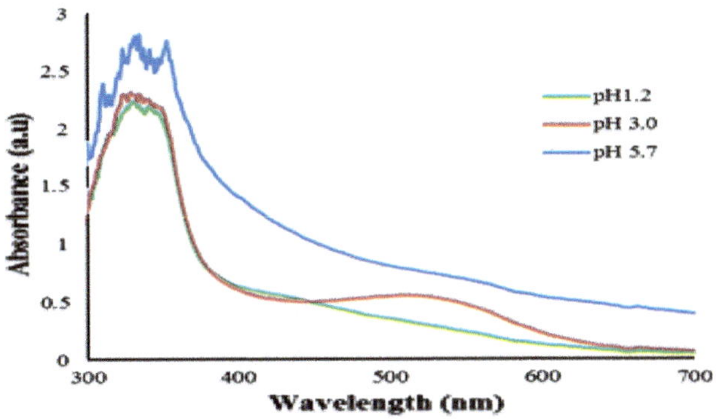

Fig. 2 Absorption spectrum of the dye extract from *red bougainvillea glabra* flower for $pH_{1.2}$, $pH_{3.0}$, and $pH_{5.7}$

Absorption spectra of a dye reflect optical transition probability between the ground state, the excited state and the solar energy range absorbed by the dye. The three dyes' extracts show absorption peaks centered between 310 and 340 nm in the UV-range, with maximum peaks at 330 nm for pH 5.7 and 327 nm for pH 3.0 and 1.2. The dye extracts at pH 1.23 and 3.0 have about the same absorption intensity in short wavelength range, with dye extract at a pH of 5.7 having the highest intensity in this range. The lowest absorption intensity in the long wavelength is observed for dye pH of 1.2, indicating a highly suppressed absorbance as a result of degradation of betanin in very strong acidic environment [20]. Dye extract at a pH of 3 displays broad absorption peak in the 480–560 nm range resulting from $\pi - \pi^*$ transitions due to the mixed contributions of the yellow–orange betaxanthins (480 nm) and of the red–purple betacyanin (540 nm) [6]. At a pH of 5.7, the extract has a highest but flattened absorption spectrum in the long-range wavelength indicating a suppressed absorption peak at this pH and a wider range of red, orange, yellow, and blue light can be absorbed. Absorbance peaks around 300 and 535 nm are characteristic absorptions for red–violet betalain group, betacyanin [14].

The betaxanthin and betanin concentration in the dye extracts in µM are estimated from the absorbance A at 482 and 536 nm from the absorbance spectra using [10],

$$[B_{betax}] = 23.8A_{482} - 7.7A_{536}$$
$$[B_{betan}] = 15.38A_{536}$$

$[B_{betax}]$ is the betaxanthin concentration and $[B_{betan}]$ the betanin concentration. The betaxanthin concentrations were 7.39, 8.79, and 15.17 µM for pH of 1.23, 3.0, and 5.7, respectively, while the betanin concentration at a pH of 1.2, 3.0 and 5.7 were 4.35, 8.15, and 11.26 µM, respectively.

Figure 3 shows the scanning electron micrograph of the TiO_2 (anatase) film. The TiO_2 film has thickness of 9 µm and a mean particle size of 20 nm. It shows a mesoporous surface of the spherical nanoparticles of TiO_2 that forms nanopores across the surface. The EDX analysis of the TiO_2 electrode in Fig. 4 shows the presence of Titanium, Oxygen, Nitrogen, and Carbon. The presence of chlorine in the compound was due to $TiCl_4$ which was applied at a final treatment of the porous TiO_2 electrodes in the anatase TiO_2 compound contained.

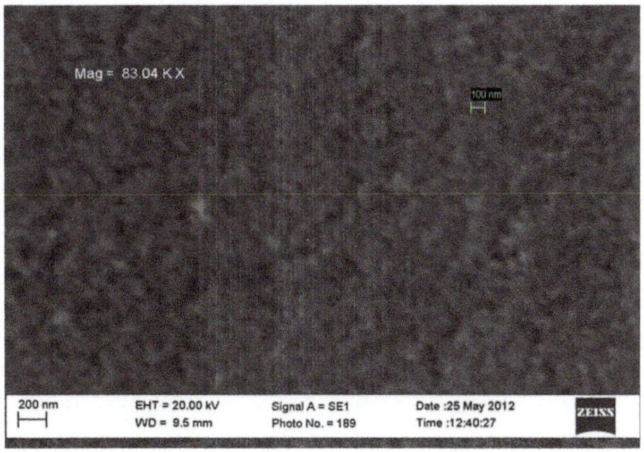

Fig. 3 SEM micrograph of a nanoporous TiO_2 (anatase) film

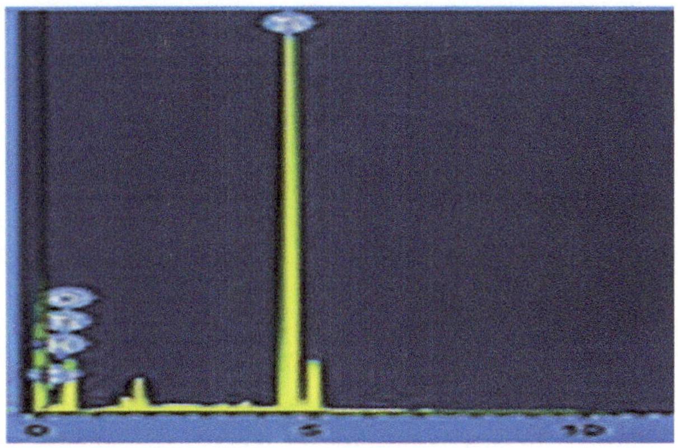

Fig. 4 Elemental composition of TiO_2 compound

The current–voltage curve of a solar cell yields important photoelectric parameters such as short-circuit current density J_{sc}, the open-circuit voltage V_{oc}, the current I_{mp}, and voltage V_{mp} at the maximum power point P_{max}. The photoelectric characteristics of the DSSCs were inspected under the simulated sunlight source (AM1.5).

The combined J–V and P–V curves of the DSSCs at different pHs are shown in Fig. 5. The open-circuit voltage V_{oc} of 0.44 V is obtained for dye pH of 1.2 and 3.0, respectively, while at a pH of 5.7 the V_{oc} is 0.41. The DSSC prepared from a dye pH of 3.0 has the highest photocurrent density J_{sc} of 3.72 mA/cm^2 and fill factor of and 0.59. At a dye sensitizer pH of 1.2, the DSSC has the lowest Photocurrent density of 1.13 mA/cm^2 and lowest fill factor of 0.43 due to betanin degradation in very strong

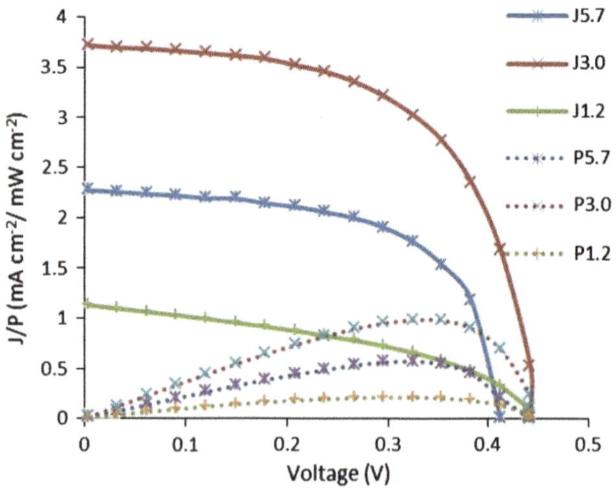

Fig. 5 Combine J–V and P–V curves of the cells for dye sensitizers at pH of 1.2, 3.0, and 5.7

Table 1 Photoelectric parameters of the cells sensitized with natural extracts

Dye pH	J_{sc} (mA/cm^2)	V_{oc} (V)	P_{max} (mW/cm^2)	FF	η
1.2	1.13	0.44	0.50	0.43	0.21
3.0	3.72	0.44	1.64	0.59	0.98
5.7	2.27	0.412	0.94	0.61	0.57

acidic environment [17, 20] with resultant inefficient light harvesting by the dye and inefficient charge injection into the TiO_2 nanoparticles; also possible cell deterioration by acid leaching is expected as the pH goes lower.

The reduction in photoelectrical property observed at pH 5.7 compared to pH 3 results because at high pH betanin is poorly absorbed onto titania due to its inhibition by indicaxanthin (a common betaxanthin in red *bougainvillea*) competing with betanin for adsorption sites, but at lower pH protonation of the betalainic carboxylic acid group from its anionic form occurs promoting its adsorption onto the TiO_2 [14].

The photoelectric parameters are listed in Table 1.

The obtained photoelectric parameters using *bougainvilla glabra* flower dye extract as sensitizer dye at pHs of 1.2, 3.0, and 5.7 are comparable to J_{sc} of 1.881–2.344 mA/cm^2 though with lower V_{oc} of 0.23–0.26 V from betalain pigments extracted from bracts of *bougainvilla glabra* and *bougainvilla spectabilis* at a pH of 5.7 [14].

Also although J_{sc} of up to 9.5 mA/cm^2 have been reported at a dye pH of 1.0 from other sources of betalain, corresponding V_{oc} was just between 0.23–0.3 V [6]. The obtained photoelectric parameters at dye pHs of 1.23 and 5.7 are much lower than that of black dye. The black dye has a certified J_{sc} of 20.5 mA/cm^2 and a V_{oc} of 0.72 V and is superior to all charge-transfer sensitizers on the basis of photovoltaic performance under AM 1.5 [21]. The J_{sc} 3.72 mA/cm^2 at dye sensitizer pH of 3.0 is encouraging.

Conclusion

Betalain raw extracts of red *bougainvillea glabra* flower were used as natural dye sensitizers for DSSCs at different dye sensitizer pH values. At a pH of 1.23, the DSSC had the lowest parameters as a result of betalain degradation in very acidic environment and leaching of the dye from the TiO_2 surface. The DSSC from dye sensitizer pH of 3.0 offered the highest J_{sc} of 3.72 mA/cm^2, a FF of 0.59, and highest P_{max} of 1.64 mW/cm^2. The poor absorption of betanin onto the titania at high pH lead to decreased photoelectrical parameter of the cell at pH of 5.7. The obtained photoelectric parameters are much lower than that of black dye which is superior to all charge-transfer sensitizers on the basis of photovoltaic performance under AM 1.5 presently, with a certified J_{sc} of 20.5 mA/cm^2 and a V_{oc} of 0.72 V [21]. Nevertheless, natural *bougainvillea glabra* flower betalain raw extracts present promising low-cost, non-toxic, and environmentally friendly alternative photosensitizers in DSSCs.

Acknowledgments This study was supported by grant from the University Board of Research (UBR) of the Federal University of Technology, Minna, Nigeria.

References

1. O'Regan, B., Grätzel, M.: A low cost high-efficiency solar cell based on dye-sensitized colloidal TiO_2 films. Nature **353**, 737–740 (1991)
2. Zhou, H., Wu, L., Gao, Y., Ma, T.: Dye-sensitized solar cells using 20 natural dyes as sensitizers. J. Photochem. Photobiol. A Chem **219**, 188–194 (2011)
3. Chang, H., Wu, H.M., Chen, T.L., Huang, K.D., Jwo, C.S., Lo, Y.J.: Dye-sensitized solar cell using natural dyes extracted from spinach and ipomoea. J. Alloys Compd. **495**(2), 606–610 (2010)
4. Erten-Ela, S.: Photovoltaic effect of nanostructured TiO_2 layers in dye sensitized solar cells. J. Optoelectron. Adv. Mater. **14**(9–10), 758–762 (2012)
5. Dai, Q., Rabani, J.: Photosensitization of nanocrystalline TiO_2 films by anthocyanin dyes. J. Photochem. Photobiol. A Chem. **148**, 17–24 (2002)
6. Calogero, G., Di Marco, G., Cazzanti, S., Caramori, S., Argazzi, R., Di Carlo, A., Alberto Bignozzi, C.: Efficient dye-sensitized solar cells using red turnip and purple wild Sicilian prickly pear fruits. Inter. J. Mol. Sci. **11**, 254–267 (2010)
7. Gao, F.G., Bard, A.J., Kispert, L.D.: Photocurrent generated on a carotenoid-sensitized TiO_2 nanocrystalline mesoporous electrode. J. Photochem. Photobiol. A Chem. **130**, 49–56 (2000)

8. Zhang, D., Lanier, S.M., Downing, J.A., Avent, J.L., Lum, J., McHale, J.L.: Betalain, pigments for dye-sensitized solar cells. J. Photochem. Photobiol. A Chem. **195**, 72–80 (2008)
9. Koboyashi, N., Schmidt, J., Wray, V., Schliemann, W.: Formation and occurrence of dopamine-derived betacyanins. Phytochem. **56**, 429–436 (2001)
10. Strack, D., Vogt, T., Schliemann, W.: Recent advances in betalain research. Phytochem. **62**, 247–269 (2003)
11. Azeredo, H.M.C.: Betalains, properties, sources, applications, and stability—a review. Inter. J. Food Sci. Technol. **44**, 2365–2376 (2009)
12. Mabberley, D.F.: A Portable Dictionary of Vascular Plants. Cambridge University Press, UK (1997)
13. Impellizzeri, G., Piattelli, M., Sciuto, S.: Acylated betacyanins from *Drosanthemum floribundum*. Phytochem. **12**, 2295–2296 (1973)
14. Hernández-Martínez, A. R.,, Vargas S., Estevez, M., Rodríguez, R.: Dye-sensitized solar cells from extracted bracts *Bougainvillea* betalain pigments. In: 1st International Congress on Instrumentation and Applied Sciences, Cancun, Q. R., Mexico, 26–29 Oct 2010
15. Piattelli, M., Imperato, F.: Pigments of *Bougainvillea glabra*. Phytochem. **9**, 2557–2560 (1970)
16. Jackman, R.L., Smith, J.L.: Anthocyanins and betalains. In: Hendry, G.F., Houghton, J.D. (eds.) Natural Food Colourants, pp. 244–309. Blackie Academic & Professional, London (1996)
17. Schwartz, S.J., von Elbe, J.H.: Identification of betanin degradation products. Eur. Food Res. Technol. **176**, 448–453 (1983)
18. Sze, S.M., Lee, M.: Physics of Semiconductor Devices. Wiley, New York (1969)
19. Kimpa, M.I., Momoh, M., Isah, K.U., Yahya, H.N., Ndamitso, M.M.: Photoelectric characterization of dye sensitized solar cells using natural dye from pawpaw leaf and flame tree flower as sensitizers. Mater. Sci. Appl. **3**, 281–286 (2012)
20. Pedreno, M.A., Escribano, J.: Correlation between antiradical activity and stability of betanine from Beta vulgaris L roots under different pH, temperature and light conditions. J. Sci. Food Agric. **81**, 627–631 (2001)
21. Qin, Y., Peng, Q.: Ruthenium sensitizers and their applications in dye-sensitized solar cells. Inter. J. Photoenerg. **2012**, 1–21 (2012)

22

A review on the application of iron(III) fluorides as positive electrodes for secondary cells

Donato E. Conte · Nicola Pinna

Abstract Among the many industrial and research level applications of fluorides like microelectronics, polymers, agronomy, imaging, dental composites, uranium enrichment processes or catalysis, there is a recent and growing interest for these ionic compounds in the field of energy storage and conversion. Li-ion technology in particular, has been attracted to fluorides due to the possibility of more-than-one electrode reaction. In fact, these reactions can potentially enhance the energy stored in commercial cells (conversion reactions). By reaction with lithium, transition metal fluorides such as MnF_3, FeF_3, CoF_3 or CuF_2, can be reduced to the respective zero valent metal/LiF mixture with resulting high theoretical capacities ranging from 3 to 4.2 times the amount currently delivered by commercial $LiFePO_4$ (170 mAh g^{-1}). Since, in practice, the most of the research efforts are being put into iron trifluoride due to chemical (relatively mild synthesis conditions), electrochemical (lower polarization than MnF_3 for example) and environmental (Fe is more environmentally friendly than Co) reasons, in this review we will focus on the electrochemistry, synthesis strategies and amelioration techniques for the high oxidation state ferric fluoride and its derivatives.

Keywords Iron fluoride · Li-ion · Batteries

D. E. Conte (✉) · N. Pinna
Institut für Chemie, Humboldt-Universität zu Berlin,
Brook-Taylor-Str. 2, 12489 Berlin, Germany
e-mail: donato.e.conte@gmail.com

Introduction

Metal fluorides are ionic compounds that have been known since the late Renaissance era when they were used to render aluminum slags less viscous during the smelting process (CaF_2 fluxes) [1]. More recent applications, both at the industrial and research levels include: etching of silicon wafers for microelectronics, production of fluorinated polymers, agronomic compounds, medical imaging products and dental composites for caries control, uranium enrichment processes, catalysis and energy storage and conversion [2–4].

In the latter field, in particular, fluorides have attracted attention due to the conversion reaction with lithium ions. In conversion reactions, the transition-metal, M, of an initial MX species (X = H, N, O, F, P or S) is reduced by lithium to give LiX and the pure metal [5]. Since the first report of such kind of reaction in oxides [6], it has been clear that the involved materials could greatly enhance the energy stored in Li-ion cells. The reason for such high energy densities lies in the possibility for the transition metal to pass through various oxidation states involving more than one electron. For example, put aside the reaction potential, FeF_3 would reach 712 mAh g^{-1} for its complete 3 electron reaction, while in $LiFePO_4$ only the Fe(III)/Fe(II) couple will be active, delivering a theoretical maximum of 170 mAh g^{-1} (4.2 times less than iron trifluoride).

Recently, more and more research targets the optimization of high energy anode materials (Si, Sn, oxides, etc.) [7]. Some prototypes could be brought to the market in the near future. However, at the current state of research, cathodes still represent the main cell limitation [8]. In fact it is not possible to assemble a performing full cell using a high energy anode if the cathode material cannot match that energy. Considering a cell that would be assembled with a Si anode, a high amount of lithium can theoretically react delivering around 4,000 mAh g^{-1} ($Si + 4.4Li^+ + 4.4\ e^- \rightarrow Li_{4.4}Si$). If, in a purely speculative way, this high capacity can be reached and sustained for several hundreds of cycles, such an achievement would be completely useless when the coupling would be done with $LiFePO_4$-based cathodes for example. In fact since in full Li-ion cells, cathodes constitute the Li^+ reservoir (this is the reason why new portable gadgets have to be charged during the very first use), the 170 mAh g^{-1} of 1 g of $LiFePO_4$ will never be enough for the full reaction of Si. This leaves the manufacturer with two options: assemble the commercial cell with a Si-based anode without using its full energy potential or balance the amount of cathode using 4.4 times more of cathodic powder than anodic one. Such an increase in the amount of active material has both an economic and an engineering impact on the final cells. For these reasons, the development of cathode materials exhibiting reversible, more-than-one electron reactions would constitute one of the major breakthroughs for the realization of the next generation of accumulators, some other requirements being high voltage reactions and flat reaction plateaus.

In this review, we will focus mainly on the electrochemical reaction and synthesis strategies for the high oxidation state ferric fluoride and its derivatives as cathodes for the next generation of electrochemical accumulators.

Early research, characteristics and electrochemical mechanism of FeF_3

Fluoride-based cathodes have been investigated very early during the development of Li-based cells. Already in the 1960s, Lockheed investigated CuF_2 and CoF_3 to develop long-lasting Li-based primary cells for use in space applications [9–11]. Later, in the 1970s, AgF_2, NiF_2 and BiF_3 were also investigated for primary cells [12]. The first prototypes of the modern Li-ion cells were developed following the proposal (around 1980 [13]) and commercial application (in 1990 [14]) of the "rocking chair" systems, effectively clarifying that research in cathode materials would be a prominent field for the following years.

The first investigation of FeF_3 was made by Arai et al. in a study published in 1997 [15]. The study was performed on a commercially available product in the region 4.5–2 V versus Li^+/Li. The authors identified the topotactic insertion reaction during the first part of the discharge with a maximum reaction of 0.5 Li^+. This behavior accounts for half of the currently demonstrated insertion of $1Li^+$ per formula unit. The most probable reason is the lack of nanostructuration of the commercial FeF_3.

After the above-mentioned study, fluoride-based cathodes were somehow left aside for some years to the profit of more immediately interesting compounds like oxides or phosphates. The reason for this slowing down of the research lies in the intrinsic, poor electronic conductivity of the fluoride which is due to the large bandgap induced by the highly ionic character of the metal-halogen bond. Badway et al. [16, 17] extensively studied the FeF_3 reaction mechanism in two publications that appeared in 2003 and later in 2007 [18]. Again the study was carried on a commercially available product but this time, high energy ball milling was used both to decrease the fluoride particle sizes down to 25–30 nm and to intimately mix it with carbon-based conducting agents. Moreover, the studied voltage regions were varied from 4.5–2 V versus Li^+/Li to 4.5–1.5 V versus Li^+/Li which permitted the description of the full electrochemical reaction of FeF_3.

Pure, anhydrous iron trifluoride possesses a hexagonal lattice belonging to the R-*3c* space group. This structure lies between that of ReO_3 (regular corner-shared octahedrons resulting in oxygen ccp array) and that of PdF_3 (where octahedrons are also corner-shared but with fluorine in hcp array). These two structures can be derived from each other by simple rotation of the octahedrons. For the sake of simplicity then, we will describe FeF_3 as having a "ReO_3-type structure" where an alpha angle ($\alpha = 58°$) transform the cubic structure into a hexagonal one. Open tunnels are present along the [012] (Fig. 1).

Resulting from the presence of these channels, the electrochemical reaction can be described as an initial insertion of Li^+, although not completely topotactic, with a first electrochemical plateau at around 3.4 V followed by a conversion reaction producing Fe^0 and LiF with a second plateau at around 2.1 V (Fig. 1). Shortly, FeF_3 electrochemical discharge reaction with Li^+ can be written as:

$$Li^+ + e^- + FeF_3 \rightarrow \text{Intercalated ``}LiFeF_3\text{''}\ (3.4\,V) \quad (1)$$

$$2Li^+ + 2e^- + \text{Intercalated ``}LiFeF_3\text{''} \rightarrow 3LiF + Fe(2.1\,V) \quad (2)$$

This behavior was documented by numerous techniques among which in situ and ex situ X-ray diffraction (XRD), ex situ selected area electron diffraction (SAED), ex situ electron energy loss spectroscopy (EELS) and galvanostatic intermittent titration technique (GITT).

Based on XRD studies, it was concluded that since a topotactic insertion should result in no appreciable change in the intensities of XRD reflections, reaction 1 proceeds through two separate mechanisms [17]. The topotactic reaction can be observed up to a composition of roughly $Li_{0.5}FeF_3$ while large variations in unit cell volume can be observed for the remaining part of the curve up to around 1 Li^+. The conversion reaction (reaction 2) then follows. On

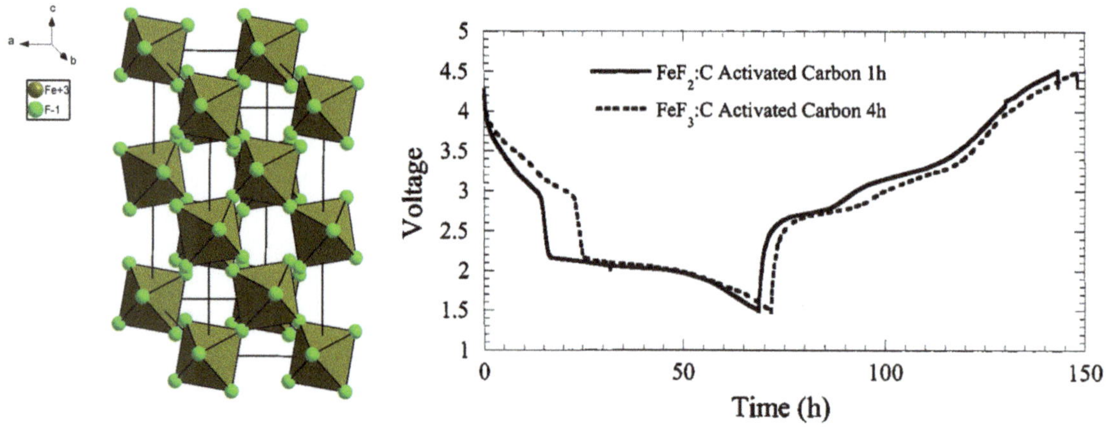

Fig. 1 Hexagonal unit cell of anhydrous FeF_3 viewed along the (012) plane and its galvanostatic response (*dotted line*) vs. Li^+/Li. Reprinted with permission from Ref [17]. Copyright 2003, The Electrochemical Society

the other hand, the authors could not conclusively account for the charge behavior: unexpectedly identifying a compound related to FeF_2 rutile structure at the end of charge, they were only able to exclude the presence of side reactions such as the formation of FeC_x or FeO_xF_{2-2x} by electrolyte decomposition.

Through density functional theory (DFT), Doe et al. [19] identified several possible intermediate Li–Fe–F phases to explain both the discharge and charge behavior. Interestingly, they concluded that discharge and charge do not follow the same reaction path with the formation of a higher number of intermediate phases during charge than during discharge. Also, the final charge product resulted to be obtained from the delithiation of a rutile-like $Li_{0.75}Fe_{0.75}F_3$ phase thus forming a rutile-distorted FeF_3 with cell parameters very close to those of FeF_2.

Experimental proof of such a behavior was given almost concomitantly by Yamakawa et al. [20] through ex situ ^6Li, ^7Li and ^{19}F magic-angle-spinning nuclear magnetic resonance (MAS-NMR) among other techniques. The analyzed commercially available iron trifluoride was ball-milled with carbon. In order to clarify the end-of-charge mechanism, the authors used an interesting approach consisting in the comparison of an electrochemically cycled FeF_3 sample with a FeF_2 sample cycled with the same conditions. The reported results confirm the structural transformation of the initial ReO_3–FeF_3 phase into a phase structurally resembling a lithiated rutile with a stoichiometry close to $Li_{0.5}FeF_3$. The comparison with pure rutile-FeF_2 showed that the Li local environments were similar to those of the FeF_3 system lending further support to the experimental data. The transformation evolves initially through a two-phase region (up to $x \sim 0.5$) and then through a solid-solution reaction that involves insertion of Li into $Li_{0.5}FeF_3$. Furthermore, the material can accommodate more lithium by progressively reducing the Fe and extruding LiF from the matrix reaching a final state where LiF and α-Fe are intimately mixed. Interestingly, NMR spectra showed that the obtained α-Fe is superparamagnetic and not ferromagnetic as one would expect. When the reconversion was studied, it was found that the LiF-Fe mix re-forms a lithiated rutile phase at 4.0 V with a composition close to $Li_{0.5}FeF_3$. The authors conclude then that the conversion reaction is largely, but not completely, reversible.

It is worth noting that the first insertion reaction (reaction 1) was very recently reinvestigated with the proposal of a simpler mechanism other than the separated two-phase region up to $Li_{0.5}FeF_3$ and solid-solution up to $LiFeF_3$. Through in situ XRD, Tan et al. [21] proposed that lithiation involves a single, continuous, topotactic insertion of Li-ions in the range 0–1 Li^+. The insertion would take place along the body diagonal of the trifluoride unit cell only resulting in limited distortion of the lattice parameters as the shifts of the observed diffraction peaks seem to suggest.

While the electrochemical reactions are, at this point, largely understood, the reaction kinetics remained to be addressed, notably the evolution of the polarization during cycling and the interfacial behaviors. Liu et al. [22] published a complete electrochemical study coupling GITT with electrochemical impedance spectroscopy (EIS) in order to clarify the above-mentioned points. In order to compare with previous studies, a high energy ball-milled FeF_3/carbon composite was analyzed in the 4.5–1 V versus Li^+/Li potential window. The authors observed that after the initial Li insertion, the formation of the LiF/Fe nanocomposite resulted in a phase separation. This nanocomposite exhibits a large internal interface between LiF and Fe and its EIS signature is characterized by both high resistance (due to LiF insulating character) and high capacitance (due to the high interfacial surface). It is important to stress that this internal interface does not involve the electrolyte in any case and is purely limited to the interface inside the solid state composite. As one can expect, the phase separation results in a large overpotential which is experimentally observed as a strong polarization of FeF_3 electrochemical curve. During charge, recombination involves migration of heavy atoms such as Fe and F to recover the rutile Li_xFeF_3 ($x \sim 0.5$). This step is initially fast at the interface but soon slows down resulting in a limitation to the reaction rate.

As a final observation, the authors also implied that such a behavior arises when the reaction kinetics are faster at higher rates. In other words, rapidly reversing the LiF/Fe nanocomposite is much more effective than let it rest for some time. This observation has, of course, important implications on the cell design should a commercial FeF_3-based cathode see the daylight.

Apart the electrochemical mechanisms involved in the conversion reaction, studies devoted to analyzing other problems involving iron trifluoride in secondary cells appeared. For example, Zhou et al. [23–25] studied the thermal stability of the usually ball-milled commercial iron trifluoride/carbon mixture by thermogravimetry (TGA-DSC). They analyzed the thermal behavior of samples cycled limiting the discharge depths to 1 Li, 2 Li and 3 Li. It was demonstrated that FeF_3 is thermally stable and that reaction peaks were observed only due to electrolyte decomposition. No appreciable reaction was observed for 1 and 2 Li reactions except for the formation of a surface film due to the contact electrode–electrolyte (decomposing in the range 100–160 °C). It is interesting to point out, however, that the iron formed at the end of the complete conversion reaction (3 Li) was found to react with PF_5 present in the electrolyte (decomposition of $LiPF_6$) at temperatures around 300 °C.

Synthesis techniques, new electroactive phases and approaches for the improvement of FeF_3

Besides the well-known mechanotreatment of commercial FeF_3 powder, authors have been increasingly interested in the direct synthesis of the electroactive species by soft-chemistry approaches such as sol–gel or hydrothermal/solvothermal methods. Iron trifluoride is generally obtained by treating the anhydrous trichloride in a flow of HF. The reaction at RT yields amorphous FeF_3 while crystalline powders can be obtained after heat treatment (at least 600 °C) [26]. This is obviously an expensive and complex procedure if one wants to produce larger quantities at the industrial scale.

Hydrated phases

The first report of a full synthesis of a FeF_3 based material at low T appeared in 2009 with the publication of Wu et al. [27, 28]. By reacting a $FeCl_3$ water solution with NaOH, the hydroxide $Fe(OH)_3$ precipitates. This precipitate easily reacts with an excess of HF water solution to give a number of hydrated iron trifluorides some of which can be partially dehydrated in mild conditions (70–150 °C at atmospheric pressure or under vacuum). The obtained fluoride was then ball-milled together with MoS_2 or V_2O_5 in order to promote the formation of a mixed conducting matrix to provide easier transport of both ions and electrons. Both samples exhibited interesting capacities (around 160 and 200 mAh $g^{-1}$for the MoS_2 and the V_2O_5 composites, respectively, at 0.1 °C rate) for at least 30 cycles. However, the authors limited the cycles to a relatively narrow potential window of 4.5–2 V vs Li^+/Li thus exploiting only the part of the curve relative to the insertion reaction (*cf.* Eq. 1). As one can imagine, some discrepancies in the electrochemical curves arise since in the reported composites FeF_3 is actually hydrated and structurally different from the ReO_3-type FeF_3 phase. Thus, the intercalation potential is relatively lower decreasing from the 3.4–3 V for the anhydrous ReO_3-type FeF_3 phase [17] to the 3–2.6 V for the reported hydrated phase.

Recently Louvain et al. [3] published a comprehensive overview of the materials obtained when even small traces of water are present in the reaction media. Since FeF_3 is highly hygroscopic, a number of H_2O-containing structures have been studied since the late 1960s resulting in a quite variegated picture. In general, both anhydrous FeF_3 and all its partially hydrated forms are not stable in air at RT and will slowly accept water molecules until the $FeF_3{\cdot}3H_2O$ stoichiometry is eventually reached. If the hydration

Table 1 Crystallographic parameters of the main iron trifluoride hydrated forms

Structure	Space group	*a* (pm)	*b* (pm)	*c* (pm)	α (°)	β (°)	γ (°)	Z
α-$FeF_3{\cdot}3H_2O$ [29]	*R-3m*	951.4	951.4	478.8	90	90	120	3
β-$FeF_3{\cdot}3H_2O$ JCPDS 32-0464	*P4/n*	784.6	784.6	387.7	90	90	90	2
$FeF_3{\cdot}0.5H_2O$ [30] (pyrochlore)	*Fd-3m*	104.1	104.1	104.1	90	90	90	4
$FeF_3{\cdot}0.33H_2O$ CPDS 76-1216 (hexagonal tungsten bronze)	*Cmcm*	742.3	1273.0	752.6	90	90	90	12

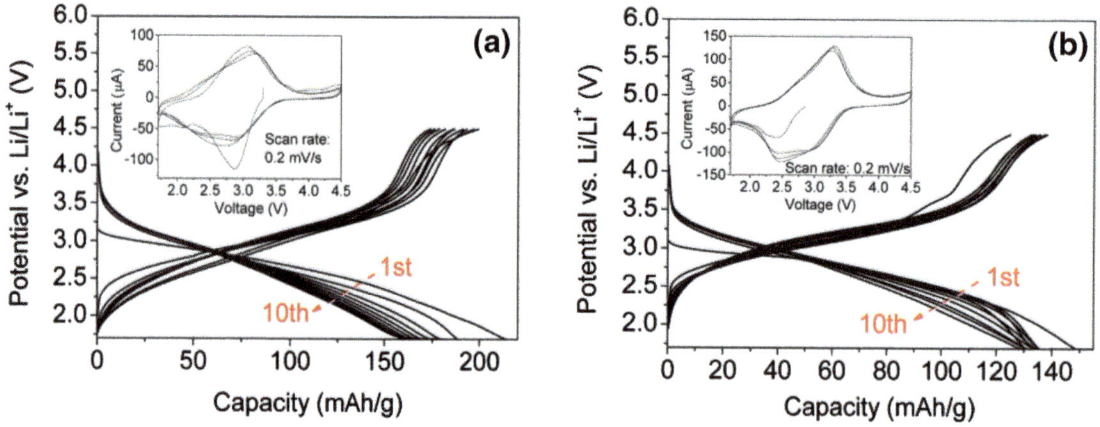

Fig. 2 Galvanostatic cycles of 5 wt % SWCNTs composites of HTB-$FeF_3{\cdot}0.33H_2O$ (**a**) and pyrochlore–$FeF_3{\cdot}0.5H_2O$ (**b**) in the voltage window 4.5–1.7 V vs. Li^+/Li (0.1 °C rate). The inset displays the respective cyclic voltammograms at a constant scanning rate of 0.2 mV s^{-1}. Adapted with permission from Ref [36]. Copyright 2011, American Chemical Society

process is observed starting from the anhydrous form, a large series of stoichiometries can be obtained the most common of which are reported in Table 1. However, solution-based syntheses can also lead to mixed valence compounds such as: $Fe_2F_5{\cdot}7H_2O$, $Fe_{1.9}F_{4.75}{\cdot}0.95H_2O$ or $Fe_3F_8{\cdot}2H_2O$ [3].

The picture is complicated also by the fact that both the pyrochlore and the hexagonal tungsten bronze (HTB) structures (see Table 1) can be obtained either in the fully dehydrated FeF_3 form [3] or in the partially hydrolyzed form with substitution of F anions by OH groups. Considering for example the HTB-$FeF_3{\cdot}0.33H_2O$ structure, it is possible to obtain it through dehydration of $FeF_3{\cdot}3H_2O$ ($T > 120$ °C [3]). Further thermal treatment yields anhydrous HTB-FeF_3 [3] or an OH substituted structure such as HTB-$FeF_{3-x}(OH)_x{\cdot}yH_2O$ ($y < 0.33$) [31–33]. By conventional structural analysis techniques, like XRD, it cannot be clearly distinguished between these three forms so other means have to be deployed to analyze the material in detail. In particular Mößbauer spectroscopy can help due to its sensitivity to the atomic environment as well as by infrared (IR) spectroscopy after adsorption of small molecules (generally NH_3 or pyridine).

In a series of studies, the group of Maier, demonstrated the electrochemical activity of both HTB and pyrochlore hydrated iron trifluorides [30, 34–38]. The synthesis involves $Fe(NO_3)_3{\cdot}9H_2O$ in 1-butyl-3-methylimidazolium tetrafluoroborate ($BmimBF_4$) ionic liquid at 50 °C; the ionic liquid serving multiple roles at a time such as: solvent, soft template and fluorine source (BF_4^- anion degrades to BF_3 and F^- in presence of water). Carbonaceous conductivity enhancers, such as single-walled carbon nanotubes (SWCNTs), can be added to the reaction medium to increase the overall electrochemical performances. The HTB-$FeF_3{\cdot}0.33H_2O$ structure is built around very large hexagonal cavities (around 550 pm across) forming channels running in the [001] direction. Water molecules occupy 1/3 of the total free positions in the channels. This water serves as a strong structural stabilizer during electrochemical lithiation with the possibility to accommodate Li^+ up to the full 2/3 of the remaining vacant positions. HTB-$FeF_3{\cdot}0.33H_2O$ thus shows an intercalation behavior (solid solution) in the region 4.5–1.6 V versus Li^+/Li with the possibility to reach the theoretical capacity of 150 mAh g^{-1} (Fig. 2a). The large cavity also permits fast Li^+ diffusion with good capacities retained even at 1 °C rates [35].

The pyrochlore–$FeF_3{\cdot}0.5H_2O$ is also built around slightly smaller hexagonal cavities compared to the HTB homologue (around 500 pm across), but the channels run in the [110] direction instead. However, pyrochlore is also a more unstable structure and it undergoes irreversible amorphisation at potentials below 1.6 V. This phenomenon is due to the loss of structural water dislodged from the channels and has secondary consequences on partial electrolyte degradation [38]. However, the pyrochlore structure seems also to produce much better distributed composites with carbonaceous additives leading to a longer-lived cathode material [30, 36].

The same working group also demonstrated the capability of both structures through sodium intercalation with comparable 140–150 mAh g^{-1} in the region 4–1.2 V versus Na^+/Na independent of the chosen compound [30, 37].

Following the direction traced with the synthetic method developed by the Maier's group, Li et al. [39] obtained a HTB-$FeF_3{\cdot}0.33H_2O$/graphene nanosheets (GNS) composite. During synthesis, GNS are stabilized in ethanol by the addition of $BmimBF_4$ ionic liquid which helps to avoid restacking due to the attraction between the $[Bmim]^+$cation and the π-electrons of graphitic layers. The resulting

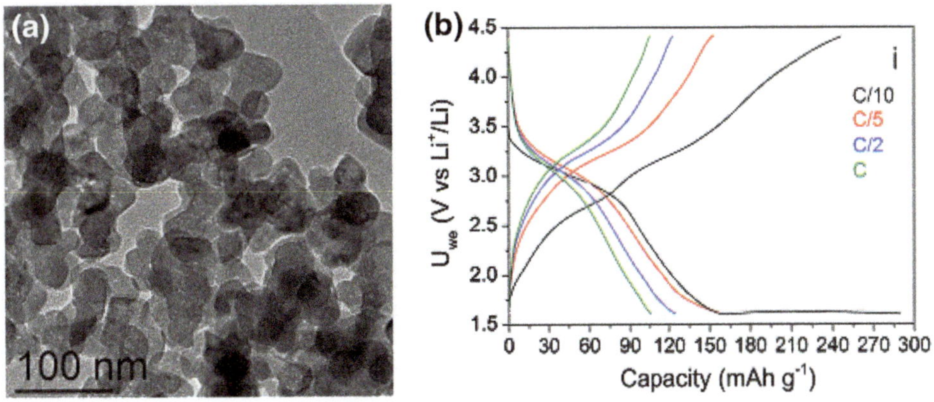

Fig. 3 TEM (**a**) and galvanostatic cycling (**b**) of $FeF_3 \cdot 0.33H_2O$/RGO composite at various C rates. Voltage window 4.4–1.6 V vs. Li^+/Li as reported in Ref [41]

composite material shows much spaced fluoride particles, 10 nm in average, anchored on the graphene sheets even if a small percentage of the carbonaceous source remains after synthesis (less than 10 wt % from elemental analysis). In the insertion reaction electrochemical window (4.5–1.7 vs. Li^+/Li) a very good performance at the higher C rates was recorded (90 mAh g^{-1} at 20 °C rate) confirming the importance of a careful address of the electrical insulating nature of fluoride particles. Prolonged cycling showed a stability lasting for over 200 cycles at 10 °C rate.

Lu et al. [40] employed the same ionic liquid-based synthesis, but implemented it by microwave irradiation (80 °C for 5 min). The authors obtained the pyrochlore type $FeF_3 \cdot 0.5H_2O$ regardless of the amount of ionic liquid added to the reaction medium. However, the sample morphology is clearly influenced by the amount of ionic liquid and in particular with its increase, the resulting particle morphology changes from octahedrons to spheres with porous surface. These latter are most probably secondary particles (around 200 nm in diameter) formed by the agglomeration of smaller primary particles (as already observed by Li et al. [35, 38]). Electrochemical results for these porous spheres in the region 4.5–1.6 versus Li^+/Li are comparable to those reported for hydrated fluoride phases with an average of 140 mAh g^{-1} for 100 cycles at around 0.07 °C rate.

Microwave irradiation has been also exploited by Di Carlo et al. [41] who demonstrated an ionic-liquid-free approach to HTB-$FeF_3 \cdot 0.33H_2O$. In this study, the $BmimBF_4$ ionic liquid was substituted by a methanolic solution of HF and a high boiling point alcohol acting as a solvent (benzyl alcohol). By microwave irradiation (150 °C for 10 min), both a pure product and a reduced graphene oxide-based composite have been successfully synthesized with the formation of C–F bonds between fluoride particles and the carbonaceous substrate (Fig. 3).

The latter in particular, was easily capable of cycling at its full theoretical capacity after more than 50 cycles but, in a more interesting way, it showed the possibility of delivering a quite stable 100 mAh g^{-1} at 1 °C rate in the potential window 4.4–1.6 versus Li^+/Li.

Very recently, also ball-milling has been revisited for the direct synthesis of HTB-$FeF_3 \cdot 0.33H_2O$-acetylene black composites [42]. $Fe(NO_3)_3 \cdot 9H_2O$ and acetylene black react with NH_4F in an ethanol/polyethylene glycol mixture (1:1 volume) for 10 h in stainless steel milling vessels. A mild heat treatment to remove NH_4NO_3 follows (120 °C, 24 h). However, unlike in previous examples where a link between the carbon and the fluoride particles can be identified in the form of C–F bonds [30, 41], there is no evidence of chemical reaction for this composite, suggesting a behavior more related to intimate mixing/coating than to real bonding between the fluoride particles and the carbonaceous matrix. Electrochemical results are in line with the pure phase HTB-$FeF_3 \cdot 0.33H_2O$ with 150 mAh g^{-1} delivered at 0.1 °C rate in the potential window 4.4–1.5 versus Li^+/Li.

Besides carbonaceous additives, another approach to improve the electronic conductivity of iron trifluoride is doping with other transition metals. As it has been predicted that Co substitution can extensively decrease the band gap from 4.49 eV for the pure fluoride to 1.46 eV for $Co_{0.5}Fe_{0.5}F_3$ [43], experimental proof soon followed. In the study of Liu et al. [44], Co-doped FeF_3 was prepared by precipitation in water ($FeCl_3 \cdot 6H_2O$, $CoCl_2 \cdot 6H_2O$ and NaOH). The resulting hydroxide was then dissolved in concentrated HF (40 wt % in H_2O) at RT and heat-treated at 170 °C (12 h under argon). Ball milling with acetylene black (15 wt %) yielded composites of general formula: $Fe_{1-x}Co_xF_3$/C, which underwent further annealing at 150 °C for 3 h. The resulting fluoride-based composite does not show signs of phase separation by XRD. The primary phase is ascertained to be a doped HTB-$FeF_3 \cdot 0.33H_2O$ with slightly smaller cell volume consistent with the smaller ionic radius of Co(III) compared to Fe(III). The best electrochemical response was obtained for the composition $Fe_{0.95}Co_{0.05}F_3$/C although only the 4.5–2 V versus Li^+/Li window was analyzed (Fig. 4). The

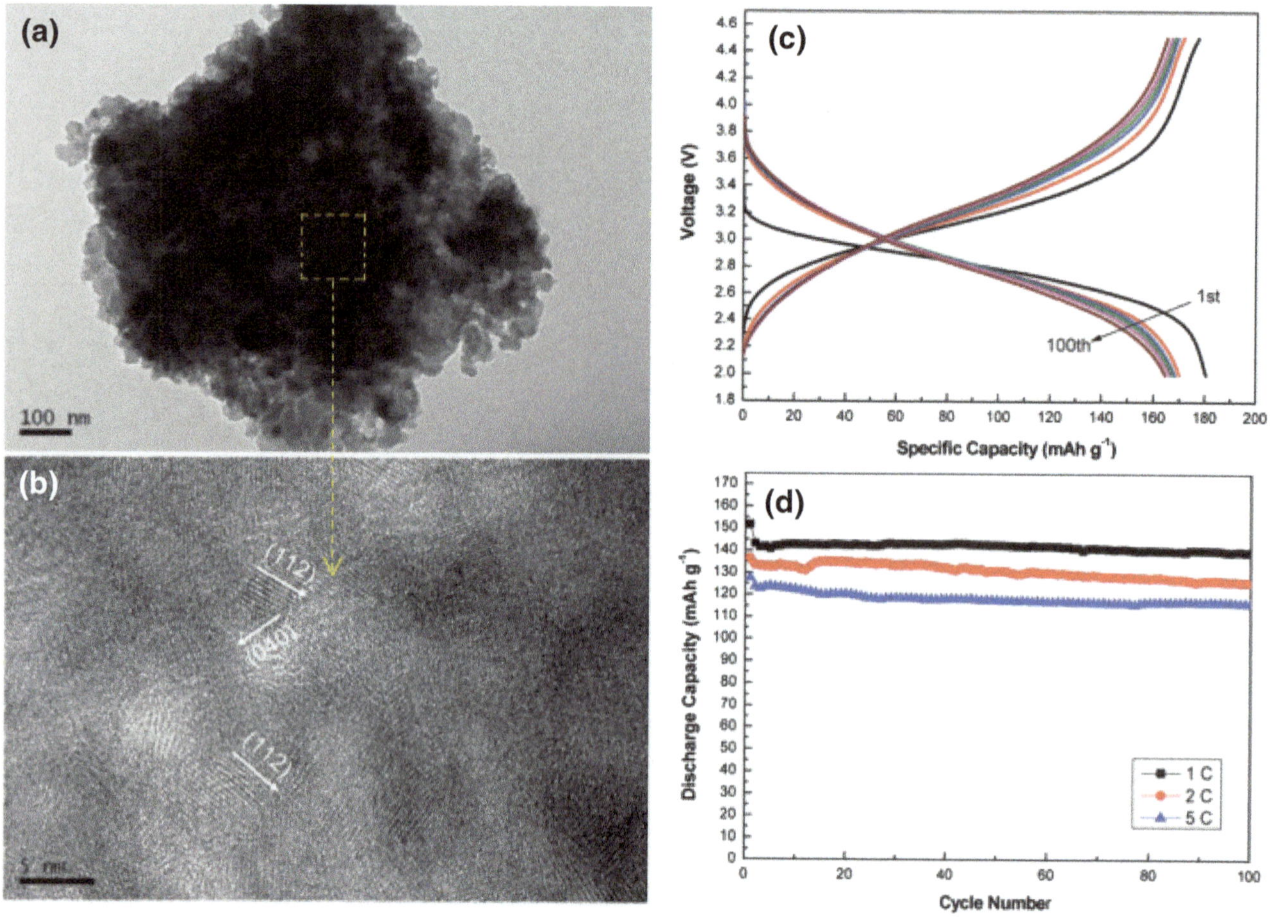

Fig. 4 TEM (**a**) and HR-TEM (**b**) images of $Fe_{0.95}Co_{0.05}F_3/C$ composite. Primary particles of around 10 nm. Galvanostatic cycling of the composite (0.5 °C rate) and its cycling behavior at various C rates are shown in (**c**, **d**), respectively. Voltage window 4.5–2 V vs. Li^+/Li as reported in Ref [44]. Images are courtesy of X. Wang, adapted with permission

composite was able to deliver around 160 mAh g^{-1} after 100 cycles at 0.5 °C and no fading was recorded for the same prolonged cycling at increased C rates. Further analysis based on the GITT technique confirmed the beneficial effect of doping demonstrating an increased Li^+ diffusion coefficient of 2×10^{-12} cm^2 s^{-1} at around 2.7 V versus Li^+/Li (midway of 1st discharge). For the sake of comparison, Li et al. reported an average value close to 10^{-14} cm^2 s^{-1} for the pure, non-milled HTB-$FeF_3{\cdot}0.33H_2O$ considering aggregates of around 350 nm [35].

In a recent publication, Liu et al. [45] compared the electrochemical response of anhydrous FeF_3, of HTB-$FeF_3{\cdot}0.33H_2O$ and of the highly hydrated $FeF_3{\cdot}3H_2O$. Interestingly, the three phases have similar behaviors in the enlarged voltage region 4.5–1 V versus Li^+/Li with an initial plateau appearing in the region 3–2 V related to the insertion reaction and a second plateau below 1.7 V. The authors conclude that the HTB structure shows the best electrochemical performances during the intercalation process, but loss of structural order leads to increased performance degradation when the conversion reaction is included.

Regardless, some studies focused on the fully hydrate phase: $FeF_3{\cdot}3H_2O$. The $FeF_3{\cdot}3H_2O$ structure is built around chains of apex-sharing octahedrons running along the [001] direction. In the apex positions, two fluorine atoms are arranged, while the four basal positions are statistically occupied by either fluorine or water (two of each). Each octahedron chain is then surrounded by four chains of exclusively water molecules, which create large spaces between each octahedron sequence originating roughly rectangular cavities around 46 pm long and 26 pm large. Since, unlike in HTB-$FeF_3{\cdot}0.33H_2O$, water molecules do not occupy the free positions in the channels, the structure is stabilized without further effect on Li^+ diffusion. Thus, in principle, during the intercalation process Li^+ diffusivity should be higher in $FeF_3{\cdot}3H_2O$ than in HTB-$FeF_3{\cdot}0.33H_2O$.

Shi et al. [46] obtained the tri-hydrate by simple precipitation ($FeCl_3$–cetyl-trimethylammonium bromide surfactant, CTAB, -HF mixed water solution), while

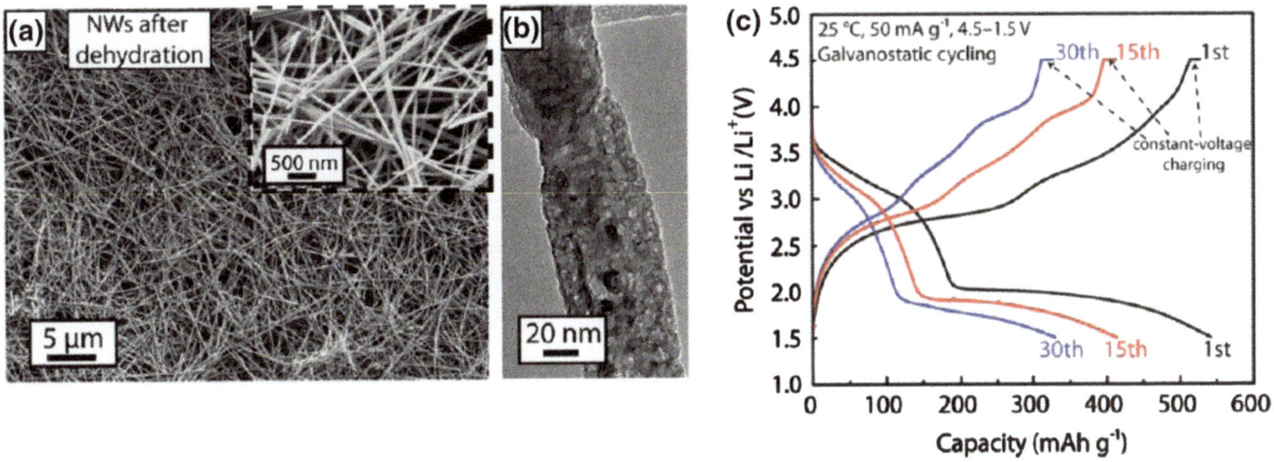

Fig. 5 SEM (**a**) and TEM (**b**) images of the iron fluoride wires after annealing with their relative galvanostatic cycles (**c**).Voltage window 4.5–1.5 V vs. Li^+/Li (~0.07 °C rate). Adapted with permission from Ref [48]. Copyright 2012, American Chemical Society

composites were obtained after ball milling with carbon black. Analysis based on cyclic voltammetry results (CV) confirmed the assumption of a better Li^+ diffusivity demonstrating an increased Li^+ diffusion coefficient of 7×10^{-13} cm^2 s^{-1} at RT. For the sake of comparison, Li et al. [35] reported an average value close to 10^{-14} cm^2 s^{-1} for the pure, non-milled HTB-FeF_3·0.33H_2O considering aggregates of around 350 nm. The tri-hydrated fluoride also delivers a robust 210–200 mAh g^{-1} in the region 4.5–1.5 V versus Li^+/Li at 0.1 °C rate for at least 30 cycles. However, unless the structure can be stabilized in some way, the instability due to the loss of structural water results in a tendency to steady performance degradation with cycling.

Synthesis followed by a post-treatment

For the synthesis of water-free ReO_3-type FeF_3 except for solid state reactions, a post-treatment to eliminate structural water is needed. In most of the cases, a second treatment is also involved, allowing particle size reduction, intimate mix with carbon additives or the formation of specific structures. In this section, we will examine the various reports which have appeared on this specific subject.

Simple heat treatment for water removal

One common synthesis method is the simple precipitation usually involving an ethanol or water solution of $FeCl_3$ or $Fe(NO_3)_3$ mixed with an ethanol or water solution of HF or other fluorine source (NH_4HF_2 for example). Surfactants, like CTAB, or carbon additives (carbon nanotubes, graphene oxide or reduced graphene oxide, etc.) are sometimes added to the reaction mixture in order to limit particle growth or provide the needed electronic conduction enhancement. Products are then submitted to a dehydration step by annealing (generally not above 400 °C).

In their report, Li et al. [47, 48] employed a precipitation method followed by a dehydration step (2 °C min^{-1}, 350 °C under Ar). The obtained material consists of ReO_3–FeF_3 with only trace amounts of FeF_2 impurity. The peculiarity about the mentioned work is that a large excess of HF is used in respect to the Fe precursor (roughly 420 times). This large difference favors the growth of fluoride wires several hundreds of micrometers long and, in average, 80 (pristine) to 50 nm thick (after annealing, Fig. 5a, b). The authors attribute the preferential 1D growth to the formation of screw dislocations [49] also helped by the fact that F^- is a good complexing ligand for Fe^{3+} and stabilizes the FeF_6^{3-} complexes. The wires could be successfully cycled with only the customary addition of carbon black during electrode preparation (20 wt %, Fig. 4c).

This publication [48] reports also an important insight in the electrochemical reaction of these fibers. From TEM observations on fully discharged electrodes, single-crystal α-Fe nanodomains (3–5 nm) are formed consistently with the already described conversion reaction. However, these nanoparticles clearly appear to be interconnected through their common [110] family of planes to form a continuous network. Such a behavior, although never reported for the trifluoride, was recently demonstrated to occur for a FeF_2/C composite and was ascribed to the small diffusivity of Fe ions thus producing a metallic iron backbone encapsulated in LiF [50]. On the other hand, upon charge the usual reconversion to defective trirutile is observed but the reaction is constantly hindered up to complete electrode failure. At the 50th charge, significant amounts of unconverted LiF and α-Fe are present but the exact reason of such degradation is still to be reported.

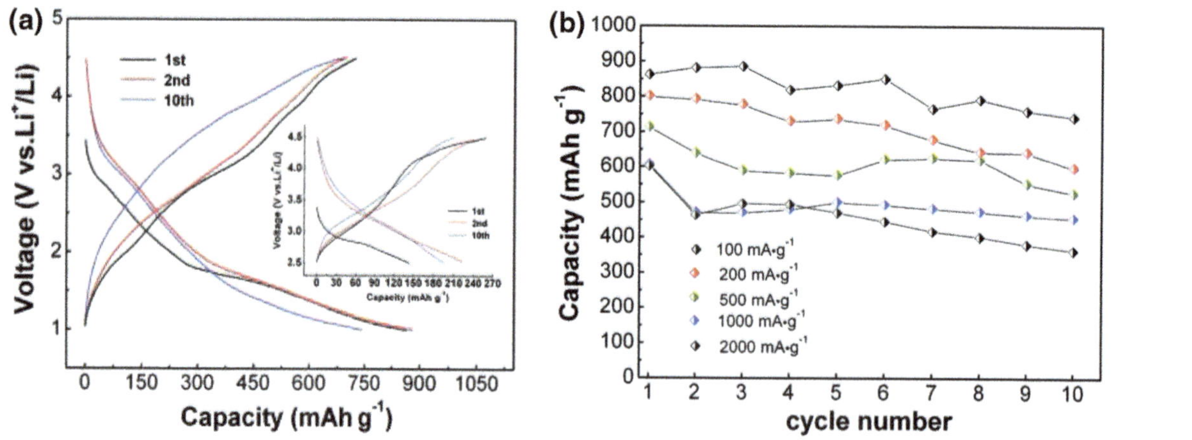

Fig. 6 **a** Galvanostatic cycles of one of the trifluoride composite electrodes described by Li et al. The *inset* displays the galvanostatic behaviors of the composite in the voltage window 4.5–2.5 V vs. Li^+/Li while the relative cycling performances at various rates are reported in **b**. Adapted with permission from Ref [54]. Copyright 2010, American Chemical Society

Other papers involve drying after precipitation either without [51] or with the presence of carbonaceous additives (for example reduced graphene oxide [52] or carbon nanotubes [53]). In the three above cases, however, although the authors provide XRD signatures suggesting full dehydration of $FeF_3{\cdot}3H_2O$ to ReO_3-type FeF_3, the reported electrochemical signatures in the region 4.5–1.5 versus Li^+/Li are very close to that of the hydrated phases (HTB or pyrochlore) suggesting either water presence in the compounds or partial dehydration/hydrolysis as discussed for the hydrated phases in the previous section [31–33]. The group of Maier [37, 38] clearly observed that the final fluoride product is highly influenced both by the initial reaction T (for example, carrying on a reaction at 0 °C yielded a completely amorphous fluoride) and by the reaction medium (for example, the dehydration process carried in a hydrophobic ionic liquid yielded an almost amorphous product). So far, no comprehensive study has appeared on this particular issue which certainly is worth exploring.

Li et al. [54] examined the influence of three different synthetic routes involving $Fe(NO_3)_3$ and NH_4HF_2 mixed with different surfactants. In any case, annealing (400 °C under Ar) and ball milling (with graphite) were necessary in order to obtain a viable material. However, well dispersed 15 nm particles showed a very high capacity of around 800 mAh g^{-1} stable for at least 10 cycles at roughly 0.14 °C rate (Fig. 6). Extra capacity could be partially attributed to interfacial storage (see "Thin films" for a more complete explanation of this phenomenon), while the impressive features at high and very high current densities can be the effect of an important carbon load (52 % of the electrode mass) diluting the active material and increasing the electroactive surface area as postulated by the authors themselves.

Coating: carbonaceous additives and protective oxides

The most common coating process involves the addition carbonaceous additives which are known to greatly enhance electronic conductivity of the electroactive material. For fluorides, graphene oxide/reduced graphene oxide (or graphene nanosheets as they are sometimes called) are the most widely used additives. Although carbonization of a carbon source (such as citric acid, a sugar or other compound) is often used to coat cathode materials such as $LiFePO_4$ at high temperatures (e.g., 650 °C [55]), these temperatures are not compatible with transition metal fluorides. HF evolution (resulting in structure decomposition) for the HTB structure can be recorded as low as 170 °C under Ar atmosphere, practically as soon as water is removed from the hexagonal channels [32]. For this reason, it is important to study coating procedures that can be carried on at lower temperatures.

Addition of graphene/reduced graphene oxide to the precipitation mixture was reported by Liu et al. [56]. The results in terms of XRD signatures, both before and after the heat treatment (150 °C for 10 h in N_2 flow), are very similar to the ones reported by Kim et al. [53] but the electrochemical signature (4.5–1.5 versus Li^+/Li) is close to that of amorphous FeF_3 reported by Li et al. [38] suggesting either water presence in the compounds or a partially amorphous state as previously discussed. However, in the wider potential window, the graphene-based composite seems to withstand cycling in a much better way. For example, the carbon nanotube-based composite reported by Kim delivers around 400 mAh g^{-1} after 10 cycles at 0.07 °C, while the performance of the amorphous FeF_3 reported by Li steadily degrades up to the 20th cycle where 250 mAh g^{-1} are registered at 0.1 °C rate. In comparison, the capacity of the graphene/reduced graphene oxide-based

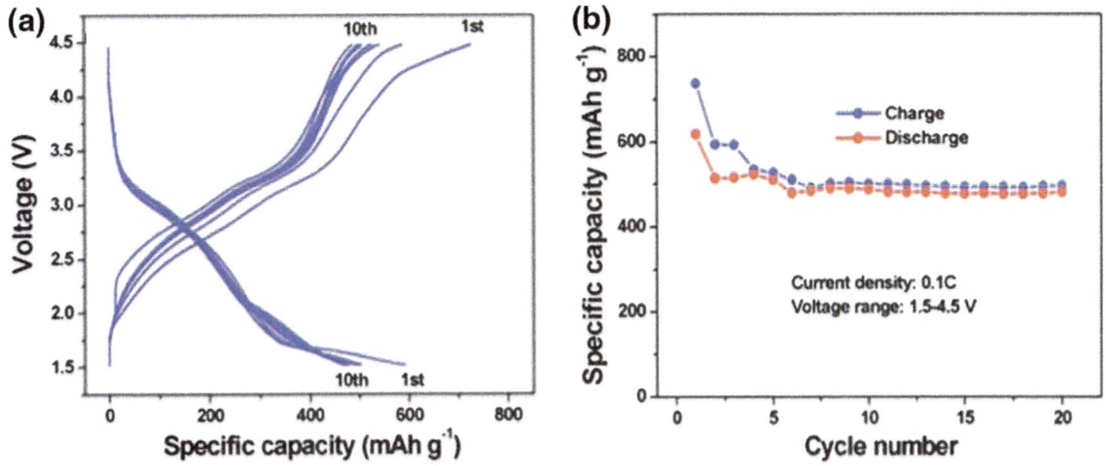

Fig. 7 Galvanostatic cycling (**a**) and cycling behavior (**b**) of the iron fluoride graphene/reduced graphene oxide composite. Voltage window 4.5–1.5 V vs. Li^+/Li as reported in Ref [56] (0.1 °C rate). Images are courtesy of Y. Zhou, adapted with permission

composite stabilizes at 500 mAh g^{-1} delivered for at least 20 cycles at 0.1 °C rate (Fig. 7).

Another graphene/reduced graphene oxide composite was reported by Ma et al. [57] through a rather convoluted synthesis involving various steps. In a first step, Fe_3O_4/C composites were obtained from graphene oxide, sodium acetate (NaAc) and $FeCl_3{\cdot}6H_2O$ dissolved in ethylene glycol (EG) and solvothermally treated at 200 °C for 12 h. An annealing step (500 °C, 2 h, Ar/H_2 atmosphere) was carried out in order to reduce (at least partially) graphene oxide. The composite was solvothermally treated a second time with HF (48 wt % H_2O solution) to transform Fe_3O_4/C into $FeF_3{\cdot}3H_2O/C$. Further annealing (200 °C, 4 h, Ar atmosphere) yields FeF_3/C. The first annealing step, promotes the formation of few hundreds nanometer iron oxide particles. This greatly hinders the electrochemical reaction with a fast degradation from 600 to 230 mAh g^{-1} after the first five cycles even at very low current density (4.5–1.5 versus Li^+/Li at 0.03 °C). However, the benefic effect of the intimate contact and, possibly, the anchoring on the reduced graphene oxide additive can be observed on the long period when the large crystals have been repeatedly ground by the protracted conversion reaction. In fact, the composites are stable at around 200 mAh g^{-1} for at least 100 cycles. A similar result can be obtained using commercial Fe_2O_3 [58] instead of magnetite nanoparticles.

It is worth noting that other examples of addition of carbonaceous additives have been reported in "Hydrated phases", namely, in the works of Maier [30, 34, 36] and Di Carlo [41] but we chose to illustrate them in a different section due to the differences in the obtained fluoride product.

Another interesting concept is the coating of cathodic materials with oxides. The reason for coating is the fact that since direct contact with the electrolyte is prevented, dissolution of the cathode material due to the decomposition of $LiPF_6$ is also hindered. Moreover, other benefits have been observed such as: maintenance of ionic conduction pathways, suppression of the release of oxygen and of phase transitions (in case of high-energy materials such as the solid solution $Li_2MnO_3/LiMO_2$, with M = transition metal), and decrease of cationic disorder [59].

Following this approach, Zhang et al. [60] used the oxide coating concept but with the added benefit of increased electronic conductivity. Anhydrous FeF_3 was thus coated with a thin layer of semiconductive α-Fe_2O_3 (hematite) and showed improved electrochemical properties without further addition of carbonaceous additives and/or ball milling. The synthesis of such a composite material involves the synthesis of hematite first (highly agglomerated particles of 100–150 nm) by a sol–gel method (a water solution of $Fe(NO_3)_3{\cdot}9H_2O$, citric acid and NH_4OH is left to gelify and then heat-treated at 500 °C, 5 h under air). The resulting oxide is then fully fluorinated by further heat treatment (475 °C, 5 h under inert gas containing F_2) and rapidly re-oxidized under air (500 °C for a duration varying from 15 to 480 s). The electrochemistry reported in the potential window 4.5–1.2 V versus Li^+/Li (0.07C) is the result of the complex interaction between various active species: the primary FeF_3, the Fe_2O_3 coating and possibly interfacial oxyfluoride formed between the two contiguous layers. Even though, the authors state that: "[...] the improvement of electrochemical performance of FeF_3 resulting from an in situ Fe_2O_3 coating is inferior to that by mixing with carbon-based conductive additives", it is remarkable that such large particles can sustain some cycling without any further treatment to reduce size and/or further enhance electronic conductivity (roughly 260 mAh g^{-1} are delivered after 10 cycles).

Fig. 8 a TEM image of the final FeF_3/PEODT composite. Pore size distribution, from BET analysis, is bimodal and centered at around 7 and 30 nm. b Cycling behavior at various temperatures (1.4 °C rate). Voltage window 4.5–2 V vs. Li^+/Li as reported in Ref [62]. Images are courtesy of X. Zhang, adapted with permission

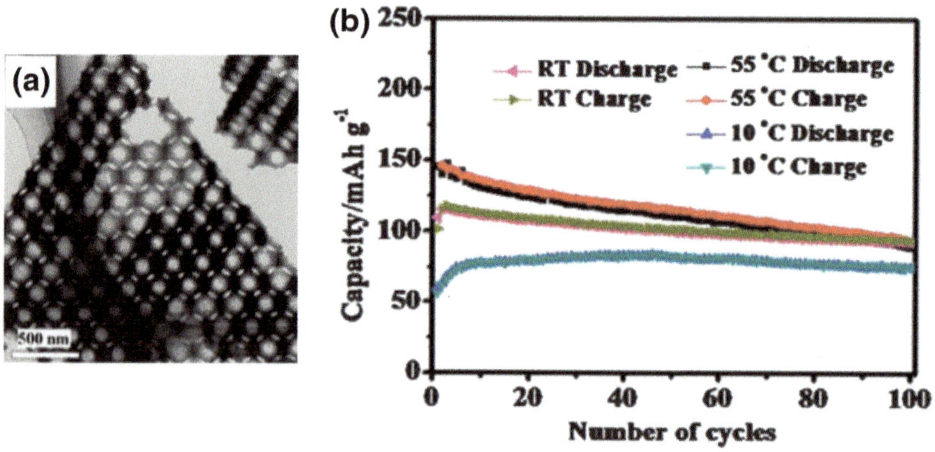

Structuration

In recent years the benefits of ordered three-dimensional (3D) porous nanostructures over nanopowders have widely taken root in the energy-materials community because they result in higher surface areas and more favorable structural stability over simple non-ordered particles. Moreover, the ameliorated diffusivity of the electrolyte through channels and pores would ensure a high specific surface area in contact with the electrode and hence an increased possibility for Li-ions to react. In the specific case of fluorides, there could also be an additional positive effect due to the minimization of nanoparticle aggregation induced by the conversion reaction [61].

An approach to the 3D structuration of the composite material has been reported by Ma et al. [62] who proposed three-dimensionally ordered macroporous (3DOM) structures coated with a conductive material (in this case, polyethylene-dioxythiophene, PEODT). Synthesis-wise, however, the steps necessary in order to obtain such a structure are quite complex. In a first step, polystyrene sphere colloidal crystals were prepared in order to form a template for further reaction. Then $FeF_3 \cdot 3H_2O$ was obtained through the well-known precipitation synthesis involving $FeCl_3$, NaOH and HF in water. In this particular case, the fluoride was purified by freeze drying. In the next step, iron fluoride colloidal suspension in methanol–water (90:10 vol %) was prepared. To the suspension were added polystyrene beads as templating agent (1–2 h). After drying, soaking in toluene (12 h) to remove the polystyrene template, and heat treatment in order to dehydrate the $FeF_3 \cdot 3H_2O$ (140 °C, Ar, 15 h) a porous three-dimensionally ordered iron fluoride structure was obtained. Finally, an ethanolic solution of lithium bis(trifluoromethanesulfonyl)imide (LiTFSI), 3,4-ethylenedioxythiophene (EDOT) and 3DOM FeF_3 structure was heated at 60 °C for 2 h. This results in the polymer coating of the fluoride nanostructure (Fig. 7a) via the in situ polymerization of EODT through the reaction:

$$FeF_3 + LiTFSI + EODT \rightarrow LiFeF_3 + PEODT \quad (3)$$

While FT-IR spectroscopy provides a clear evidence of the PEODT formation, there is no certainty about the lithiation of iron fluoride since the product appears to be amorphous to XRD. Moreover, if Li was already present in the structure (Eq. 3), the material could start its electrochemical cycling with a charge step, however, this has not been tested. The electrochemical response in the 4.5–1.5 V versus Li^+/Li window demonstrates that the 3D architecture is capable of delivering 540 mAh g^{-1} at the 1st cycle (0.07 °C rate) with galvanostatic cycles typical of iron fluoride, i.e., a sloping curve at around 3 V followed by a long plateau at around 1.6 V. Unfortunately, the authors do not concentrate on this enlarged potential window, but instead chose to analyze only the insertion part of the reaction in the 4.5–2 V versus Li^+/Li potential window. The material delivers a stable 200 mAh g^{-1} for 30 cycles at a low rate of 0.03 °C. More interestingly, at high rate (1.4 °C) it delivers 100 mAh g^{-1} for 100 cycles rate (Fig. 8b). This demonstrates that the structuration plays undeniably a key role in the cycling of iron fluoride-based materials.

Impregnation of ordered mesoporous carbon (OMC) was employed by Jung et al. [63]. The initial $FeCl_3$/HF water solution was added, dropwise and for several times, onto OMC with a following drying process (80 °C under vacuum). Calcination (200 °C, for 36 h under Ar) yielded the final composite material. XRD results evidence the formation of pure ReO_3-type FeF_3 with trace amounts of $FeF_3 \cdot 0.33H_2O$. Moreover, because of capillary diffusion during synthesis, most of the fluoride particles appear to be on the inside of the OMC pores rather than being formed onto the external surface (Fig. 9). From the electrochemistry point of view, the authors also choose to analyze only the insertion part of the reaction taking place in the 4.5–2 V versus Li^+/Li window. The main advantage of a carbonaceous-based structuring agent is the certain

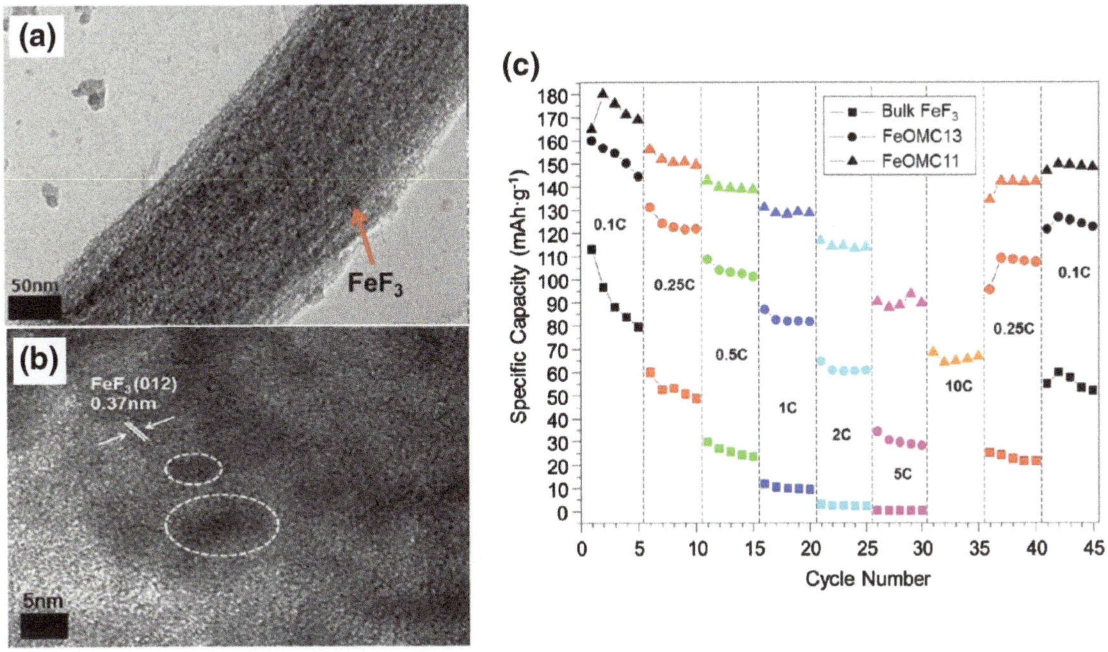

Fig. 9 TEM image of the fluoride/OMC composite (**a**) and an enlargement on a fluoride particle (**b**) along with its cycling behavior at various C rates (**c**). The composite described in this review is labeled FeOMC11 (*triangles*). Voltage window 4.5–2 V vs. Li^+/Li. Adapted with permission from Ref [63]. Copyright 2013, American Chemical Society

increase in electronic conductivity, which permits delivering 70 mAh g^{-1} at 10 °C rate. On the other hand, the high rate achieved could also be attributed to the HTB-type fluoride that has been demonstrated to easily sustain fast rates in respect to the other iron fluoride phases [35].

Among carbon additives, excluding the already mentioned carbon nanotubes, graphene/graphene oxide and ordered mesoporous carbon, activated carbon microbead (ACMB) can be also used as a structuring agent [64]. After the synthesis of ACMB from glucose, the composite material was obtained via a CTAB-assisted precipitation synthesis: a CTAB–$FeCl_3$ mixed water solution was slowly added to an ACMB-HF water solution and stirred for 24 h at RT. The obtained material was then heat-treated at 170 °C for 10 min, Ar atmosphere. The resulting composite is constituted of a mixture of $FeF_3{\cdot}3H_2O$ and HTB-$FeF_3{\cdot}0.33H_2O$ which seems to be grafted at the surface of ACMB spherules (about 30 μm in diameter) thus keeping the spherical morphology. The authors do not show wide area SEM images, so it is impossible to appreciate how homogeneous the coverage is, also taking into account the small initial amount of ACMB (2 wt % in respect to the overall fluoride quantity). The delivered capacity in the 4.5–2 V versus Li^+/Li window was slightly lower than the theoretical 150 mAh g^{-1} with a stabilization at around 120 mAh g^{-1} after 50 cycles (at 0.1 °C rate).

As it has been pointed out in "Simple heat treatment for water removal", surfactants are employed in order to limit particle growth and aggregation and, among them, the most employed in iron fluoride synthesis is CTAB. However, the molecule itself can be also employed as a structuring agent directing the assembly of the particles. In the study of Tan et al. [65], CTAB was dissolved with $Fe(NO_3)_3{\cdot}9H_2O$ in an ethanol/HF mixed solution and solvothermally treated for 2 h at 80 °C. The obtained precipitate was then washed and dried at 180 °C for 12 h. The resulting product, a mixture of $FeF_3{\cdot}3H_2O$ and HTB-$FeF_3{\cdot}0.33H_2O$, consists of quite uniform prismatic/cylindric hollow rods around 3 μm long and 500 nm thick. Unfortunately, the bare structures were not able to meet the requirement of prolonged cycling with a marked capacity fade after 20 cycles (4.5–2 V vs. Li^+/Li at 0.5 °C rate) and so a ball-milling treatment with acetylene black was carried out. The treatment completely destroyed the hollow structures and lead to small particles intimately mixed with the carbonaceous additive with a consequently ameliorated electrochemical response (150 mAh g^{-1} after 100 cycles in the 4.5–2 V vs. Li^+/Li voltage window at 0.5 °C rate) but effectively eliminating every effort initially put into the formation of self-assembled coherent structures.

Another concept has been investigated by Martha et al. [66] who approached structuration from the point of view of electrode preparation instead of synthesis. The generally accepted method consists of pasting an electrode slurry (active material/carbonaceous additive and plasticizer) onto aluminum foils. Instead of Al, the authors used non-graphitic carbon fibers as current collectors and petroleum pitch as a binding agent between iron fluoride and the fibers, thus eliminating the need of an inactive plasticizer (providing only dead weight to the final electrode). After ball milling of commercial FeF_3 and a graphene-based electron conductor, the composite powder was mixed with petroleum pitch and the resulting slurry was coated onto the carbon fibers.The electrodes were then directly used after carbonization (around 450 °C, 5 h under Ar). In the voltage window, 4.5–1 V versus Li^+/Li, 600 mAh g^{-1} have been recorded for the composite at low current density (0.05 °C rate) with good performances also retained at 1 °C rate (a stable 400 mAh g^{-1} for 10 cycles). Moreover, it is noteworthy that the replacement of aluminum by carbon fibers contributes to further reduce the weight of the electrodes permitting higher overall energy per weight unit and thus highlighting the interest of such a technological approach.

Thin films

The current, rechargeable battery technology implies the use of a liquid electrolyte, which needs to be imbibed into a separator. This characteristic brings to the commercial models several restrictions in terms of design, size and leak-proof assembly. In recent years, the concept of all-solid-state (micro)batteries has found large support from the scientific community due to the increased flexibility of the design of stand-alone microelectronic devices and the enhanced applicability brought by the complete absence of leakage (for example medical implants would be much safer than what they are today) [67]. In order to advance on such a concept, it is very important to be able to produce thin, homogeneous films of active material/separator so that miniaturization is possible without losing the advantages of energy densities and cycle life, which are now achieved by Li-ion cells.

Unlike the extensive work conducted on oxides and phosphates [67, 68], only few works have been published on iron fluoride thin films applied to Li-ion batteries. In one of them, Makimura et al. [69, 70] used pulsed laser deposition (PLD) to grow iron fluoride thin films from 80 to 150 nm in thickness. The technique consists in pulsing a laser source with a 2 Hz frequency onto FeF_3 or FeF_2 target. The target is then vaporized creating a thin film on a stainless steel substrate. Typical electrochemical curves for FeF_2 samples deposited at the high T (600 °C) and RT are recorded. All the reported films presented a very large reversible over-capacity at low potentials (between 1 and 0.05 V vs. Li^+/Li) acting as the main electrochemical phenomenon and granting the very interesting performances reported (around 500 mAh g^{-1} for 20 cycles). Such a behavior has been explained as an interfacial interaction of lithium within the Fe/LiF matrix leading to local charging (interfacial storage).

More recently, direct formation of a fluoride-based layer has also been achieved by electrodeposition. Guitian et al. [71] anodized iron strips in an ethylene glycol/water NH_4F electrolyte with a potential of 50 V for duration of 15 min. The resulting electrodes are formed of a highly porous layer (Fig. 10) constituted by a mixture of iron oxy-/hydroxy-fluorides. The electrochemical curve shows many features correlated to the various fluoride species and, in particular to FeO_xF_y [72]. The more interesting characteristic, however, is the enhanced cyclability of the layer which retains both its overall structure and porosity after 200 cycles.

Prelithiation

Current, commercial Li-ion technology comprises a carbon-based anode and lithium-containing cathode material. Adopting such a configuration was inevitable after the problems deriving from Li dendritic growth during charge [73]. As such, the first generation Li/TiS_2 cells were replaced by the so-called "rocking chair" systems as we know them nowadays. At the current state of technology, FeF_3 suffers from a more debilitating plague than electronic insulation or reaction kinetics: the lack of lithium in the initial material. One way to solve the problem would be the use of prelithiated anodes which are difficult to manufacture and handle in practice. Research is focusing on how to introduce lithium in the structure so as to actually give a chance to fluorides in the next generation of accumulators.

The direct synthesis of the charge end-product (the $Li_{0.5}FeF_3$ trirutile structure, see "Early research, characteristics and electrochemical mechanism of FeF3") has been reported in 1969 [74]. Through heat treatment at high temperature (1000 °C) of an equimolar mixture of LiF, FeF_2 and FeF_3 powders, it was possible to obtain the pure $LiFe_2F_6$ trirutile phase. A successive study pointed out that single-crystals can be obtained through hydrothermal synthesis in fluorolytic medium (5 M HF solution at 650 °C) starting from LiF and FeF_3 [75]. A more recent study, however, introduces the possibility of mechanical milling of the same equimolar mixture of fluorides although the process requires very high milling times (40 h for the initial product and another hour after addition of carbon black) [76, 77]. The pure product showed a poor intercalation behavior (less than 120 mAh g^{-1} after 15 cycles in the 4.5–2 V vs. Li^+/Li voltage window at 0.08C rate) which could be ameliorated when the synthesis was conducted with a slight excess of lithium ($Li_{1.2}Fe_2F_{6.2}$). Mößbauer spectroscopy pointed out that the mechanical treatment is rough enough to promote loss of fluorine, resulting in an under stoichiometric compound bearing a

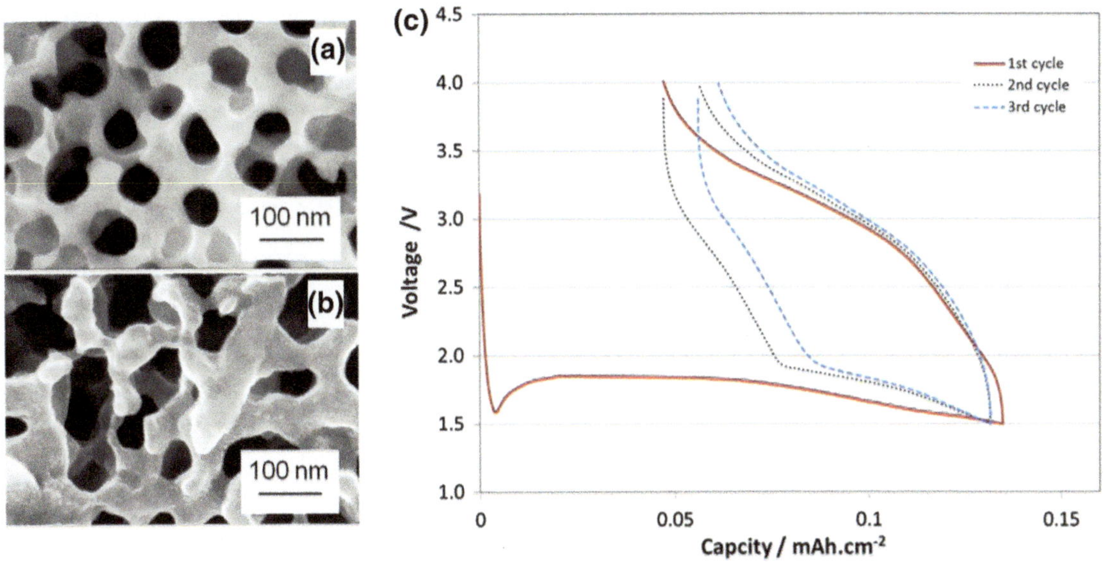

Fig. 10 SEM images of the oxy-/hydroxy-fluoride layer obtained at 50 V before (**a**) and after (**b**) 200 cycles (4–1.5 vs. Li^+/Li at 30 μA cm^{-2}). The first few galvanostatic cycles are reported in (**c**). Figure adapted form Ref [71]. Copyright 2013, with permission from Elsevier

large number of anionic vacancies. Thus, the authors ascribe the ameliorated electrochemical performance (around 140 mAh g^{-1} after 30 cycles in the 4.5–2 V vs. Li^+/Li voltage window at 0.08 °C rate) to the vacancy effect over structural order.

Since the trirutile $Li_{0.5}FeF_3$ requires both harsh conditions for its synthesis and some defective structure in order to work properly, other approaches are being explored with the aim of producing a prelithiated fluoride cathode. The most direct approach consists in obtaining the end products of the conversion reaction, LiF and Fe (see Eq. 2), as an intimate mix.

The first exploration of the concept was performed by Amatucci around 2004 and later in 2011 [78, 79]. Prelithiation was achieved through a solid state route involving high-energy milling of Li_3N, FeF_3 and carbon powders. For the sake of comparison, the same technique was employed also for $BiF_3$and FeF_2. The choice of reactants lies at the base of the following reaction (Eq. 4):

$$Li_3N + FeF_3 \rightarrow 3LiF + Fe + 1/2N_2 \quad (4)$$

Even if the theoretical reaction can seem simple, the authors showed that the final products have a deep dependency on the Li_3N to FeF_3 initial ratio. Although the reaction seems to proceed as expected for the stoichiometric 1:1 ratio, a wide variety of products can be obtained for a lower or higher amount of nitride. For amounts of Li_3N up to 0.5 Li, a rutile-like phase can be obtained which cannot, however, be identified as $Li_{0.5}FeF_3$. For higher amounts up to 1 Li, Li_3FeF_6 and Fe are formed, while for even higher amounts (above the 1:1 ratio), various iron nitride phases were also observed. The Bi-based composites showed a more satisfactory electrochemistry than the Fe-based ones for two main reasons: on the one hand, during the mechanotreatment there can be formation of Li_xBi alloys, which can help the diffusivity of Li into the resulting particles, while Fe does not produce any alloy with Li. On the other hand, iron fluorides have a greater sensitivity to crystallite size due to their slower kinetics in respect to BiF_3 and the mechanotreatment was able to produce particles of roughly 10 nm for the Fe-based products, while half that size was detected for the Bi-based product.

Based on the above results, Liao et al. [80, 81] used magnetron co-sputtering to create a 64-sample library of $LiF_{(1-x)}/Fe_x$ $(0 < x < 1)$ compositions. The principle of the experiment was to rapidly examine a series of possible compositions in order to quantify the optimal LiF/Fe ratio for an acceptable electrochemical result. Without surprise, the best performances were obtained when the ratio of LiF/Fe was 3/1 but the interesting result turned out to be a confirmation of the appearance of interfacial storage (see previous section) for cutoff voltages of 1.2 V versus Li^+/Li. In fact, co-deposited LiF/Fe thin layers already possess an interfacial region between LiF domains and Fe domains that enables overcapacity. Moreover, by Mößbauer spectroscopy, the 'excess' of Li ions could be univocally proved to be located at the LiF side of the interface with the corresponding electrons enriching the Fe side as a consequence.

Further ball milling techniques were then carried directly on the LiF/Fe 3/1 ratio. Li et al. [82] used a two-step milling adding to the commercial LiF and iron powders a nanometric TiN. In the second step, further milling was conducted on the LiF/Fe/TiN with addition of graphite. Instead of promoting the formation of LiF with release of a gas, as described in the work of Amatucci (Eq. 4), here the TiN nanoparticles act as both a grinding-enhancer powder and as a structuring agent forming a conductive nanometric core capable of fixing the electroactive components in order to prevent phase segregation and/or aggregation. Electrodes were prepared with the customary addition of conductive agent and plasticizer (10 wt % each) and good performances were recorded in the enlarged potential window 4.5–1 V versus Li^+/Li with around 300 mAh g^{-1} for 20 cycles at 0.7 °C rate.

A third approach, explored by Kim et al. [83], consists not in the direct milling of the end products of the conversion reaction (such as LiF/Fe in the 3/1 ratio, see Eq. 2) but in mimicking the composition at the end of the insertion reaction (such as LiF/FeF_2 in the 1/1 ratio, see Eq. 1). A two-step milling was again used, first adding FeF_2 to LiF (both commercial powders) and then further milling after addition of graphite. Electrodes were prepared by slurry (20 wt % of conductive agent and 10 wt % of plasticizer) and around 200 mAh g^{-1} were delivered in the potential window 4.8–2 V versus Li^+/Li. The interesting point here is the peculiar reaction of the FeF_2–LiF composite which, although similar in composition to the electrochemically formed $LiFeF_3$, reacts via a different pathway. With the help of X-ray absorption fine structure and near edge structure (EXAFS/XANES), XPS and DRX studies, the authors were able to prove that, during the first charge, the excess of fluorine resulting from Li^+ release, would be incorporated into the remaining FeF_2 to form a higher oxidation state fluoride analogous to ReO_3-type FeF_3 although highly defective. The following reaction would follow the scheme reported in Eq. 1 with an insertion inside the newly formed iron trifluoride.

Ma et al. [84] produced a composite starting from hydrothermally synthesized Fe_3O_4/reduced graphene oxide ball-milled with LiF. The resulting LiF/Fe_3O_4/reduced graphene oxide powder was then heat-treated in a reducing environment (550 °C, 4 h, Ar/H_2). After the heat treatment, TEM images reveal that magnetite is successfully reduced to α-Fe particles (50–60 nm), grafted onto the reduced graphene oxide surface. At the same time, while no evolution of average particle size is recorded for LiF (100–150 nm), the particles do not seem to be well homogenized with iron, and large void spaces can be observed. However, a good cycling stability with a certain decrease in polarization can be observed in the potential window 4.5–1.5 V versus Li^+/Li (0.03–0.14 °C rates) certainly due to the benefic effect of the conductive reduced graphene oxide layers.

The synthetic method employed by Prosini et al. [85] sits in the midway between the formation of LiF with release of a gas proposed by Amatucci's group (Eq. 4) and the previously illustrated oxide reduction to obtain Fe^0. In their study, a mixture of hematite and lithium hydride was placed in an oven at 700 °C for 1 h in N_2/H_2 flow to obtain a Fe/Li_2O composite. A second heat treatment in the presence of NH_4F (500 °C for 1 h, N_2/H_2) transformed the Fe/Li_2O composite into a Fe/LiF composite. Overall, the following reactions occur (Eqs. 5, 6):

$$6LiH + Fe_2O_3 \rightarrow 3Li_2O + 2Fe + 3H_2 \quad (5)$$

$$3Li_2O + 2Fe + 6NH_4F \rightarrow 6LiF + 2Fe + 6NH_3 + 3H_2O \quad (6)$$

The synthesis does not seem to be optimized at the moment as FeF_2 and Li_5FeO_4 impurities have been detected by XRD; morphology is also an issue as the high temperatures favor particle coalescence and growing deeply affecting the electrochemical results.

An original approach was published by Prakash et al. [86, 87] in which carbon nanotubes and other carbon-based structures were directly grown on a LiF/Fe composite by decomposition of ferrocene deriving from a dried dispersion with LiF in diethyl ether. The resulting powders were sealed (under Ar) into a specially designed stainless steel/quartz rotating reactor and heated at 700 °C (35 °C min^{-1}, 10 rpm for 5 h). The concept behind this procedure is that the iron produced from the ferrocene decomposition, will act as catalyst for the growth of nanotubes and core-shell structures deriving from various carbon-based reactive gases (Fig. 11a, b).

In addition to the LiF/Fe/C composite, however, iron carbide (Fe_3C) was also obtained as a result of the deactivation of the Fe catalyst. Electrochemistry studies showed very encouraging results with a stable capacity of 220 mAh g^{-1} for 150 cycles in the broad electrochemical window 4.5–1.3 V versus Li^+/Li, but at a low current density of 0.03 °C. It is noteworthy that the reported electrodes were used as pure powders without further additives (carbon black or PVdF binder) and that the remarkable properties could be related to the high amount of carbon (around 45wt % from elemental analysis) unfortunately, the inactive iron carbide contributes to some hindering of the electrochemical performance due to the consumption of iron.

Another ferrocene-based synthesis was recently introduced by Zhang et al. [88]. The authors produced fibers from electrospun solutions of LiF, ferrocene and polyacrylonitrile (PAN) which were then heat-treated in two steps: a first annealing at 280 °C in air for 5 h followed by a carbonization at 900 °C (8 h under Ar). The resulting composite yielded freestanding and flexible electrodes with some degrees of 3D structuration. Electrochemical tests were conducted in the window 4.2–0.5 V versus Li^+/Li

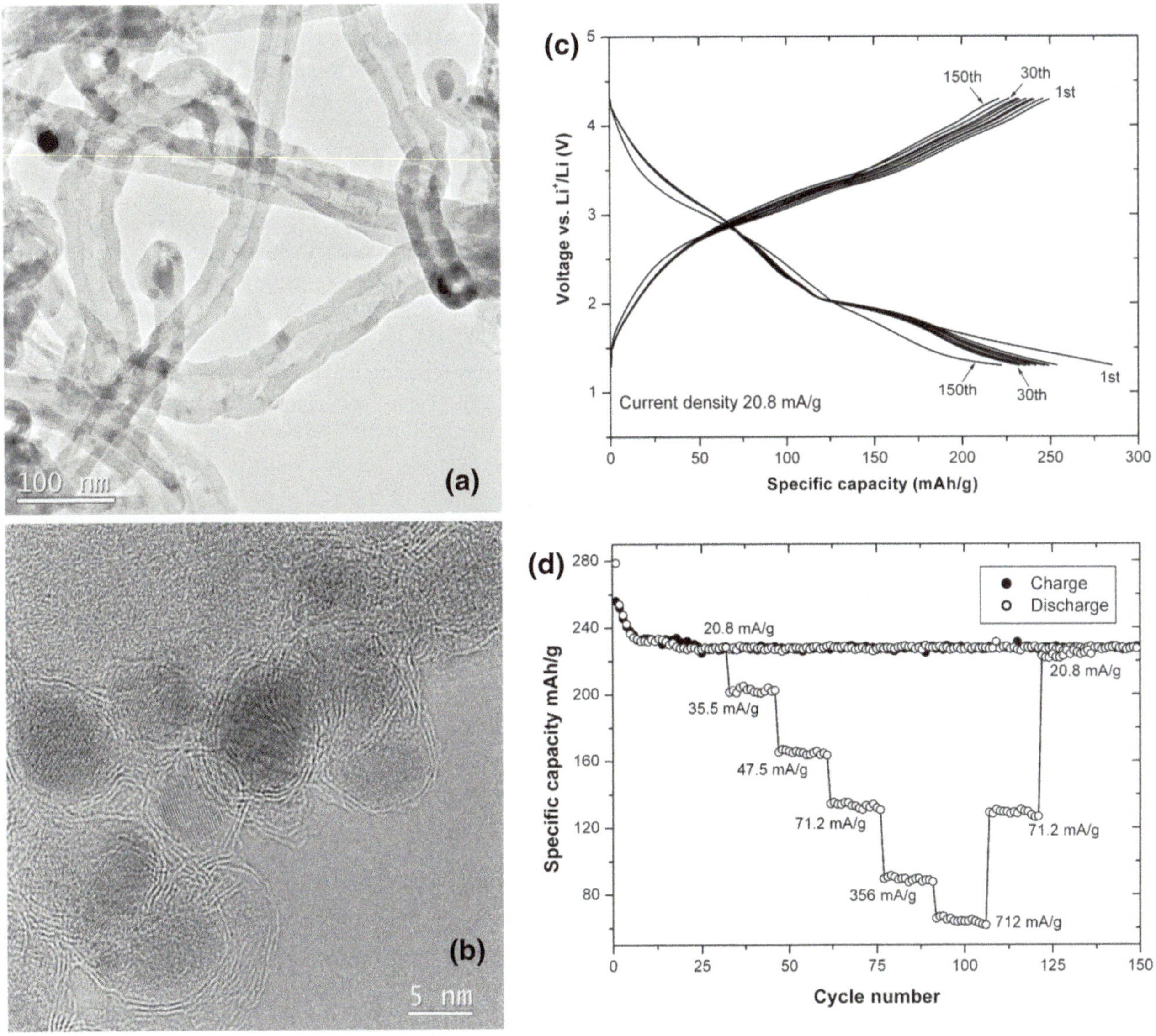

Fig. 11 TEM images of the structures obtained from the ferrocene-based reaction: nanotubes with encapsulated iron particles (**a**) and onion-like coating of iron particles (**b**). Galvanostatic cycles are reported in (**c**) while the cycling behavior at various C rates is shown in (**d**). Voltage window: 4.5–1.3 vs. Li^+/Li. Adapted from Ref [87]. Copyright 2011, with permission from Elsevier

resulting in around 400 mAh·g^{-1} delivered after 50 cycles (current density of 0.05 °C) even if a marked tendency to performance degradation can be observed after the 30th cycle.

Other derivatives

Other iron fluoride compounds could be employed for the development of cathodic materials for the next generation of accumulators. It is the case of lithium fluoroferrate or sodium-iron fluorides which can open new perspectives for the research and development sectors.

Lithium fluoroferrate

Lithium fluoroferrate (Li_3FeF_6) possesses two polymorphs (monoclinic α-phase and orthorhombic β-phase) with structures closely related to cryolite (Na_3AlF_6). Both structures consist of isolated and slightly distorted FeF_6 octahedra which under tilt rotation produce either one or the other polymorph. These octahedra are linked together by interconnected lithium polyhedra in the tri-dimensional structure. The most active group on the electrochemical study of this compound is that of Garcia-Alvarado [89–92] who deeply studied the behavior of both phases. They found a better activity of the α-polymorphin in respect to the β-form and a correlation between particle size and delivered capacity. As a general remark, the α-polymorph can be obtained with smaller particle size, and the synthesis was optimized until the full theoretical capacity was obtained (140 mAh g^{-1}). This optimized synthesis involves stoichiometric amounts of reagents ($Fe(NO_3)_3{\cdot}9H_2O$, HF water solution and Li_2CO_3), which lead to the precipitated fluoroferrate product after addition of 2-propanol at 0 °C. A further ball milling treatment (12 h) yielded the final fluoride/carbon composite. The composite material obtained by this synthetic route was able to deliver 140 mAh g^{-1} in the voltage window 4.5–2.25 V versus Li^+/Li at 0.06 °C rate. More recently [92], simulations pointed out that defects are energetically favoured especially for small crystallites with the most probable defect types being Li-excess (with consequent local reduction of Fe^{III} to Fe^{II}) and/or Frenkel defects (lithium insertion into interstitial sites). Moreover, lithium-ion transport in Li_3FeF_6 seems to be closely related to interstitial migration (low calculated activation energies of 0.4 eV) paving the way for a doping strategy consisting in the introduction of different ions into iron sites, such as divalent Ni, Mg, Co or Mn, to promote Li interstitial compensation.

At the current state, however, the fluoroferrate cannot represent a viable electrode material due the low capacity associated with a 1e⁻ reaction and to problems in lithium extraction. In fact, the charge reaction of Li_3FeF_6 involves Fe^{IV} metal centers and, even if it has been recently calculated that Li_2FeF_6 is a stable compound, there is currently no experimental observation for the latter also due to the prohibitive reaction potential of 6.1 V versus Li^+/Li [93].

Sodium-iron fluorides ($NaFeF_3$ and Na_3FeF_6)

Sodium-ion batteries have recently drawn the attention of the scientific community as low cost and low environmental impact devices with widely available starting materials and low toxicity [94]. As it has been highlighted in previous sections, some iron fluorides have found an application also in Na-ion cells; one example is the HTB-$FeF_3{\cdot}0.33H_2O$ structure. As it concerns the ReO_3-type FeF_3, its structure has the potential to accommodate also Na-ions during the intercalation reaction (Eq. 1) although with slower kinetics than Li-ions [95]. In the case of sodium, however, the intermediate $NaFeF_3$ is well known and can be easily synthesized. Although technically not a Fe^{III} compound, $NaFeF_3$ finds easily its place in this review due to the concept it represents. There is not much research concerning this material [96–98]. An optimized synthesis method, based on the observation that capacity increases with the crystallinity of the final $NaFeF_3$ compound, involves a stoichiometric mixture of $FeF_2$and $NaHF_2$. The reagents are then submitted to a roll-quenching technique where the melt obtained by heating at 1,000 °C (40 s under Ar) is injected into a fast spinning copper roller (2,000 rpm). Ball-milling with acetylene black provides the final composite material which can deliver the full capacity associated with the 1e⁻ reaction (200 mAh g^{-1} at 0.014 °C rate in the potential window 4.5–1.5 V vs. Na^+/Na). A more significative finding was, however, that among the isostructural $NaMF_3$ series (with M = Mn, Fe, Co and Ni), $NaFeF_3$ is the only phase capable of reversible Na^+ deinsertion/insertion. This occurrence finds its explanation mainly in the chemical stability of the de-intercalated fluorides, MF_3 (with M=Mn, Co and Ni) being known fluorinating agents. Only the affinity of iron toward fluorine ensures the ability of $NaFeF_3$ to sustain cycling.

Yamada et al. [99] prepared dispersed and uniform particles by liquid-phase synthesis involving the thermal decomposition of iron and sodium trifluoroacetates in a mixture of oleic acid and oleylamine (10/10 molar ratio) at 280 °C for 30 min under Ar. The resulting particles were mixed with acetylene black by ball milling to yield the final composite material. TEM microscopy studies evidence an average particle size 10–20 nm. The composite delivers 115 mAh g^{-1} at 1° C rate in the potential window 4.5–1.5 V versus Na^+/Na.

Just like lithium fluoroferrate, the sodium analogue, Na_3FeF_6, is related to cryolite structure. Shakoor et al. [100] prepared sodium fluoroferrate through dry ball milling using stoichiometric amounts of sodium and iron fluorides (3NaF:1FeF_3). The composite delivers around 110 mAh g^{-1} (0.1 °C rate) in the potential window 4.25–0.5 V versus Na^+/Na, but capacity faded progressively from the 10th cycle on. The probable reason has been identified by the authors in the difficult-to-reverse conversion reaction since NaF and Fe have been found in the discharged electrode. As a final remark, unless the same concept of interstitial charge compensation, which has been postulated for the lithium fluoroferrate is also valid for the sodium analogue (see Ref [92]), the charge reaction of Na_3FeF_6 to Na_2FeF_6 cannot be the initial step, so a hypothetical device would require a discharge step in order to start its operative life.

Conclusions

Before concluding this review, we would like to present the reader with an overview of the most interesting points the research on iron fluorides applied to batteries has elucidated:

- Pure FeF_3 is highly hygroscopic so that its stability in air at RT is very limited. The compound will accept water molecules producing a series of H_2O-stabilized structures until reaching the $FeF_3{\cdot}3H_2O$ stoichiometry. This hydration process is fairly common in solution-based syntheses and generally yields HTB-$FeF_3{\cdot}0.33H_2O$ or pyrochlore-$FeF_3{\cdot}0.5H_2O$ structures;
- Water molecules act as strong structural stabilizers during the electrochemical lithiation promoting an intercalation behavior in the same potential window where anhydrous FeF_3 would react with the full conversion. This occurrence limits the theoretical capacity (150 mAh g^{-1} for HTB-$FeF_3{\cdot}0.33H_2O$ instead of 712 mAh g^{-1} for FeF_3 at 1.5 V vs. Li^+/Li) but permits faster Li^+ diffusion with good capacities retained even at 1C rates;
- These water molecules can be highly mobile so it is not uncommon to observe loss of structural water resulting in a tendency to steady performance degradation with cycling. One of the possible reasons is that water can hydrolize the $LiPF_6$ salt into LiOH, PF_5 and HF. Therefore, careful structural assessment, coating or other structure stabilization has to be envisaged for such materials;
- When envisaging the production of a water-free FeF_3, the effect of the initial F^- concentration has to be taken into account. A large excess of HF compared to Fe, favors the growth of amorphous fluoride wires. The wire morphology favors the formation of contiguous and interconnected 1D Fe domains at the end of the conversion reaction potentially paving the way for the full $3e^-$ reaction;
- It is possible to create ordered three-dimensional structures for fluoride-based materials, which would result in ameliorated diffusivity of the electrolyte and possibly in the reduction of the nanoparticles aggregation originating from the conversion reaction [61];
- The absence of preliminary Li content is a major obstacle for the development of fluoride-based cathode materials. Prelithiation is thus a point to extensively study;
- An alternative to prelithiation could be Li^+ compensation induced by divalent cation doping. This effect has, so far, been predicted only for the Li_3FeF_6 compound but it is not impossible that a similar behavior could be induced in structures such as the HTB-$FeF_3{\cdot}0.33H_2O$ or pyrochlore-$FeF_3{\cdot}0.5H_2O$;
- Open structures and certain derivatives also suggest a viable exploration of the sodiation reaction for the development of cheaper cells.

As far as it is possible to observe in the actual state of research, the most promising iron fluoride-based cathodic materials are obviously represented by the hydrated HTB and pyrochlore phases. In fact, it has been widely proven that these two structures are easy enough to synthesize through different soft-chemistry-based techniques using low-cost or reusable starting materials. The main hindrances to the development of these materials are represented by the relatively low capacity (150 mAh g^{-1}), the lower, sloped reaction potential (3.2–3.3 V vs. Li^+/Li) compared to the higher, flatter potential of, for example, $LiFePO_4$ (3.4 V vs. Li^+/Li) and the lack of lithium/sodium in the initial compound.

While Fe-based materials offer advantages with respect to environmental and economic considerations, the major difficulty in the electrochemistry of Fe is the lower voltage obtained when compared to other transition metals such as Co or Ni. A way to overcome this issue was demonstrated in 1989 on a comparative study between the three iron analogues $Fe_2(MoO_4)_3$, $Fe_2(WO_4)_3$ and $Fe_2(SO_4)_3$ [101]. Since all three compounds possess very similar metal–oxygen distances and formal charge, the increased redox potential with lithium from around 3.0 V for the molybdate and tungstate to 3.6 V for the sulfate, came as a surprising effect. The "inductive affect", as it became known, was originating from the significantly larger electronegativity of the sulfate groups resulting in an increase of the ionic character of the Fe–O bonds. Thus, it is thanks to the inductive effect that sulfide-based electrode materials always display a lower potential than the corresponding oxides, as well as why the fluoride analogues show even higher potentials. In this last case in particular, manipulations axed on the increase of Fe–F bond ionic character by replacement of the anion would be complex since fluorine is already the most electronegative element. Instead, one way to act on the iron-fluorine bond could be the application of chemical pressure [102]. When smaller cations are replaced by larger ones in a close-packed structure, the result from the point of view of the smaller ion is comparable to an elongation of the bond thereby increasing its ionic nature. This is why accurate doping with Co, Ni or Cu would probably give some result in this direction.

Both capacity and prelithiation could be ameliorated by the careful application of blending, combining different active materials. Just like spinel/layered mixtures ($xLi_2MnO_3{\cdot}(1-x)LiMO_2$ with M = Mn, Co, Ni or their mixtures) [103] orphosphate/silicate mixtures ($xLiFePO_4{\cdot}(1-x)Li_2FeSiO_4$) [104] have demonstrated incremented capacity in regular half-cell tests (usually delivering at least 200 mAh g^{-1}), fluoride mixtures could be easily analyzed in this regard. For example the system $xLiFe_2F_6{\cdot}(1-x)Li_3FeF_6$ has not been investigated so far, while the mixture could certainly represent a valid candidate as a cathodic material.

Besides the more advanced concepts expressed hereby, some fundamental questions still remain to be answered for the iron fluoride hydrated phases in particular. It is noteworthy that since the detailed description of the FeF_3 behavior by Badway, Doe and Yamakawa, no in-depth structural/mechanistical studies have appeared. Except for the Maier's group, who carefully assessed not only the synthesis procedure but also the electrochemical characteristics of both the HTB and pyrochlore systems, all other publications have been focused for the best part on the synthesis procedures and/or the effects of addition of various carbon additives. For example, it has been pointed out that in ionic liquid-based syntheses, the final fluoride products are influenced both by the initial reaction temperature and by the reaction medium (nature of the ionic liquid). So far, no study presents a careful and rationalized assessment of the dehydration products for other synthetic mediums or studies their influence on the final fluoride material. Moreover, as the same structure can exist with different hydrated/hydrolyzed states, the relation to water content-electrochemical behavior has still to be elucidated.

Another valid concept is the study of the electrode preparation/optimisation instead of synthesis. Effects such as current collector influence, separator influence, electrolyte decomposition and slurry composition have found very limited and non-organized reports in the current literature.

Finally, we would like to stress again that probably no practical materials would see the light of day if a solution to the prelithiation issue or the exploration of Na-based derivatives is not extensively investigated.

References

1. Agricolae, G.: Bermannus sive de re metallica (on the nature of metals, or minerals). Dover Publications NY (1986)
2. Tressaud, A.: Functionalized inorganic fluorides: synthesis, characterization & properties of nanostructured solids. Wiley, New York (2010)
3. Louvain, N., Fakhry, A., Bonnet, P., El-Ghozzi, M., Guerin, K., Sougrati, M.-T., Jumas, J.-C., Willmann, P.: One-shot versus stepwise gas-solid synthesis of iron trifluoride: investigation of pure molecular F_2 fluorination of chloride precursors. CrystEngComm **15**, 3664–3671 (2013)
4. Fedorov, P.P., Luginina, A.A., Kuznetsov, S.V., Osiko, V.V.: Nanofluorides. J. Fluor. Chem. **132**, 1012–1039 (2011)
5. Cabana, J., Monconduit, L., Larcher, D., Palacin, M.R.: Beyond intercalation-based Li-ion batteries: the state of the art and challenges of electrode materials reacting through conversion reactions. Adv. Mater. **22**, E170–E192 (2010)
6. Poizot, P., Laruelle, S., Grugeon, S., Dupont, L., Tarascon, J.M.: Nano-sized transition-metal oxides as negative-electrode materials for lithium-ion batteries. Nature **407**, 496–499 (2000)
7. Liang, C., Gao, M., Pan, H., Liu, Y., Yan, M.: Lithium alloys and metal oxides as high-capacity anode materials for lithium-ion batteries. J. Alloys Compd. **575**, 246–256 (2013)
8. Tarascon, J.M.: Key challenges in future Li-battery research. Phil. Trans. R. Soc. A **368**, 3227–3241 (2010)
9. Bauman, H.F.: Limited-cycle secondary battery using lithium anode. In: center, D.D. (ed.). p. 29. Lockheed Missiles & Space Company, Palo Alto (1964)
10. Bauman, H.F., Chiton, J.E., Mauri, R.: Lithium anode limited cycle battery investigation. In: center, D.d. (ed.). p. 81. Lockheed Missiles & Space Company, Palo Alto (1966)
11. Bauman, H.F., Chilton, J.E., Hultquist, A.E., Adams, G.B.: Lithium anode limited cycle battery investigation. In: center, D.D. (ed.). p. 40. Lockheed Missiles & Space Company, Palo Alto (1966)
12. Fiordiponti, P., Panero, S., Pistoia, G., Temperoni, C.: Non-aqueous batteries with BiF_3 cathodes. J. Electrochem. Soc. **125**, 511–515 (1978)
13. Guyomard, D., Tarascon, J.M.: Rocking-chair or lithium-ion rechargeable lithium batteries. Adv. Mater. **6**, 408–412 (1994)
14. Whittingham, M.S.: Lithium batteries and cathode materials. Chem. Rev. **104**, 4271–4301 (2004)
15. Arai, H., Okada, S., Sakurai, Y., Yamaki, J.: Cathode performance and voltage estimation of metal trihalides. J. Power Sources **68**, 716–719 (1997)
16. Badway, F., Pereira, N., Cosandey, F., Amatucci, G.G.: Carbon-metal fluoride nanocomposites: structure and electrochemistry of FeF_3:C. J. Electrochem. Soc. **150**, A1209–A1218 (2003)
17. Badway, F., Cosandey, F., Pereira, N., Amatucci, G.G.: Carbon metal fluoride nanocomposites: high-capacity reversible metal fluoride conversion materials as rechargeable positive electrodes for Li batteries. J. Electrochem. Soc. **150**, A1318–A1327 (2003)
18. Cosandey, F., Al-Sharab, J.F., Badway, F., Amatucci, G.G., Stadelmann, P.: EELS spectroscopy of iron fluorides and FeF_x/C nanocomposite electrodes used in Li-ion batteries. Microsc. Microanal. **13**, 87–95 (2007)
19. Doe, R.E., Persson, K.A., Meng, Y.S., Ceder, G.: First-principles investigation of the Li-Fe-F phase diagram and equilibrium and nonequilibrium conversion reactions of iron fluorides with lithium. Chem. Mater. **20**, 5274–5283 (2008)
20. Yamakawa, N., Jiang, M., Key, B., Grey, C.P.: Identifying the local structures formed during lithiation of the conversion material, iron fluoride, in a Li ion battery: a solid-state NMR, X-ray diffraction, and pair distribution function analysis study. J. Am. Chem. Soc. **131**, 10525–10536 (2009)
21. Tan, H.J., Smith, H.L., Kim, L., Harding, T.K., Jones, S.C., Fultz, B.: Electrochemical cycling and lithium Insertion in nanostructured FeF_3 cathodes. J. Electrochem. Soc. **161**, A445–A449 (2014)
22. Liu, P., Vajo, J.J., Wan, J.S., Li, W., Liu, J.: Thermodynamics and kinetics of the Li/FeF_3 reaction by electrochemical analysis. J. Phys. Chem. C **116**, 6467–6473 (2012)
23. Zhou, M.J., Zhao, L.W., Doi, T., Okada, S., Yamaki, J.: Thermal stability of FeF_3 cathode for Li-ion batteries. J. Power Sources **195**, 4952–4956 (2010)
24. Zhou, M.J., Zhao, L.W., Okada, S., Yamaki, J.: Thermal characteristics of a FeF_3 cathode via conversion reaction in comparison with $LiFePO_4$. J. Power Sources **196**, 8110–8115 (2011)
25. Zhou, M.J., Zhao, L.W., Kitajou, A., Okada, S., Yamaki, J.: Mechanism on exothermic heat of FeF_3 cathode in Li-ion batteries. J. Power Sources **203**, 103–108 (2012)
26. Brauer, G.: Handbook of preparative inorganic chemistry, vol. 1, 2 ed. Academic Press, NY (1963)
27. Wu, W., Wang, X., Wang, X., Yang, S., Liu, X., Chen, Q.: Effects of MoS_2 doping on the electrochemical performance of FeF_3 cathode materials for lithium-ion batteries. Mater. Lett. **63**, 1788–1790 (2009)

28. Wu, W., Wang, Y., Wang, X., Chen, Q., Wang, X., Yang, S., Liu, X., Guo, J., Yang, Z.: Structure and electrochemical performance of FeF_3/V_2O_5 composite cathode material for lithium-ion battery. J. Alloys Compd. **486**, 93–96 (2009)
29. Karraker, D.G., Smith, P.K.: Alpha-FeF_3*$3H_2O$ and beta-FeF_3*$3H_2O$ revisited: crystal structure and Fe^{57} Mossbauer spectra. Inorg. Chem. **31**, 1118–1120 (1992)
30. Li, C., Yin, C., Gu, L., Dinnebier, R.E., Mu, X., van Aken, P.A., Maier, J.: An FeF_3*$0.5H_2O$ polytype: a microporous framework compound with intersecting tunnels for Li and Na batteries. J. Am. Chem. Soc. **135**, 11425–11428 (2013)
31. Demourgues, A., Francke, L., Durand, E., Tressaud, A.: Chemistry and key structural features of oxyhydroxy-fluorides: relationships with the acidic character, thermal stability and surface area. J. Fluor. Chem. **114**, 229–236 (2002)
32. Francke, L., Durand, E., Demourgues, A., Vimont, A., Daturi, M., Tressaud, A.: Synthesis and characterization of Al^{3+}, Cr^{3+}, Fe^{3+} and Ga^{3+} hydroxyfluorides: correlations between structural features, thermal stability and acidic properties. J. Mater. Chem. **13**, 2330–2340 (2003)
33. Vimont, A., Lavalley, J.C., Francke, L., Demourgues, A., Tressaud, A., Daturi, M.: Infrared study of the surface properties of HTB-type Al-, Cr-, Fe-hydroxyfluorides. J. Phys. Chem. B **108**, 3246–3255 (2004)
34. Li, C.L., Gu, L., Tsukimoto, S., van Aken, P.A., Maier, J.: Low-temperature ionic-liquid-based synthesis of nanostructured iron-based fluoride cathodes for lithium batteries. Adv. Mater. **22**, 3650 (2010)
35. Li, C.L., Gu, L., Tong, J.W., Tsukimoto, S., Maier, J.: A mesoporous iron-based fluoride cathode of tunnel structure for rechargeable lithium batteries. Adv. Funct. Mater. **21**, 1391–1397 (2011)
36. Li, C.L., Gu, L., Tong, J.W., Maier, J.: Carbon nanotube wiring of electrodes for high-rate lithium batteries using an imidazolium-based ionic liquid precursor as dispersant and binder: a case study on iron fluoride nanoparticles. ACS Nano **5**, 2930–2938 (2011)
37. Li, C.L., Yin, C.L., Mu, X.K., Maier, J.: Top-down synthesis of open framework fluoride for lithium and sodium batteries. Chem. Mater. **25**, 962–969 (2013)
38. Li, C.L., Mu, X.K., van Aken, P.A., Maier, J.: A high-capacity cathode for lithium batteries consisting of porous microspheres of highly amorphized iron fluoride densified from Its open parent phase. Adv. Energy Mater. **3**, 113–119 (2013)
39. Li, B., Rooney, D.W., Zhang, N., Sun, K.: An in situ ionic-liquid-assisted synthetic approach to iron fluoride/graphene hybrid nanostructures as superior cathode materials for lithium ion batteries. ACS Appl. Mater. Interfaces **5**, 5057–5063 (2013)
40. Lu, Y., Wen, Z., Rui, K., Wu, X., Cui, Y.: Worm-like mesoporous structured iron-based fluoride: facile preparation and application as cathodes for rechargeable lithium ion batteries. J. Power Sources **244**, 306–311 (2013)
41. Di Carlo, L., Conte, D.E., Kemnitz, E., Pinna, N.: Microwave-assisted fluorolytic sol-gel route to iron fluoride nanoparticles for Li-ion batteries. Chem. Commun. **50**, 460–462 (2014)
42. Xu, X., Chen, S., Shui, M., Xu, L., Zheng, W., Shu, J., Cheng, L., Feng, L., Ren, Y.: One step solid state synthesis of FeF_3*$0.33H_2O$/C nano-composite as cathode material for lithium-ion batteries. Ceram. Int. **40**, 3145–3148 (2014)
43. Yang, Z.H., Pei, Y., Wang, X.Y., Liu, L., Su, X.P.: First principles study on the structural, magnetic and electronic properties of Co-doped FeF_3. Comput. Theor. Chem. **980**, 44–48 (2012)
44. Liu, L., Zhou, M., Yi, L., Guo, H., Tan, J., Shu, H., Yang, X., Yang, Z., Wang, X.: Excellent cycle performance of Co-doped FeF_3/C nanocomposite cathode material for lithium-ion batteries. J. Mater. Chem. **22**, 17539–17550 (2012)
45. Liu, L., Guo, H., Zhou, M., Wei, Q., Yang, Z., Shu, H., Yang, X., Tan, J., Yan, Z., Wang, X.: A comparison among FeF_3*$3H_2O$, FeF_3*$0.33H_2O$ and FeF_3 cathode materials for lithium ion batteries: structural, electrochemical, and mechanism studies. J. Power Sources **238**, 501–515 (2013)
46. Shi, Y.L., Wu, N., Shen, M.F., Cui, Y.L., Jiang, L., Qiang, Y.H., Zhuang, Q.C.: Electrochemical behavior of iron(III) fluoride trihydrate as a cathode in lithium-ion batteries. ChemElectroChem. (2014, in press)
47. Li, L.S., Yu, Y.H., Meng, F., Tan, Y.Z., Hamers, R.J., Jin, S.: Facile solution synthesis of alpha-FeF_3*$3H_2O$ nanowires and their conversion to alpha-Fe_2O_3 nanowires for photoelectrochemical application. Nano Lett. **12**, 724–731 (2012)
48. Li, L., Meng, F., Jin, S.: High-capacity lithium-ion battery conversion cathodes based on iron fluoride nanowires and insights into the conversion mechanism. Nano Lett. **12**, 6030–6037 (2012)
49. Morin, S.A., Bierman, M.J., Tong, J., Jin, S.: Mechanism and kinetics of spontaneous nanotube growth driven by screw dislocations. Science **328**, 476–480 (2010)
50. Wang, F., Robert, R., Chernova, N.A., Pereira, N., Omenya, F., Badway, F., Hua, X., Ruotolo, M., Zhang, R.G., Wu, L.J., Volkov, V., Su, D., Key, B., Whittinghan, M.S., Grey, C.P., Amatucci, G.G., Zhu, Y.M., Graetz, J.: Conversion reaction mechanisms in lithium ion batteries: study of the binary metal fluoride electrodes. J. Am. Chem. Soc. **133**, 18828–18836 (2011)
51. Chu, Q., Xing, Z., Tian, J., Ren, X., Asiri, A.M., Al-Youbi, A.O., Alamry, K.A., Sun, X.: Facile preparation of porous FeF_3 nanospheres as cathode materials for rechargeable lithium-ion batteries. J. Power Sources **236**, 188–191 (2013)
52. Chu, Q., Xing, Z., Ren, X., Asiri, A.M., Al-Youbi, A.O., Alamry, K.A., Sun, X.: Reduced graphene oxide decorated with FeF_3 nanoparticles: Facile synthesis and application as a high capacity cathode material for rechargeable lithium batteries. Electrochim. Acta **111**, 80–85 (2013)
53. Kim, S.W., Seo, D.H., Gwon, H., Kim, J., Kang, K.: Fabrication of FeF_3 nanoflowers on CNT branches and their application to high power lithium rechargeable batteries. Adv. Mater. **22**, 5260–5264 (2010)
54. Li, T., Li, L., Cao, Y.L., Ai, X.P., Yang, H.X.: Reversible three-electron redox behaviors of FeF_3 nanocrystals as high-capacity cathode-active materials for Li-ion batteries. J. Phys. Chem. C **114**, 3190–3195 (2010)
55. Zaghib, K., Guerfi, A., Hovington, P., Vijh, A., Trudeau, M., Mauger, A., Goodenough, J.B., Julien, C.M.: Review and analysis of nanostructured olivine-based lithium recheargeable batteries: status and trends. J. Power Sources **232**, 357–369 (2013)
56. Liu, J., Wan, Y., Liu, W., Ma, Z., Ji, S., Wang, J., Zhou, Y., Hodgson, P., Li, Y.: Mild and cost-effective synthesis of iron fluoride-graphene nanocomposites for high-rate Li-ion battery cathodes. J. Mater. Chem. A **1**, 1969–1975 (2013)
57. Ma, R., Lu, Z., Wang, C., Wang, H.-E., Yang, S., Xi, L., Chung, J.C.Y.: Large-scale fabrication of graphene-wrapped FeF_3 nanocrystals as cathode materials for lithium ion batteries. Nanoscale **5**, 6338–6343 (2013)
58. Myung, S.T., Sakurada, S., Yashiro, H., Sun, Y.K.: Iron trifluoride synthesized via evaporation method and its application to rechargeable lithium batteries. J. Power Sources **223**, 1–8 (2013)
59. Li, X., Liu, J., Meng, X., Tang, Y., Banis, M.N., Yang, J., Hu, Y., Li, R., Cai, M., Sun, X.: Significant impact on cathode performance of lithium-ion batteries by precisely controlled metal oxide nanocoatings via atomic layer deposition. J. Power Sources **247**, 57–69 (2014)

60. Zhang, W., Ma, L., Yue, H., Yang, Y.: Synthesis and characterization of in situ Fe_2O_3-coated FeF_3 cathode materials for rechargeable lithium batteries. J. Mater. Chem. **22**, 24769–24775 (2012)
61. Tiwari, J.N., Tiwari, R.N., Kim, K.S.: Zero-dimensional, one-dimensional, two-dimensional and three-dimensional nanostructured materials for advanced electrochemical energy devices. Prog. Mater Sci. **57**, 724–803 (2012)
62. Ma, D., Cao, Z., Wang, H., Huang, X., Wang, L., Zhang, X.: Three-dimensionally ordered macroporous FeF_3 and its in situ homogenous polymerization coating for high energy and power density lithium ion batteries. Energy Environ. Sci. **5**, 8538–8542 (2012)
63. Jung, H., Shin, J., Chae, C., Lee, J.K., Kim, J.: FeF_3/Ordered Mesoporous Carbon (OMC) nanocomposites for lithium ion batteries with enhanced electrochemical performance. J. Phys. Chem. C **117**, 14939–14946 (2013)
64. Liu, L., Zhou, M., Wang, X.Y., Yang, Z.H., Tian, F.H.: Synthesis and electrochemical performance of spherical FeF_3/ACMB composite as cathode material for lithium-ion batteries. J. Mater. Sci. **47**, 1819–1824 (2012)
65. Tan, J., Liu, L., Hu, H., Yang, Z., Guo, H., Wei, Q., Yi, X., Yan, Z., Zhou, Q., Huang, Z., Shu, H., Yang, X., Wang, X.: Iron fluoride with excellent cycle performance synthesized by solvothermal method as cathodes for lithium ion batteries. J. Power Sources **251**, 75–84 (2014)
66. Martha, S.K., Nanda, J., Zhou, H., Idrobo, J.C., Dudney, N.J., Pannala, S., Dai, S., Wang, J., Braun, P.V.: Electrode architectures for high capacity multivalent conversion compounds: iron (II and III) fluoride. Rsc Adv. **4**, 6730–6737 (2014)
67. Oudenhoven, J.F.M., Baggetto, L., Notten, P.H.L.: All-solid-state lithium-ion microbatteries: a review of various three-dimensional concepts. Adv. Energy Mater. **1**, 10–33 (2011)
68. Knoops, H.C.M., Donders, M.E., van de Sanden, M.C.M., Notten, P.H.L., Kessels, W.M.M.: Atomic layer deposition for nanostructured Li-ion batteries. J. Vac. Sci. Technol. A **30** (2012)
69. Makimura, Y., Rougier, A., Tarascon, J.M.: Pulsed laser deposited iron fluoride thin films for lithium-ion batteries. Appl. Surf. Sci. **252**, 4587–4592 (2006)
70. Makimura, Y., Rougier, A., Laffont, L., Womes, M., Jumas, J.C., Leriche, J.B., Tarascon, J.M.: Electrochemical behaviour of low temperature grown iron fluoride thin films. Electrochem. Commun. **8**, 1769–1774 (2006)
71. Guitian, B., Lascaud, S., Novoa, X.R., Ribeaucourt, L., Vidal, E.: On the growth of nanostructured iron hydroxy-fluorides for Li-ion batteries. J. Power Sources **241**, 567–571 (2013)
72. Pereira, N., Badway, F., Wartelsky, M., Gunn, S., Amatucci, G.G.: Iron oxyfluorides as high capacity cathode materials for lithium batteries. J. Electrochem. Soc. **156**, A407–A416 (2009)
73. Li, Z., Huang, J., Liaw, B.Y., Metzler, V., Zhang, J.: A review of lithium deposition in lithium-ion and lithium metal secondary batteries. J. Power Sources **254**, 168–182 (2014)
74. Portier, J., Tressaud, A., Depape, R., Hagenmuller, P.: Crystallography and magnetic study of an unpublished fluoride of trirutile type. C. R. Acad. Sc. Ser. C **267**, 1711–1713 (1968)
75. Fourquet, J.L., Lesamedi, E., Calage, Y.: Crystal growth and crystal structure of trirutile $LiFe_2F_6$. J. Solid State Chem. **77**, 84–89 (1988)
76. Liao, P., Li, J., Dahn, J.R.: Lithium intercalation in $LiFe_2F_6$ and $LiMgFeF_6$ disordered trirutile-type phases. J. Electrochem. Soc. **157**, A355–A361 (2010)
77. Liao, P., Dunlap, R.A., Dahn, J.R.: In situ Mossbauer effect study of lithium intercalation in $LiFe_2F_6$. J. Electrochem. Soc. **157**, A1080–A1084 (2010)
78. Amatucci, G.G.: Electrode composition useful as positive electrode material of electrochemical cell e.g. lithium battery cell, comprises lithium fluoride compound demonstrating high specific capacity. WO2004051772-A2; US2004121235-A1; AU2003295927-A1; AU2003295927-A8; US7371338-B2; WO2004051772-A3
79. Amatucci, G.G., Pereira, N., Badway, F., Sina, M., Cosandey, F., Ruotolo, M., Cao, C.: Formation of lithium fluoride/metal nanocomposites for energy storage through solid state reduction of metal fluorides. J. Fluor. Chem. **132**, 1086–1094 (2011)
80. Liao, P., MacDonald, B.L., Dunlap, R.A., Dahn, J.R.: Combinatorially prepared $[LiF]_{1-x}Fe_x$ nanocomposites for positive electrode materials in Li-ion batteries. Chem. Mater. **20**, 454–461 (2008)
81. Liao, P., Dunlap, R.A., Dahn, J.R.: A Mossbauer effect study of combinatorially prepared Al_2O_3/Fe and LiF/Fe multilayers. J. Phys.: Condens. Matter **20**, 055203 (2008)
82. Li, T., Chen, Z.X., Ai, X.R., Cao, Y.L., Yang, H.X.: LiF/Fe nanocomposite as a lithium-rich and high capacity conversion cathode material for Li-ion batteries. J. Power Sources **217**, 54–58 (2012)
83. Kim, S.W., Nam, K.W., Seo, D.H., Hong, J., Kim, H., Gwon, H., Kang, K.: Energy storage in composites of a redox couple host and a lithium ion host. Nano Today **7**, 168–173 (2012)
84. Ma, R., Dong, Y., Xi, L., Yang, S., Lu, Z., Chung, C.: Fabrication of LiF/Fe/graphene nanocomposites as cathode material for lithium-ion batteries. ACS Appl. Mater. Interfaces **5**, 892–897 (2013)
85. Prosini, P.P., Masci, A., Cento, C., Gislon, P.: A composite electrode based on sub-micrometric iron metal and lithium fluoride. Electrochim. Acta **106**, 312–319 (2013)
86. Prakash, R., Mishra, A.K., Roth, A., Kuebel, C., Scherer, T., Ghafari, M., Hahn, H., Fichtner, M.: A ferrocene-based carbon-iron lithium fluoride nanocomposite as a stable electrode material in lithium batteries. J. Mater. Chem. **20**, 1871–1876 (2010)
87. Prakash, R., Wall, C., Mishra, A.K., Kuebel, C., Ghafari, M., Hahn, H., Fichtner, M.: Modified synthesis of [Fe/LiF/C] nanocomposite, and its application as conversion cathode material in lithium batteries. J. Power Sources **196**, 5936–5944 (2011)
88. Zhang, S., Lu, Y., Xu, G., Li, Y., Zhang, X.: LiF/Fe/C nanofibres as a high-capacity cathode material for Li-ion batteries. J. Phys. D Appl. Phys. **45**, 395301–395307 (2012)
89. Gonzalo, E., Kuhn, A., Garcia-Alvarado, F.: On the room temperature synthesis of monoclinic Li_3FeF_6: a new cathode material for rechargeable lithium batteries. J. Power Sources **195**, 4990–4996 (2010)
90. Gonzalo, E., Kuhn, A., Garcia-Alvarado, F.: A comparative study of alpha- and beta-Li_3FeF_6: structure and electrochemical behavior. J. Electrochem. Soc. **157**, A1002–A1006 (2010)
91. Basa, A., Gonzalo, E., Kuhn, A., Garcia-Alvarado, F.: Reaching the full capacity of the electrode material Li_3FeF_6 by decreasing the particle size to nanoscale. J. Power Sources **197**, 260–266 (2012)
92. Gonzalo, E., Kuhn, A., Garcia-Alvarado, F., Islam, M.S.: Defect and dopant properties of the alpha- and beta-polymorphs of the Li_3FeF_6 lithium battery material. J. Mater. Chem. A **1**, 6588–6592 (2013)
93. Schroeder, M., Eames, C., Tompsett, D.A., Lieser, G., Islam, M.S.: Li_xFeF_6 ($x = 2, 3, 4$) battery materials: structural, electronic and lithium diffusion properties. Phys. Chem. Chem. Phys. **15**, 20473–20479 (2013)
94. Palomares, V., Casas-Cabanas, M., Castillo-Martinez, E., Han, M.H., Rojo, T.: Update on Na-based battery materials. A

growing research path. Energy Environ. Sci. **6**, 2312–2337 (2013)
95. Nishijima, M., Gocheva, I.D., Okada, S., Doi, T., Yamaki, J.-I., Nishida, T.: Cathode properties of metal trifluorides in Li and Na secondary batteries. J. Power Sources **190**, 558–562 (2009)
96. Gocheva, I.D., Nishijima, M., Doi, T., Okada, S., Yamaki, J.-I., Nishida, T.: Mechanochemical synthesis of $NaMF_3$ (M = Fe, Mn, Ni) and their electrochemical properties as positive electrode materials for sodium batteries. J. Power Sources **187**, 247–252 (2009)
97. Kitajou, A., Komatsu, H., Chihara, K., Gocheva, I.D., Okada, S.: Yamaki, J.-i.: Novel synthesis and electrochemical properties of perovskite-type $NaFeF_3$ for a sodium-ion battery. J. Power Sources **198**, 389–392 (2012)
98. Dimov, N., Nishimura, A., Chihara, K., Kitajou, A., Gocheva, I.D., Okada, S.: Transition metal $NaMF_3$ compounds as model systems for studying the feasibility of ternary Li-M-F and Na-M-F single phases as cathodes for lithium-ion and sodium-ion batteries. Electrochim. Acta **110**, 214–220 (2013)
99. Yamada, Y., Doi, T., Tanaka, I., Okada, S.: Yamaki, J.-i.: Liquid-phase synthesis of highly dispersed $NaFeF_3$ particles and their electrochemical properties for sodium-ion batteries. J. Power Sources **196**, 4837–4841 (2011)
100. Shakoor, R.A., Lim, S.Y., Kim, H., Nam, K.-W., Kang, J.K., Kang, K., Choi, J.W.: Mechanochemical synthesis and electrochemical behavior of Na_3FeF_6 in sodium and lithium batteries. Solid State Ionics **218**, 35–40 (2012)
101. Manthiram, A., Goodenough, J.B.: Lithium insertion into $Fe_2(SO_4)_3$ frameworks. J. Power Sources **26**, 403–408 (1989)
102. Melot, B.C., Tarascon, J.M.: Design and preparation of materials for advanced electrochemical storage. Acc. Chem. Res. **46**, 1226–1238 (2013)
103. Wu, F., Li, N., Su, Y., Shou, H., Bao, L., Yang, W., Zhang, L., An, R., Chen, S.: Spinel/layered heterostructured cathode material for high-capacity and high-rate Li-ion batteries. Adv. Mater. **25**, 3722–3726 (2013)
104. Gao, H., Hu, Z., Zhang, K., Cheng, F., Chen, J.: Intergrown Li_2FeSiO_4*$LiFePO_4$-C nanocomposites as high-capacity cathode materials for lithium-ion batteries. Chem. Commun. **49**, 3040–3042 (2013)

23

Trials for oxide photo-thermoelectrics

Ichiro Terasaki · Ryuji Okazaki ·
Partha Sarathi Mondal · Yu-Chin Hsieh

Abstract Thermoelectrics is an energy conversion technology from heat into electricity, and vice versa, through the thermoelectric phenomena in solids, while photovoltaics is an energy conversion technology from solar photon energy into electricity using the photo-excitations in solids. We are trying to find a way to combine thermoelectrics with photovoltaics to establish a new method to generate renewable energy with high efficiency. In this article, we show two approaches for this purpose using oxide materials: thermoelectric energy conversion by photo-excited carriers and the thermoelectric power generation using a focused light as a heat source.

Keywords Thermoelectric power generation · Oxide · Photo-doping · Photo-thermoelectrics

Introduction

No one can deny that our modern society is based on vast consumption of electric/chemical energies. Since all the developing countries have the right to enjoy life as comfortable as the advanced countries do, energy demands are increasing year by year in spite of serious shortage of petroleum. Thus, a search for sufficient energy resources is a responsibility of researchers in all areas of science and technology. Best energy resources are of no doubt renewable energies, which preliminarily come from the solar energy.

The energy conversion technique using the solar energy is classified into two; The one is photovoltaics in which an electron–hole pair created by an incident photon is separated by an internal electric field at the *p–n* junction [27]. This technology is now commercially available as solar battery cells. The other is solar-thermal energy conversion, where heat generated by focused sunlight vaporizes water to rotate a gas turbine [34]. Although these two techniques are matured, there remain issues to be addressed. In photovoltaics, the conversion efficiency is close to a theoretical limit, and raw materials of silicon of high quality are about to run out. Of course alternative materials are being developed, but the cost and natural abundance are still issues. In the case of solar-thermal conversion, the conversion efficiency is not satisfactory except for some areas around the equators.

We have studied thermoelectric energy conversion using oxide materials, which are superior at high temperatures in air [9, 13]. A serious drawback of thermoelectrics is poor efficiency [28]. A good thermoelectric material requires high electrical conductivity, large Seebeck coefficient, and low thermal conductivity at the same time, which is very difficult to be realized in real materials. In fact, reliable calculations of materials parameters do not give promising results. Since such calculations are done near equilibrium states, we hope that the thermoelectric performance may go beyond theoretical limitations in non-equilibrium states [40, 41]. As such, we are trying to find ways to break through the poor efficiency by focusing non-equilibrium states. In this article, we show our preliminary results for two types of photo-excited thermoelectrics. One is the thermoelectrics using the photo-Seebeck effects (Fig. 1a), and the other is the thermoelectric energy conversion in a

This work was partially supported by ALCA, Japan Science and Technology Agency, and by The Mitsubishi Foundation.

I. Terasaki (✉) · R. Okazaki · P. S. Mondal · Y.-C. Hsieh
Department of Physics, Nagoya University, Nagoya 464-8602, Japan
e-mail: terra@cc.nagoya-u.ac.jp

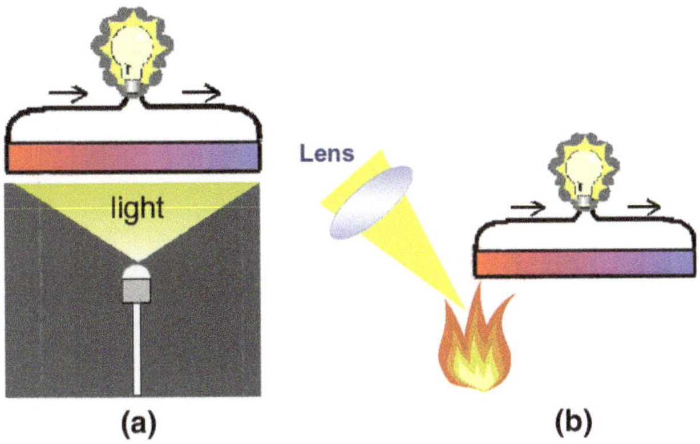

Fig. 1 Schematics of (**a**) power generation using photo-Seebeck effect and (**b**) power generation from focused light

large temperature gradient using solar light as a heat source (Fig. 1b). One may associate the two with photovoltaic and solar-thermal energy conversions.

To emphasize our originality, we will briefly summarize the preceding works. The photo-Seebeck effect was first reported by Tauc [31] in 1955, and was later examined in conventional semiconductors [7, 15]. Note that the word "photo-thermoelectric" is confusing; it stood for the photo-Seebeck effect before the 80s, but is now used as photo-thermal energy conversion through the Seebeck effect (for example, see [2, 12]). To our knowledge, our work is the first trial for the thermoelectric energy conversion using the photo-Seebeck effect. In contrast, the second trial shown in Fig. 1b has been examined by many groups. Originally Telkes [32] examined the concept of solar thermoelectric generator in 1954. Naito et al. [18] designed a power converter using concentrated solar light and achieved a high temperature of 2,200 K in vacuum. Suter et al. [30] fabricated the solar thermoelectric generator and analyzed the efficiency and the maximum power. Fan et al. [3] reported an efficiency of 3 % using commercially available Peltier modules. Thus, our originality lies on the fact that our device is made of *single crystals of transition metal oxide*, which can work at 800 K in air.

Photo-Seebeck effect in oxide single crystals

Although the first observation of the photo-Seebeck effect was reported in the mid 50s [31], there have been very few reports on the photo-Seebeck measurements since the 80s. Thus, we had to newly establish measurement procedure with recently available equipments such as a light-emitting diode (LED). Here, we elaborate on the measurement and analysis details for the photo-Seebeck coefficient, which was not included in the published papers [17, 20].

Commercially available substrates were used as ZnO single crystals, and flux method was employed for making single crystals of PbO. The photoconductivity of ZnO and PbO single crystals was measured with a two-probe method. The thermoelectric voltage of ZnO was measured with two-probe technique with several temperature difference and several photon intensities [20]. The resistance of PbO was too high to use the same measurement setup as in the case of ZnO. Instead, the thermoelectric current was measured with several temperature difference and several photon intensities [17].

The photo-Seebeck effect is a change in the Seebeck coefficient with light illumination, which can be evaluated by comparing the thermoelectric voltage before and after illumination. However, the light illumination affects various properties at the same time; It causes the photovoltaic voltage at the contacts, increases the sample temperature, and changes the temperature difference. To be more quantitative, the measured thermoelectric voltage V_{TE} can be expressed by temperature difference ΔT at dark as

$$V_{TE} = S\Delta T + V_0, \tag{1}$$

where S is the Seebeck coefficient at dark and V_0 is the offset voltage. When illuminated, the voltage can be written as

$$V_{TE} = (S + \delta S)(\Delta T + \delta T) + (V_0 + \delta V_0), \tag{2}$$

where δS is the photo-induced change in the Seebeck coefficient. δT and δV_0 are the photo-induced temperature difference and photovoltaic component, respectively. Note that we measured the voltage in a cryostat (Quantum Design PPMS), where the sample temperature was strictly controlled.

The best way to measure the photo-Seebeck coefficient is to measure V_{TE} as a function of $\Delta T + \delta T$. When V_{TE} is found to be a linear function of $\Delta T + \delta T$, the photo-Seebeck coefficient of $S + \delta S$ and the photovoltaic component $V_0 + \delta V_0$ are evaluated from the slope and the intercept, respectively. Figure 2 shows a typical example for the thermoelectric voltage with various photon intensities. The value of I_{LED} represents the input current to an LED of 365 nm and 0 mA corresponds to the dark state. As is clearly shown, the slope systematically decreases with increasing I_{LED}, which means that the Seebeck coefficient decreases in magnitude with light illumination. This is reasonable because the light illumination generates carriers in the sample (photo-doping). Note that the intercept also changes with I_{LED}, which indicates a significant photovoltaic component induced by light illumination. We further comment that the temperature difference changes with I_{LED}. This implies that the absorbed photon is converted into heat, and the heat flows through the sample to the heat bath. If the heat flow is completely homogeneous, the temperature

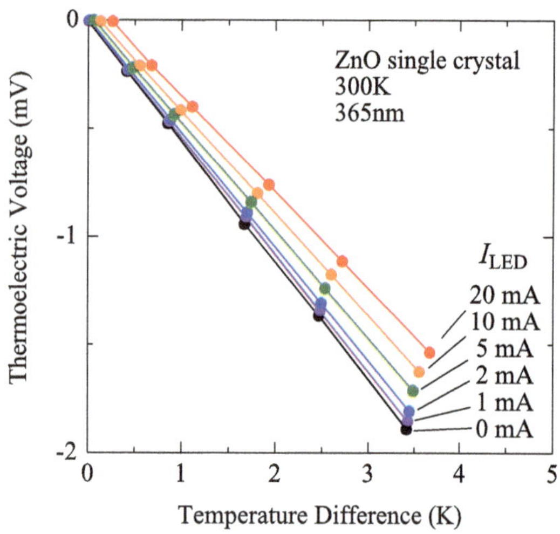

Fig. 2 Thermoelectric voltage in a ZnO single crystal plotted as a function of temperature difference. The LED current I_{LED} corresponds to the photon intensity

difference will be unchanged. Owing to unwanted asymmetry, a finite value of δT is observed.

In the case of PbO, the sample resistance is too high to measure V_{TE} because of the huge Johnson–Nyquist noise. Instead, the thermoelectric current I_{TE} was measured with a pico/femto-ampere meter as a function of temperature difference expressed by

$$I_{TE} = (S + \delta S)(\Delta T + \delta T)/(R - \delta R) + I_0, \quad (3)$$

where R is the sample resistance at dark and δR is the photo-resistance of the sample. Thus, if the photo-resistance is measured in different runs, the thermoelectric voltage is evaluated as $(R - \delta R)I_{TE}$. When $(R - \delta R)I_{TE}$ is a linear function of $\Delta T + \delta T$, the photo-Seebeck coefficient is evaluated from the slope.

Figure 3 shows a typical experimental result in a PbO single crystal. The thermoelectric current I_{TE} is measured as a function of ΔT in Fig. 3a. With increasing I_{LED}, the slope increases rapidly, which is mainly determined by photoconduction. Figure 3b is the thermoelectric voltage plotted as a function of temperature difference, where the slope slightly changes with I_{LED}, indicating photo-Seebeck effect in this oxide. Similarly to ZnO, the results in Fig. 3 can be understood in terms of photo-doping. The type of the carrier is hole in PbO, whereas it is electron in ZnO. It should be noted that chemical doping is quite difficult in PbO, and that light illumination is an effective tool for carrier doping in PbO.

After a phenomenological analysis based on the two-layer model, we have obtained the intrinsic photo-Seebeck coefficient and the intrinsic values of photo-doped carrier density [17, 20]. We find a maximum value for the photo-doped electron density is of the order of 10^{19} cm^{-3}, which is nearly the same value of the optimally doped

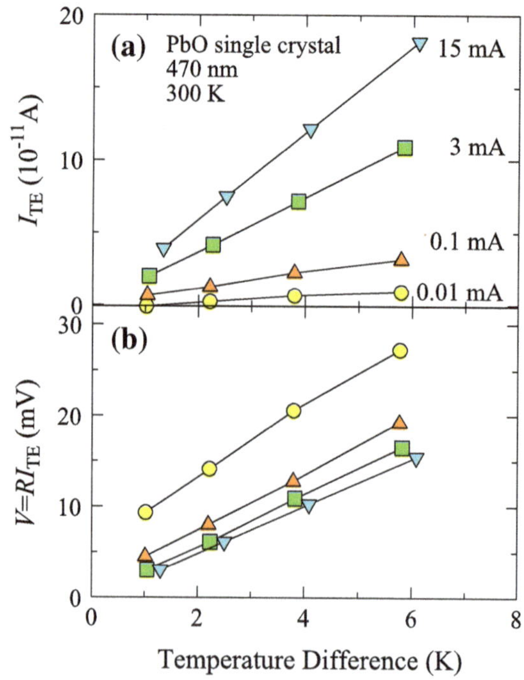

Fig. 3 (**a**) Thermoelectric current I_{TE} and (**b**) thermoelectric voltage RI_{TE} in a PbO single crystal plotted as a function of temperature difference. The *numbers* in mA represent the LED current I_{LED} corresponding to the photon intensity

thermoelectric semiconductors [28]. Thus, we can conclude that the ultraviolet LED can be an alternative carrier source. In the case of PbO, a maximum value is of the order of 10^{18} cm^{-3}, one-order of magnitude smaller than in the case of ZnO.

Thermoelectric uni-couple made of single-crystal oxides

Since the discovery of good thermoelectric properties in the layered cobalt oxide Na_xCoO_2 [33], oxide thermoelectrics has been explored extensively. The layered cobalt oxides are now ready to be applied for thermoelectric power generation, and a prototype power generator is now commercially available at a venture company [1]. All the layered cobalt oxides show good thermoelectric performance at high temperatures. Above 700 K, the thermoelectric performance exceeds a qualifying value (the dimensionless figure of merit larger than unity) in single-crystal samples [5, 6, 25], while it remains a somewhat lower value in ceramic samples [11, 19, 37, 39]. Thus, a device using single crystals is expected to show a better performance than ceramic samples.

We have made a trial device of a uni-couple consisting of single-crystal $Bi_2Sr_2Co_2O_y$ (p-type) and single-crystal $CaMnO_3$ (n-type). Single crystals were grown through a floating-zone technique, and the p–n junction was made

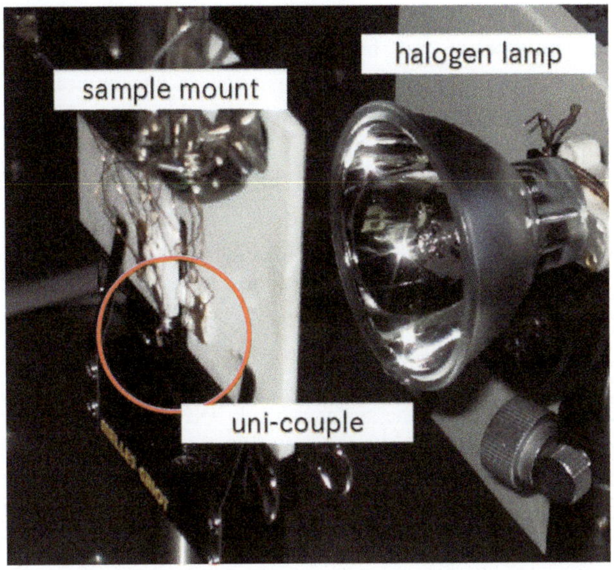

Fig. 4 Power curve measurement setup using halogen lamp as a heat source

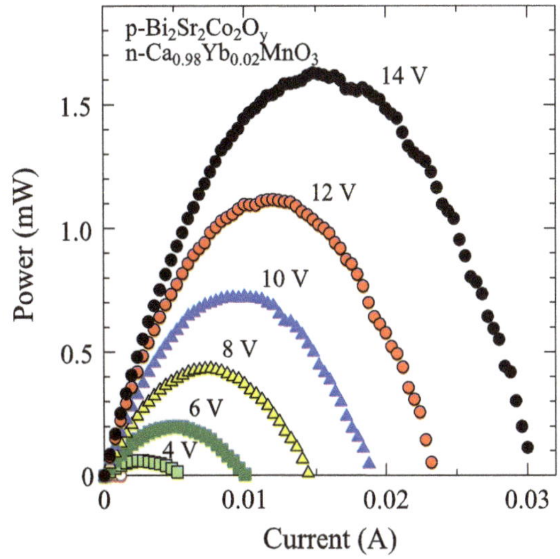

Fig. 5 Power curves of the uni-couple device made of single-crystal oxides. The *numbers* in volt represent the input voltage for the halogen lamp

using silver paint. The dimensions were 1.4 × 1.2 × 9.4 mm^3 for *p*-type and 1.2 × 1.1 × 11 mm^3 for *n*-type. The current–voltage characteristics were measured with a home-made characterization system. A focused halogen light was used as a heat source, and three platinum resistance thermometers were attached at the ends of the device (one at the hot junction and the other two at the cold edges). Figure 4 shows a photographic image of our measurement setup. The device supported on a machinable ceramic plate is set at the focal point of the halogen lamp.

Figure 5 shows typical power curves of the single-crystal uni-couple. Since the device design is not yet optimized, the maximum power output is only 1.5 mW. Although both *p*- and *n*-legs are made of single crystals, the thermoelectric properties are not yet optimized either. In addition, this measurement includes the contact resistance and circuit resistance, which underestimates the power output.

Let us compare the device performance with other oxide devices. Table 1 lists the oxide device performance reported so far. The materials, the operating temperature difference ΔT, and the number of pairs N are different from one to another. The maximum output per pair P_p in our device is comparable with some devices.

One characteristic feature of this device is that a large temperature difference is generated across the sample. Figure 6a shows the hot-side temperature T_H and the cold-side temperature T_C of the uni-couple plotted as a function of the input voltage of the halogen lamp V_{light}. With increasing V_{light}, T_H increases monotonically, while the two T_C's almost stay near room temperature. At a maximum voltage of 14 V, the temperature difference reaches 500 K.

Table 1 Comparison with oxide thermoelectric modules

Reference	Materials	N	ΔT (K)	V_{oc} (mV)	P_{max} (mW)	P_p (mW)
Matsubara et al. [16]	CaGdCoO/ CaMnO	8	773	988	63.5	7.9
Urata et al. [36]	CaCoO/CaMnO	8	973	700	340	42.5
Shin et al. [26]	NaCoCuO/ BaSrPbO	1	506	140	12.0	12.0
Ono et al. [21]	NaCoO/ CaLaMnO	2	166	63	2.3	1.15
Hayashi et al. [8]	LaSrCuO/ NdCeCuO	25	360	1,300	26.3	1.05
Reddy et al. [23]	CaCoO/ CaSmMnO	2	925	400	31.5	17.0
Tomes et al. [35]	LaSrCuO/ CaMnNbO	2	622	46	88.8	44.4
Inagoya et al. [10]	NdCaCoO/ LaCoMnO	10	170	1,010	10.8	1.08
This work	BiSrCoO/ CaYbMnO	1	500	200	1.6	1.6

The number of pairs N, the temperature difference ΔT, the open-circuit voltage V_{oc}, and the maximum power P_{max} are listed. The maximum power per pair $P_p = P_{max}/N$ is also listed

Since the device length is approximately 1 cm, the temperature gradient reaches a large value of 500 K/cm by air-cooling without blower.

Owing to the large temperature difference, thermoelectric parameters change within the device though their temperature dependence. Snyder and Ursell [29] have found the exact expression for the conversion efficiency that is valid even in such conditions. Surprisingly, the efficiency is determined only by the values at the high and

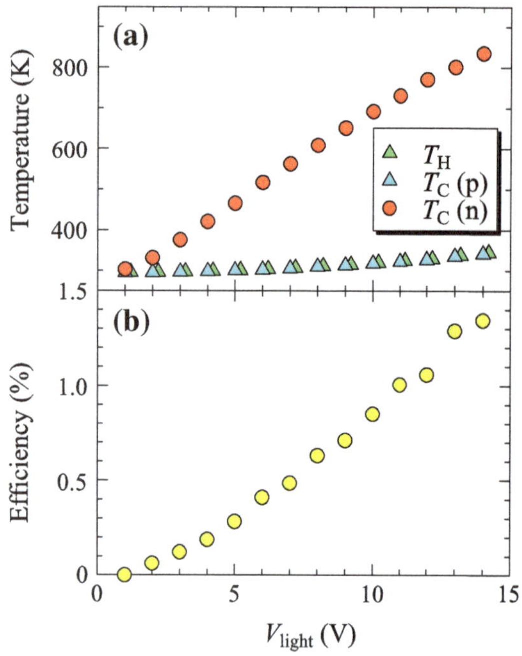

Fig. 6 (**a**) The temperatures at the hot edge T_H and the temperatures at the cold edges T_C's (**b**) the conversion efficiency calculated by Eq. (4) plotted as a function of the input voltage for the halogen lamp V_{light}

low temperature edges of the device. A concrete form of the expression is given by

$$\eta = 1 - \frac{S_C T_C + 1/u_C}{S_H T_H + 1/u_H}, \tag{4}$$

where the subscripts of H and C represent the values at the hot and cold edges. The quantity of $u \equiv j/\kappa \nabla T$ is called the relative current density, which is the ratio of the electrical current density to the heat flux density by thermal conduction.

The temperatures (T_H and T_C) at the cold and hot edges were monitored in the power curve measurements. The Seebeck coefficients (S_H and S_C) at the cold and hot edges were known by measuring the Seebeck coefficients of *p*- and *n*-type crystals in advance. To calculate the relative current density, the resistivity and the thermal conductivity are necessary. The resistivity values were measured for the prepared crystals, and the thermal conductivity data were taken from the literature [4, 24]. Figure 6b shows the conversion efficiency plotted as a function of V_{light}. The maximum efficiency is a reasonably large value of 1.4 % in spite of the fact that all the conditions were far from ideal. The present measurement is a first step for this kind of thermoelectric power generator, and we notice many issues to improve the device performance. Kraemer et al. [14] have reported an efficiency of 4.6 % in a solar thermoelectric generators using Bi_2Te_3-based device. Although the device concept is completely different from our device, we preliminarily obtained a similar efficiency in another modified uni-couple (not shown).

An advantage of oxide device is seen in the smooth evolution of the power curves in Fig. 5. This implies that the device resistance is almost constant from 300 to 800 K, and the device is sufficiently stable at 800 K in air. Considering that the surface temperature at the focal point on the device is much higher, we think that this experiment demonstrates the high-temperature stability of oxide devise in air.

We are considering a possibility to use this type of device to the black-body radiation from high-temperature furnace. Recalling that the sun is a black body of 6,000 K, we can use black-body radiation from other hot sources as well. For example, a black body of 1,000 K will give a large photon flux of 57 kW/m^2, being 50 times larger than that of the sun. Heat is difficult to control; It is hardly stored, transported or accumulated at our disposal, but electromagnetic wave can be done so.

Summary and future prospects

In this article, we have reported our attempts to find a better thermoelectric power generation using light. One approach is the thermoelectrics using the photo-Seebeck effect. We have shown the *n*-type photo-Seebeck effect in ZnO and the *p*-type one in PbO. Although the thermoelectric energy conversion is not yet done, the photo-doping gives reasonably good thermoelectric parameters. The second approach is the thermoelectric energy conversion using focused light as a heat source. The device structure and thermoelectric materials are not yet optimized, and the efficiency is 1.4 % at the present stage. Nevertheless, we have successfully shown the high-temperature stability of the oxide device, and also shown that the oxide thermoelectric uni-couple can be made from single crystals. The former approach uses ultraviolet light for a doping tool, while the latter uses infrared light as a heat source. Since these two lights are not used in Si-based solar cells, our two approaches can live with solar cells to increase the total energy conversion efficiency. Actually, Park et al. [22] and Wang et al. [38] already have combined an existing photovoltaic cell with a thermoelectric module. We look forward to developing a new device making full use of solar energy by combining photovoltaics with thermoelectrics, even though we cannot see the concrete shape at the moment.

Acknowledgments We would like to thank A. Horikawa for her tenacious efforts to establish measurement procedure for the photo-Seebeck effect. We also wish to thank Y. Yasui, H. Taniguchi, M. Fujita, and Y. Shiraishi for close collaboration of this work. We also appreciate R. Funahashi and H. Ohta for fruitful discussion.

References

1. TES New Energy corporation (2010).
2. Basko, D.: A photothermoelectric effect in graphene. Science **334**(6056), 610–611 (2011).
3. Fan, H., Singh, R., Akbarzadeh, A.: Electric power generation from thermoelectric cells using a solar dish concentrator. J. Electr. Mater. **40**(5), 1311–1320 (2011).
4. Flahaut, D., Mihara, T., Funahashi, R., Nabeshima, N., Lee, K., Ohta, H., Koumoto, K.: Thermoelectrical properties of A-site substituted $Ca_{1-x}Re_xMnO_3$ system. J. Appl. Phys. **100**(8), 084911 (2006).
5. Fujita, K., Mochida, T., Nakamura, K.: High-temperature thermoelectric properties of $Na_xCoO_{2-\delta}$ single crystals. Jpn. J. Appl. Phys. **40**(Part 1, No. 7), 4644–4647 (2001).
6. Funahashi, R., Shikano, M.: $Bi_2Sr_2Co_2O_y$ whiskers with high thermoelectric figure of merit. Appl. Phys. Lett. **81**(8), 1459–1461 (2002).
7. Harper, J.G., Matthews, H.E., Bube, R.H.: Two-carrier photothermoelectric effects in GaAs. J. Appl. Phys. **41**(7), 3182–3184 (1970).
8. Hayashi, S.F., Nakamura, T., Kageyama, K., Takagi, H.: Monolithic thermoelectric devices prepared with multilayer cofired ceramics technology. Jpn. J. Appl. Phys. **49**(9), 096505 (2010).
9. He, J., Liu, Y., Funahashi, R.: Oxide thermoelectrics: the challenges, progress, and outlook. J. Mater. Res. **26**(15), 1762–1772 (2011).
10. Inagoya, A., Sawaki, D., Horiuchi, Y., Urata, S., Funahashi, R., Terasaki, I.: Thermoelectric module made of perovskite cobalt oxides with large thermopower. J. Appl. Phys. **110**(12), 123712 (2011).
11. Ito, M., Nagira, T., Furumoto, D., Katsuyama, S., Nagai, H.: Synthesis of $Na_xCo_2O_4$ thermoelectric oxides by the polymerized complex method. Scripta Mater. **48**(4), 403–408 (2003).
12. Kasymakhunova, A., Nabiev, M.: Photothermoelectric converters of concentrated radiation. Tech. Phys. Lett. **29**(3), 253–255 (2003).
13. Koumoto, K., Wang, Y., Zhang, R., Kosuga, A., Funahashi, R.: Oxide thermoelectric materials: a nanostructuring approach. Ann. Rev. Mater. Res. **40**(1), 363–394 (2010).
14. Kraemer, D., Poudel, B., Feng, Hp, Caylor, J.C., Yu, B., Yan, X., Ma, Y., Wang, X., Wang, D., Muto, A., McEnaney, K., Chiesa, M., Ren, Z., Chen, G.: High-performance flat-panel solar thermoelectric generators with high thermal concentration. Nat. Mater. **10**, 532–538 (2011).
15. Kwok, H.B., Bube, R.H.: Thermoelectric and photothermoelectric effects in semiconductors: CdS single crystals. J. Appl. Phys. **44**(1), 138–144 (1973).
16. Matsubara, I., Funahashi, R., Takeuchi, T., Sodeoka, S., Shimizu, T., Ueno, K.: Fabrication of an all-oxide thermoelectric power generator. Appl. Phys. Lett. **78**(23), 3627–3629 (2001).
17. Mondal, P.S., Okazaki, R., Taniguchi, H., Terasaki, I.: Photoseebeck effect in tetragonal PbO single crystals. J. Appl. Phys. **114**(17), 173710 (2013).
18. Naito, H., Kohsaka, Y., Cooke, D., Arashi, H.: Development of a solar receiver for a high-efficiency thermionic/thermoelectric conversion system. Sol. Energy **58**(4–6), 191–195 (1996).
19. Ohtaki, M., Nojiri, Y., Maeda, E.: Improved thermoelectric performance of sintered $NaCo_2O_4$ with enhanced 2-dimensional microstructure. In: Rowe, M. (ed.) Proceedings of The 19th International Conference on Thermoelectrics (ICT2000), pp. 190–195. Babrow, Cardiff (2000)
20. Okazaki, R., Horikawa, A., Yasui, Y., Terasaki, I.: Photo-seebeck effect in ZnO. J. Phys. Soc. Jpn. **81**(11), 114722 (2012).
21. Ono, Y., Kimura, D., Kawano, S.H.O., Watanabe, R., Kajitani, T.: Fabrication and performance of an oxide thermoelectric power generator. In: Proceedings of The 21st International Conference on Thermoelectrics (ICT2002), pp. 454–457. IEEE (2002)
22. Park, K.T., Shin, S.M., Tazebay, A.S., Um, H.D., Jung, J.Y., Jee, S.W., Oh, M.W., Park, S.D., Yoo, B., Yu, C., Lee, J.H.: Lossless hybridization between photovoltaic and thermoelectric devices. Sci. Rep. **3**, 2123 (2013).
23. Reddy, E.S., Noudem, J.G., Hebert, S., Goupil, C.: Fabrication and properties of four-leg oxide thermoelectric modules. J. Phys. D **38**(19), 3751–3755 (2005).
24. Satake, A., Tanaka, H., Ohkawa, T., Fujii, T., Terasaki, I.: Thermal conductivity of the thermoelectric layered cobalt oxides measured by the Harman method. J. Appl. Phys. **96**(1), 931–933 (2004).
25. Shikano, M., Funahashi, R.: Electrical and thermal properties of single-crystalline $(Ca_2CoO_3)_{0.7}CoO_2$ with a $Ca_3Co_4O_9$ structure. Appl. Phys. Lett. **82**(12), 1851–1853 (2003).
26. Shin, W., Murayama, N., Ikeda, K., Sago, S., Terasaki, I.: Thermoelectric device of $Na(Co, Cu)_2O_4$ and $(Ba, Sr)PbO_3$. J. Cer. Soc. Jpn. **110**(8), 727–730 (2002).
27. Singh, G.: Solar power generation by PV (photovoltaic) technology: a review. Energy **53**, 1–13 (2013).
28. Snyder, G.J., Toberer, E.S.: Complex thermoelectric materials. Nat. Mater. **7**, 105–114 (2008).
29. Snyder, G.J., Ursell, T.S.: Thermoelectric efficiency and compatibility. Phys. Rev. Lett. **91**(14), 148301 (2003).
30. Suter, C., Tomes, P., Weidenkaff, A., Steinfeld, A.: Heat transfer and geometrical analysis of thermoelectric converters driven by concentrated solar radiation. Materials **3**(4), 2735–2752 (2010).
31. Tauc, J.: The thermal photo-electric phenomenon in semi-conductors. Czech. Phys. J. **5**(4), 528–535 (1955).
32. Telkes, M.: Solar thermoelectric generators. J. Appl. Phys. **25**(6), 765–777 (1954).
33. Terasaki, I., Sasago, Y., Uchinokura, K.: Large thermoelectric power in $NaCo_2O_4$ single crystals. Phys. Rev. B **56**(20), R12685–R12687 (1997).
34. Thirugnanasambandam, M., Iniyan, S., Goic, R.: A review of solar thermal technologies. Renew. Sustain. Energy Rev. **14**(1), 312–322 (2010).
35. Tomes, P., Trottmann, M., Suter, C., Aguirre, M.H., Steinfeld, A., Haueter, P., Weidenkaff, A.: Thermoelectric oxide modules (TOMs) for the direct conversion of simulated solar radiation into electrical energy. Materials **3**(4), 2801–2814 (2010).
36. Urata, S., Funahashi, R., Mihara, T., Kosuga, A., Sodeoka, S., Tanaka, T.: Power generation of a p-type $Ca_3Co_4O_9$/n-type $CaMnO_3$ module. Int. J. Appl. Cer. Technol. **4**(6), 535–540 (2007).

37. Van Nong, N., Pryds, N., Linderoth, S., Ohtaki, M.: Enhancement of the thermoelectric performance of *p*-type layered oxide $Ca_3Co_4O_{9+d}$ through heavy doping and metallic nanoinclusions. Adv. Mater. **23**(21), 2484–2490 (2011).
38. Wang, N., Han, L., He, H., Park, N.H., Koumoto, K.: A novel high-performance photovoltaic–thermoelectric hybrid device. Energy Environ. Sci. **4**(9), 3676–3679 (2011).
39. Wang, Y., Sui, Y., Cheng, J., Wang, X., Su, W.: Comparison of the high temperature thermoelectric properties for Ag-doped and Ag-added $Ca_3Co_4O_9$. J. Alloys Comp. **477**(1–2), 817–821 (2009).
40. Whitney, R.S.: Nonlinear thermoelectricity in point contacts at pinch off: a catastrophe aids cooling. Phys. Rev. B **88**(6), 064302 (2013).
41. Zebarjadi, M., Esfarjani, K., Shakouri, A.: Nonlinear peltier effect in semiconductors. Appl. Phys. Lett. **91**(12), 104122 (2007).

24

Materials science aspects of zinc–air batteries: a review

Vincenzo Caramia · Benedetto Bozzini

Abstract Metal–air batteries are becoming of particular interest, from both fundamental and industrial viewpoints, for their high specific energy density compared to other energy storage devices, in particular the Li-ion systems. Among metal–air batteries, the zinc–air option represents a safe, environmentally friendly and potentially cheap and simple way to store and deliver electrical energy for both portable and stationary devices as well as for electric vehicles. Zinc–air batteries can be classified into primary (including also the mechanically rechargeable), electrically rechargeable (secondary), and fuel cells. Research on primary zinc–air batteries is well consolidated since many years. On the contrary, research on the electrically rechargeable ones still requires further efforts to overcome materials science and electrochemical issues related to charge and discharge processes. In addition, zinc–air fuel cells are also of great potential interest for smart grid energy storage and production. This review aims to report on the latest progresses and state-of-the-art of primary, secondary and mechanically rechargeable zinc–air batteries, and zinc–air fuel cells. In particular, this review focuses on the critical aspects of materials science, engineering, electrochemistry and mathematical modeling related to all zinc–air systems.

Keywords Zinc air batteries · Primary batteries · Secondary batteries · Fuel cells · Zinc · Air cathode

V. Caramia · B. Bozzini (✉)
Dipartimento di Ingegneria dell'Innovazione, Università del Salento, via Monteroni, 73100 Lecce, Italy
e-mail: benedetto.bozzini@unisalento.it

Introduction

In the modern industrialized society, the electrical energy demand is increasing exponentially [1]; but environmental pollution, due to the usage of fossil fuels for power generation, is a very well known and urgent problem [1, 2]. Renewable and sustainable energy, such as solar and wind [3–6], could replace hydrocarbons, but it is also important to find a safe, reliable and efficient way to store such energy and use it in transportation systems and large-scale applications, for instance. For this reason, researchers and industry are looking for new strategies for better electrical energy storage devices. Among these, the batteries represent the right key for the next generation of green vehicles and grid energy storage, due to their relatively high energy density compared to supercapacitors that have, instead, a higher power density [7–9].

Nowadays, the need for energy storage in a robust and reliable electric grid is increasing as a result of the growing and worldwide use of renewable power generation [10]. In Japan, for instance, due to the potential decommissioning of nuclear fleet, it is becoming crucial the use of alternative and smart ways to generate, store and distribute electric energy [10]. Italy, instead, has substantial renewable capacity relative to grid size, and the grid is currently struggling with reliability issues; moreover, in the USA, Canada and Germany there is a keen interest to invest much more in renewable power sources (wind, solar, thermal, etc.,) [10–13]. For these reasons, it is important to develop more efficient grid storage systems. The currently available mature and reliable technologies for high system power ratings (100 MW–1 GW) are pumped hydro-storage (PHS) and compressed air energy storage. However, for smaller and smarter grids, electrochemical devices, chiefly batteries, play an important role for power rating in the

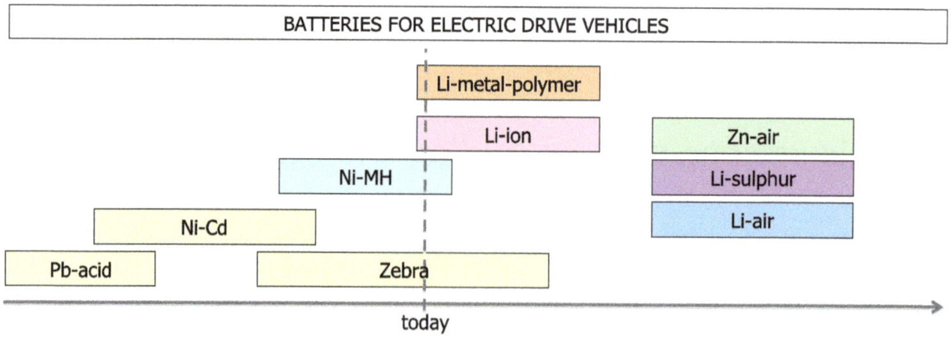

Fig. 1 Batteries for electric drive vehicles. Reproduced with permission from [34]

range from c.a. 1 kW to 10 MW [9]. Among batteries, the following are the most investigated ones: the well known and widely used lead–acid ones, that have also recently undergone notable technical improvements; the more recent lithium ion-based technology concept; the redox-flow batteries (e.g., Zn/Br; Zn/Cl, Vanadium-redox), and finally the sodium–sulfur and metal–air devices that are being actively studied by several industrial and academic institutions [14–19]. In this framework, metal–air batteries potentially represent the method of choice for smart and green grid storage, since their peculiarity is the high theoretical energy density and the low environmental impact of their components [12, 20]. Recently (2012), the EOS Energy company began its mission to employ zinc–air batteries for grid storage, believing in their potentialities over other types of batteries where less safe, scarce and expensive materials, such as lithium, are employed [21].

Even if the technology of lithium ion batteries (LIBs) is well established for portable applications such as mobile telephones, digital cameras, etc., their energy density is still not entirely satisfactory for electric vehicles (EVs) applications and driving ranges comparable to those typically obtained with fossil fuels. Nevertheless, electric vehicles totally based on Li-ion technology are commercially available with a full charge autonomy up to c.a. 200 km, such as: Renault ZEO and Kangoo [22], BMW 1 series ActiveE [23], Ford Focus Electric [24], Mercedes Benz Smart Electric Drive [25], Mitsubishi Innovative Electric Vehicle ("I-MiEV") [26], Nissan Leaf [27], Tesla Model S [28], Toyota RAV4 EV [29], and Volkswagen e-up [30]. Furthermore, some automobile companies are selling vehicles with a hybrid technology; Toyota hybrid system has been in the market since 1997 using Ni–MH batteries, and since 2011 using Li-ion technology. Apart from Toyota, also Opel (Opel Ampera) [31], Ford [32] and Volkswagen [30] are in the market of hybrid electric vehicles powered by a Li-ion technology and by an internal combustion engine.

Therefore, in the near future, the so-called metal–air batteries could replace the ion-based technology as shown in Fig. 1 [33, 34]. Among all the different types of metal—air batteries—such as those based on Ca, Mg, Li, Zn, Al, Fe and Cd–, Li–air and Zn–air batteries are the most promising ones. A comparison between theoretical and practical gravimetric energy density (Wh/kg) for different electrochemical energy storage devices is reported in Fig. 2 [35]. In detail, the practical energy density of lead–acid and Ni–Cd batteries is c.a. 40 Wh/kg; that of Ni–MH systems is 50 Wh/kg, that of devices based on Li-ion chemistry is 160 Wh/kg and that of Li–S concepts is about 370 Wh/kg. The practical energy density of metal–air systems, instead, can be summarized as follows: Zn–air 350 Wh/kg, Al–air (not shown in Fig. 2) 1,300 Wh/kg and Li–air 1,700 Wh/kg [36]. A very recent review on metal–air batteries, including the zinc–air ones, provides the reader with a general overview and comparison on the different metal–air technologies [37]. To the best of our knowledge, a rich and recent scientific literature is available for Li–air cells, but not for Zn–air systems. Nevertheless, apart from some debates on the last two technologies, which are summarized in the quite recent review of Lee et al. [38], from a materials science point of view it is very challenging to obtain an efficient electrically rechargeable metal–air battery employable for EVs.

Zinc as a metal is safer than lithium and can be fully recycled [39]. The availability of zinc is not a major problem, compared to lithium, since the number of zinc reserves is high around the world, currently the largest producers being China, Australia, Canada and the USA [39, 40]. Some properties of the electrochemical zinc–air cells and other metals are listed in the Table 1 [41], where it is possible to notice that Li, Ca, Mg and Al have the highest specific energy, but they have very negative electrode potentials and, as a result, these electrodes corrode easily in aqueous electrolytes with the production of hydrogen gas.

A zinc–air battery, as schematically illustrated in Fig. 3, is composed of three main components: a zinc anode, an alkaline (KOH) electrolyte and an air cathode (usually a porous and carbonaceous material). Oxygen diffuses through the porous air cathode, and the catalyst layer on the cathode allows the reduction of oxygen to hydroxide ions

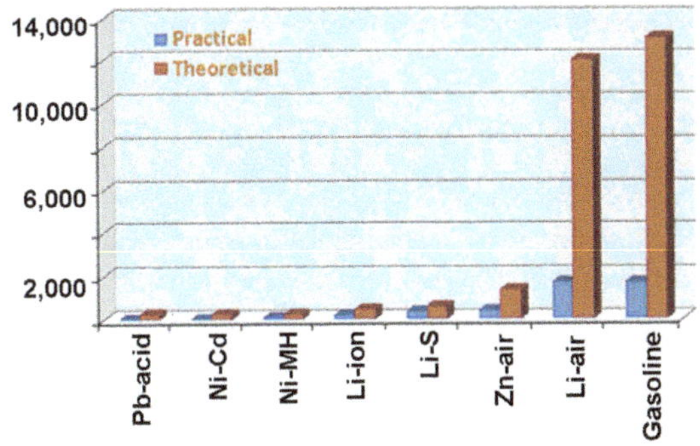

Fig. 2 Comparison of practical and theoretical gravimetric energy density (Wh/kg) for different batteries. Reproduced with permission from [35]

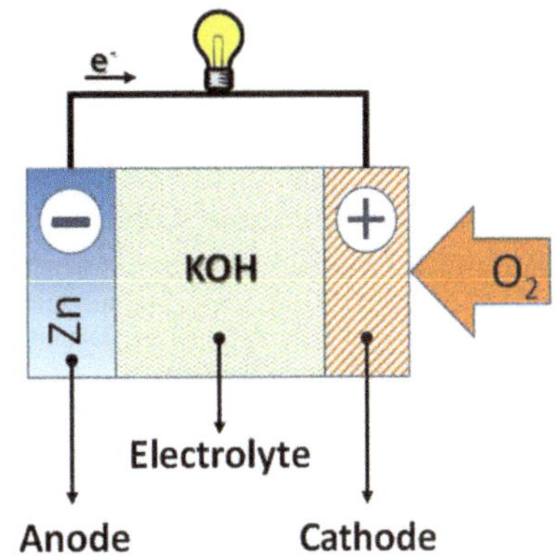

$$Zn + \frac{1}{2}O_2 \rightarrow 2ZnO$$

Fig. 3 Schematic illustration of a primary, not electrically rechargeable, zinc–air battery

in the alkaline electrolyte with the electrons obtained from the electrochemical oxidation of the zinc anode. On recharge, oxygen will be evolved and diffuses out of the positive electrode, whereas zinc is deposited at the negative electrode. An oxygen electrode with dual catalytic activity (for ORR and OER) (not shown in Fig. 3) must be used for a rechargeable battery. This is a very hot topic and some key advancements have been recently published [42].

Zinc–air batteries can be classified into primary and electrically rechargeable. The former ones are commonly employed in hearing aid devices since the 1960s and represent a very well established technology without the need of further developments in terms of research on new materials and engineering improvements. The latter ones, on the contrary, still require research efforts to improve the efficiency and minimize obnoxious side effects occurring during the charging process [42], such as, for instance, zinc dendrites formation, limited number of charge–discharge cycles due to deterioration of the air cathode and carbonation of KOH electrolyte. In addition, in an electrically rechargeable zinc–air system, it is also important to deeply study and understand the chemistry of zincate solubility in the alkaline electrolyte as well as to develop new catalyst materials [42].

This review provides a comprehensive summary of the latest developments in zinc–air battery and fuel cell science and technology, covering, in particular, the materials used for the anode, the cathode, and the electrolyte as well as all the problems currently limiting the widespread success of electrically rechargeable zinc–air batteries.

Primary zinc–air batteries

This section concentrates on anodic, electrolytic and cathodic materials employed in primary zinc–air battery. From the point of view of applications, primary Zn–air batteries are well known for use in hearing aid devices (button-type cells), as hinted at in the introduction. Nevertheless, large primary Zn–air batteries have been also used to provide low rate and long-life power for applications such as seismic telemetry, railroad signaling and navigational buoys as well as remote communications. The theoretical specific energy density of Zn–air batteries is 1,084 Wh/kg. The theoretical voltage of a zinc–air cell is

Table 1 Characteristics of metal–air cells

Metal anode	Electrochemical equivalent of metal (Ah/g)	Theoretical cell voltage[a] (V)	Valence charge	Theoretical specific energy (of metal) (kWh/kg)	Practical operating voltage (V)
Li	3.86	3.4	1	13.0	2.4
Ca	1.34	3.4	2	4.6	2.0
Mg	2.20	3.1	2	6.8	1.2–1.4
Al	2.98	2.7	3	8.1	1.1–1.4
Zn	0.82	1.6	2	1.3	1.0–1.2
Fe	0.96	1.3	2	1.2	1.0

Data derived from [41]

[a] Cell voltage with oxygen cathode

1.667 V, but, in practice, the open circuit voltage is about 1.35 V. In addition, the discharge curve of a Zn–air cell is flat with minimum potential decay—compared to Li–air systems—when a constant current density (50–100 mA cm^{-2}) is applied at the battery, and the voltage is measured and monitored over discharge time.

Anode materials

The anode materials of any Zn–air battery is simply zinc, employed in different forms such as powder or fibers, to increase efficiency. During discharge of the battery, the following chemical reaction scheme describes the processes involved at the zinc anode in alkaline solutions:

$$Zn \rightarrow Zn^{2+} + 2e^- \quad (1)$$

$$Zn^{2+} + 4OH^- \rightarrow Zn(OH)_4^{2-} \quad (2)$$

$$Zn(OH)_4^{2-} \rightarrow ZnO + H_2O + 2OH^- \quad (3)$$

$$Zn + 2H_2O \rightarrow Zn(OH)_2 + H_2\uparrow \quad (4)$$

In particular, during discharge, the oxidation of the zinc electrode can involve several processes including oxidation of surface zinc atoms, ion solvation in the solution, ion diffusion in the electrolyte and precipitation into a solid phase when the solubility limit is reached.

Some interesting properties of zinc relevant to battery applications (at room temperature) are listed in the Table 2 [43]. One of the most important characteristics of Zn metal is its low electrical resistivity, equal to 5.96 μΩ cm. In addition, the corrosion rate of zinc in alkaline solutions is not very fast compared, for instance, to aluminum; but during the anodic discharge of the battery, as it is possible to conclude from reaction (4), the hydrogen evolution (HER) in the form of gas is problematic. In particular, due to the basicity of the concentrated (4–9 M, with pH ~ 14 in order to ensure a high ionic conductivity) KOH electrolyte in aqueous solutions—usually employed in Zn–air batteries—, zinc corrodes producing hydrogen, as summarized in the Pourbaix diagram (Fig. 4a) [44], giving rise to charge losses and potentially dangerous hydrogen gas buildup. It is

Table 2 Properties of zinc relevant to battery applications (at room temperature)

Atomic weight	65.38
Crystalline structure	Closed packed hexagonal
Density	7.14 g/cm3
Melting temperature	419.5 °C
Resistivity	5.96 μΩ cm
Ion valence	2
Ionic radius	0.074–0.083 nm
Stable dissolved form in KOH	$Zn(OH)_4^{2-}$
Standard potential	−0.763 V_{SHE}
Reversible potential in 35 % KOH, 8.2 mol/L	−1.35 V_{SHE}

Adapted from [43]

SHE standard hydrogen electrode

also worth noticing that, when the pH is lower than 13, passivation of zinc results in the formation of stable (known as type II), compact, insoluble and n-type semiconductor zinc oxide (ZnO), that is inconvenient for battery operation [45].

The rest of this section will be devoted to the discussion of the different types and configurations of Zn anodes for primary Zn–air batteries. The use of zinc foil is inconvenient, because it reduces the performance of the electrochemical cell due to the low surface area. In order to increase the surface of the zinc anode particular strategies can be adopted, resulting in an enhancement of the battery lifecycle; among these solutions, the use of nano- and microstructured materials plays an important role. Figure 5 shows the different forms of zinc materials that have been employed as anodes in both primary and secondary zinc–air batteries [43, 46, 47]. All these types of zinc materials have different values of porosity. A suitable value of porosity is crucial to discharge the battery at large current densities with minimal voltage losses and long durability. A porosity value in the range between 60 and 80 % is commonly found at the anode of a Zn–air battery, which corresponds to a capacity of 1.2–2.2 Ah/cm^3. Moreover, it is worth noticing that, as reported in Fig. 6 [47], the

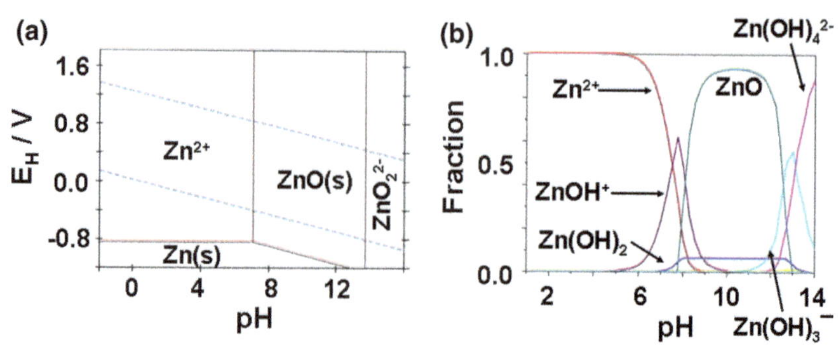

Fig. 4 **a** Pourbaix diagram of zinc [44]; **b** fractions of different Zn^{2+} species formed as a function of pH for a zinc–water system. Reproduced with permission from [45]

Fig. 5 Different forms of zinc materials produced by different methods. Reproduced with permission from [43]

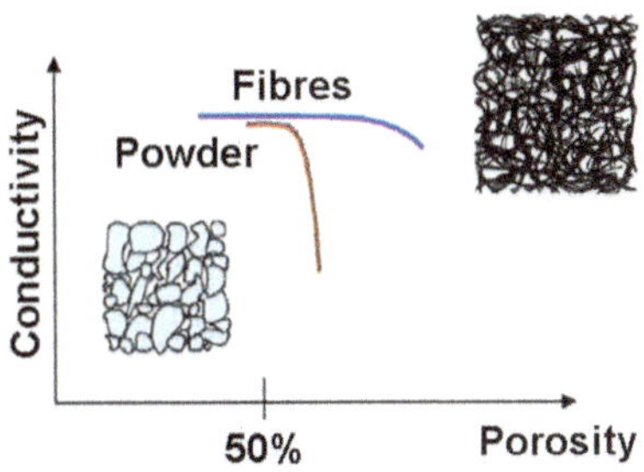

Fig. 6 Qualitative relationship between conductivity and porosity for powder electrodes and fibrous electrodes. Reproduced with permission from [47]

conductivity of the zinc anode behaves in different ways, depending on the morphology of the material (powder or fiber) and on its porosity. Fibrous zinc has a better electrical performance in the range of porosities required for an efficient Zn–air battery. On the contrary, an atomized zinc powder has a low porosity (about 50 %), and, for this reason, a gelling agent is used to make a mix of zinc powder and electrolyte, such that the former is suspended in the electrolyte gel and, then, increases the porosity of the anode material to c.a. 70 % [48, 49]. Different types of gels or binders have been reported, such as polytetrafluoroethylene (PTFE) [50, 51], sago [49, 52], carbopol gel [53], and a tapioca binder [54]. Notwithstanding its advantages in terms of porosity enhancement, the use of gelling agents increases both electrolyte resistance and manufacturing costs.

It is important to note that, when the zinc anode is employed in dispersed forms such as powder or fiber, mercury is typically added to enhance the electrical conductivity and to reduce the evolution of H_2 [55]. Of course, mercury ought to be replaced in sustainable Zn–air technologies. Four chief strategies have been developed for Hg replacement, mainly aiming to increase the overpotential for hydrogen evolution.

(i) Alloying zinc with other metals (Pb, Cd, Ni) with high hydrogen evolution overvoltage (a potentiodynamic polarization analysis is used to measure the relevant overpotential). However, all these elements are either highly toxic or not environmental friendly. For this reason, Zhang et al. [56] suggested the introduction of metallic bismuth to a pasted zinc electrode. In addition, alloys of Zn and Al are also used not only to inhibit corrosion of Zn, but also to increase the capacity of the anode material and decrease its weight [57] by harvesting both the reactivity of Al in alkaline solutions and its density. Special attention has been devoted to the more environmentally acceptable Zn–Ni alloy system. The comparative corrosion of Zn and Zn–Ni alloys in different conditions (temperature and concentration of the KOH electrolyte) has been studied in [58] by potentiodynamic and impedance methods. By plotting the corrosion current vs. concentration of KOH electrolyte (log[KOH]) for Zn and Zn–Ni alloys, a linear trend is observed for each alloy. Of course, by alloying Zn with Ni, the corrosion resistance of the pure metal is enhanced.

(ii) Addition of Al_2O_3 either by mixing of Zn and Al_2O_3 particles or by surface modification of the Zn powders with an Al_2O_3 coating, deposited via a chemical solution process (see also Fig. 7 for details on the respective effects) [59]. The effectiveness of the approach was tested by comparing the volumetric amount of hydrogen spontaneously evolved on the surface of bare Zn, Al_2O_3 mixed with Zn, and Zn powders coated with Al_2O_3 in a 9 M KOH electrolyte at 60 °C. The hydrogen evolution is almost suppressed in the presence of the alumina coating, owing to the formation of a passivation layer that prevents direct contact of Zn with the KOH electrolyte.

(iii) Coating the zinc metal particles with other materials such as $Li_2O–2B_2O_3$ in a core–shell structure. This configuration prevents the zinc particles from contacting directly the alkaline electrolyte, thus hindering corrosion side-reactions.

(iv) Mixing the alkaline electrolyte with organic corrosion inhibitors such as (a) anions of organic acids [60], (b) $HCO_2–CH_2–(OCH_2CH_2)–CH_2–CO_2H$ dicarboxylic acid-modified poly(ethylene glycol) [61] and phosphoric acid esters (e.g., GAFAC RA600) [62].

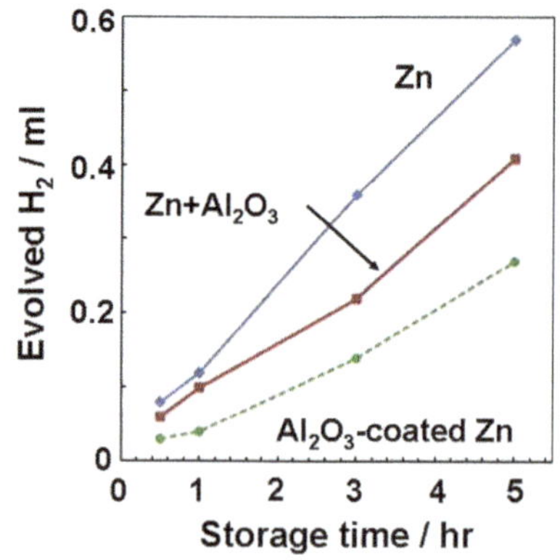

Fig. 7 Comparison of volumetric amount of hydrogen spontaneously evolved on the surface of bare Zn, Al_2O_3 mixed with Zn, and Al_2O_3-coated Zn gel anodes in a 9 M KOH electrolyte at 60 °C. Reproduced with permission from [59]

Cathode materials

The cathode is, of course, an "air cathode" and performs the reduction of O_2 to complete the redox reaction of the battery. Being a gas electrode, the cathode has to be porous and electrocatalytically active. The cathodic electrochemical reaction can be written as follows:

$$O_2 + 2H_2O + 4e^- \rightarrow 4OH^- \quad (5)$$

Since an alkaline solution is typically employed in Zn–air batteries, Eq. (5) represents the typically favored mechanism [63, 64]. As far as the catalysts are concerned, Zn–air batteries do not require precious metal catalysts, such as platinum, ruthenium or palladium, for the relevant chemistry; moreover, it has also been shown that noble catalysts tend to decrease the electrochemical performance since noble metal traces, released from the cathode, tend to diffuse to the zinc anode, lower its hydrogen overvoltage thus enhancing corrosion and gas production. References [33, 65] review the electrochemistry of the oxygen reduction reaction in metal–air batteries (including Zn–air ones) as well as the relevant catalysts. Moreover, Ref. [66] comprehensively compiles the published catalyst options and cathode configurations, with special attention to the patent literature. Among air cathode materials, Fe- and Co-based catalysts obtained by pyrolysis of N- and C-containing precursors (such as polypyrrole and polyaniline) [67] as well as graphene loaded with Mn_3O_4 nanoparticles [68] and CNT mixed with CoO [69], have been shown to be efficient for primary Zn–air batteries and to exhibit better performance than noble metals. The role of the carbonaceous materials is to support the catalyst and to create a porous and electronically conductive path. However, to avoid leakage of the electrolyte from the air cathode (see, e.g., [70]), superhydrophobic materials, such as polytetrafluoroethylene (PTFE), are also employed as fillers in the fabrication of air cathodes [64].

References [71–90] provide a comprehensive description of the available air cathode fabrication concepts that can be summarized as follows: (i) Teflon® membrane on the air side with a high air permeability of 2,000–4,000 s (Gurley method). This membrane protects the air cathode against possible leakage of electrolyte to the current collector. (ii) Current collector, typically a Ni mesh, a cheaper woven copper mesh with Ni coating, or an expensive Ni foam. Ni foam has been reported to exhibit a typically higher surface area, ensuring a better performance of the air cathode. (iii) Gas diffusion layer (GDL) chiefly composed of acetylene blacks (AB) with low Brunauer–Emmett–Teller (BET) surface area and wettability mixed with polytetrafluoroethylene (PTFE). The use of acetylene blacks with higher hydrophobic properties promotes more hydrophobic GDL properties. (iv) Catalyst layer: composed by the catalyst mixed with carbon blacks exhibiting high BET surface area and PTFE for hydrophobization.

Electrolyte

The most suitable electrolyte for Zn–air batteries is potassium hydroxide (KOH, some relevant properties of this electrolyte have been listed in Table 3), owing to its high ionic conductivity and low cost. Other electrolytes have been considered, chiefly NaOH, but these have proved less effective. The concentration of KOH is selected in view of minimizing Zn corrosion and H_2 evolution. 30 % KOH is a typical compromise among good ionic conductivity, maximum solubility of zinc oxide and minimal zinc corrosion. One of the main problems with these aqueous alkaline electrolytes is carbonation, resulting from the presence of carbon dioxide in the atmospheric air. Not only does the formation of potassium carbonate lower the I–V curves and adversely affect the capacity of Zn–air batteries, but—since potassium carbonate is less soluble than potassium hydroxide—it also causes pore clogging of air cathodes [70].

Table 3 Properties of KOH electrolyte relevant to battery applications (at room temperature)

Density of 35 % KOH, 8.2 mol/L	1.34 g/cm^3
Resistivity of 8 mol/L KOH	2.3 Ω cm
Solubility of ZnO in 35 % KOH, 8.2 mol/L	1 M
Supersaturation of zincate in KOH	2–4 times of solubility
Ratio of diffusion coefficient of $Zn(OH)_4^{2-}$/ OH^-	c.a. 0.1
Ratio of diffusion coefficient of $Zn(OH)_4^{2-}/K^+$	c.a. 0.25

Adapted from [43]

Mechanically rechargeable Zn–air batteries

Among primary Zn–air batteries, the mechanically rechargeable ones, schematically illustrated in Fig. 8, consist of a replaceable anode cassette comprising a slurry of zinc particles in contact with a potassium hydroxide solution and brought in contact—typically by gravity—with a current collector. The anode cassette is confined by a separator envelope and put into contact with the air electrode via a free-electrolyte compartment. Of course, the mechanical recharging policy allows fast regeneration of a spent cell in a configuration that preserves the benefits of a primary cell concept, but allows the body of the cell to be hosted in a more rugged way to the device exploiting the battery, in a way that is typical of secondary batteries. Of course, mechanically rechargeable batteries do not exhibit the degradation problems related to morphology and shape changes taking place during electric charging and discharge that are typical of secondary batteries. This type of Zn–air battery concept has been chiefly studied by Electric Fuel Ltd. for defense applications, consumer electronic products and electric vehicles (buses in particular). In particular, the best-documented mechanically rechargeable zinc–air battery module employed is the one tested in electric buses, exhibiting a high specific energy density (>200 Wh/kg), able to power the vehicle for c.a. 400 km [91–94]. A typical regeneration route, schematically illustrated in Fig. 9, is described in [93] and includes the following operations: (i) disassembly, in which separator bags are removed from the anodes, and the zinc oxide discharge product (along with residual, undischarged zinc) is removed from the current collector frames; (ii) dissolution, in which zinc oxide is dissolved in a KOH solution to form a zincate-rich feed, according to the following reaction: $ZnO + 2KOH + H_2O \rightarrow K_2Zn(OH)_4$; (iii) electrowinning, in which the zincate solution is electrolyzed according to the following reaction: $K_2Zn(OH)_4 \rightarrow Zn + 2KOH + H_2O + \frac{1}{2}O_2$; (iv) reassembly, in which the electrowon zinc, together with residual (charged) metallic zinc, is compacted onto the current collector frame and the anode is inserted into a separated bag.

Electrically rechargeable Zn–air batteries

In this section, a discussion of the anode, cathode and electrolyte materials employed in electrically rechargeable Zn–air batteries (ERZAB) will be provided. In addition, the main problems, limiting their commercialization and widespread application, will be discussed. It is worth underlining that most of the materials employed in primary Zn–air batteries are also used in ERZAB, especially the anode materials, the electrolyte and the catalysts for the oxygen reduction reaction (ORR). Of course, the main research challenges and difficulties in materials selection stem from the requirement of reversibility and efficiency in

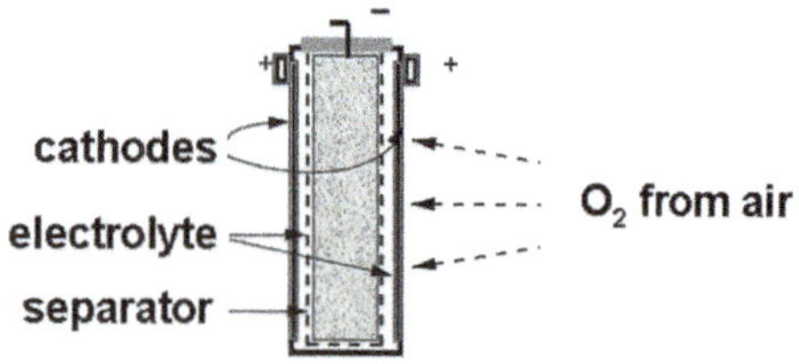

Fig. 8 Schematic diagram of a mechanically rechargeable zinc–air battery (Electric Fuel company). Reproduced with permission from [93]

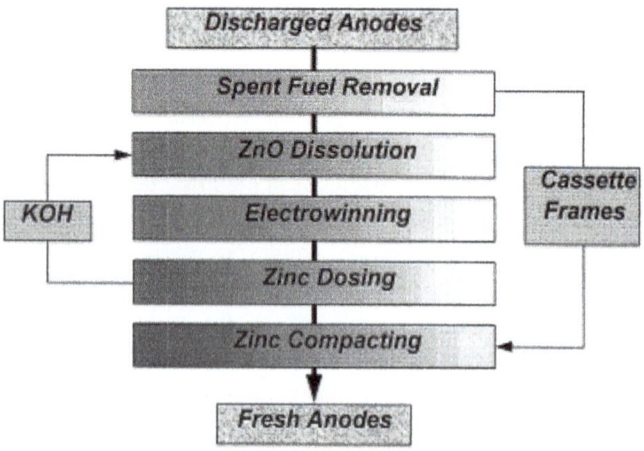

Fig. 9 Regeneration process of zinc anodes. Reproduced with permission from [93]

both discharge and charge processes. Li et al. [69], in a very recent paper (May 2013) demonstrated an excellent bifunctional electrocatalyst, exhibiting optimal ORR and OER activities during discharge and charging processes, respectively.

Anode materials

As in the primary Zn–air battery, in the electrically rechargeable one, the anode material is zinc. However, in a secondary battery, this component changes structure and shape during repeated charge/discharge cycles, corresponding to zinc dissolution and re-deposition for several reasons, chiefly inhomogeneous current distribution and presence of concentration gradients in the electrolyte. As far as the shape changes are concerned, the typical shape change problems that have been reported to take place during charging are dendrite formation (leads to the loss of active material and to short-circuiting) and electrode densification (causing the originally porous zinc to leave active locations and agglomerate into self-screening compacts) [42]. It is worth noting that the different types of shape change phenomena can be strongly interconnected. A comprehensive account of shape change phenomena can be found in Ref. [43]. In the quest for reversible electrical recharging processes, effective dendrite suppression was achieved by applying a zeolite film over the zinc electrode [94]. This film keeps the Zn^{2+} discharge product close to the electrode, preventing it from being flushed away into the electrolyte.

Cathode materials

As hinted at above, at the cathode of a rechargeable Zn–air battery, both the oxygen reduction reaction (ORR) and the oxygen evolution reaction (OER) are key factors for the efficiency of the system. Presently, a cathode material that is efficient and durable for both ORR and OER is not available, but research in the field seems active. Reference [42] reports on a novel core-corona bifunctional catalyst (CCBC) consisting of lanthanum nickelate centers supporting nitrogen-doped carbon nanotubes (NCNT). Reference [95] reports on the preparation of MnO_2 nanotubes functionalized with Co_3O_4 nanoparticles for bifunctional air cathodes. These hybrid MnO_2/Co_3O_4 nanomaterials exhibit enhanced catalytic reactivity toward oxygen evolution reaction under alkaline conditions compared with MnO_2 nanotubes or Co_3O_4 nanoparticles alone. Reference [96] reports on novel silver nanoparticles-decorated MnO_2 nanorods as an air electrode bifunctional catalyst. An alternative to the use of bifunctional catalysts is the approach of employing two specialized air cathodes for ORR and OER, respectively [69].

Electrolyte

The electrolyte most commonly employed in an electrically rechargeable Zn–air battery is again KOH. However, in the oxidation–reduction cycling, the electrochemistry of zinc in potassium hydroxide gives some problems because the precipitation of zinc oxide is irreversible and reduces the availability of Zn^{2+} ions upon cycling. The use of chelating ionic liquids has been proposed to circumvent precipitation issues [97].

Mathematical modeling of electrically rechargeable Zn–air batteries

To the best of our knowledge, just one paper has appeared on the mathematical modeling of the thickness of the anode and the optimal KOH concentration. Reference [98] presents a numerical model for a single-cell rechargeable Zn–air batteries able to estimate the I–V response of the cell, the individual Zn and O_2 electrode potentials as well as the potential and concentration profiles (see Fig. 10). The model proposed in [98] consists of two partial differential equations for the material balance of ions moving in the electrolyte and solid species locked in the anode.

Zinc–air fuel cell

Zinc–air fuel cell systems are potentially attractive for stationary and large-scale production and storage of electric energy. At variance with primary and electrically rechargeable zinc–air batteries, the zinc–air fuel cells (ZAFCs) are based on the flow of a mixture of Zn particles and electrolyte. Zn particles produced by different methods have been tested, including cut wire chunks, compacted electrodeposited dendrites, Zn spheroids reconstructed electrochemically in a spouted bed reactor (see, e.g., Fig. 11); zinc pellets are preferred to compacted dendrites owing to their better fluid dynamic behavior [93, 99–102]. The spent Zn particles, along with the KOH electrolyte, are replaced hydraulically by pumping from an external tank. Moreover, during discharge potassium zincate is continuously pumped away with the electrolyte. A schematic representation of a regenerative zinc–air fuel cell is shown in Fig. 12 [43].

Conclusions

In this review, the different types of zinc–air batteries described in the literature and their functional components are examined from the point of view of electrochemical materials science. Among all metal–air batteries, the zinc–air ones are of particular interest since in principle they can be safer, cheaper and more environmentally friendly than other competing technologies. Furthermore, zinc–air batteries, both primary and electrically rechargeable, can meet the requirements of the whole range of applications: portable electronics, medium-scale energy production and storage and eventually grid storage. Fully engineered secondary zinc–air batteries are not yet available: research and development is still needed, especially in the fields of: (i) shape changes of the Zn electrode during charge/discharge cycles, (ii) durable and dual air cathode catalysts, (iii) KOH-based electrolyte chemistry. As far as high-power applications are concerned, the mechanically rechargeable and the flowing anode concepts are particularly appealing for the possibility of achieving essentially continuous operation as well as of implementing an off-line electrical generation process.

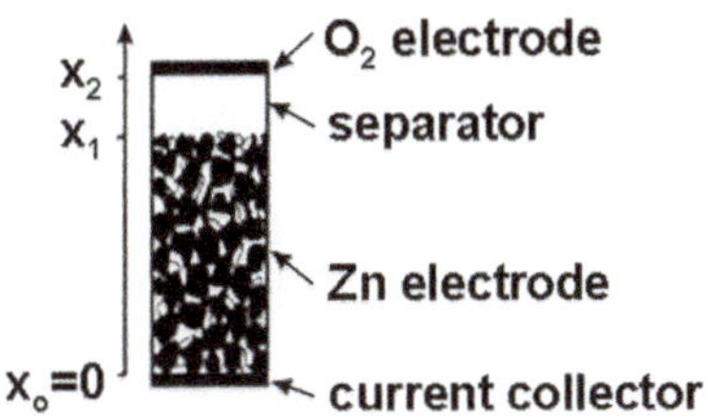

Fig. 10 Model of the one-dimension single-cell Zn–air battery. Reproduced with permission from [98]

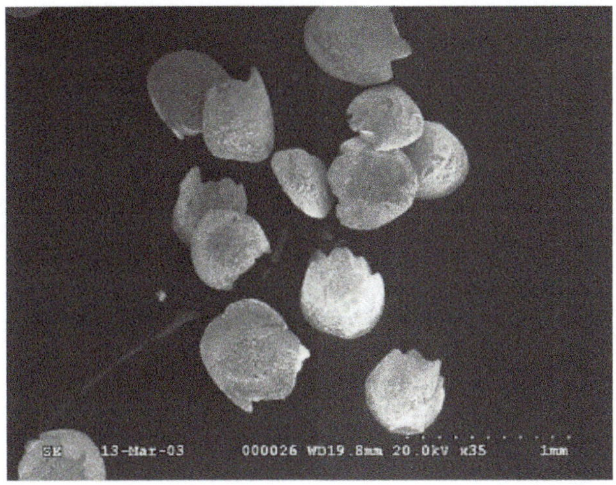

Fig. 11 SEM image of the zinc pellets obtained after regeneration. Reproduced with permission from [99]

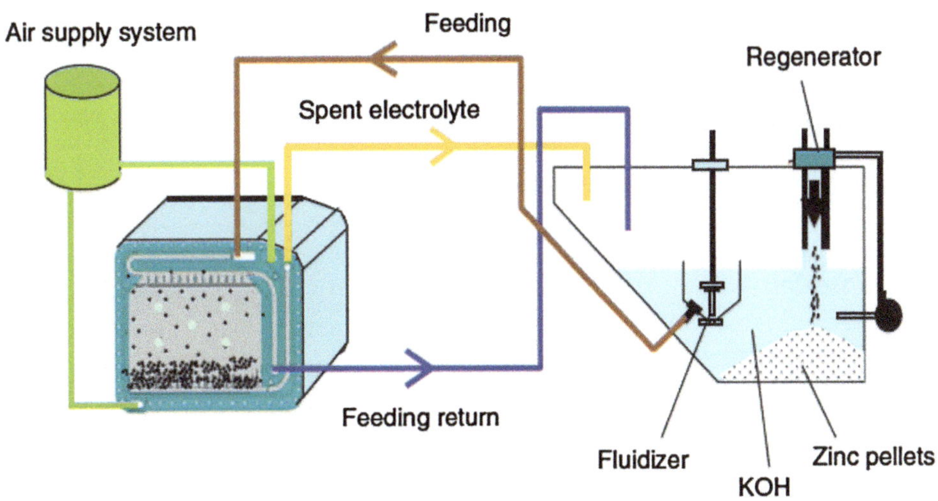

Fig. 12 Schematic illustration of the flow of material in a zinc–air fuel cell system. Reproduced with permission from [43]

References

1. Omer, A.M.: Energy, environment and sustainable development. Renew. Sustain. Energy Rev. **12**, 2265–2300 (2008)
2. Bose, B.K.: Global energy scenario and impact of power electronics in 21st century. IEEE Trans. Ind. Electron. **60**, 2638–2651 (2013)
3. Omer, A.M.: Green energies and the environment. Renew. Sustain. Energy Rev. **12**, 1789–1821 (2008)
4. Solangi, K.H., Islam, M.R., Saidur, R., Rahim, N.A., Fayaz, H.: A review on global solar energy policy. Renew. Sustain. Energy Rev. **15**, 2149–2163 (2011)
5. Islam, M.R., Mekhilef, S., Saidur, R.: Progress and recent trends of wind energy technology. Renew. Sustain. Energy Rev. **21**, 456–468 (2013)
6. Devabhaktuni, V., Alam, M., Shekara Sreenadh Reddy Depuru, S., Green Ii, R.C., Nims, D., Near, C.: Solar energy: trends and enabling technologies. Renew. Sustain. Energy Rev. **19**, 555–564 (2013)
7. Tie, S.F., Tan, C.W.: A review of energy sources and energy management system in electric vehicles. Renew. Sustain. Energy Rev. **20**, 82–102 (2013)
8. Pollet, B.G., Staffell, I., Shang, J.L.: Current status of hybrid, battery and fuel cell electric vehicles: from electrochemistry to market prospects. Electrochim. Acta **84**, 235–249 (2012)
9. Dunn, B., Kamath, H., Tarascon, J.-M.: Electrical energy storage for the grid: a battery of choices. Science **334**, 928–935 (2011)
10. http://www.energy.gov/sites/prod/files/2013/12/f5/Grid%20Energy%20Storage%20December%202013.pdf. Accessed Feb 2014
11. Koohi-Kamali, S., Tyagi, V.V., Rahim, N.A., Panwar, N.L., Mokhlis, H.: Emergence of energy storage technologies as the solution for reliable operation of smart power systems: a review. Renew. Sustain. Energy Rev. **25**, 135–165 (2013)
12. Rastler, D.: Electricity energy storage technology options: a white paper primer on applications, costs and benefits, p. 1020676. EPRI, Palo Alto (2010)
13. Steinke, F., Wolfrum, P., Hoffmann, C.: Grid vs. storage in a 100% renewable Europe. Renew. Energy **50**, 826–832 (2013)
14. http://www.sbc.slb.com/SBCInstitute/Publications/ElectricityStorage.aspx. Accessed Feb 2014
15. Yang, Z., Zhang, J., Kintner-Meyer, M.C.W., Lu, X., Choi, D., Lemmon, J.P., Liu, J.: Electrochemical energy storage for green grid. Chem. Rev. **111**, 3577–3613 (2011)
16. Alotto, P., Guarnieri, M., Moro, F.: Redox flow batteries for the storage of renewable energy: a review. Renew. Sustain. Energy Rev. **29**, 325–335 (2014)
17. Ding, C., Zhang, H., Li, X., Liu, T., Xing, F.: Vanadium flow battery for energy storage: prospects and challenges. J. Phys. Chem. Lett. **4**, 1281–1294 (2013)
18. Leung, P., Li, X., Ponce de Leon, C., Berlouis, L., Low, C.T.J., Walsh, F.C.: Progress in redox flow batteries, remaining challenges and their applications in energy storage. RSC Adv. **2**, 10125–10156 (2012)
19. Leadbetter, J., Swan, L.G.: Selection of battery technology to support grid-integrated renewable electricity. J. Power Sources **216**, 376–386 (2012)
20. http://www.technologyreview.com/news/512206/years-in-the-making-promising-rechargeable-metal-air-batteries-head-to-market/. Accessed Feb 2014
21. http://www.eosenergystorage.com/zinc-air-batteries-for-grids/. Accessed Feb 2014
22. http://www.renault-ze.com. Accessed Nov 2013
23. http://www.bmw.com/com/en/newvehicles/1series/activee/2011/showroom/index.html. Accessed Nov 2013
24. http://www.ford.com/cars/focus/trim/electric/. Accessed Nov 2013
25. http://tools.mercedes-benz.co.uk/current/smart/brochures/smart-electric-drive.pdf. Accessed Nov 2013
26. http://www.mitsubishi-motors.com/special/ev/whatis/index.html. Accessed Nov 2013
27. http://www.nissan.co.uk/vehicles/electric-vehicles/electric-leaf/leaf.htm#vehicles/electric-vehicles/electric-leaf/leaf. Accessed Nov 2013
28. http://www.teslamotors.com/models. Accessed Nov 2013
29. http://www.toyota.com/rav4ev/. Accessed Nov 2013
30. http://emobility.volkswagen.it/it/it/private/technology.html. Accessed Nov 2013
31. http://www.opel.com/microsite/ampera/#/ampera_charging_daily_use. Accessed Nov 2013
32. http://social.ford.com/our-articles/cuvs/c-max/latest-ford-c-max-news-from-the-2011-north-american-international-auto-show/. Accessed Nov 2013
33. Cheng, F., Chen, J.: Metal–air batteries: from oxygen reduction electrochemistry to cathode catalysts. Chem. Soc. Rev. **41**, 2172–2192 (2012)
34. Catenacci, M., Verdolini, E., Bosetti, V., Fiorese, G.: Going electric: expert survey on the future of battery technologies for electric vehicles. Energy Policy. **61**, 403–413 (2013)
35. Wang, J., Li, Y., Sun, X.: Challenges and opportunities of nanostructured materials for aprotic rechargeable lithium–air batteries. Nano Energy **2**, 443–467 (2013)
36. Girishkumar, G., McCloskey, B., Luntz, A.C., Swanson, S., Wilcke, W.: Lithium–air battery: promise and challenges. J. Phys. Chem. Lett. **1**, 2193–2203 (2010)
37. Rahman, M.A., Wang, X., Wen, C.: High energy density metal–air batteries: a review. J. Electrochem. Soc. **160**, A1759–A1771 (2013)
38. Lee, J.-S., Tai Kim, S., Cao, R., Choi, N.-S., Liu, M., Lee, K.T., Cho, J.: Metal–air batteries with high energy density: Li–air versus Zn–air. Adv. Energy Mater. **1**, 34–50 (2011)
39. http://www.zinc.org/. Accessed Nov 2013
40. http://www.ilzsg.org/static/statistics.aspx. Accessed Nov 2013
41. Hamlen, R.P., Atwater, T.B.: Metal/air batteries, chapter 38. In: Linden, D., Reddy, T.B. (eds.) Handbook of batteries, 3rd edn. McGraw-Hill, New York (2002)
42. Chen, Z., Yu, A., Higgins, D., Li, H., Wang, H., Chen, Z.: Highly active and durable core-corona structured bifunctional catalyst for rechargeable metal–air battery application. Nano Lett. **12**, 1946–1952 (2012)
43. Zhang, X.G.: Secondary batteries—zinc systems|zinc electrodes: overview. In: Jürgen, G. (ed.) Encyclopedia of electrochemical power sources, pp. 454–468. Elsevier, Amsterdam (2009)
44. http://commons.wikimedia.org/wiki/File:Zn-pourbaix-diagram.svg. Accessed Nov 2013
45. Thomas, S., Birbilis, N., Venkatraman, M.S., Cole, I.S.: Self-repairing oxides to protect zinc: review, discussion and prospects. Corros. Sci. **69**, 11–22 (2013)
46. Yang, C.-C., Lin, S.-J.: Improvement of high-rate capability of alkaline Zn–MnO_2 battery. J. Power Sources **112**, 174–183 (2002)

47. Zhang, X.G.: Fibrous zinc anodes for high power batteries. J. Power Sources **163**, 591–597 (2006)
48. Othman, R., Basirun, W.J., Yahaya, A.H., Arof, A.K.: Hydroponics gel as a new electrolyte gelling agent for alkaline zinc–air cells. J. Power Sources **103**, 34–41 (2001)
49. Mohamad, A.A.: Zn/gelled 6 M KOH/O_2 zinc–air battery. J. Power Sources **159**, 752–757 (2006)
50. Lee, C.W., Eom, S.W., Sathiyanarayanan, K., Yun, M.S.: Preliminary comparative studies of zinc and zinc oxide electrodes on corrosion reaction and reversible reaction for zinc/air fuel cells. Electrochim. Acta **52**, 1588–1591 (2006)
51. Müller, S., Holzer, F., Haas, O.: Optimized zinc electrode for the rechargeable zinc–air battery. J. Appl. Electrochem. **28**, 895–898 (1998)
52. Masri, M.N., Mohamad, A.A.: Effect of adding potassium hydroxide to an agar binder for use as the anode in Zn–air batteries. Corros. Sci. **51**, 3025–3029 (2009)
53. Wu, G.M., Lin, S.J., Yang, C.C.: Alkaline Zn–air and Al–air cells based on novel solid PVA/PAA polymer electrolyte membranes. J. Membr. Sci. **280**, 802–808 (2006)
54. Masri, M.N., Nazeri, M.F.M., Ng, C.Y., Mohamad, A.A.: Tapioca binder for porous zinc anodes electrode in zinc–air batteries. J. King Saud Univ. Eng. Sci. 2013.
55. Devyatkina, T.I., Gun'ko, Y.L., Mikhalenko, M.G.: Development of ways to diminish corrosion of zinc electrode. Russ. J. Appl. Chem. **74**, 1122–1125 (2001)
56. Zhang, C., Wang, J.M., Zhang, L., Zhang, J.Q., Cao, C.N.: Study of the performance of secondary alkaline pasted zinc electrodes. J. Appl. Electrochem. **31**, 1049–1054 (2001)
57. Lan, C.J., Chin, T.S., Lin, P.H., Perng, T.P.: Zn–Al alloy as a new anode-metal of a zinc–air battery. J. New Mater. Electrochem. Syst. **9**, 27–32 (2006)
58. El-Sayed, A.-R., Mohran, H., Abd El-Lateef, H.: Corrosion study of zinc, nickel, and zinc-nickel alloys in alkaline solutions by Tafel plot and impedance techniques. Metall. Mater. Trans. A **43**, 619–632 (2012)
59. Lee, S.-M., Kim, Y.-J., Eom, S.-W., Choi, N.-S., Kim, K.-W., Cho, S.-B.: Improvement in self-discharge of Zn anode by applying surface modification for Zn–air batteries with high energy density. J. Power Sources **227**, 177–184 (2013)
60. Kim, K., Cho, Y.-H., Eom, S.W., Kim, H.-S., Yeum, J.H.: Anions of organic acids as gas suppressants in zinc–air batteries. Mater. Res. Bull. **45**, 262–264 (2010)
61. Auinat, M., Ein-Eli, Y.: Enhanced inhibition of zinc corrosion in alkaline solutions containing carboxylic acid modified PEG. J. Electrochem. Soc. **152**, A1158–A1164 (2005)
62. Ein-Eli, Y., Auinat, M., Starosvetsky, D.: Electrochemical and surface studies of zinc in alkaline solutions containing organic corrosion inhibitors. J. Power Sources **114**, 330–337 (2003)
63. Kinoshita, K.: Electrochemical Oxygen Technology. Wiley, New York (1992)
64. Harting, K., Kunz, U., Turek, T.: Zinc–air batteries: prospects and challenges for future improvement. Z. Phys. Chem. **226**, 151–166 (2012)
65. Cao, R., Lee, J.-S., Liu, M., Cho, J.: Recent progress in non-precious catalysts for metal–air batteries. Adv. Energy Mater. **2**, 816–829 (2012)
66. Neburchilov, V., Wang, H., Martin, J.J., Qu, W.: A review on air cathodes for zinc–air fuel cells. J. Power Sources **195**, 1271–1291 (2010)
67. Chen, Z., Choi, J.-Y., Wang, H., Li, H., Chen, Z.: Highly durable and active non-precious air cathode catalyst for zinc air battery. J. Power Sources **196**, 3673–3677 (2011)
68. Lee, J.-S., Lee, T., Song, H.-K., Cho, J., Kim, B.-S.: Ionic liquid modified graphene nanosheets anchoring manganese oxide nanoparticles as efficient electrocatalysts for Zn–air batteries. Energy Environ. Sci. **4**, 4148–4154 (2011)
69. Li, Y., Gong, M., Liang, Y., Feng, J., Kim, J.-E., Wang, H., Hong, G., Zhang, B., Dai, H.: Advanced zinc–air batteries based on high-performance hybrid electrocatalysts. Nat. Commun. 4 (2013).
70. Liu, X.-Y., Xu, X.-Z.: Mesoscopic numerical computation model of air-diffusion electrode of metal/air batteries. Appl. Math. Mech. **34**, 571–576 (2013)
71. Deborski, G.A.: US Patent 4,256,545 (1978)
72. Armstrong, W.A.: US Patent 3,948,684 (1976)
73. Passaniti, J., Dopp, R.B.: US Patent 5,378,562 (1995)
74. Passaniti, J.L., Dopp, R.B.: US Patent 5,308,711 (1994)
75. Bach, S., Baffier, N., Henry, M., Livage, J.: FR Patent 2,659,075 (1991)
76. Hoge, W.H.: US Patent 4,906,535 (1988)
77. Hoge, W.H.: US Patent 5,032,473 (1991)
78. Dopp, R.B.: US Patent 5,656,395 (1997)
79. Golovin, N.: US Patent 6,444,609 (2002)
80. Golovin, N.: US Patent 6,428,931 (2002)
81. Ndzebet, R.: US Patent 6,780,347 (2002)
82. Koshiba, N., Hayakawa, H., Momose, K., Ohta, A.: US Patent 4,595,643 (1984)
83. Sassa, R.L., Zuckerbrod, D., Buerger, W., Bacino, J.E.: US Patent 6,921,606 (2005)
84. Sun, F., Wang, F.: US Patent 6,248,476 (2001)
85. Borbelly, L., Wang, F.: US Patent 6,248,476 (1990)
86. McEvoy, J.J.: US Patent 4,585,710 (1986)
87. Kelsey, G.S., Chalilpoyil, P., Trainer, P.D., Kaplan, A., Cintra, G., Vu, V.H., Sillesky, J.D.: US Patent 6,207,322 (1998)
88. Shun, Y-K., Lou, C-L.: US Patent 6,127,061 (2000)
89. Blanchart, A.P.O., Van Der Poorten, C.J. E.: US Patent 4,696,872 (1987)
90. Henry, G.: US Patent 4,333,993 (1982)
91. http://www.electric-fuel.com/index.html. Accessed Nov 2013
92. Koretz, B., Goldstein, J.R.: Regeneration of zinc anodes for the electric fuel(R) zinc–air refuelable EV battery system. In: Energy Conversion Engineering Conference. IECEC-97., Proceedings of the 32nd Intersociety, vol. 2, pp. 877–882 (1997)
93. Goldstein, J., Brown, I., Koretz, B.: New developments in the Electric Fuel Ltd. zinc/air system. J. Power Sources **80**, 171–179 (1999)
94. Bergman, M.: Synthesis, preparation and characterization of materials for metal air battery applications. Master of Science Thesis in Applied Physics, Department of Applied Physics—Division of Condensed Matter Physics. Chalmers University of Technology (2013)
95. Du, G., Liu, X., Zong, Y., Hor, T.S.A., Yu, A., Liu, Z.: Co_3O_4 nanoparticle-modified MnO_2 nanotube bifunctional oxygen cathode catalysts for rechargeable zinc–air batteries. Nanoscale **5**, 4657–4661 (2013)
96. Goh, F.W.T., Liu, Z., Ge, X., Zong, Y., Du, G., Hor, T.S.A.: Ag nanoparticle-modified MnO_2 nanorods catalyst for use as an air electrode in zinc–air battery. Electrochim. Acta **114**, 598–604 (2013)
97. Kar, M., Winther-Jensen, B., Forsyth, M., MacFarlane, D.R.: Chelating ionic liquids for reversible zinc electrochemistry. Phys. Chem. Chem. Phys. **15**, 7191–7197 (2013)
98. Deiss, E., Holzer, F., Haas, O.: Modeling of an electrically rechargeable alkaline Zn–air battery. Electrochim. Acta **47**, 3995–4010 (2002)
99. Smedley, S.I., Zhang, X.G.: A regenerative zinc–air fuel cell. J. Power Sources **165**, 897–904 (2007)
100. Smedley, S.I., de Tezanos Pinto, M., des Jardins, S.R., Novkov, D.J., Gulino, R.: Discrete particle electrolyzer cathode and

method of making same. US Patent Application Publication No. 2004/0168922 A1, (September 2nd, 2004)

101. Sapkota, P., Kim, H.: Zinc–air fuel cell, a potential candidate for alternative energy. J. Ind. Eng. Chem. **15**, 445–450 (2009)

102. Cooper, J.F., Krueger, R.: The refuelable zinc–air battery: alternative techniques for zinc and electrolyte regeneration. Final report UCRL-TR-218414, ILZRO Program ZB-1 Zinc–air batteries, (April 15th, 1996)

A new method of pretreatment of lithium manganese spinels and high-rate electrochemical performance of $Li[Li_{0.033}Mn_{1.967}]O_4$

A. V. Potapenko · S. I. Chernukhin · S. A. Kirillov

Abstract Lithium manganese spinels tend to aggregate upon annealing and do not allow for attaining high discharge rates when used as cathodes in lithium-ion batteries. To obtain spinel samples of lower aggregation and better high-rate properties, precursors synthesized by means of a citric acid-aided route are suggested to be pyrolyzed in an inert atmosphere, instead of pyrolysis in air. The synthesis of nanosized $Li[Li_{0.033}Mn_{1.967}]O_4$ is described, and its characteristics including X-ray diffraction, scanning electron microscopy, and porosity, as well as electrochemical test results are presented. The particle size of the materials obtained is smaller, the degree of aggregation is lower, and high-rate properties are better than for analogues pyrolyzed in air. In particular, sample Li||$Li[Li_{0.033}Mn_{1.967}]O_4$ cells deliver ~60 mAh g^{-1} at the current loads of 4,000 mA g^{-1} (30 C). After a sequence of 62 charge/discharge cycles with the currents growing from 0.1 to 100 C the specific capacity retains its initial value.

Keywords Lithium battery · Cathode material · $LiMn_2O_4$ spinel · High-rate properties

Introduction

Lithium manganese spinel $LiMn_2O_4$ as a cathode material for lithium-ion batteries has been suggested in 1983 [1] and since that time attracts much attention of experimentalist [2]. Compared to currently employed $LiCoO_2$ it is cheap, non-toxic, has a high potential against lithium electrode (3.0–4.5 V) and the theoretical capacity of 148 mAh g^{-1} [1, 3, 4]. Properties of this and related spinel compounds and advances in their synthesis are reviewed in a recent paper [5].

$LiMn_2O_4$ belongs to the cubic $Fd\bar{3}m$ space group, $Z = 8$ [4, 6]. In this cubic structure, oxygen ions occupy the tetrahedral positions (32e) and form the cubic dense packing. The octahedral positions (16d) are inhabited by manganese ions (Mn^{3+}/Mn^{4+}). Lithium ions hold the tetrahedral (8a) positions. Each of them is separated from four neighboring ions by voids (16c). This means that three-dimensional channels (8a–16c–8a–16c) exist for possible insertion and transport of lithium ions.

Electrochemical intercalation/deintercalation of Li^+ into/from this material occurring in the potential range of 3.0–4.5 V can be expressed as

$$LiMn_2O_4 \leftrightarrow Li_{1-x}Mn_2O_4 + xLi^+ + xe^-. \qquad (1)$$

In the composition range of $0 \leq x \leq 0.5$ and $0.5 \leq x \leq 1.0$, reaction (1) occurs in two stages easily discernible in both galvanostatic and potentiodynamic regime, however, the spinel structure of $LiMn_2O_4$ remains cubic keeping the lattice parameter a close to ca. 8.248 Å.

Much work has been done so as to understand the nature of the $Li_{0.5}Mn_2O_4$ phase and positioning of the lithium ions in it. Careful electrochemical, X-ray and neutron diffraction studies have shown that the charge/discharge process occurs in three steps, hence up to three phases coexist in the $0 \leq x \leq 1$ range [7–9]. The structure of $Li_{0.5}Mn_2O_4$ has been determined as belonging to the modulated $R\bar{3}m$ space group, $Z = 6$, with the O^{2-} ions in the 6c and 18h positions, the Mn^{3+} ions in the 3b positions, the Mn^{4+} ions in

A. V. Potapenko · S. I. Chernukhin · S. A. Kirillov
Joint Department of Electrochemical Energy Systems,
38A Vernadsky Ave., Kiev 03142, Ukraine

S. A. Kirillov (✉)
Institute for Sorption and Problems of Endoecology,
13 Gen. Naumov St., Kiev 03164, Ukraine
e-mail: kir@i.kiev.ua

the 9e positions, and the Li^+ ions partially filling 6c sites. In a cubic setting, this cell has $a = 8.142$ Å. Its remarkable feature is an opportunity to derive superlattices corresponding to the two "missing" phases having the predicted $Li_{0.25}Mn_2O_4$ and $Li_{2/3}Mn_2O_4$ compositions [9].

If x exceeds 1.0, i.e. at potentials of ~3.0 V against Li/Li^+, a tetragonal $LiMnO_2$ phase with a slightly increased unit volume ($a = 8.007$ Å, $c = 9.274$ Å) appears,

$$Li + LiMn_2O_4 \leftrightarrow 2LiMnO_2. \quad (2)$$

Furthermore, tetragonal distortions in lithium–manganese spinels are favored by the Jahn–Teller effect which facilitates lowering the symmetry of the structure if more than 50 % of Mn^{3+} is accumulated in the product of reaction (2) [10].

An effect of solid state synthesis conditions and annealing of $LiMn_2O_4$ on its stoichiometry and electrochemical properties is studied in detail [11, 12]. It is found that annealing at temperatures not exceeding 850 °C and slow cooling ensure the minimal loss of oxygen and the maximal specific capacity. However, major drawbacks of the material are inability to gain the theoretical capacity values and a significant capacity fading on cycling (up to 10 % after 50 cycles).

There are at least two possible directions in improving the electrochemical properties of the lithium manganese spinel electrode materials. One of them is based on the replacement of a part of the manganese ions in the 16d positions by excess lithium ions and obtaining substituted spinels of a general formula $Li[Li_qMn_{2-q}]O_4$ [13, 14]. It should be mentioned that annealing in air often leads to an oxygen loss and to a non-stoichiometricity of the spinels whose formula can be written as $Li[Li_qMn_{2-q}]O_{4-\delta}$. A detailed investigation by Paulsen and Dahn [15] reveals that between 400 and 880 °C only the mentioned spinel is stable having a negligible oxygen nonstoichiometric range, $\delta < 0.02$. The lithium stoichiometry can vary from $0.05 < q < 0.13$ and increases with decreasing temperature and increasing oxygen partial pressure. Upon the substitution, the fraction of Mn^{3+} decreases and the specific capacity diminishes (e.g., to 117 mAh/g for $Li_{1.063}Mn_{1.932}O_4$ [13] and to 98 mAh/g for $Li_{1.12}Mn_{1.88}O_4$ [14]). However, in this case, the mean degree of oxidation of manganese exceeds 3.5 and thus prevents of arising the Jahn–Teller effect and capacity-fading phenomena.

Another way for improving the electrochemical properties of electrode materials is obtaining them in the form of nanosized crystals. Decreasing the size of particles leads to their better cycling ability and makes it possible to discharge lithium ion batteries with higher currents, i.e. increases their rate [16]. To do so, significant efforts are undertaking in the field of sol–gel synthesis of electrode materials with improved electrochemical parameters [17–26], in particular, using oxyacid metal salts (citrates on the first place) as precursors [21–27]. High homogeneity of final products and lower thermal treatment temperatures decrease the probability of particle growth and are considered to be advantages of such approaches.

In a previous paper [28], we have described the high-rate properties of the lithium substituted spinel of the $Li[Li_{0.033}Mn_{1.967}]O_4$ composition with the lithium to manganese ratio equal to 1.05:2 and the theoretical capacity of 133 mAh g^{-1}. It has been obtained according to the aforementioned directions by means of an optimized citric acid-aided route. In terms of Paulsen and Dahn [15], this composition is structurally stable and does not tend to lose oxygen upon heating. As follows from Ref. [28], sample cells with cathodes made of the material in question evidence the absence of discharge capacity fading within 100 cycles and the capability to be discharged with great currents. In particular, at the current density of 1,480 mA g^{-1} (~11 C) the discharge capacity equals to 58 mAh g^{-1} or almost the half of the theoretical capacity of the material, and at the highest current load of 2,220 mA g^{-1} (~16.5 C) applied, the capacity of 33 mAh g^{-1} has been retained.

On the other hand, the data presented in Ref. [15] and our practice have shown that even being synthesized at mild conditions, these and other lithium manganese spinel samples demonstrate a clear trend to the fast aggregation of particles upon the thermal treatment of a pyrolyzed material. This trend levels the advantages of the citric acid-aided route where pyrolysis gives nanosized powders with crystallites of ~15 nm size.

To avoid this obstacle and to obtain spinel samples of the lower degree of aggregation and better high-rate properties the following idea is suggested in this paper. Instead of the pyrolysis in air (a 'standard' scheme), citrate precursors are pyrolyzed in an inert atmosphere (a proposed scheme). In this case, two circumstances are considered to favor the smaller degree of aggregation of the target product. First of all, no chemical reactions between lithium carbonate and manganese-II oxide obtained in the course of pyrolysis shall occur. Second, a great amount of carbon shall be formed covering the products of pyrolysis. This means that upon further annealing, carbon-coated Li_2CO_3 and MnO will react, and the coating, even being burnt, will prevent the rapprochement of the spinel particles. In what follows, we describe the physicochemical and electrochemical properties of the lithium-substituted spinel of the $Li[Li_{0.033}Mn_{1.967}]O_4$ composition supporting this idea. Its particle size is smaller, the degree of aggregation is lower, and high-rate properties are better that for its analogue pyrolyzed in air. In particular, model Li||$Li[Li_{0.033}Mn_{1.967}]O_4$ cells deliver ~60 mAh g^{-1} at the current loads of 4,000 mA g^{-1} (30 C); at the highest current load of 13,300 mA g^{-1} (100 C), the capacity 11 mAh g^{-1}

is retained; after a sequence of 62 charge/discharge cycles with the currents varied from 0.1 to 100 C the specific capacity returns at its initial value.

Experimental

Citrate precursors have been obtained in the following way [27–31]. Solutions of lithium nitrate, manganese-II nitrate, and citric acid (all of analytical grade) of approx. 1 M concentration have been analyzed with the four significant digits accuracy, mixed using calibrated pipettes in the 1.05:2:6 ratio, and evaporated up to a resin state (80 °C). The resins have been dried at 120 °C giving the precursors for all further operations.

According to the 'standard' scheme, the precursors have been pyrolyzed and annealed in air. By the proposed scheme, pyrolysis has been performed in the argon atmosphere and after that, pyrolyzed samples have been annealed in air. The temperature of pyrolysis in both cases has been equal to 400 °C; the duration of pyrolysis has been taken equal to 0.5 h. Subsequent operations with the samples pretreated in air have been made according to Ref. [28] and consisted of 24 h annealing in air at 700 °C with heating and cooling rates taken equal to 5 and 2 $°C \cdot min^{-1}$, respectively. The samples pretreated in argon have been annealed in air at two different regimes: for 24 h at 700 °C and for 10 h at 800 °C; the heating and cooling rates have been kept as the same as in the case of the samples pretreated in air.

Since the content of lithium and manganese is crucial for the correct comparison of the properties of the samples pretreated at different conditions, it has been controlled by means of atomic absorption spectrophotometry (AA-8500, Nippon Jarrell–Ash, Japan). The mean Li:2Mn ratio of 1.0503 has been determined for thermally treated samples with standard deviation of 0.00082 and standard error of 0.00033 thus signifying that no lithium loss occurs. This Li:2Mn ratio perfectly matches the title composition, $Li[Li_{0.033}Mn_{1.967}]O_4$. To check the thermal decomposition process, the differential thermal analysis has been employed (Q-1500 D, MOM, Hungary). Porosity studies have been utilized on an ASAP 2000 device (Micromeritics, USA). The phase composition of the samples, their morphology and the particle size have been studied by means of X-ray diffraction (DRON-4-07, LOMO, Russia, CoK_α and CuK_α radiation) and scanning electron microscopy (JSM 6700F, JEOL, Japan). To qualitatively determine the phase composition of samples JCPDS-ICDD database has been used.

Electrochemical tests have been performed in sample Li||$Li[Li_{0.033}Mn_{1.967}]O_4$ cells on a home-made versatile workstation. Working electrodes have consisted of 82 wt. % of the studied material, 10 wt. % of a conductive additive (soot) and 8 wt. % of a binder (poly (vinyliden) difluoride). A 20 μm layer of the slurry has been put on an aluminum foil with a doctor blade and then dried at 120 °C for 18-21 h under an IR radiator. The quantity of $Li[Li_{0.033}Mn_{1.967}]O_4$ in a dried remainder was of 3–8 mg cm^{-2}. The electrodes were rolled so as to decrease the thickness of the layer from 70–80 to 25–35 μm. Sample coin-type cells have been assembled in 2016 cases with Celgard 2500 separators and an electrolyte containing the 1 M solution of $LiPF_6$ in a mixture of ethylene carbonate and dimethyl carbonate taken in the 1:1 mass ratio. The reference and counter electrode has been made of lithium metal. Potentiodynamic tests have been performed within the voltage range of 3.4–4.5 V at the scanning speed of 0.1 mV s^{-1}. High rate tests have been carried out in a galvanostatic mode within the same voltage range using a CCCV regime with the charge current of 1C and a trickle charge at 0.1 C; discharge currents varied from 0.1C to 100C. As noticed above, the theoretical specific capacity of the compound studied, $Li[Li_{0.033}Mn_{1.967}]O_4$ is 133 mAh g^{-1}, therefore, 1C = 133 mA g^{-1}.

Results and Discussion

As follows from Refs. [29–31], citrate precursors after drying at 120 °C (dried resins) consist of stoichiometric hydrous citrates and upon pyrolysis, decompose giving the target products. In the case of lithium manganese spinels, the precursors contain the mixture of dihydrates of lithium citrate, manganese-II citrate, and two excess moles of citric acid, and their decomposition in air is described by the following reaction [28],

$$LiC_6H_7O_7 \cdot 2H_2O + 2MnC_6H_6O_7 \cdot 2H_2O + 2C_6H_8O_7 + 23.25O_2 \rightarrow LiMn_2O_4 + 30CO_2 + 23.5H_2O. \quad (3)$$

Unlike pyrolysis in air, the decomposition of the spinel precursor in argon atmosphere should proceed according the following scheme,

$$LiC_6H_7O_7 \cdot 2H_2O + 2MnC_6H_6O_7 \cdot 2H_2O + 2C_6H_8O_7 \rightarrow {}^1\!/_2Li_2CO_3 + 2MnO + 15.5C + 23.5H_2O + 14CO. \quad (4)$$

In this case, the reaction products are lithium carbonate and manganese-II oxide. On further annealing in air, the excess carbon should serve as a coating preventing the aggregation of the spinel particles. The calculated mass loss in reaction (4) is 60.9 %. In accord with this estimate, pyrolysis at 400 °C for 0.5 h leads to the experimental mass loss of 62.0 %. X-ray diffraction data evidence the formation of MnO. A diffraction pattern obtained (Fig. 1a) well agrees with existing data (JCPDS 07-0230). The lithium salt formed is not recognizable due to low scattering abilities of lithium, carbon and oxygen.

Upon heating the mixture of Li_2CO_3, MnO and carbon in air, the following reaction accompanied by the mass loss of 51.0 % should occur,

$$Li_2CO_3 + 4MnO + 31C + 32.5O_2 \rightarrow 2LiMn_2O_4 + 32CO_2. \quad (5)$$

Respective thermal analysis data are given in Fig. 2. The mass loss of 47.0 % registered is in a good agreement with the calculations by reaction (5). On the other hand, the burning of carbon and accompanying exothermic effect makes any attempts to interpret the DTA data difficult. An X-ray analysis proves that the product obtained is the lithium manganese spinel (Fig. 2b).

To determine the crystallite size d of spinels the Scherrer formula has been employed [32],

$$d = 0.9\lambda/(B\cos\theta), \quad (6)$$

where λ is the wavelength of X-ray radiation, B is the line width at the half maximum of the diffraction peak, and θ is the angle of diffraction. The lattice parameters a (Å) for the spinel samples have been calculated by the Holland-Redfern method using a UnitCell 1997 software [33]. The values of d and a are summarized in Table 1. As follows from these data, the samples consist of nanosized crystallites of comparable dimensions.

Scanning electron microscopy supports a conclusion regarding the nanosized nature of the spinels under investigation (Fig. 3). Comparing the micrographs one easily reassures that the particle size $d_{part.}$ in the sample pyrolyzed in argon is at least twice smaller than in the sample pyrolyzed in air (50–70 nm against 70–100 nm, respectively). Moreover, the former sample demonstrates a less significant tendency to aggregation. Combining these data with the crystallite size d obtained by means of X-ray diffraction, Table 1, one can speculate that the particles in the argon-pyrolyzed samples are single crystals, whereas the particles in the air-pyrolyzed samples are aggregates of single crystals.

Unlike lithium manganese spinels obtained by means of nanochemistry routes [14, 34] and showing high surface areas and clear hysteresis loops on nitrogen adsorption/desorption curves, porosity data for our samples are much simpler (Fig. 4). They reveal low surface areas ($\sim$11 m^2/g) and do not enable one to perform the calculation of pore size distributions. This means that the materials obtained have no micro- and mesopores, and their porosity, as follows from SEM micrographs, is determined by voids between particles.

After physico-chemical characterization the samples have been subjected to electrochemical tests. In Fig. 5, stationary voltammetric curves corresponding to the 3rd cycle for sample Li||Li[$Li_{0.033}Mn_{1.967}$]O_4 cells are presented. The term 'stationary' means the complete reproducibility of voltammetric data in preceding and following measurements. Peaks in the cathodic and anodic regions reflect processes occurring upon insertion/deinsertion of lithium ions into/from the lithium manganese spinel in the composition range of $0 \le x \le 0.5$ and $0.5 \le x \le 1.0$. It should be mentioned that the low-voltage insertion and deinsertion peaks are clearly asymmetric possibly

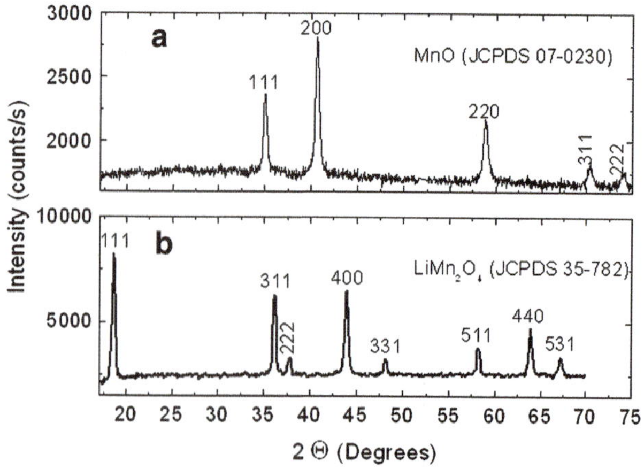

Fig. 1 X-ray diffraction pattern of the decomposition products of the precursor treated in argon atmosphere (**a**) and the spinel obtained upon further thermal treatment (**b**)

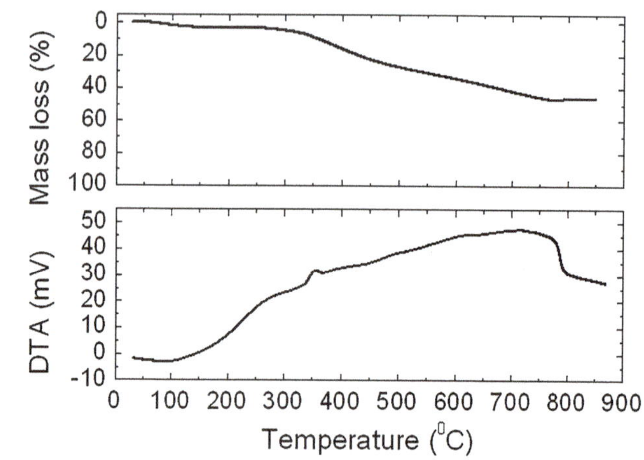

Fig. 2 Thermal analysis data for the mixture of Li_2CO_3, MnO and carbon obtained in the course of its heating in air

Table 1 X-ray diffraction and scanning electron microscopy data for Li[$Li_{0.033}Mn_{1.967}$]O_4

Sample treatment	d (400), nm	a, Å	V, Å^3	$d_{part.}$, nm
Air	30	8.2356	559	70–100
Argon	65	8.2405	560	50–70

Fig. 3 Scanning electron micrographs (×100,000) of $Li[Li_{0.033}Mn_{1.967}]O_4$ treated in air (*left*) and in argon (*right*)

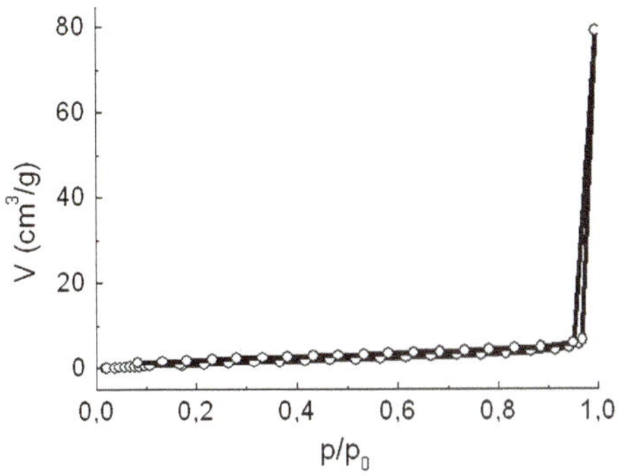

Fig. 4 Nitrogen adsorption/desorption isotherm for $Li[Li_{0.033}Mn_{1.967}]O_4$ treated in argon

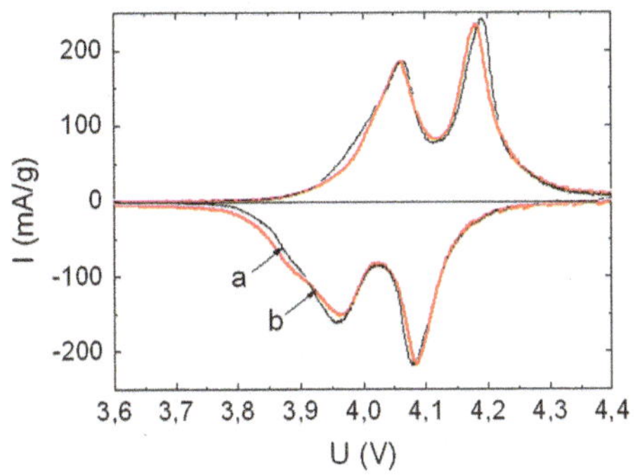

Fig. 5 Stationary voltammetric characteristics of $Li[Li_{0.033}Mn_{1.967}]O_4$–based electrodes. Scanning speed 0.1 mV/s. **a** *Thin black curve* $Li[Li_{0.033}Mn_{1.967}]O_4$ treated in air. **b** *Thick red curve* $Li[Li_{0.033}Mn_{1.967}]O_4$ treated in argon

signifying the presence of a third phase, as described in Introduction. These features, as well as an analysis of diffusion coefficients in $Li[Li_qMn_{2-q}]O_4$ spinels ($0 \leq q \leq 0.065$) determined from voltammetric and impedance measurements will be described elsewhere.

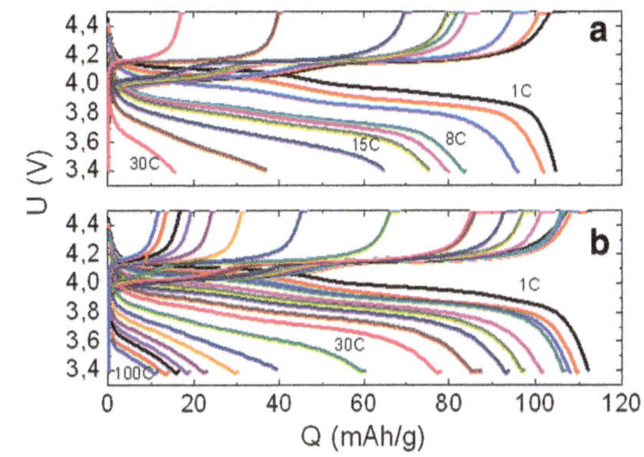

Fig. 6 Discharge curves for sample Li|Li$[Li_{0.033}Mn_{1.967}]O_4$ cells at various discharge current densities: **a** $Li[Li_{0.033}Mn_{1.967}]O_4$ treated in air; **b** $Li[Li_{0.033}Mn_{1.967}]O_4$ treated in argon

Discharge curves for sample Li|Li$[Li_{0.033}Mn_{1.967}]O_4$ cells at various discharge current densities are given in Fig. 6, and dependences of specific capacities on current densities are shown in Fig. 7. At minimal current loads, the specific capacities of materials under investigation are lower than the theoretical values. For the samples fired in air discharge capacities $Q_d^{max.}$ at minimal discharge currents $I_{min.} = 0.1C$ are equal to 103 mAh g^{-1}; for those fired in argon $Q_d^{max.} = 113$ mAh g^{-1}. As follows from Figs. 6 and 7, the cells with the cathode materials fired in argon retain their working ability at the currents growing from 14.8 to 14,800 mA g^{-1} (from 0.1 to 100 C). After heavy-duty tests during 62 charge/discharge cycles their specific capacity returns to its initial value. The materials treated in air remain stable at much lower current loads and do not retain their initial capacity after loading.

It is tempting to compare these results with those existing in the literature. Since the rate capability of Li-substituted spinels has not been widely studied we present

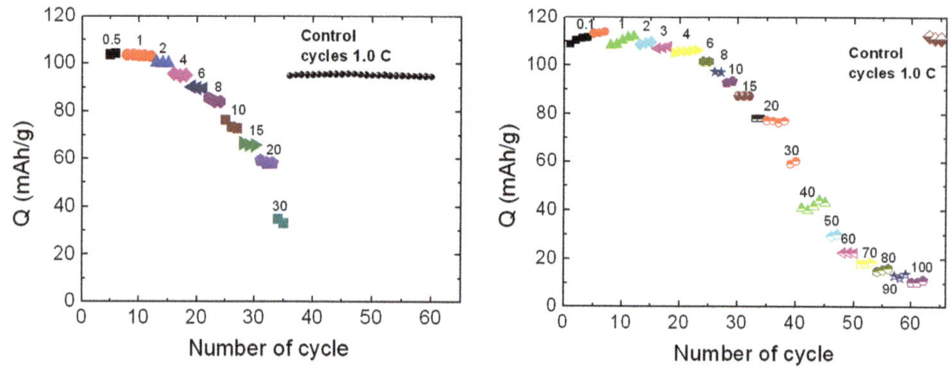

Fig. 7 Dependences of discharge capacity on cycle number for Li||Li[$Li_{0.033}Mn_{1.967}$]O_4 cells. *Left* Li[$Li_{0.033}Mn_{1.967}$]O_4 treated in air. *Right* Li[$Li_{0.033}Mn_{1.967}$]O_4 treated in argon. Numeric values mean discharge current densities in C units

Table 2 Parameters of materials obtained by various routes and demonstrating good high-rate properties

Route	Initial capacity $Q_d^{min.}$, mAh g^{-1}	dQ_d/dI	References
Literature data			
Resorcinol-formaldehyde	137	0.6 (60 C)	[20]
Solid state	105	1.9 (10 C)	[35]
Solid state	119	1.45 (20 C)	[36]
Sol–gel	100	2.1 (20 C)	[17]
Sol–gel	115	0.8 (40 C)	[18]
Hydrothermal	108	0.12 (90 C)	[37]
Hydrothermal	118	0.23 (95 C)	[38]
Nanowires	120	0.22 (135 C)	[39]
Pechini	120	0.8–2.1 (60 C)	[40]
This work			
Pechini/air	103	4.2 (16.7 C)	–
Pechini/argon	113	2.3 (40 C)–1.2 (100 C)	–

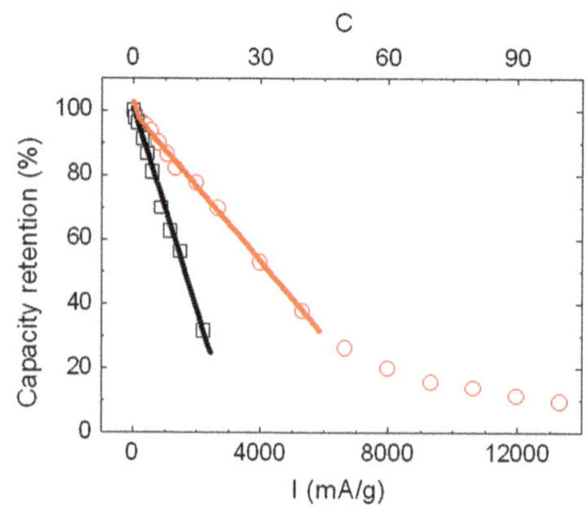

Fig. 8 Capacities retained by Li||Li[$Li_{0.033}Mn_{1.967}$]O_4 cells upon various current loads. *Squares, black* Li[$Li_{0.033}Mn_{1.967}$]O_4 treated in air. *Circles, red* Li[$Li_{0.033}Mn_{1.967}$]O_4 treated in argon

Table 2 where selected latest data on $LiMn_2O_4$ are summarized. This seems appropriate because the lithium excess in the sample is very small. More detailed structural and electrochemical information about these and other $LiMn_2O_4$ samples demonstrating good high-rate properties can be found in Ref. [5]. For the sake of comparison, the high-rate parameters are characterized by the minimal capacity losses per the unit current rate, $dQ_d/dI = (Q_d^{max.} - Q_d^{min.})/(I_{max.} - I_{min.})$, where $Q_d^{max.}$ and $Q_d^{min.}$ are the discharge capacities at the minimal and maximal discharge currents applied, $I_{min.}$ and $I_{max.}$. For our samples these values have been determined from Fig. 8.

The inspection of Table 2 reveals that electrode materials demonstrating good high-rate properties have low initial capacities, except the sample obtained by means of a resorcinol-formaldehyde route and exhibiting exceptional results. It should be mentioned parenthetically that, to our surprise, numerous materials with high initial capacities (130–133 mAh g^{-1} [19, 41], 135 mAh g^{-1} [42–45], 145 mAh g^{-1} [46]) have never been treated at high current loads. On the other hand, it is well-seen that pyrolyzing the citrate precursor in argon one can attain better high-rate capabilities and significantly improve the initial capacity of the spinel material.

In our opinion, this promising result is bound not only to a decrease in the size and degree of aggregation of the spinel particles, but to the nature of the crystallites in the argon-treated sample, each of which, according to XRD and SEM data, can be considered as a separate single crystal. As follows from Ref. [5], just the perfectness of the particles, together with their size and degree of aggregation, is crucial for the accessibility of theoretical capacities and high speed of discharge of lithium batteries.

Conclusion

In this paper, we have synthesized two samples of the nanosized lithium-doped lithium manganese spinel, Li[$Li_{0.033}Mn_{1.967}$]O_4 using a citric acid-aided route, performed their characterization by means of X-ray diffraction, scanning electron microscopy, and porosity studies, and carried out their electrochemical tests in Li||Li[$Li_{0.033}Mn_{1.967}]O_4$ cells. The difference between these samples lies in the pyrolysis regime, viz., one has been pyrolyzed in air and another in argon. The particle size in the sample pyrolyzed in argon (50–70 nm) is at least twice smaller than in the sample treated in air (70–100 nm), and the former demonstrates a smaller tendency to aggregation. In high-rate tests, the argon-fired samples overwhelm the air-fired ones retaining the working ability at the currents growing from 14.8 to 14,800 mAh g^{-1} (from 0.1 to 100C), and after 63 charge/discharge cycles their specific capacity returns to its initial value with no capacity fading. This result may by prospective for lithium-ion batteries for heavy-duty applications.

References

1. Thackeray, M.M., David, W.I.F., Bruce, P.G., Goodenough, J.B.: Lithium insertion into manganese spinels. Mater. Res. Bull. **18**, 461–472 (1983)
2. Whittingham, M.S.: Lithium batteries and cathode materials. Chem. Rev. **104**, 4271–4302 (2004)
3. Thackeray, M.M., Johnson, P.J., de Piciotto, L.A., Bruce, P.G., Goodenough, J.B.: Electrochemical extraction of lithium from $LiMn_2O_4$. Mater. Res. Bull. **19**, 179–187 (1984)
4. Rossouw, M.H., de Kock, A., de Piciotto, L.A., Thakeray, M.M., David, W.I.F., Ibberson, R.M.: Structural aspects of lithium-manganese-oxide electrodes for rechargeable lithium batteries. Mater. Res. Bull. **25**, 173–182 (1990)
5. Potapenko, A.V., Kirillov, S.A.: Lithium manganese spinel materials for high-rate electrochemical application. J. Energy Chem. **23**, 543–588 (2014)
6. Ohzuku, T., Kitagawa, M., Hirai, T.: Electrochemistry of manganese dioxide in lithium nonaqueous cell III. X-Ray diffractional study on the reduction of spinel-related manganese dioxide. J. Electrochem. Soc. **137**, 769–775 (1990)
7. Liu, W., Kowal, K., Farrington, G.C.: Mechanism of the electrochemical insertion of lithium into $LiMn_2O_4$ spinels. J. Electrochem. Soc. **145**, 459–465 (1998)
8. Lee, Y.J., Wang, F., Mukerjee, S., McBreen, J., Grey, C.P.: ^{6}Li and ^{7}Li magic-angle spinning nuclear magnetic resonance and in situ x-ray diffraction studies of the charging and discharging of LixMn2O4 at 4 V. J. Electrochem. Soc. **147**, 803–812 (2000)
9. Björk, H., Gustafsson, T., Thomas, J.O., Lidin, S., Petříček, V.: Long-range ordering during delithiation of $LiMn_2O_4$ cathode material. J. Mater. Chem. **13**, 585–589 (2003)
10. Yamada, A.: Lattice instability in $Li(Li_xMn_{2-x})O_4$. J. Solid State Chem. **122**, 160–165 (1996)
11. Tarascon, J.M., McKinnon, W.R., Coowar, F., Bowmer, T.N., Amatucci, G., Guyomard, D.: Synthesis conditions and oxygen stoichiometry effects on Li insertion into the spinel $LiMn_2O_4$. J. Electrochem. Soc. **141**, 1421–1431 (1994)
12. Tarascon, J.M., Wang, E., Shokoohi, F.K., McKinnon, W.R., Colson, S.: The spinel phase of $LiMn_2O_4$ as a cathode in secondary lithium cells. J. Electrochem. Soc. **138**, 2859–2864 (1991)
13. Xia, Y., Yoshio, M.: An investigation of lithium ion insertion into spinel structure Li-Mn-O compounds. J. Electrochem. Soc. **143**, 825–833 (1996)
14. Jiao, F., Bao, J., Hill, A.H., Bruce, P.G.: Synthesis of ordered mesoporous Li–Mn–O spinel as a positive electrode for rechargeable lithium batteries. Angew. Chem. **47**, 9711–9716 (2008)
15. Paulsen, J.M., Dahn, J.R.: Phase diagram of Li-Mn-O spinel in air. Chem. Mater. **11**, 3065–3079 (1999)
16. Aricò, A.S., Bruce, P., Scrosati, B., Tarascon, J.M., Van Schalkwijk, W.: Nanostructured materials for advanced energy conversion and storage devices. Nat. Mater. **4**, 366–377 (2005)
17. Tang, W., Wang, X.J., Hou, Y.Y., Li, L.L., Sun, H., Zhu, Y.S., Bai, Y., Wu, Y.P., Zhu, K., Van Ree, T.: Nano $LiMn_2O_4$ as cathode material of high rate capability for lithium ion batteries. J. Power Sources **198**, 308–311 (2012)
18. Mao, J., Dai, K.H., Zhai, Y.C.: High rate capability and cycling stability of $Li_{1.07}Mn_{1.93}O_4$ nanoflakes synthesized via gel-combustion method. Acta Phys. Chim. Sin. **28**, 349–354 (2012)
19. Feng, C., Li, H., Zhang, C., Guo, Z., Wu, H., Tang, J.: Synthesis and electrochemical properties of non-stoichiometric Li–Mn-spinel ($Li_{1.02}M_xMn_{1.95}O_{4-y}F_y$) for lithium ion battery application. Electrochim. Acta **61**, 87–93 (2012)
20. Chen, Y., Xie, K., Pan, Y., Zheng, C.: Effect of calcination temperature on the electrochemical performance of nanocrystalline $LiMn_2O_4$ prepared by a modified resorcinol–formaldehyde route. Solid State Ionics **181**, 1445–1450 (2010)
21. Liu, W., Farrington, C.C., Chaput, F.: Synthesis and electrochemical studies of spinel phase $LiMn_2O_4$ cathode materials prepared by the Pechini process. J. Electrochem. Soc. **143**, 879–884 (1996)
22. Hwang, B.J., Santhanam, R., Liu, D.G.: Characterization of nanoparticles of $LiMn_2O_4$ synthesized by citric acid sol–gel method. J. Power Sources **97–98**, 443–446 (2001)
23. Hwang, B.J., Santhanam, R., Liu, D.G.: Effect of various synthetic parameters on purity of $LiMn_2O_4$ spinel synthesized by a sol–gel method at low temperature. J. Power Sources **101**, 86–89 (2001)
24. Hwang, B.J., Santhanam, R., Hu, S.G.: Synthesis and characterization of multidoped lithium manganese oxide spinel, $Li_{1.02}Co_{0.1}Ni_{0.1}Mn_{1.8}O_4$, for rechargeable lithium batteries. J. Power Sources **108**, 250–255 (2002)
25. Wang, X., Chen, X., Gao, L., Zheng, H., Ji, M., Shen, T., Zhang, Z.J.: Citric acid-assisted sol–gel synthesis of nanocrystalline $LiMn_2O_4$spinel as cathode material. Cryst. Growth. **256**, 123–127 (2003)
26. Yi, T., Dai, C., Gao, K., Hu, X.: Effects of synthetic parameters on structure and electrochemical performance of spinel lithium manganese oxide by citric acid-assisted sol–gel method. J. Alloys Compd. **425**, 343–347 (2006)
27. Potapenko, A.V., Chernukhin, S.I., Romanova, I.V., Rabadanov, KSh, Gafurov, M.M., Kirillov, S.A.: Citric acid aided synthesis, characterization, and high-rate electrochemical performance of $LiNi_{0.5}Mn_{1.5}O_4$. Electrochim. Acta **134**, 442–449 (2014)
28. Potapenko, A.V., Chernukhin, S.I., Romanova, I.V., Kirillov, S.A.: Physico-Chemical and Electrochemical Properties of Nanosized Li[Li0.033Mn1.967]O4. Chem. Phys. Tech. Surf. (In Russian) **2**, 175–181 (2011)
29. Farbun, I.A., Romanova, I.V., Terikovskaya, T.E., Dzanashvili, D.I., Kirillov, S.A.: Complex formation in the course of synthesis of zinc oxide from citrate solutions. Russ. J. Appl. Chem. **80**, 1798–1803 (2007)
30. Romanova, I.V., Farbun, I.A., Khainakov, S.A., Kirillov, S.A., Zazhigalov, V.A.: Investigaon fcatlyic proetis ofmaterils baed ontrasion metals oxides and Ce oxide. Rep. Nat. Acad. Sci. Ukraine. (In Russian) **10**, 154–159 (2008)

31. Farbun, I.A., Romanova, I.V., Kirillov, S.A.: Optimal design of powdered nanosized oxides of high surface area and porosity using a citric acid aided route, with special reference to ZnO. J. Sol Gel Sci. Technol. **68**, 411–422 (2013)
32. Patterson, A.: The diffraction of X-rays by small crystalline particles. Phys. Rev. **56**, 978 (1939)
33. Holland, T.J.B., Redfern, S.A.T.: Unit cell refinement from powder diffraction data: the use of regression diagnostics. Mineral. Mag. **61**, 65–77 (1997)
34. Wang, Y., Shao, X., Xu, H., Xie, M., Deng, S., Wang, H., Liu, J., Yan, H.: Facile synthesis of porous $LiMn_2O_4$ spheres as cathode materials for high-power lithium ion batteries. J. Power Sources **226**, 140–148 (2013)
35. Huang, S., Wen, Z., Yang, X., Zhu, X., Lin, B.: Synthesis and the improved high-rate performance of $LiMn_2O_4$/Ag composite cathode for lithium-ion batteries. Electrochem. Solid State Lett. **9**, A443–A447 (2006)
36. Lanz, M., Kormann, C., Steininger, H., Heil, G., Haas, O., Novák, P.: Large-agglomerate-size lithium manganese oxide spinel with high rate capability for lithium-ion batteries. J. Electrochem. Soc. **147**, 3997–4000 (2000)
37. Tang, W., Liu, L.L., Tian, S., Li, L., Li, L.L., Yue, Y.B., Bai, Y., Wu, Y.P., Zhu, K., Holze, R.: $LiMn_2O_4$ nanorods as a super-fast cathode material for aqueous rechargeable lithium batteries. Electrochem. Commun. **13**, 1159–1162 (2011)
38. Kim, J.S., Kim, K.S., Cho, W., Shin, W.H., Kanno, R., Choi, J.W.: A truncated manganese spinel cathode for excellent power and lifetime in lithium-ion batteries. Nano Lett. **12**, 6358–6365 (2012)
39. Hosono, E., Kudo, T., Honma, I., Matsuda, H., Zhou, H.: Synthesis of single crystalline spinel $LiMn_2O_4$ nanowires for a lithium ion battery with high power density. Nano Lett. **9**, 1045–1051 (2009)
40. Chen, Y.C., Xie, K., Pan, Y., Zheng, C.M., Wang, H.L.: High power nano-$LiMn_2O_4$ cathode materials with high-rate pulse discharge capability for lithium-ion batteries. Chin. Phys. B **20**, 028201–028206 (2011)
41. Sun, Y.K., Kim, D.W., Jin, S.H., Hyung, Y.E., Moon, S.I., Park, D.K.: Synthesis and cycling behavior of $LiMn_2O_4$cathode materials prepared by glycine-assisted sol-gel method for lithium secondary batteries. Korean J. Chem. Eng. **15**, 64–70 (1998)
42. Suryakala, K., Marikkannu, K.R., Paruthimal Kalaignan, G., Vasudevan, T.: Synthesis and rlectrochemical characterization of LiMn2O4 and LiNd0.3Mn1.7O4 as cathode for lithium ion batteryIntern. J. Electrochem. Sci. **3**, 136–144 (2008)
43. Thirunakaran, R., Sivashanmugam, A., Gopukumar, S., Dunnill, C.W., Gregory, D.H.: Studies on chromium/aluminium-doped manganese spinel as cathode materials for lithium-ion batteries-A novel chelated sol-gel synthesis. J. Mater. Proc. Tech. **208**, 520–531 (2008)
44. Thirunakaran, R., Sivashanmugam, A., Gopukumar, S., Dunnill, C.W., Gregory, D.H.: Phthalic acid assisted nano-sized spinel $LiMn_2O_4$ and $LiCr_xMn_{2-x}O_4$ ($x = 0.00$–0.40) via sol–gel synthesis and its electrochemical behaviour for use in Li-ion-batteries. Mater. Res. Bull. **43**, 2119–2129 (2008)
45. Li, R., Li, M.: Enhancement of the electrochemical properties of $LiMn_2O_4$ by glycolic acid-assisted sol-gel method. Ionics **15**, 215–219 (2009)
46. Vivekanandhan, S., Venkateswarlu, M., Satyanarayana, N., Suresh, P., Nagaraju, D.H., Munichandraiah, N.: Effect of calcining temperature on the electrochemical performance of nanocrystalline $LiMn_2O_4$ powders prepared by polyethylene glycol (PEG-400) assisted Pechini process. Mater. Lett. **60**, 3212–3216 (2006)

Permissions

All chapters in this book were first published in MRSE, by Springer; hereby published with permission under the Creative Commons Attribution License or equivalent. Every chapter published in this book has been scrutinized by our experts. Their significance has been extensively debated. The topics covered herein carry significant findings which will fuel the growth of the discipline. They may even be implemented as practical applications or may be referred to as a beginning point for another development.

The contributors of this book come from diverse backgrounds, making this book a truly international effort. This book will bring forth new frontiers with its revolutionizing research information and detailed analysis of the nascent developments around the world.

We would like to thank all the contributing authors for lending their expertise to make the book truly unique. They have played a crucial role in the development of this book. Without their invaluable contributions this book wouldn't have been possible. They have made vital efforts to compile up to date information on the varied aspects of this subject to make this book a valuable addition to the collection of many professionals and students.

This book was conceptualized with the vision of imparting up-to-date information and advanced data in this field. To ensure the same, a matchless editorial board was set up. Every individual on the board went through rigorous rounds of assessment to prove their worth. After which they invested a large part of their time researching and compiling the most relevant data for our readers.

The editorial board has been involved in producing this book since its inception. They have spent rigorous hours researching and exploring the diverse topics which have resulted in the successful publishing of this book. They have passed on their knowledge of decades through this book. To expedite this challenging task, the publisher supported the team at every step. A small team of assistant editors was also appointed to further simplify the editing procedure and attain best results for the readers.

Apart from the editorial board, the designing team has also invested a significant amount of their time in understanding the subject and creating the most relevant covers. They scrutinized every image to scout for the most suitable representation of the subject and create an appropriate cover for the book.

The publishing team has been an ardent support to the editorial, designing and production team. Their endless efforts to recruit the best for this project, has resulted in the accomplishment of this book. They are a veteran in the field of academics and their pool of knowledge is as vast as their experience in printing. Their expertise and guidance has proved useful at every step. Their uncompromising quality standards have made this book an exceptional effort. Their encouragement from time to time has been an inspiration for everyone.

The publisher and the editorial board hope that this book will prove to be a valuable piece of knowledge for researchers, students, practitioners and scholars across the globe.

List of Contributors

M. Z. Rahman
Department of Electrical and Electronic Engineering, Ahsanullah University of Science and Technology, 141-142 Love Road, Tejgaon I/A, Dhaka 1208, Bangladesh

S. I. Khan
Department of Electrical and Electronic Engineering, Bangladesh University of Engineering and Technology, Dhaka 1000, Bangladesh

A. Yamamoto
Graduate School of Engineering, University of Fukui, 3-9-1 Bunkyo, Fukui 910-8507, Japan

A. Hashimoto
Graduate School of Engineering, University of Fukui, 3-9-1 Bunkyo, Fukui 910-8507, Japan

K. Sugita
HISAC, University of Fukui, 3-9-1 Bunkyo, Fukui 910-8507, Japan

A. G. Bhuiyan
Khulna University of Engineering and Technology, Khulna 9203, Bangladesh

N. Narita
Energy Use R&D Centre, The Kansai Electric Power Co., Inc, 3-11-20 Wakoji, Amagasaki, Hyogo 661-0974, Japan

Md. A. Hossain
Department of Electrical and Electronic Engineering, Khulna University of Engineering and Technology, Khulna, Bangladesh

Md. R. Islam
Department of Electrical and Electronic Engineering, Khulna University of Engineering and Technology, Khulna, Bangladesh

M. K. Hossain
Center of Research Excellence in Renewable Energy, King Fahd University of Petroleum and Minerals (KFUPM), Dhahran 31261, Kingdom of Saudi Arabia

F. Deganello
Consiglio Nazionale delle Ricerche (CNR), Istituto per lo Studio dei Materiali Nanostrutturati (ISMN), UOS Palermo, Via Ugo La Malfa 153, 90146 Palermo, Italy

L. F. Liotta
Consiglio Nazionale delle Ricerche (CNR), Istituto per lo Studio dei Materiali Nanostrutturati (ISMN), UOS Palermo, Via Ugo La Malfa 153, 90146 Palermo, Italy

G. Marcı
Dipartimento di Ingegneria Elettrica Elettronica e delle Telecomunicazioni (DIEET), Universita` di Palermo (UNIPA), 90128 Palermo, Italy

E. Fabbri
National Institute for Materials Science (NIMS), International Research Center for Materials Nanoarchitectonics (MANA), Tsukuba 305-0044, Japan
Electrochemistry Laboratory, General Energy Research Department, Paul Scherrer Institut, 5232 Villigen, Switzerland

E. Traversa
International Research Center for Renewable Energy, State Key Laboratory of Multiphase Flow in Power Engineering, Xi'an Jiaotong University, Xi'an 710049, Shaanxi, China

A. K. Baral
High Temperature Energy Materials Center, Korea Institute of Science and Technology, 39-1, Hawolgok-dong, Seongbuk-gu, Seoul, South Korea

S. Choi
High Temperature Energy Materials Center, Korea Institute of Science and Technology, 39-1, Hawolgok-dong, Seongbuk-gu, Seoul, South Korea

B. K. Kim
High Temperature Energy Materials Center, Korea Institute of Science and Technology, 39-1, Hawolgok-dong, Seongbuk-gu, Seoul, South Korea

J.-H. Lee
High Temperature Energy Materials Center, Korea Institute of Science and Technology, 39-1, Hawolgok-dong, Seongbuk-gu, Seoul, South Korea

A. K. Baral
Department of Chemistry, University of Calgary, 2500 University Dr NW, Calgary, AB T2N1N4, Canada

K. Annou
Institut Charles Gerhardt, UMR 5253, Universite´ de Montpellier II, Sciences et Techniques du Languedoc, CC015, 2 Place Euge`ne Bataillon, 34095 Montpellier Cedex 5, France

M. Pelosi
Institut Charles Gerhardt, UMR 5253, Universite´ de Montpellier II, Sciences et Techniques du Languedoc, CC015, 2 Place Euge`ne Bataillon, 34095 Montpellier Cedex 5, France

F. Favier
Institut Charles Gerhardt, UMR 5253, Universite´ de Montpellier II, Sciences et Techniques du Languedoc, CC015, 2 Place Euge`ne Bataillon, 34095 Montpellier Cedex 5, France

M. Tillard
Institut Charles Gerhardt, UMR 5253, Universite´ de Montpellier II, Sciences et Techniques du Languedoc, CC015, 2 Place Euge`ne Bataillon, 34095 Montpellier Cedex 5, France

K. Annou
Institut d'Electronique du Sud, UMR5214, Universite´ de Montpellier II, Sciences et Techniques du Languedoc, 2 Place Euge`ne Bataillon, 34095 Montpellier Cedex 5, France

Y. Cuminal
Institut d'Electronique du Sud, UMR5214, Universite´ de Montpellier II, Sciences et Techniques du Languedoc, 2 Place Euge`ne Bataillon, 34095 Montpellier Cedex 5, France

G. Gershinsky
Department of Chemistry, Bar Ilan Institute of Nanotechnology and Advanced Materials, Bar-Ilan University, 52900 Ramat Gan, Israel

D. Zitoun
Department of Chemistry, Bar Ilan Institute of Nanotechnology and Advanced Materials, Bar-Ilan University, 52900 Ramat Gan, Israel

T. W. Day
Department of Materials Science, California Institute of Technology, Pasadena, CA 91106, USA

F. Drymiotis
Department of Materials Science, California Institute of Technology, Pasadena, CA 91106, USA

D. R. Brown
Department of Materials Science, California Institute of Technology, Pasadena, CA 91106, USA

G. J. Snyder
Department of Materials Science, California Institute of Technology, Pasadena, CA 91106, USA

K. A. Borup
Department of Chemistry and iNANO, Center for Materials Crystallography, Aarhus University, Aarhus 8000, Denmark

B. B. Iversen
Department of Chemistry and iNANO, Center for Materials Crystallography, Aarhus University, Aarhus 8000, Denmark

T. Zhang
CAS Key Laboratory of Energy-conversion Materials, Shanghai Institute of Ceramics, Chinese Academy of Sciences, Shanghai 200050, China

X. Shi
CAS Key Laboratory of Energy-conversion Materials, Shanghai Institute of Ceramics, Chinese Academy of Sciences, Shanghai 200050, China

L. Chen
CAS Key Laboratory of Energy-conversion Materials, Shanghai Institute of Ceramics, Chinese Academy of Sciences, Shanghai 200050, China

J. H. Lee
Graduate School of EEWS (Energy, Environment, Water, and Sustainability) and Center for Nature-inspired Technology (CNiT), KAIST Institute NanoCentury, Korea Advanced Institute of Science and Technology (KAIST), 291 Daehakro, Yuseong-gu, Daejeon 305-701, Republic of Korea

W. H. Shin
Graduate School of EEWS (Energy, Environment, Water, and Sustainability) and Center for Nature-inspired Technology (CNiT), KAIST Institute NanoCentury, Korea Advanced Institute of Science and Technology (KAIST), 291 Daehakro, Yuseong-gu, Daejeon 305-701, Republic of Korea

S. Y. Lim
Graduate School of EEWS (Energy, Environment, Water, and Sustainability) and Center for Nature-inspired Technology (CNiT), KAIST Institute NanoCentury, Korea Advanced Institute of Science and Technology (KAIST), 291 Daehakro, Yuseong-gu, Daejeon 305-701, Republic of Korea

B. G. Kim
Graduate School of EEWS (Energy, Environment, Water, and Sustainability) and Center for Nature-inspired Technology (CNiT), KAIST Institute NanoCentury, Korea Advanced Institute of Science and Technology (KAIST), 291 Daehakro, Yuseong-gu, Daejeon 305-701, Republic of Korea

J. W. Choi
Graduate School of EEWS (Energy, Environment, Water, and Sustainability) and Center for Nature-inspired Technology (CNiT), KAIST Institute NanoCentury, Korea Advanced Institute of Science and Technology (KAIST), 291 Daehakro, Yuseong-gu, Daejeon 305-701, Republic of Korea

C. Gupta
Physics and Engineering of Carbon, CSIR-National Physical Laboratory, New Delhi, India

P. H. Maheshwari
Physics and Engineering of Carbon, CSIR-National Physical Laboratory, New Delhi, India

R. B. Mathur
Physics and Engineering of Carbon, CSIR-National Physical Laboratory, New Delhi, India

S. Sasikala
CSIR-Central Electrochemical Research Institute, Chennai, India

D. R. Abd El-Hafiz
Egyptian Petroleum Research Institute, Cairo 11727, Egypt

M. A. Ebiad
Egyptian Petroleum Research Institute, Cairo 11727, Egypt

R. A. El-salamony
Egyptian Petroleum Research Institute, Cairo 11727, Egypt

U. Gangopadhyay
Meghnad Saha Institute of Technology, Nazirabad, Kolkata 700107, India

S. Jana
Meghnad Saha Institute of Technology, Nazirabad, Kolkata 700107, India

S. Das
Meghnad Saha Institute of Technology, Nazirabad, Kolkata 700107, India

P. Ghosh
Meghnad Saha Institute of Technology, Nazirabad, Kolkata 700107, India

Sushant Kumar
Center for the Study of Matter at Extreme Conditions, College of Engineering and Computing, Florida International University, Miami, FL 33199, USA

Surendra K. Saxena
Center for the Study of Matter at Extreme Conditions, College of Engineering and Computing, Florida International University, Miami, FL 33199, USA

Vadym Drozd
Center for the Study of Matter at Extreme Conditions, College of Engineering and Computing, Florida International University, Miami, FL 33199, USA

Andriy Durygin
Center for the Study of Matter at Extreme Conditions, College of Engineering and Computing, Florida International University, Miami, FL 33199, USA

Paris A. Fokaides
School of Engineering and Applied Sciences, Frederick University, Cyprus, 7., Y. Friderickou Str., 1036 Nicosia, Cyprus

Angeliki Kylili
School of Engineering and Applied Sciences, Frederick University, Cyprus, 7., Y. Friderickou Str., 1036 Nicosia, Cyprus

Soteris A. Kalogirou
Department of Mechanical Engineering and Materials Science and Engineering, Cyprus University of Technology, Limassol, Cyprus

E. H. Fontes
Instituto de Pesquisas Energéticas e Nucleares, IPEN/CNEN-SP, Av. Prof. Lineu Prestes, 2242 Cidade Universitaŕia, CEP 05508-900 São Paulo, SP, Brazil

R. M. Piasentin
Instituto de Pesquisas Energéticas e Nucleares, IPEN/CNEN-SP, Av. Prof. Lineu Prestes, 2242 Cidade Universitaŕia, CEP 05508-900 São Paulo, SP, Brazil

J. M. S. Ayoub
Instituto de Pesquisas Energéticas e Nucleares, IPEN/CNEN-SP, Av. Prof. Lineu Prestes, 2242 Cidade Universitaŕia, CEP 05508-900 São Paulo, SP, Brazil

J. C. M. da Silva
Instituto de Pesquisas Energéticas e Nucleares, IPEN/CNEN-SP, Av. Prof. Lineu Prestes, 2242 Cidade Universitaŕia, CEP 05508-900 São Paulo, SP, Brazil

M. H. M. T. Assumpc¸a˜o
Instituto de Pesquisas Energéticas e Nucleares, IPEN/CNEN-SP, Av. Prof. Lineu Prestes, 2242 Cidade Universitaŕia, CEP 05508-900 São Paulo, SP, Brazil

E. V. Spinace
Instituto de Pesquisas Energéticas e Nucleares, IPEN/CNEN-SP, Av. Prof. Lineu Prestes, 2242 Cidade Universitaŕia, CEP 05508-900 São Paulo, SP, Brazil

A. O. Neto
Instituto de Pesquisas Energéticas e Nucleares, IPEN/CNEN-SP, Av. Prof. Lineu Prestes, 2242 Cidade Universitaŕia, CEP 05508-900 São Paulo, SP, Brazil

R. F. B. de Souza
Instituto de Pesquisas Energéticas e Nucleares, IPEN/CNEN-SP, Av. Prof. Lineu Prestes, 2242 Cidade Universitaŕia, CEP 05508-900 São Paulo, SP, Brazil

Haikun Li
School of Engineering, Grand Valley State University, Allendale, MI 49401, USA

Lindsay M. Corneal
School of Engineering, Grand Valley State University, Allendale, MI 49401, USA

Charles R. Standridge
School of Engineering, Grand Valley State University, Allendale, MI 49401, USA

S. Jegadheeswaran
Department of Mechanical Engineering, Bannari Amman Institute of Technology, Sathyamangalam, Erode (DT) 638401, India

S. D. Pohekar
Mechanical Engineering Area, Tolani Maritime Institute, Induri, Pune 410507, India

T. Kousksou
Laboratoire des Sciences de l'Inge´nieur Applique´es a` la Me´canique et au Génie Electrique (SIAME), Université de Pauet des Pays de l'Adour, IFR, A. Jules Ferry, 64000 Pau, France

V. Kumar
Hydrogen Energy Lab, Department of Physics, Deenbandhu Chhotu Ram University of Science and Technology, Murthal (Sonipat), Haryana, India

D. Pukazhselvan
Hydrogen Energy Lab, Department of Physics, Deenbandhu Chhotu Ram University of Science and Technology, Murthal (Sonipat), Haryana, India
Department of Physics and Astronomy, Padova University, Padua, Italy

S. K. Singh
Hydrogen Energy Lab, Department of Physics, Deenbandhu Chhotu Ram University of Science and Technology, Murthal (Sonipat), Haryana, India

A. K. Tyagi
Solid State Chemistry Section, Babha Atomic Research Centre, Trombay, Mumbai, India

L. A. Kosyachenko
Chernivtsi National University, Chernivtsi 58012, Ukraine

Daniel Symes
Centre for Hydrogen and Fuel Cell Research, School of Chemical Engineering, University of Birmingham, Edgbaston, Birmingham, West Midlands B15 2TT, UK

Connie Taylor-Cox
Centre for Hydrogen and Fuel Cell Research, School of Chemical Engineering, University of Birmingham, Edgbaston, Birmingham, West Midlands B15 2TT, UK

Leighton Holyfield
Centre for Hydrogen and Fuel Cell Research, School of Chemical Engineering, University of Birmingham, Edgbaston, Birmingham, West Midlands B15 2TT, UK

Bushra Al-Duri
Centre for Hydrogen and Fuel Cell Research, School of Chemical Engineering, University of Birmingham, Edgbaston, Birmingham, West Midlands B15 2TT, UK

Aman Dhir
Centre for Hydrogen and Fuel Cell Research, School of Chemical Engineering, University of Birmingham, Edgbaston, Birmingham, West Midlands B15 2TT, UK

C. Lecoeur
Centre Interuniversitaire de Recherche et d'Ingénierie des Matéiaux CIRIMAT, UMR CNRS 5085, Université Paul Sabatier, 118 route de Narbonne, 31062 Toulouse Cedex 09, France
Réseau sur le Stockage Electrochimique de l'Energie (RS2E), FR CNRS 3459, Toulouse, France

B. Daffos
Centre Interuniversitaire de Recherche et d'Ingénierie des Matéiaux CIRIMAT, UMR CNRS 5085, Université Paul Sabatier, 118 route de Narbonne, 31062 Toulouse Cedex 09, France
Réseau sur le Stockage Electrochimique de l'Energie (RS2E), FR CNRS 3459, Toulouse, France

R. Lin
Centre Interuniversitaire de Recherche et d'Ingénierie des Matéiaux CIRIMAT, UMR CNRS 5085, Université Paul Sabatier, 118 route de Narbonne, 31062 Toulouse Cedex 09, France
Réseau sur le Stockage Electrochimique de l'Energie (RS2E), FR CNRS 3459, Toulouse, France

L. Divay
Thales Research and Technology, Campus Polytechnique,1 avenue Augustin Fresnel, 91767 Palaiseau Cedex, France

P. Le Barny
Thales Research and Technology, Campus Polytechnique,1 avenue Augustin Fresnel, 91767 Palaiseau Cedex, France

M. Pham Thi
Thales Research and Technology, Campus Polytechnique,1 avenue Augustin Fresnel, 91767 Palaiseau Cedex, France

P.-L. Taberna
Centre Interuniversitaire de Recherche et d'Ingénierie des Matéiaux CIRIMAT, UMR CNRS 5085, Université Paul Sabatier, 118 route de Narbonne, 31062 Toulouse Cedex 09, France
Réseau sur le Stockage Electrochimique de l'Energie (RS2E), FR CNRS 3459, Toulouse, France

P. Simon
Centre Interuniversitaire de Recherche et d'Ingénierie des Matéiaux CIRIMAT, UMR CNRS 5085, Université Paul Sabatier, 118 route de Narbonne, 31062 Toulouse Cedex 09, France
Réseau sur le Stockage Electrochimique de l'Energie (RS2E), FR CNRS 3459, Toulouse, France

Kasim Uthman Isah
Department of Physics, Federal University of Technology, Minna, Nigeria

Umar Ahmadu
Department of Physics, Federal University of Technology, Minna, Nigeria

Adamu Idris
Department of Physics, Federal University of Technology, Minna, Nigeria

Mohammed Isah Kimpa
Department of Physics, Federal University of Technology, Minna, Nigeria

Uno Essang Uno
Department of Physics, Federal University of Technology, Minna, Nigeria

Muhammed Muhammed Ndamitso
Department of Chemistry, Federal University of Technology,Minna, Nigeria

Noble Alu
Physics Advanced Laboratory, Sheda Science and Technology complex Sheda, Abuja, Nigeria

Donato E. Conte
Institut für Chemie, Humboldt-Universität zu Berlin,Brook-Taylor-Str. 2, 12489 Berlin, Germany

Nicola Pinna
Institut für Chemie, Humboldt-Universität zu Berlin,Brook-Taylor-Str. 2, 12489 Berlin, Germany

I. Terasaki
Department of Physics, Nagoya University, Nagoya 464-8602, Japan

R. Okazaki
Department of Physics, Nagoya University, Nagoya 464-8602, Japan

P. S. Mondal
Department of Physics, Nagoya University, Nagoya 464-8602, Japan

Y.-C. Hsieh
Department of Physics, Nagoya University, Nagoya 464-8602, Japan

V. Caramia
Dipartimento di Ingegneria dell'Innovazione, Universitá del Salento, via Monteroni, 73100 Lecce, Italy

B. Bozzini
Dipartimento di Ingegneria dell'Innovazione, Universitá del Salento, via Monteroni, 73100 Lecce, Italy

A. V. Potapenko
Joint Department of Electrochemical Energy Systems, 38A Vernadsky Ave., Kiev 03142, Ukraine

S. I. Chernukhin
Joint Department of Electrochemical Energy Systems, 38A Vernadsky Ave., Kiev 03142, Ukraine

S. A. Kirillov
Joint Department of Electrochemical Energy Systems, 38A Vernadsky Ave., Kiev 03142, Ukraine
Institute for Sorption and Problems of Endoecology, 13 Gen. Naumov St., Kiev 03164, Ukraine